Glass and Glass Production in the Near East during the Iron Age

Evidence from objects, texts and chemical analysis

Katharina Schmidt

Archaeopress Archaeology

Archaeopress Publishing Ltd
Summertown Pavilion
18-24 Middle Way
Summertown
Oxford OX2 7LG

www.archaeopress.com

ISBN 978-1-78969-154-2
ISBN 978-1-78969-155-9 (e-Pdf)

Cover image: Inlaid glass bowl from the Amman Citadel, © École biblique/ Department of Antiquities, Jordan, photo: Johannes Kramer

To my family and Samar

Contents

List of Figures

List of Tables

Sources for Plates Illustrations

1. Foreword and Acknowledgements

The colour tones of the lands between the Euphrates and the Tigris are very light, dusty, and dull. They need for contrast pure strong colours. (...) But the general effect through the long eight months of yearly drought, and the prevalence in the landscape of the endless, barren, parched plains and little hills between and beside the rivers - all that is indeed dusty, and veiled in yellowish white insipidity, which spreads over villages and towns, over houses and streets, over palms and stepped flora (...). Men have there unconsciously a strong need for expressing themselves in arrangements of colours. (Andrae 1925: 1)

Today glass is an everyday commodity, often even considered a substitute, similar to plastic. In ancient Mesopotamia, where mud-brick and ceramics were the common materials, the situation was different. The quote of the archaeologist Walter Andrae above, who spent a considerable part of his life in Iraq, is an impressive testimony to the little diverse and colourful monotonous landscape of Mesopotamia, which did not differ much in its current appearance from the Bronze and Iron Age. But unlike today, coloured, shiny and smooth surfaces were rare in Iron Age Mesopotamia, on which this study focuses. Objects made of stone, metal and glazed materials that create a shiny and colourful effect were for the majority of people not part of their daily experience. That is why they were valuable. Due to its deep, bright colours and shiny surface, glass is one of the materials that share these highly appreciated properties and, as the youngest of the materials artificially produced in antiquity, ranks among the most admired materials of that time.

In the Late Bronze Age, when glass production was first introduced in Mesopotamia and Egypt, glass was a rare commodity and was used as a material for prestige objects available only to the elite. This book examines the history of glass in the first half of the 1st millennium BCE in Iron Age Mesopotamia, a time that has been underrepresented in research to date. In many cases, a hiatus in glass production was assumed for this period. It was only in Roman times, with the invention of glass blowing technology, that glass became an everyday commodity.

This monograph is intended to contribute to the history of glass and close the gap between the Late Bronze Age and the Hellenistic period, both of which have been well studied. It becomes clear that many glass technological developments that are of great importance for the following periods began and were prepared in Iron Age Mesopotamia. This is the reason why the presentation of the glass material from this period is of such great importance.

This monograph represents the revised version of the author's doctoral thesis, which was submitted in May 2016, and defended in June 2016 at the Department of Near Eastern Archaeology (Vorderasiatische Archäologie) at the Faculty of Cultural Studies (Fakultät für Kulturwissenschaften) at the Ludwig-Maximilians-University in Munich.

This monograph would not have been possible without the support of many individuals and institutions. I would therefore like to express my sincere gratitude to everyone who has contributed to this book:

First, I would like to thank the Graduate School 'Distant Worlds' at the Ludwig-Maximilians-University, Munich for its generous financial and structural support in the framework of my dissertation scholarship. Only by the support of the school was it possible for me to carry out the many research trips that were necessary to study the glass objects in the different museums and collections worldwide. In this context, I would like to particularly thank all principal investigators, post-docs and colleagues of the Graduate school for fruitful discussions and helpful advice during the numerous seminars.

I would like also to express my gratitude to the Research Training Group 'Value and Equivalence' of the Goethe University, Frankfurt, for their structural support with regard to my association with their training group, and for the financial support to obtain some of the publication rights for photographs.

My sincere gratitude goes to Professor Adelheid Otto, Professor Michael Roaf and Professor Walther Sallaberger from the Ludwig-Maximilians-Universität München, and Professor Jan-Waalke Meyer (University of Frankfurt), who supervised this doctoral dissertation. In this regard, I would also like to thank Professor Ian Freestone from University College London, who supervised the archaeometric part of this study. Without their critical comments, suggestions and support in their particular fields of research, the completion of the dissertation would not have been possible.

The know-how of glassmakers and people with practical experience in handling glassy materials contributed significantly to this work. In this context I would like to thank sincerely the 'Roman Glassmakers' Mark Taylor and David Hill for their comments and advice with regard to the manufacturing processes described in this monograph and their critical remarks on several parts of this text. I would like to also thank William Gudenrath from the Corning Museum of Glass, for various conversations with regard to the manufacturing process of glass, and Dr Zahed Taj-Eddin for stimulating discussions and many critical comments on the text.

The book is the result of collaboration between several museums, collections, institutions and individuals. I would therefore like to thank Professor Markus Hilgert and Dr Lutz Martin (Vorderasiatisches Museum, Berlin), who granted me access to the collection of finds and also gave me permission to publish photographs. My thanks also go to Dr Helen Gries for her support in accessing the objects and Olaf Teßmer for the excellent photographs of the glass objects, and to Professor Johannes Renger for granting me access to the Aššur database and allowing me to use the information. I would like to thank Nigel Tallis, Dr St. John Simpson and Alex Truscott (British Museum, London) for the opportunity to work on the artefacts in the British Museum collection and their helpful comments and advice on the collection. I also thank Dr Joan Aruz for granting me access to the collection of the Metropolitan Museum of Art, New York, and in particular Dr Kim Benzel, Tim Healing, and Dr Christopher S. Lightfoot for their support in examining the glass objects and the generous permission to publish the photographs taken during my research visit. Thanks also go to Katherine Blanchard (University of Pennsylvania Museum of Archaeology and Anthropology, Philadelphia) for her support in the study of the objects of the museum collection, and to Professor Richard Zettler for permission to publish the photographs taken during my research stay. In this context I would like to also thank Professor Stephan Kroll, who made the Hasanlu documentation available to me. My sincere gratitude goes to Dr Kiersten Neumann, who enabled me to examine the finds in the Oriental Institute Museum in Chicago and provided me with new photographs for this monograph. I would also like to thank in this regard Dr Helen McDonald (Oriental Institute Museum, Chicago) for making the finds available for me in the collection. My thanks also go to John Kelly (The Field Museum, Chicago), Dr Jutta-Annette Page (Toledo Museum of Art, Toledo), Terry Bloxham (The Victoria and Albert Museum, London) and Dr Rachel Sparks (Institute of Archaeology Collection, London) for providing access to their collections and the opportunity to study and publish objects in their collections. I would also like to thank Dr Karol B. Wight, who gave me access to the collection of the Corning Museum of Glass, and I would like to particularly thank Alexandra M. Ruggiero, who not only helped me with the data collection but also connected me with all the museum staff and made my stay in Corning very enriching. Finally, I am grateful that I was able to study the material from the collection of the Musée du Louvre in Paris, and I would like to particularly thank Dr Ariane Thomas for the granting me access to the finds and allowing me to publish the photography, but also Jorge Vasquez for providing the objects. I thank Professor Jean-Babtiste Humbert (École Biblique et Archéologique) and the Department of Antiquities in Jordan for permission to publish the extraordinary glass bowl from the Citadel in Amman, which also serves as the cover motif for this book. In this context I would like to thank Dr Barbara Porter (ACOR Jordan), who gave me permission to study and photograph the piece on her premises, and also to Johannes Kramer, who took the excellent photographs. I also thank Yael Barschak (Israel Museum, Jerusalem) for permission to publish a bead from their collection, and Professor Tim Matney and Professor Dirk Wicke for permission to publish one vessel from Ziyaret Tepe.

With regard to the writing and finishing process of this book, my sincere gratitude goes to Dr Jutta Häser and Dr Jean Evans for proof-reading and the English corrections of the dissertation manuscript. My thanks goes also to Dominik Blöse, and Luisa Goldammer for providing the digitized drawings of the objects and Dr Simon Halama for the plates section. With regard to the archaeometric chapter, I would like to thank Dr Matt Phelps most sincerely for his critical remarks. Finally, I am very grateful to all my colleagues at the Ludwig-Maximilians-Universität Munich, the Goethe-Universität Frankfurt, and University College London for innumerable enriching discussions and advice relating to this study.

1.1. Aims and objectives

This book examines the history of glass in the first half of the 1st millennium BCE with a focus on Iron Age Mesopotamia (1000–539), a period that was previously underrepresented in glass research, and it is therefore the first to cover this topic comprehensively and in detail. The necessity of working on this subject resulted from the observations made with regard to the glass object corpora of the Late Bronze Age and the Hellenistic period, which compared with one another consist of very different types of glass objects and which were also appreciated in a different way within the society. This gave rise to the question of how the situation was in the intermediate period, the Iron Age, and the idea was born to write a study focusing on this period and region with the aim of presenting a compilation that was as far-reaching as possible and largely complete with regard to the existing types of glass.

The overall aim of this monograph is therefore to determine which glass finds date to the Iron Age period, and, as a result, to identify the different types of glass objects that exist and their respective manufacturing techniques. Based on this fundamental question, another objective of this study is to understand how raw glass (primary production) and glass objects (secondary production) were manufactured at that time, and how both these industries were organised. This raises the further question of how widespread glass objects were in Mesopotamian society and how important, both the material glass and the individual glass objects were for the people in the Iron Age period.

In this monograph the principle of '*chaîne opératoire*' is not only employed with regard to archaeological data, but also with relation to cuneiform texts, archaeometric analyses and experimental-archaeological investigations. This enables the reconstruction of every single step of the operational sequence of primary and secondary glass production, to recreate a reasonable picture of the production of raw glass and glass objects in Iron Age Mesopotamia. The identified technological processes then serve as a starting point for considerations on the organisation of production and on further questions, such as the distribution of the objects, their function, and assumptions about how these objects were appreciated within society.

With regard to the different disciplines incorporated into this study, an attempt was made not to view them in isolation from each other but to establish connections between these areas, for example, to identify different manufacturing techniques, detailed investigations of the objects themselves must be combined with the practical experience of glassworking, since a fundamental understanding of the material is necessary. With regard to cuneiform texts, for example, a basic understanding of the chemical composition and physical properties of the material itself is required to comprehend the processes described in the texts. This combination and integration of disciplines is the second step. This is preceded by a detailed subject-specific analysis which is provided at the beginning of each chapter in this book.

Chapters 1 and 2 are to be understood as introductory chapters. The first chapter gives an overview of the geographical and chronological framework, and also explains the concept of *chaîne opératoire*, the history of research and the beginning of glass production in the Late Bronze Age. The second chapter deals with the physical and chemical properties of glass to understand the material.

A detailed discussion of the sites and archaeological contexts in which glass finds were made is provided in Chapter 3. This overview is of fundamental importance with regard to the date of the objects and their distribution, both geographically and contextually, and forms the basis for the evaluation of the entire corpus discussed in Chapter 5.

In Chapter 4 the material under study will be presented. The glass finds were compiled from museum collections, literature and excavation databases. With regard to most of the objects, the author was able to examine the artefacts visually within the limitations of museum accessibility (see Acknowledgements), in some cases only photographs could be used. Almost all types of objects made of glass during the Iron Age period are incorporated, only beads and seals are omitted, as their large number would have far exceeded the scope of the work.[1] First, the glass objects will be described in detail, listed in the catalogue and depicted in the plate section. Second, these glass objects were divided into different typological groups according to their specific manufacturing technique.

The following Chapter 5 deals with the geographical distribution of the finds, their specific finding contexts and the chronological range of the different types, including information discussed in detail in Chapters 3 and 4.

Philological evidence forms an intrinsic part of the monograph, which is described in Chapter 6. The texts represent the most extensive source for the reconstruction of glass production in Mesopotamia and thus form a central part of this monograph. As a case study, the recipe on the production of blue *zagindurû*-glass was selected, a text preserved in six different versions. The text is presented both in an edition and in a content-related commentary (Appendix 1). An intimate knowledge of the chemical and physical composition of glass, as well as of the contemporaneous glass finds, contributes essentially to the interpretation of the text.

Results based on chemical analysis are provided in Chapter 7, which is divided into two parts. The first part contains a general discussion of chemical components of glass and their occurrences. In the second part, the chemical analyses taken from glass at different sites are discussed. For this purpose, the chemical raw data were collected, standardised and compiled in a table (Appendix 2). In the course of this work, only published chemical raw data were integrated.

The main results of the study are comprised in the final conclusion. With the exception of the concept of the *chaîne opératoire*, no further theoretical approach to the topic is made. In the course of her dissertation work, the author dealt, however, extensively with the concepts of 'exchange' and 'network analyses', as well as with the various aspects of 'value and equivalence' in connection with glass, which are definitely worthwhile

[1] Detailed study of the glass seals would, however, be worthwhile for future research, especially with regard to a combined iconographic and archaeometric evaluation.

aspects for further studies, and which therefore deserve separate detailed works in the future.

At this point, some further remarks with regard to this monograph are made. The year dates are omitted in this study but refer to BCE unless otherwise specified. The study refers to the common site names and their spellings throughout the text, and are listed in Chapter 3. The spelling of king names follows 'The Royal Inscriptions of the Neo-Assyrian Period' (RIMA) guidelines. Special terms used in the study are marked with an asterisk and are explained and further defined in a technical index.

1.2. Primary and secondary production and the principles of *chaîne opératoire*

Glass production is divided into a 'primary production' which means the fusion of raw materials into a batch, and a 'secondary production' concerning the fashioning of the material into objects (Henderson 2013: 307). Objects associated with primary production are ingots, lumps and pieces of raw glass, as well as waste material. In contrast, glass objects that are not supposed to be further processed such as vessels, inlays, beads or pendants are the outcome of the secondary production process. Both branches of these industries differ considerably and require varying working conditions. First, different ranges of temperatures are needed. Glassmaking (primary production) demands very high temperatures which could either have been held for a short period of time at a very high level or for a longer period of time at a lower level. Glassworking (secondary production), in contrast, requires lower temperatures (Shortland 2012: 27–28). Some secondary processes are even only carried out by the use of cold-working techniques. The need for different temperatures has a direct effect on fuel, kiln construction, tools needed for manipulation, as well as on the shape and constitution of the crucibles in which glass was melted.

The primary and secondary production of every glass object presumes a selection of (raw) materials, tools, energy sources and techniques. In this regard, craftsmen had to choose from a range of different alternatives (Sillar and Tite 2000: 3). To be able to reconstruct the past production process and the choice for one specific technique, it is necessary to look at each step of the manufacturing sequence. This sequence of steps is defined as 'chaîne opératoire' and 'refers to the range of processes by which naturally occurring raw materials are selected, shaped and transformed into usable cultural products' (Schlanger 2005: 25).[2]

Regarding primary and secondary glass production, the requirements (materials, tools, energy sources, techniques) of the two production branches presupposes different knowledge and skills, and consequently result in a different configuration of the workshop itself. The choice for a specific alternative was in most of the cases a response to functional necessities that are directly connected with environmental, economic, social, political and also ideological factors (Sillar and Tite 2000: 5). For instance, environmental factors affect the availability of raw materials, tools and energy sources, whereas the economic and political situation provides information about possible trade routes and exchange systems that existed in a specific time period and region (Sillar and Tite 2000: 9). Therefore, it is assumed that technological developments are not only the cause of technological factors but also result from social, political, and ideological ones. These are crucial for technological innovation since they underlie every decision to adopt or reject a particular technological process. The procedures involved in the production of a glass artefact must therefore be understood and interpreted in their cultural context, and the object itself must be situated in the broader historical and political context of the time in which it was produced. During the Iron Age, the region under study underwent a fundamental change, i.e. the emergence of the first empire, which put an end to the political and cultural fragmentation of the preceding Late Bronze Age. These changes may also have an impact on the material culture and the production principles of glass, which is to be investigated in this study.

1.3. Previous studies on glass in the ancient Near East

The existence of pre-Hellenistic glass production was long denied by scientists concerned with the general history of glass (Kisa 1908: 102). Various excavators, on the other hand, who uncovered Late Bronze Age and Iron Age glass objects in their excavations, referred – in response to the assumption of Kisa – to these 'early' glass finds in their excavation publications, such as the excavators of Babylon (Koldewey 1913: 249–250), Nippur (Peters 1898: 134–135) or Nuzi (Starr 1939: 157–159). On this basis, Koldewey (1913) suggested that Mesopotamian glass production existed independently of Egypt already in the Late Bronze Age period. Starr (1939) even dedicated a separate chapter to the glass finds from Nuzi (15th/14th century) in his monograph and thus emphasised their significance. Similarly, the opinion was also held in Assyriology that there was an early glass industry in Mesopotamia that was independent of Egypt, in this context Meissner (1920: 235) in particular is to be mentioned (for details, see Chapter 6.1.1).

Enhanced interest in Mesopotamian glass finally developed out of the intensive research on Mediterranean core-formed vessels, which was

[2] For a literary review on this concept, see Sellet 1993: 106.

promoted by Fossing in 1940. In his monograph *Glass vessels before glass-blowing*, Fossing established from the scant number of objects known at that time, four major chronological and typological groups which are broadly still accepted today (for details, see Chapter 4.3.1). He also identified an early group of Mesopotamian vessels and pointed towards a Mesopotamian glass production which existed as early as 1300 (Fossing 1940: 31–41). Thus, pre-Hellenistic glass from Mesopotamia emerged as a field of interest in the discipline of glass research. As a consequence, glass became acknowledged in the field and was more frequently published in excavations reports.

Regarding Iron Age glass in particular, Nimrud yielded the majority of glass finds, which have been well discussed in a number of articles (for instance by Barag 1983; Brill 1978; Curtis 1999; Orchard 1978; Saldern 1966a). This can partly be explained through the broad discussions of the decorated ivories found at the site, which were also inlaid with glass (see Chapter 4.2.2.9). Similarly, glass inlays for ivories were also published from Samaria (Crowfoot and Crowfoot 1938: 44–45), and Arslan Taş (Thureau-Dangin 1931: 138). Articles devoted with great attention to glass finds can only be found sporadically at sites such as Hasanlu (de Schauensee 2001; Marcus 1991; Saldern 1966b), Aššur (Haevernick 1968) and Gordion (Jones 2009).

In the 1950s, technological and chemical questions became major concerns, not only in the research of glass, but also regarding other archaeological materials. This enhanced interest in materials in general was driven by the rapid development of scientific and technological methods which also promoted cooperation between the field of archaeology and natural science. The discipline of 'archaeometry' was finally coined by Hawkes, Professor of European Archaeology in Oxford, in the mid 1950s, which promptly also became the title of a journal (*Archaeometry*). Archaeometry denotes the application of physical, chemical, biological and earth sciences to archaeological problems and, since its emergence, has played an integral part in the field of archaeology (Wagner 2007: 5).[3]

Since the 1950s, chemical analysis has become increasingly important in glass research. Today, it has become an integral part of glass research. Of particular importance for this development are the three volumes of *Chemical Analysis of Early Glasses* published by Brill (1999; 1999a) and Brill and Stapleton (2012). The volumes contain the largest compilation of chemical raw data and their interpretation (vol. 3) for the eastern Mediterranean that have so far been published. These monographs are among the few in which data from Mesopotamian sites, including the Iron Age period, have been recognised. This is particularly important against the background that the amount of published archaeometric data from Iron Age Mesopotamian sites is very small compared, for example, to Late Bronze Age glass or analyses from other regions and periods. So far, archaeometrical research on Iron Age glass material was only sporadically published from sites such as Nimrud (Bimson and Freestone 1985; Brill 1978; Brill 1999b: 47–49; Cable and Smedley 1987; Reade *et al.* 2005; Turner 1955; 1956), Hasanlu (Brill 1999b:44; Stapleton and Swanson 2002a; 2002b; Stapleton 2003; 2011), Gordion (Privat *et al.* 2014; Reade *et al.* 2012) and Pella (Privat *et al.* 2014).

The foundation of the Corning Museum of Glass in 1951 is another milestone in the history of glass research. The museum is the world's largest institution in the field of glass research, as it houses not only an extensive collection of antique and modern glass objects, but also laboratories for analytical and experimental studies and an extensive library on glass history. Of particular importance is the publication series *The Corning Museum of Glass Monographs* and the journal *Journal of Glass Studies*, published by the Corning Museum, which contribute to glass research.

One of the most important monographs on Mesopotamian glass yet published is *Glass and glassmaking in ancient Mesopotamia* by Oppenheim (1970). The approach pursued in this study is the joint work of philology (Oppenheim 1970), archaeometry (Brill 1970) and archaeology (Barag 1970; Saldern 1970), which has led to the most comprehensive reconstruction of ancient glass production in Mesopotamia to date.

Publications on ancient Near Eastern glass were primarily presented in the form of catalogues. An important contribution was published in 1985 by Barag, who first presented the collection of Mesopotamian glass objects from the Bronze and Iron Ages in the British Museum (1985). Harden (1981) focused specifically on the Mediterranean core-formed vessels from the British Museum, including earlier Mesopotamian pieces in his typology. An important study, which not only catalogues the finds from the Toledo Museum of Art, but also makes typological and chronological considerations and contributes to the discussion about the production techniques of the different glass vessels was published by Grose in 1989. This was followed in a similar way in *Frühes Glas der alten Welt* by Stern and Schlick-Nolte (1994), which includes the objects from the private Ernesto Wolf collection and also considers their manufacturing techniques. The discussions about

[3] The first issue of Archaeometry was published in 1958. It is still one of the most important journals in the field, see http://www.arch.ox.ac.uk/rlaha.html (accessed: 2.3.2016). A cooperation between archaeology, assyriology and the natural sciences can sporadically be observed already in the early 20th century and was led by the assyrologist Oppenheim, among others, for details see Chapter 6.1.

production methods from both publications mentioned above are partly based on experimental studies.

Moorey (1994) also included a chapter on glass in his comprehensive work on all materials existing in ancient Mesopotamia – *Ancient Mesopotamian Materials and Industries*. He lists all glass objects of the Bronze and Iron Age known at that time, including their archaeological contexts and dating. He also presents an overview of the manufacturing techniques and gives technological considerations on glass and its connections to other materials.

One of the most comprehensive studies on the history of ancient glass in general was provided by Saldern in 2004, incorporating glass finds from the Late Bronze Age to the Roman period. This monograph includes a broad range of glass objects, including their occurrence and development in an historical context.

The research of glass in the Near East has gained increasing importance in recent years and was driven, as previously stated, by a rising interest and study of chemical composition. *Lapis Lazuli from the kiln* by Shortland is concerned with Late Bronze Age glass from the Near East and Egypt (2012). On a large scale, the interaction between these two early centres for glassmaking and forming are discussed, and questions about technology, workshops, and the use of raw materials are also considered.

Henderson's monograph *Ancient Glass* incorporates data from the Near East, Europe, Asia and Africa from the Late Bronze Age to the Islamic period (2013). This study draws its focus particularly from chemical compositions and their relations, but also includes case studies of individual sites of this vast region.

Despite the growing interest in studies on glass from Iron-Age Mesopotamia, which is evident from the growing number of articles, a monograph that takes typological, chronological and technological considerations for the entire region into account has not yet been published. The present study therefore attempts to take a first step in this direction on the basis of the research history presented here and with the incorporation of previously unexplored artifacts.

1.4. Geographical and chronological framework

The glass objects included in this study are distributed over a geographically broad area, with Mesopotamia (Assyria and Babylonia) as the core region and including its adjoining land masses. The easternmost sites comprise Susa in southwest and Hasanlu in northwest Iran. The southern boundary is Babylonia, which comprises the territory of southern Iraq. Towards the west, the area of interest stretches over northern Mesopotamia, which includes the upper Tigris region and the Khabur triangle, as well as the region of the Middle Euphrates, situated in the modern states of Syria and Turkey. Further west, the area of the Levantine coast, here divided into northern and southern Levant, is also included in this study and incorporates the land that stretches from the Amanus Mountains in the north to the eastern fringes of the Egyptian Delta in the south. Singular glass objects found in western central Anatolia (Gordion), Cyprus (Idalion), Rhodes (Kameiros), Crete (Fortetsa), Italy (Praeneste), and Tunisia (Carthage) are also incorporated, since these pieces represent key objects in glass research.

A brief outline of the history of Assyria and Babylonia, as well as the northern and southern Levant, will be provided in this chapter. Singular sites in central Anatolia and the Mediterranean mentioned above will be discussed separately.

1.4.1 Assyria and Babylonia

The most important political units in Mesopotamia during the Iron Age were the Assyrian and Babylonian empires.

1.4.1.1 The Neo-Assyrian period

The way the Neo-Assyrian empire was politically and socially structured, and, in particular, how its conquered territories were organised reflects greatly on the way objects and ideas travel. The following section largely takes the study by Radner (2014) as its basis, in order to focus on this aspect in particular. The core region of Assyria (northern Mesopotamia) is situated in the north of modern Iraq, from which the state controlled most of the Near East during that time, governing its territory of influence either directly or indirectly (see below). The beginning of the Neo-Assyrian period is difficult to determine. Different rulers are considered the first Neo-Assyrian kings, such as Tiglath-pileser I (1114–1076), Ashurnasirpal I (1049–1031) or Aššur-dān (911–890), amongst others. Also in respect to archaeology, the beginning of the Neo-Assyrian period cannot be determined exactly. Therefore the year 1000 is followed in this study for convenience (Liverani 2011; Roaf 2001).

In the 11th and 10th centuries the territory of Assyria was reduced to the city of Aššur and its hinterland, due to political and social disturbances caused by migrating people in the region at the end of the Late Bronze Age (see Chapter 1.4.2). By about 1100, the Assyrians faced the Arameans, who were tribal groups in the north of Syria. Later, the Arameans were incorporated into Assyria, forming an Aramean-Assyrian symbiosis, which is also visible in the material culture (Berlejung *et al.* 2017; Tadmor 1982). By the mid 9th century,

Assyria had recaptured the territories located in the north and west of the core region, with Shalmaneser III (858–824) consolidating this regained territory. This was achieved by the establishment of client states with 'royal cities', ruled by local dynasts who were bound to the Neo-Assyrian king by oath and treaty (Radner 2011: 103, 105). During the 8th century, Assyria suffered a brief phase of political decline, caused by aggressive rival states (Urartu, Upper Egypt), as well as by weak Assyrian monarchs. Finally, after 754,[4] Tiglath-pileser III (744–727), Shalmaneser V (726–722), and Sargon II (721–705) reasserted power by sending armies beyond their traditional Assyrian territories, such as Syria, Palestine and parts of Egypt, Anatolia and Iran. Finally, in the middle of the 7th century, the Assyrian Empire reached its maximum expansion, relocating strategically large amounts of people - in particular for economic exploitation. In this regard, Sennacherib (704–681) was the monarch who moved the greatest numbers of people across the Empire compared to his predecessors and successors, with the majority of people coming from Babylonia (Oded 1979: 20–21; Radner 2014: 109). In particular, experts from different fields were brought to the Assyrian heartland – in the time of Sennacherib mainly to Nineveh – to generate knowledge and wealth. Craftsmen have to be considered among this class of individuals, skilled people needed to furnish those temples and palaces that were (re-)built in the capitals (Oded 1979; Radner 2009; Radner 2014: 106, 108–109). From the 8th century onwards, the territory was organised into provinces, governed by local provincial governors appointed by the king (Radner 2006). The expansion of the Neo-Assyrian Empire, in particular towards the west, and the relocation of experts, played an important role with regard to production traditions and distribution of different types of objects. The extent to which this resettlement of experts has an impact on the spread of glass technology in the early 1st millennium is part of the investigation of this study.

The fall of the Neo-Assyrian empire occurred with the collapse of the major centres in the years 614 (Aššur) and 612 (Nineveh), which was caused by the attacks of Babylonians and Medes (Radner 2014: 111). The period after 612 is still broadly obscure and can archaeologically only be determined at some sites, for instance at Nimrud, Nineveh or Dur-Katlimmu, to mention only a few. This period is commonly referred to as post-Assyrian, and is marked by squatter occupations that exist at almost all the major sites. The post-Assyrian period is characterised by a continuity in material culture that makes a distinction between Neo-Assyrian and post-Assyrian difficult (Curtis 2003: 164; Kreppner 2006: 128).

1.4.1.2 The Neo-Babylonian period

According to Jursa (2014: 125, 140), the Neo-Babylonian empire cannot be disconnected from Assyrian rule over Babylonian regions and its fate; also because later it served as a cultural bridge between the Neo-Assyrian and Achaemenid periods. The Neo-Babylonian period (612–539), is historically marked by Babylonian domination in Mesopotamia and its bordering regions. During this period, the Babylonians gained control over southern and northern Mesopotamia, as well as over parts of south-west Iran, Syria and the Levant. This was mainly achieved by Nabopolassar (626–605), the founder of the Neo-Babylonian state (Da Riva 2008: 1–16; Jursa 2014: 124). The expansion to the west was mainly carried out by Nebuchadnezzar II (605–562), a process which also included colonisation and province formation. Unfortunately, royal inscriptions only vaguely report on this. Booty from Assyria, Syria and the Levant funded large, royally-sponsored building programs, in particular within the city of Babylon, and allowed Babylonia to experience a phase of great prosperity. Finally, in 539, Cyrus the Great defeated the last Neo-Babylonian king, Nabonidus (556–539), captured Babylon, and terminated the Neo-Babylonian empire by incorporating their territory into the Persian empire (Jursa 2014: 125–126, 140–142). Already in the previous years, Cyrus had captured those bordering regions in the east and north that had previously been known as Media and Lydia. By incorporating the territory of the Neo-Babylonian Empire, Cyrus finally created an empire of previously unknown size (Rollinger 2014: 150).

1.4.2 Levant

1.4.2.1 History and chronology of the Levant

The chronological nomenclature for the Levant is adopted from the European pre- and proto-history and its classification of Stone, Bronze and Iron Ages, and stands therefore in contrast to the historical nomenclature of Mesopotamia. The Iron Age in the Levant is estimated around 200 years earlier than in Mesopotamia, with Iron Age I therefore stretching from 1200/1150 to 980/930 BCE. Its beginning is marked by an epochal transition which kept the eastern Mediterranean Late Bronze Age world and changed it radically. Its breakdown has to be contextualised with the fall of the superpower Egypt and the Hittite Empire, the decline of the Mycenaean city-states with their palatial structures.[5] This ultimately resulted in a new geopolitical situation in the subsequent Iron Age II period, characterised by smaller and independent political units, associated with new group identities –

[4] The year of the accession of Aššur-nārārī V.

[5] For a detailed study and summary of all factors with further literature, see Cline 2014.

known as the kingdoms of Ammon, Moab, Edom, Aram, Israel, Juda, and the northern 'Phoenician' and southern 'Philistine' city-states. The emergence of these units was a gradual process, and only little is known about their political and social organisation as textual sources are largely missing (Weippert 1988: 353).[6]

The Iron Age IIA/B periodisation is heavily discussed. The main protagonists in this diverse chronological debate are Finkelstein, Mazar and Garfinkel. The discourse is based on a missing chronological anchor between the 12th and 8th centuries which would make it possible to correlate archaeological data with absolute dates. A detailed discussion of the Iron Age chronological debate is omitted here, and, for convenience, the 'Conventional Chronology' is applied.[7] In this investigation, however, the term 'Iron Age' is used in a general sense as the period of the first half of the 1st millennium BCE.

Dates for the 'Conventional Chronology' in the Levant

Iron IA	1200/1150–980/930
Iron IIA	1000–926/900
Iron IIB	830/800–730/700
Iron IIC	700–520

Historically, the 8th and 7th centuries are marked by the conquest of the region by the Neo-Assyrian Empire. The first phase of the Assyrian expansion took place under Ashurnasirpal II (883–859) and Shalmaneser III (858–824), which affected the northern Levant more than its southern part, since Aram still served as a buffer zone. Assyrian influence grew successively in the first half of the 8th century, in particular under Tiglath-pileser III (744–727). By the 7th century, the Assyrian Empire had grown into a 'world power', which was secured by Sennacherib (704–681) and Esarhaddon (680–669). Under Ashurbanipal (669–627) the empire finally reached its maximum expansion. The Assyrians left allies or vassals in the northern and southern Levant. For many parts of the region, in particular for Trans- and Cisjordan, it is not clear how the provincial system was exactly governed in the different regions (Bagg 2011; 2013: 132–135; Parpola 2004; Routledge 1997; Ussishkin 2006: 339–358). Some of the provinces in the west were, however, primarily established to generate trade between Assyria and the neighbouring regions, such as Ashdod (established in 711), and Sidon (established in 677) (Radner 2004). The Neo-Assyrian empire is furthermore characterised by massive deportations and resettlements of large groups of people and a supra-regional economic and trading system, all factors which also influenced the way objects, knowledge and ideas were distributed (George 1997; Gitin 1997; Oded 1979). To which extent Assyrians were present in the Levant is discussed controversially (Bagg 2011: 281; Parpola 2003).

By the end of the Assyrian supremacy in the region, the Babylonians rose to power. Nebuchadnezzar (605–562) conquered Syria and defeated the Egyptian king Nechos II at Carchemish in 605. In the subsequent years he captured large parts of the northern and southern Levant. However, the exact way of the execution of Babylonian domination over the Levant is a matter of debate, its presence can only be realised to a small extent in the material culture, e.g. the rock reliefs (da Riva 2008: 2–19; Jursa 2014: 124–126; Lipschitz 2005: 3–48, 66).

1.4.2.2 'Phoenicia' and related terminological difficulties

'Eastern Mediterranean', 'Phoenicia', 'Syria' or 'Syria-Phoenicia' can be either understood as geographic locations, or they are connected with a specific group of people. To avoid controversial terminology, the term 'Phoenician' is not used in this study, unless it serves as part of an established name, in favour of the more general differentiation between the northern and southern Levant.[8] The northern Levant finds its northern boundary in Cilicia and incorporates large parts of western Syria, from its coastal line up to the middle Euphrates. The southern Levant is defined as 'greater Syria', bounded by the Antilibanon in the north, and Egypt (Wādī el-Ġazze/el-'Arīš) in the south.

Regarding ancient glass, the term 'Phoenician' is particularly widely used in literature, and needs therefore to be discussed here. The Phoenicians are elusive in history because they never designated themselves as such.[9] Often they are referred to as successors of the Canaanites, and are therefore categorised regarding chronological considerations. Even if both groups are regarded as inhabitants of the same geographical region, a dividing line between them is often drawn chronologically, and marks the transition between the Bronze Age (Canaanites) and the Iron Age

[6] The relatively short phase of the Iron Age IIA ends around 918/7 or 926/5 BCE, when Pharaoh Sheshonq I carried out a number of military campaigns into the southern Levant to regain Egyptian influence in the region. This is witnessed by many destruction layers in various settlements of the Iron Age IIA period – although it is not always certain they were caused by Sheshonk I (Nakhai 2001: 183; Weippert 1988: 426).

[7] For the discussion of the 'Low Chronology' (Finkelstein) and the 'Modified Conventional Chronology' (Mazar), see, i. a., summarising Levy et al. 2005; Mazar 2011 with further literature, Finkelstein 2011 with further literature, as well as Ben-Tor 2000: 9–15.

[8] The designation 'Levanti' is also controversial; for a summary of the discussion, see Fischer 2007: 5–7.

[9] The original name *phoínikes* is of Greek origin, a word of Indo-European root indicating 'red', 'blood', or 'death'; Aubet 1993: 6–7. This word is linked by Greek lexicographers to a dark-purple colour, which is connected to the Phoenicians because of a dye they produced to create the most valuable purple fabrics by using a secretion of a sea-snail called murex; see Tubb 2014b: 132. The etymology is, however, by no means entirely clear; for a detailed discussion on etymology, see Aubet 1993: 5–11.

(Phoenicians), even though this cannot be confirmed by textual sources (Albright 1961; Aubet 1993: 10).[10] It is commonly accepted that the transition from the Late Bronze to the Iron Age period has to be seen as a subtle and regionally specific change, rather than as an abrupt event that can be connected with a specific date (Ussishkin 1985). The geographical demarcation of 'Phoenicia' – the region where 'Phoenicians' settled – is based on historical and linguistic considerations (Elayi 2000: 332). According to this, the northern border is situated around the island of Arwad, and the southern frontier is placed around the site of Akko. The eastern border is formed by the various mountains of Lebanon. The area is divided into city-states.[11]

The establishment of a chronological sequence of the Levantine coast has been moved forward only slowly, for which there are different reasons. To begin with, there is a general lack of a continuous stratigraphic sequence of the Iron Age in the coastal region, because many ancient sites like Beirut, Sidon or Tyre either have been overbuilt by modern settlements or show large Hellenistic, Roman, Byzantine, and Islamic superstructures. This makes the study of older periods almost impossible. Secondly, the exploration of the Levantine coast had already started in the 19th century CE, when archaeological techniques were non-existent or in their infancy. Many finds therefore lack information about provenance or an adequate description. Third, military conflicts in this region have consistently interrupted scientific work. Only in recent years have excavations yielded adequately documented results at sites such as Kamid el-Loz, Tell Kazel, Sarepta, Tell Arqa, Tell el-Burak, or Beirut, to mention only few.[12] Finally, and most significantly, this region is of interest for different archaeological disciplines, such as Near-Eastern Archaeology, Biblical Archaeology, Egyptology, and Classical Archaeology, which in pursuit of their specific aims and approaches has resulted in a lack of typological, sequential, chronological and terminological homogeneity.

[10] The terms 'Punic' and 'Carthaginian' instead refer to Phoenicians from the region of North Africa from the 6th century onwards; see Aubet 1993: 11–12. The fact that this dividing line is vague can, for instance, be seen on the basis of excavations at Kamid el-Loz: here the term 'frühe Phönizier' is used to refer to the Late Bronze Age occupation; see Hachmann 1983. For a summary of essential literature with regard to textual sources, see Fischer 2007: 3–4, in particular footnote 3.

[11] Tubb (2014a: 38) in this regard remarks that the part of the Levantine coast that was spared by the destructions of the 'Sea People' later became the territory where the 'Phoenicians' settled. An expansion of the 'Phoenician' territory to northern Syria can probably be recognised on the basis of pottery, which has been claimed to be 'Phoenician' pottery, as well as by 'Phoenician' inscriptions. The latter appear in larger quantities not before the 9th and 8th centuries, and show an increasing interaction during this time with the northern Syrian region; see Lehmann 2008: 241.

[12] See Nunez Calvo 2008; for a summary of most of the sites, see Fontan 2007: 267–280.

In the case of the Iron Age glass finds from the Levant that have been compiled in this book, this means that a large number of glass objects from the Levant are to be expected that have either not yet been excavated or have not been published, or are very difficult to date, since archaeological structures above them have damaged Iron Age contexts.

1.5. The beginnings of glass production in ancient Mesopotamia

Before the first glass was made artificially, naturally occurring glass, such as obsidian, was used widely.[13] Additionally, accidentally-formed glass has to be considered as the predecessor of intentionally made glass. In this regard, glassy slags need to be mentioned, which could occur in any high-temperature environment, as for example in kilns and furnaces in which metals were smelted, pots were fired, or in which faience* objects were produced. These slags were probably the first glassy materials observed by ancient craftsmen (Henderson 2013: 6).

It is also important to clearly differentiate between the very first singular glass finds that occur before the 16th century and the first regularly produced glass present from the late 16th century onwards. Glass that dates prior to the 16th century is rare and often cannot be attributed to secure archaeological contexts. Firmly datable early glass objects comprise a bead from Tell Judeideh that dates to the early 3rd millennium, and a pin-head from Nuzi (burial 5A, stratum IV), dating to the Old Akkadian period (2340–2200).[14] One of the most significant early finds is a translucent blue glass lump from Eridu and attributed to the Ur III period (2112–2004).[15] Chemical analysis has shown that the lump was coloured by a cobalt-rich material, which could indicate that the piece was produced deliberately (Garner 1956: 147–148).[16] It is debatable whether these early glass finds can be regarded as intentional or non-intentional products.[17]

Glass objects on a larger scale appear regularly in the early Kassite period (around 1595), and were promoted by the core-forming* technique (see Chapter 4.3.2) and the production of vessels of this type.

[13] Obsidian is formed from volcanic magma. Natural glass, furthermore, incorporates fulgurites and tektites formed by lightening or metorites, for example in the Sahara Desert; see therefore Henderson 2013: 6; Shortland 2012: 28–29.

[14] For literature and details on these objects, as well as on further earlier glass finds, see Moorey 1994: 190–191.

[15] The lump was found in the 'rubbish', but not directly on the pavement of the house, immediately beneath a pavement dated to the time of Amar-Sîn; see Hall 1930: 213–214.

[16] Henderson (2013: 8) refers to an unpublished find of a greenish glassy slag found in an Akkadian burial.

[17] Shortland (2012: 46) doubts the intentional production of these early finds. Moorey (1994: 193), in contrast, argues that some kind of glass production must have existed prior to 1650.

In this regard, the earliest vessels come from Syrian Alalakh, level VI (late 16th century, according to McClellan) (Moorey 1994: 193).[18] In Egypt the oldest vessels date to the reign of Tuthmosis III (middle 15th century). On this basis, a slightly earlier date to Syria for regular vessel production was therefore often assumedin the past.[19] Some of this evidence has recently been questioned by Shortland *et al.* 2017, who, based on newevidence from Nuzi's glass finds, suggest that glass production in Egypt was no later than in the Near East. Glass finds in Late Bronze Age Mesopotamia and Syria have been well studied, in particular by Barag (1970: 135–154; 1985: 35–49), Henderson (2013: 127–143), Moorey (1994: 196–202), and Shortland (2012: 47–84). Glass objects disappear almost entirely in Mesopotamia and Egypt towards 1200, in the wake of the political, social and economic changes in the region (see Chapter 1.4.2).

[18] Also early mould-made objects, for instance naked figurines, occur at Tell Atchana; see Barag 1985: pl. 2, no. 15, 16. For the latest research on glass from Alalakh, see Dardeniz 2016.

[19] For an extensive discussion on the Egyptian vessels, see Schlick-Nolte 1968; for a recent summary and comparison with Near-Eastern glass, also incorporating chemical data, see Shortland 2012: 47–62.

2. Glass and Glassy Materials: Definitions and Material Properties

Glass and glassy materials are materials made from silica (see Chapter 7.1.1), flux* (see Chapter 7.1.2) and lime (see Chapter 7.1.5) in varying amounts. While glass can only be produced at a temperature of at least 1000°C, glassy materials are sintered, i.e. the basic material mixture is heated at only up to 800°C. The different heating stages lead to different physical properties of the two substances, which are listed in Chapter 2.1.1. It is often difficult to distinguish between glass and the different glassy materials, especially when the condition is poor and there is severe weathering*. In the following chapter, the properties of glass and glass-like materials are discussed and attempts to separate these materials from each other (physical, chemical, physical) are made. The discussion of the weathering properties of glass should also make it easier for the reader to recognise this material within the archaeological context.

2.1. Glass

2.1.1 Physical properties and chemical composition

A discussion of the physical structure, i.e. the atomic arrangement of glass, is necessary to understand how glass behaves at different temperatures, and to understand how it could be manipulated in the Iron Age. Furthermore, the individual basic components of glass and their chemical behaviour are of importance, and which are explained in detail in Chapter 7.1, 7.2; for this reason only a brief reference to the chemical composition is given here.

By far the most common components of ancient glass are soda-lime-silica. Silica forms the largest percentage of the mixture. It exists in many forms on the earth's crust. The most common form is quartz (see Chapter 7.1.1). Quartz melts at a temperature of 1713°C, which was not achievable by ancient craftsmen. It was therefore necessary to lower the melting temperature, which was done with the use of soda serving as a flux* (see Chapter 7.1.2). By the addition of soda, the melting temperature of glass was lowered down to about 1000°C, a temperature that could, for instance, be reached in normal pottery kilns (Pollard and Heron 1996: 156; Shortland 2012: 24). As a third component, lime served as stabilizer. It is essential for the durability of glass, as it prevents the material from dissolving in water (see Chapter 7.1.5). All these substances had to be crushed, ground and mixed to form a so-called 'batch'* before they were heated to at least 1000°C. One of the major characteristics of ancient glass is its colour. Different colours could be accomplished by adding different components to basic composition. In this regard, raw materials that contain oxides of transition metals, for instance copper, were added to the melted batch to create different colours (see Chapter 7.2).

An important characteristic of (ancient) glass is its behaviour under transmissive light, which defines whether a glass is opaque*, translucent or transparent (Figure 2.1). The amount of light that travels through the glass could be influenced by the addition of so-called opacifiers* (see Chapter 7.1.6).

It is, however, not only the specific chemical composition, but rather the specific physical structure that defines the material. The arrangement of the atoms affects its behaviour when heated and cooled and the way the material could be manipulated (Newton and Davison 1989: 4). Concerning their atomic structure, all materials can be divided into one of the three 'classical

Figure 2.1: Opaque (left), translucent (middle) and transparent glass (right) (left, middle: Staatliche Museen zu Berlin – Vorderasiatisches Museum, photo: Olaf Teßmer; right: Courtesy of the Oriental Institute of the University of Chicago).

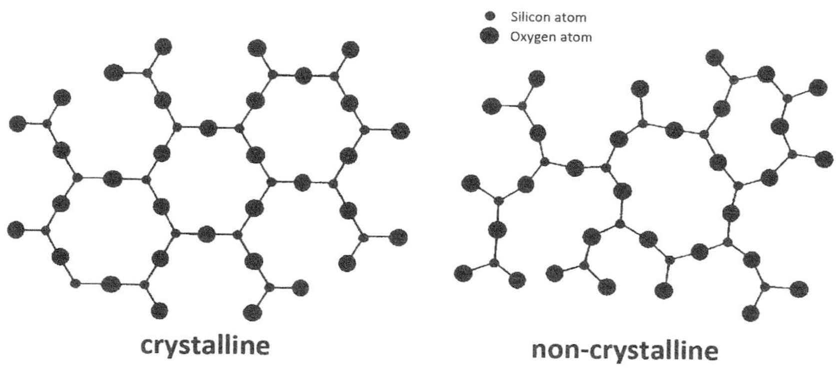

Figure 2.2: Left: Atomic structure of a solid. Right: Atomic structure of a glassy material (Shortland 2012: 23 fig. 1, 2).

states' of matter: solids, liquids and gases. In this respect, glass plays a very specific role as it lies between two of these groups: even though glass is a solid material, it possesses the atomic properties of a liquid. Whereas in solids, the atoms are arranged in a regular and ordered structure (Figure 2.2 left), the atomic structure of liquids is random and disordered. The atomic structure of glass is similar to that of a liquid but in the strict sense resembles that of plastic and rubber, which is also known as the so-called 'polymerised state' (Figure 2.2 right).[1]

The specific atomic structure of glass is a result of heating and cooling. In the process of making glass the major components must be heated to a high temperature, which causes the bonds between the atoms to break. The atoms collapse into a melted state that has the random, disordered structure of a liquid (Brill 1962: 127–129). The melt cools slowly after being heated, the atoms have enough time to find their ways back into the regularly ordered system of a solid. In contrast, if the melt cools too quickly, as is the case with glass, the atoms cannot reorganise into a regular crystalline network. This results in the typical structure of glass (Brill 1962: 131–132; Shortland 2012: 20). This irregular structure of the network causes the glass to break irregularly, resulting in a typical conchoidal fracture, similar to flint or obsidian (Freestone 1991: 38). In some cases an identification of the archaeological material glass can solely be achieved on the basis of this specific type of fracture, as in many instances glass suffers from severe corrosion.

2.1.2 Melting properties and workability

The atomic structure of a material directly influences its melting properties. While crystal melts at a specific temperature, glass does not have a defined melting point, but softens gradually. The reason for this lies, as indicated, in the atomic bonds; within crystals, the bonds between atoms are identical, as the network is

Reference point	*Definition*	*Viscosity (poise)*	*96% SiO_2 (°C)*	*Soda-lime silica (°C)*
Working point	Sufficiently soft for shaping	10^4	–	1000
Softening point	Glass tubes can be bent in a flame	$10^{7.6}$	1500	700
Annealing point	Internal stresses removed in a few minutes	$10^{13.4}$	900	510
Strain point	Highest temperature from which the glass can be rapidly cooled without serious internal stress	$10^{14.5}$	820	470

Figure 2.3: Degree of viscosity and corresponding reference points (Pollard and Heron 1996: 151).

regular. The bonds therefore break down at the same temperature. In contrast, bonds within glass vary, and therefore the breaking down of the network occurs within a wider range of temperature.[2]

Thus the different working states for glass are directly related to temperature. The relationship between temperature and stiffness is viscosity, which is a measure of resistance to flow expressed in poise. This means that the lower the viscosity, the more solvent, and the higher the viscosity, the stiffer the substance.[3]

At room temperature, glass is rigid, hard and brittle and can only be worked by using cold-working techniques. When the temperature rises, glass starts to flow slowly and becomes more and more plastic and flexible. At the so-called 'softening point' the glass can be drawn in rods. The strain levels at this point are high, making it likely to break. If the temperature is raised, the glass flows faster and becomes more and more fluid. In this state it is possible, for instance, to fill the material in a mould* (working point) (Figure 2.3).

[1] See Newton and Davison 1989: 4, and Shortland 2012: 20 for further details.

[2] The 'random network theory' was the first on the structure of glass to become widely accepted; see Zachariasen 1932; Brill 1962: 131; Pollard and Heron 1996: 152–154.

[3] For instance, honey has a higher viscosity (104) than water (101.2); see therefore in detail Newton and Davison 1989: 12; Pollard and Heron 1996: 151.

After being worked, glass must be cooled gradually and held at certain temperatures to release the stresses. This is called annealing*. If glass is not annealed properly the strain will cause the glass to crack or shatter (Cummings 1997: 161–166; Shortland 2012: 27). Today, annealing is carried out in a heated chamber called a 'lehr' or 'garage', but it is still a critical step in glassmaking (Hess and Wight 2005: 2–3). Various possibilities of annealing in ancient Mesopotamia will be discussed throughout the text.

2.1.3 Weathering effects

Weathering* is a severe problem regarding glass objects, especially from Mesopotamia. Soils in Mesopotamia are saline and humid compared, for example, to the dry and calcareous soils of Egypt. Glassy materials found in Mesopotamia are very vulnerable to weathering, in contrast to objects from Egypt, which are much better preserved (Bouquillon *et al.* 2008: 93). Therefore, glass objects from Mesopotamia and Egypt from the same period are generally very differently preserved, with Egyptian objects showing an intensive colouration and thick layer of glassy surface, whereas Mesopotamian objects are largely dull and heavily weathered (Figure 2. 7).

Weathering* is caused by the chemical reaction of water and gases in the environment. It mainly occurs on the surface of glasses, leaving behind siliceous weathering products that easily flake off. The corrosion layers have, in most cases, an iridescent appearance showing different colours (Figure 2.4) (Hess and Wight 2005: 47, 87; Pollard and Heron 1996: 173). In addition to iridescence*, also pitting*, the formation of little holes, occurs frequently on the surface (Figure 2.4).

Figure 2.4: Layer of colourful iridescence and pitting. Due to corrosion, parts of the surface are flaked off and exhibit a layer of iridescence and severe pitting (detail: Barag 1985: plate B no. 27).

Weathered layers often hinder an appropriate assessment of glass as they influence colour and light transmission. Formerly transparent glass can, for instance, appear as a whitish opaque glass, and formerly opaque blue colouration can turn into a greenish colour (see Chapter 7.2.3). An identification of the original colour and characteristics of glass have therefore to be cautiously considered and can often only be securely determined on the basis of chemical analysis.

2.2. Glassy and sintered materials: definitions and compositions

2.2.1 Faience

Faience* is a vitreous material that is made of powdered quartz, flux* and lime, which are combined with water to make a clay-like paste. The paste is formed in a mould in cold state. The faience object is afterwards heated at a temperature of around 800–900 °C (Kühne 1999: 105). During this sintering* stage, the basic components that were mixed with water become a coherent mass, but do not fuse (Hess and Wight 2005: 76). Even though faience and glass exhibit the same major components, faience differs from glass because the base material is not completely fused together to a liquid melt, but is solely sintered (Caubet and Bouquillon 2007: 13; Moorey 1994: 167; Vandiver 2008).[4] Faience is characterised by a glazed layer, which is due to alkali salts that migrate to the surface during the drying process, which happens before the objects are fired (Figure 2.5). This process is called efflorescence or self-glazing (Paynter and Tite 2001: 241). Only after the faience object has dried can it finally be burned at high temperature.

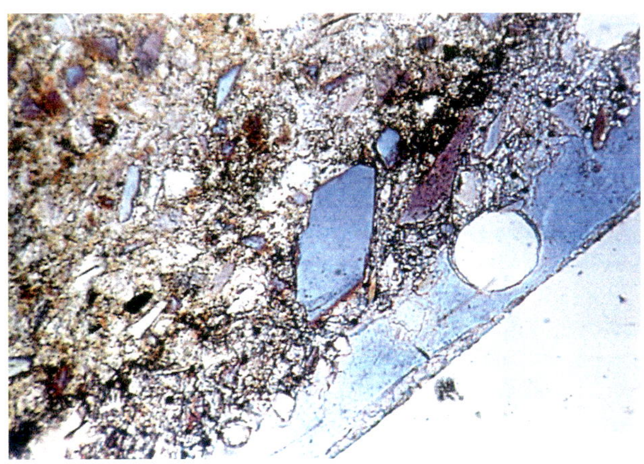

Figure 2.5: Thin section of a blue faience bead. The outer surface is covered by a fully fused glass layer (Kühne 1999: 105, fig. 2).

The thickness of the glazing layer depends on the amount of flux* used in the batch* – the higher the amount of flux in the batch, the thicker the glazing layer, and the

[4] For details and relations of the raw materials used in faience objects from Mesopotamia, see Bouquillon et al. 2008: 93–103. For components and exact mixing ratios in ancient Egyptian faience, see Kaczmarczyk and Vandiver 2008 and Taj-Eddin 2014: 83–84.

duration of the drying time. The longer the object dries the more efflorescence salts can migrate to the surface.[5] Faience* can be recognised by the fact that the colour does not run through the entire body of the object, but is only present in the glassy layer on the surface (Figure 2.6). This is a result of the migration process of the water to the surface during the drying phase, which also transports the colourants* to the surface.

Figure 2.6: Faience bowl exhibiting a blue glaze and light-brown core (Egypt, 18th dynasty) (Busz and Gercke 1999: 282, no. 17).

Faience* is often referred to as 'Egyptian faience' because 'its fullest development and best achievements were achieved in Egypt' (Noble 1969: 435).[6] Since faience objects are not only found in Egypt, but also in Mesopotamia and the northern and southern Levant, the term 'Egyptian faience' is misleading and should therefore be avoided. The term faience itself is problematic because it describes a medieval tin-glazed pottery produced in the Italian city of Faenza, also known as majolica (Noble 1969: 435). Since the English term faience is so deeply ingrained in literature, this term is still used and therefore also in this monograph (Moorey 1994: 167). In German, the terms 'Kieselkeramik', or more frequently 'Quarzkeramik', were proposed instead of 'faience' to acknowledge the actual composition of the material. This terminology is much more precise and avoids misleading geographical references and should therefore be used in German (Wartke 1999).[7]

In Mesopotamia, faience* emerges for the first time in the final phases of the Ubaid period and mainly occurs in connection with the production of beads. By the late 2nd millennium, polychrome faience objects start being produced, coinciding with the start of glass production and the emergence of glazed ceramics (Bouquillon 2008: 94). In this period, faience was still largely a 'prestige technology' associated with palatial contexts, for instance in Nuzi or Tell Rimah (Moorey 1994: 178). Faience of the 1st millennium differs considerably from the repertoire of the previous period. Almost exclusively beads, pendants and seals were manufactured during that period (Moorey 1994: 180). Faience vessels are largely absent in Iron Age Mesopotamia, while glazed pottery vessels and glazed bricks occur now frequently. The situation is different at sites situated in western Iran, such as Hasanlu, Ziwiye or Susa, where a considerable amount of faience vessels and bricks were found (Moorey 1994: 180).

With regard to its manufacturing, the pulverised raw materials (silica, lime, soda, colourants) are mixed in water to form a paste. Unlike clay, faience paste is difficult to shape, this is why moulds and cores were employed. Only a limited number of forms and sizes could therefore possible be achieved (Taj-Eddin 2014: 41).

2.2.2 Glassy faience

The term 'glassy faience*' refers to a more compact and less porous form of the material, and represents a stage between it and glass. The major difference of glassy faience to faience is that the colour is evenly spread throughout the body core, and is not restricted to the surface of the object, as it is with regard to faience objects (Brill 1970: 114). Glassy faience is cast into moulds*, similar to the manufacturing of faience objects (see Chapter 2.2.1). In the present study, glassy faience is of particular importance regarding mosaic bowls (see Chapter 4.1.2.1).

2.2.3 Frit

Moorey (1994: 167) described frit as 'is the most problematic of the terms relevant to the subject, since it is often used as if it were synonymous with faience', but the two should be kept apart from one another. Whereas faience* has a glazed surface, frit is characterised by an almost complete absence of such. Similar to faience, frit is sintered at a temperature of around 800–900C°. Frit has a specific meaning within the process of glassmaking, as the process of repeated fritting aims to produce an intermediate product which is free from impurities. This process is also described in detail in the cuneiform texts (see Chapter 6.6.3). The production of frit as an intermediate product was therefore common practice regarding ancient glassmaking.[8]

[5] For detailed descriptions on the compositions of faience from the Near East, see Bouquillon et al. 2008: 93–103.
[6] In the German language 'Faience' is frequently associated with an Egyptian origin, whereas the term 'Fritte' is used for a Mesopotamian origin of the same material.
[7] For a general overview of vitreous materials, see Freestone 1997: 306–309.
[8] Schlick-Nolte (1999: 20) refers to this as 'Glasfritte'; see also Moorey 1994: 167–168.

Figure 2.7: Glazed pottery vessels in different states of preservation. The bottle in the middle exhibits crazing (Babylon, 900–500 BCE) (Busz and Gercke 1999: 341, no. 152–154).

In addition to this technological function, the term frit is often also used to describe a polycrystalline object that has, in contrast to faience*, an unglazed surface. Therefore, if the glazed surface of a faience object is worn off completely, it is sometimes difficult or impossible to distinguish it from frit. This is because, regarding the German usage, 'Fritte' usually refers to Mesopotamian objects that are less well preserved, whereas 'Faience' is used for Egyptian objects. Coloured frit can only be distinguished from unglazed faience if the object is broken, as the colour in frit objects runs throughout the material, whereas with faience objects the colour is only present in the glazed layer of the surface (Moorey 1994: 167). In practice, a distinction between frit and unglazed faience cannot be made in most cases. For these reasons, frit is solely used in the present study to describe the intermediate product in glass production.[9]

2.2.4 Glazes

Glazes are layers of glass that are bonded on a substrate, for example on clay, silica or stone. The composition of glazes is very similar to glass, primarily containing large proportions of quartz (Brill 1970: 115).[10] Glazes occur on the substrate material either in the form of self-glazes (efflorescence), as discussed for faience*, or they can be applied to the surface of an object, for instance, by the 'application process'. The glazing material was applied onto the surface of the object as a powder or slurry.[11]

Glazed quartz and steatite objects were first produced in Mesopotamia, Egypt and in the Indus valley, at the end of the 5th millennium, for the production of beads, and form, together with faience, one of the earliest groups of glazed materials.[12]

Glazed pottery, in contrast, occurs in Mesopotamia for the first time in the middle of the 2nd millennium (Paynter 2001: 239–254), and falls together, therefore, with the beginning of glass production, showing technical correlations between these two crafts. To produce a glazing suspension, quartz and a colourant had to be added to the alluvial clay in Mesopotamia. These compounds were ground and mixed together. It is likely that the mixture was first fritted and ground and thereafter added to a water suspension.[13] The ceramic object with glazing suspension was afterwards heated to 1000°C under oxidised kiln conditions (Bouquillon 2007: 15). The crucial point of glazing pottery is the effect of crazing, which occurs because of different thermal expansions of glaze and clay (Figure 2.7). If during cooling the glaze shrinks much more than the clay, the stresses on the glaze cause it to craze. Therefore, the composition of the clay is of particular

[9] With regard to the use of terminology in German, frit should similarly only be used in connection with glass production. For any kind of sintered objects, the neutral term 'Quarzkeramik' is to be preferred.

[10] There are two different kinds of glazes which are classified by the flux used in the composition, these are alkaline glazes, fluxed with plant ash or mineral natron, or lead glazes, fluxed with lead; see Moorey 1994: 167.

[11] The cementation process involved burying the object in a glaze mixture, which was followed by firing; see Paynter and Tite 2001: 241. This method is not attested for ancient Mesopotamian glazes.

[12] In Egypt they start being made in the last quarter of the 5th millennium. For an extensive history of glazing and glazed objects, as well as the methods of glazing, see Bouquillon et al. 2008: 23–36; Moorey 1994: 169–171.

[13] This suggestion is based on ethnographic observations, therefore Paynter and Tite 2001: 245.

importance when producing glazed ceramics (Paynter and Tite 2001: 247). Mesopotamian clays have a high content of calcium oxide, which has a positive impact on the expansion coefficient. Therefore glazed ceramic objects were considerably easier to manufacture there. In contrast, glazed clay objects from Egypt were not produced until the 1st millennium because clays available in Egypt were less suitable for glazing (Paynter and Tite 2001: 247). While solely monochrome glazes on clay occur in the Late Bronze Age, polychromatic glazes are produced from the Neo-Assyrian period onwards (Bouquillon 2007: 15).

2.2.5 Summary

Glass, faience*, and glazes belong to the group of vitreous materials, which are products of high temperature processes that cause the production of a certain amount of glass in the material. The basic substance of all vitreous materials is silica. They are therefore not clay-based. Vitreous objects can be distinguished from one another when well preserved. Frequently, however, early vitreous materials have suffered weathering*, which results in the loss of glaze. This is particularly severe regarding faience objects, as the glaze layer is thin, leaving a white and powdery surface (Tite and Shortland 2008: 20). Because of the usually dry conditions in Egypt, the glazes and glass phases are much better preserved there than in Mesopotamia.

Glass and faience objects from Egypt therefore usually exhibit glazed surfaces and strong, shining colouration. In contrast, the colours of Mesopotamian glass and faience objects are rather dull, the glazing layer is often only visible in some parts (Moorey 1994: 167). A differentiation between the different glassy materials cannot be observed, in most of the cases, with the naked eye. Therefore the use of a magnifying glass of at least x 10, or even SEM or X-ray diffraction analysis, is necessary (Tite and Shortland 2008: 21).

Material Description	Core Material	Emergence in Mesopotamia
Faience	Sintered quartz	Late Ubaid
Glassy faience	Sintered quartz	No comparative data available
Frit	Sintered quartz	Intermediate product in glassmaking
Glaze	Ceramic; quartz and steatite stones	Ubaid (quartz and steatite stones); Late Bronze Age (ceramic)
Glass	Glass	Late Bronze Age

Table 2.1: The material descriptions, their core materials and the date of invention discussed in this chapter.

3. Archaeological Contexts: Sites with Iron Age Glass Finds

The following chapter provides a discussion of those sites and archaeological contexts in which glass objects were found (Figure 3.1). With regard to the excavation history, solely campaigns and monographs relevant to the glass objects and their stratigraphic context will be presented. Reference will also be drawn to the collections and museums in which the objects are stored. The glass finds included in this work represent a selection of objects that were either published or made available to the author for personal study. They are therefore to be understood as a selection of finds.

3.1. Assyria

3.1.1 Aššur

Aššur, modern Qal'at Šerqat, lies on the west bank of the Tigris, *c.* 110 km south of Mosul. The city is situated on a natural mound, 40 m above the alluvial plain, and is flanked on two sides by branches of the river Tigris.

The site of Aššur looks back over a long excavation history. A major source referring to a large number of unpublished finds, also on glass, are the so-called 'field-journals', in which all information on the finds and stratigraphy from various contexts are kept. Many of the glass finds from Aššur relevant to this study are only listed in these field-journals and were made available to the author during the preparation of this monograph in the form of an electronic database.[1] There is no summarising publication on the glass finds from Aššur yet published. The mosaic glass* and glassy faience* objects were, however, issued in an article by Haevernick 1968, and also Wartke 2012 referred to one of the mosaic bowls in detail, focussing on the material and manufacturing techniques, also including experimental studies. Beyond that, the glass objects were presented only in connection with stratigraphic works of different building structures. In this respect the following monographs are of particular importance: Andrae 1913; 1967; Haller 1954; Hauser 2012; Miglus 1996; Pedde 2015; Preusser 1955; Schmitt 2012. Today, the glass finds from Aššur are part of the collections of the Vorderasiatisches Museum in Berlin and the Istanbul Arkeoloji Müzeleri, where they arrived through divisions of finds.

In the following the different find contexts of the glass objects are reconsidered in order to suggest a date for them. Particularly to be emphasised are the mosaic objects, including bowls, inlays and tiles, since their dating has been discussed repeatedly, falling either into the Middle-Assyrian or Neo-Assyrian periods (Busz and Gercke 1999: 339; Haevernick 1968: 68; Marcus 1991; Wartke 2012). In the following sections the find contexts of the individual mosaic objects will be discussed, which contribute decisively to the suggested date of mosaic objects in Chapter 4.1.1.

3.1.1.1 Ištar and Nabû temples

The stratigraphic attribution of some of the finds to a specific layer is often not explicit. Some descriptions of the findspots of the objects often lack precision. With the mosaic glass vessel (VA Ass 3656) (Haevernick 1968: Farbtafel I, 1), this is different, and to which an exact location is assigned: 'Im Altarraum östl. der Altäre bis zur Mauer des Nabû-Tempels gefunden im Schutt unmittelbar auf dem Ziegelpflaster bis ca. 60 cm darüber.' (Haevernick 1968: 63, no. 1). Based on this description, it becomes evident that the findspot of the mosaic fragments can only refer to room 6 of the Ištar Temple, and therefore to the reign of Tukultī-Ninurta I (1243–1207). Schmitt (2012: 38), as well as Andrae (1967: 24), refer to a 'Schuttschicht' of around 60 cm thickness that served as a filling, to create an elevated floor level for room 6. The given elevation of this filling corresponds with the information in the field-journal. Furthermore, the journal indicates a findspot east of the altars ('Altäre'). The altars most likely refer to two stone pedestals (Ass19835, Ass19868) which can be identified on the plan of room 6 of the Ištar Temple (Figure 3.2). The journal further defines the findspot as between the altars and the walls of the Nabû Temple. On basis of this description, the findspot of the mosaic objects is suggested in the area between pedestal Ass19868 and the south-western outer wall of the later Nabû Temple (Figure 3.2, red circle). Finally, the find register mentions a fragment of a similar, perhaps even the same, glass vessel that was found in room 6, thus making an attribution of the pieces to this context very likely. In summary, the findspot of the mosaic glass provides the object with a Middle-Assyrian rather than a Neo-Assyrian date.

A number of sherds from glass vessels are mentioned in connection with the Nabû Temple by Schmitt (2012: 209–210, 230), but no details about the objects are provided. However, this shows that glass objects generally form part of the temple inventory of the Nabû Temple.

[1] I would like to thank Markus Hilgert sincerely for granting me access to the collection of the 'Vorderasiatisches Museum', Berlin. My gratitude goes also to Lutz Martin and Helen Gries for their great support by providing the objects for study. I would also like to thank Johannes Renger for granting me access to the Aššur database and allowing me to use the information.

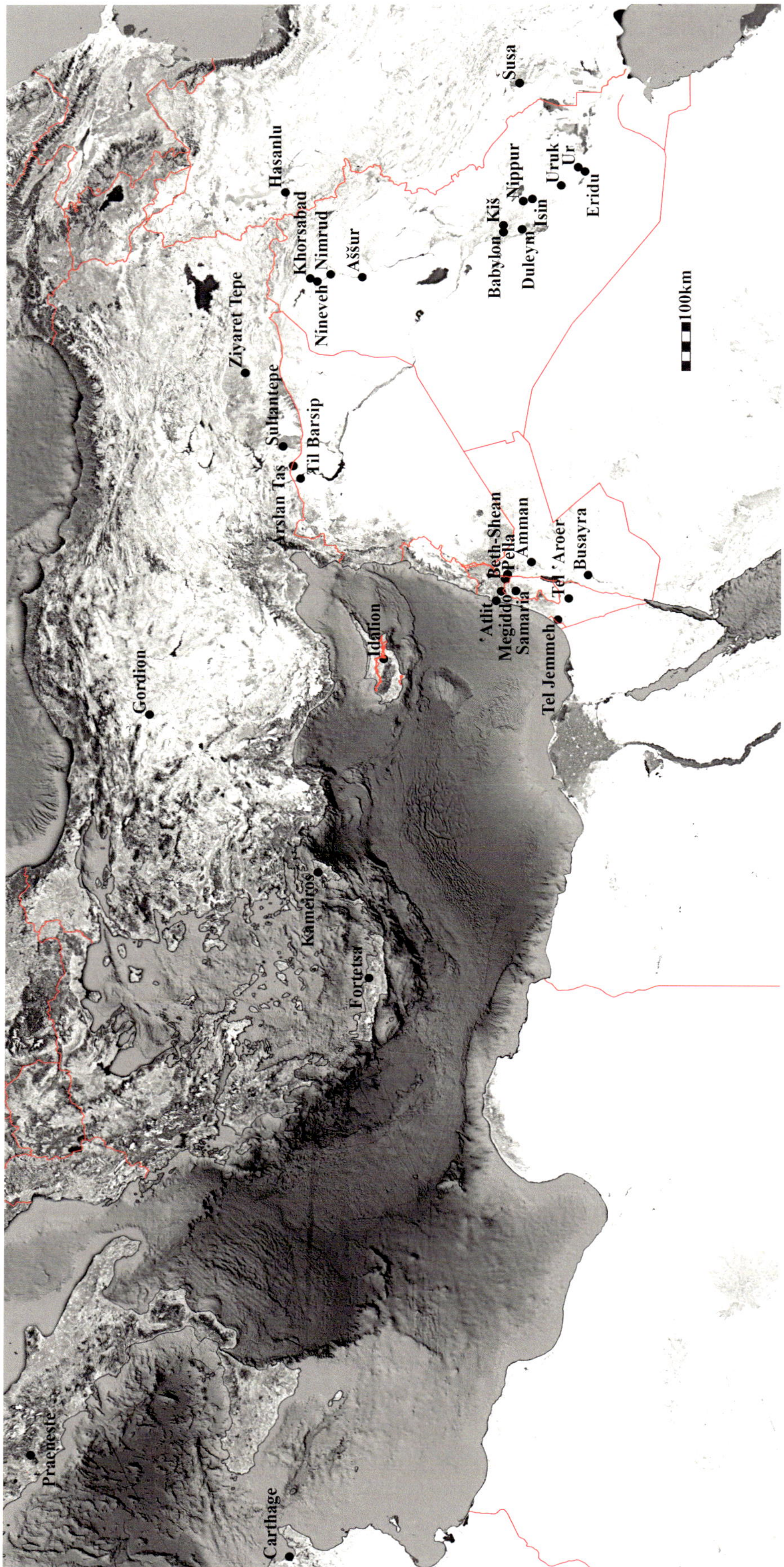

Figure 3.1: Map showing sites from which finds are included in this study.

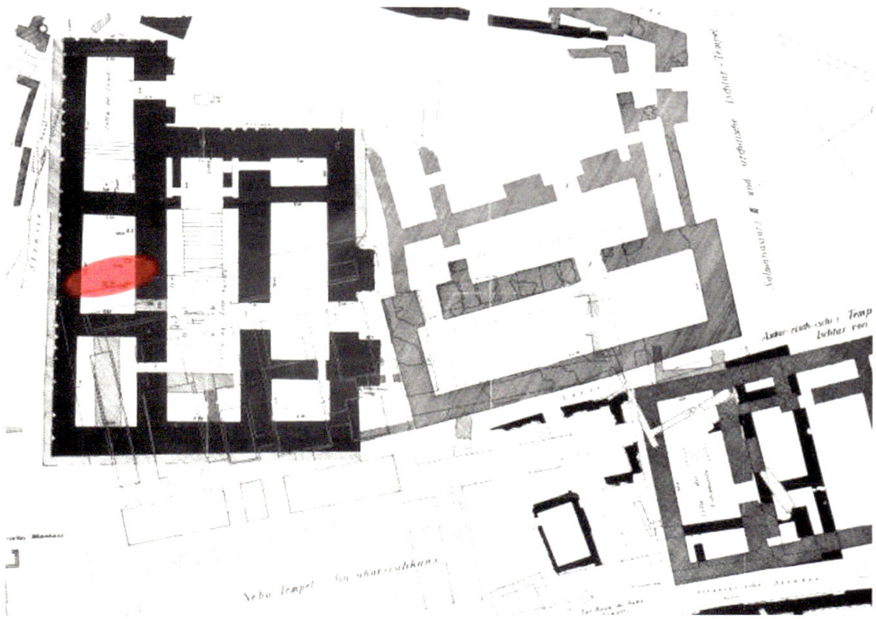

Figure 3.2: Plan of the Ištar Temple of Tukulti Ninurta I (black structures) and the Nabû Temple (white structures). The red circle indicates the findspot of the mosaic glass vessel (VA Ass 3656) (Andrae 1967: pl. 1).

3.1.1.2 Graves

The almost intact mosaic bowl As1 belongs to the tomb furniture of grave no. 311 (Ass12481b). The corpse in this burial was covered with three large pottery sherds (Pedde 2015: pl. 80).[2] The mosaic bowl was placed on the stomach, the hand on it. In addition, a rounded alabastron with lid was laid next to it (Pedde 2015: pl. 80). The burial is considered post-Assyrian by Haller (1954: 29), whereas Miglus (1996: 385) suggests, on the basis of stratigraphic considerations, a Neo-Assyrian date, and Pedde (2015: 101), on basis of the mosaic bowl and the alabastron, proposes a Middle-Assyrian date.

The core-formed vessel As9 was found in grave no. 691 ('Kompositgrab') together with six ceramic bottles, one beaker, two small glazed bottles, a terracotta head, a copper bowl with traces of food offerings, a handle, a bone needle, two silver rings, and an agate bead set in gold (Haller 1954: 88). The burial is assigned by Haller (1954: 88) to the Neo-Assyrian period, which is confirmed by Miglus (1996: 382) on the basis of stratigraphy.

3.1.1.3 Other contexts

In addition to the examples from temples and graves, discussed above, a number of other glass objects were found at Aššur with no precise contextual information. Mosaic inlay As6 can be attributed to 'Fundstelle: i 15. Südwall, O. Wallkrone' (Haevernick 1968: 64). As the given findspot incorporates a vast area (grid square i 15), neither contextual nor chronological conclusions can be drawn. The same is true for the mosaic inlay As7, as the object comes from the debris of waste material, which is indicated in the field-journal. However, As6, As7, and inlay As8 fall into the same typological group and, therefore, a chronological proximity is likely (see Chapter 4.1.3.2).

Several fragments of mosaic bowls have been found in Aššur. One of the fragments, As2, was found close to the Tukultī-Ninurta I canal ('am Tukultī-Ninurta I Kanal'), and can therefore not be assigned to a specific context; Miglus (1996: 88) points towards lead plaques which he defines as Middle-Assyrian; Wartke (2012: 405), in contrast, refers to a glazed terracotta statue of a horseman and to a glazed tile decorated with plaited bands which he attributes to the Neo-Assyrian period.[3] Therefore the accompanying finds hint either to a Middle-Assyrian or a Neo-Assyrian date.

Similarly difficult is the chronological attribution of the mosaic bowl fragment As4, which was found in the area of the terrace of the palace of Tukultī-Ninurta I (dC6II) together with terracotta figurines typical for the Neo-Assyrian period. In addition, the mosaic bowl fragments As3 and As5 originate from a context that cannot be attributed to a specific locus and period: 'e D 7 I, Suchgraben östl. der Lehmziegelkante.' (Haevernick 1968: 66, no. 13).

A large number of thick mosaic glass tiles (As13, As14, As15, As16, As17) of the same typological group come from an area around the Ziggurat, and, to a lesser extent, around the Ištar Temple, but were found not *in situ* (Haevernick 1986: 66, no. 11).[4] The accumulation of the

[2] Haller (1954: 29) classified the burial as 'Scherbengrab'.

[3] For comparable terracotta figurines, see Klengel-Brandt 1978: 85–88.

[4] Haevernick 1968: 66, no. 13 reports that there was a great number of mosaic glass tiles found around the ziggurat. It is therefore likely that only a small number of the original corpus has been preserved.

mosaic tiles around the Ziggurat and Ištar Temple are striking, and even though they lack exact localisation it is likely that they were used as decorations for these building structures (see Chapter 4.1.3.3). If an association with these monuments is assumed, the pieces date rather to the Middle-Assyrian (suggested by Haevernick 1986: 68) than to the Neo-Assyrian period (suggested by Busz and Gercke 1999: 339).

3.1.2 Khorsabad

Khorsabad, ancient Dūr Šarrukīn, is located in northern Iraq, *c.* 20 km north-east of Nineveh (today part of Mosul).

During the excavations led by Botta during the 1840s, an assortment of finds and relief slabs from the residences on the citadel was sent to Paris (Musée du Louvre), of which the majority of these objects sank in the Tigris (Caubet 1995). From 1927 until 1935, the Oriental Institute started a major field-project at the site, first under Chiera (1927–1930), and later under Frankfort (1930–1932) and Loud (1932–1935) (Loud and Altman 1938).[5] Small finds and relief slabs came to the Iraq Museum in Baghdad and to the Oriental Institute Museum in Chicago. So far, none of the glass objects has been published, the relevant pieces in this monograph were studied by the author in the Oriental Institute Collection, as well as in the Musée du Louvre.

The site was founded as a capital by Sargon II (721–705). The city remained unfinished in the year of Sargon´s death and was thereafter abandoned for a long period of time (Albenda 2003: 6–7). All architectural remains and finds, and therefore also the glass finds, can largely be attributed to Sargon II.

According to the field-register available at the Oriental Institute, Khor1, the fragment of a colourless glass bowl, was found in room 51/ 52 in Residence K, which is one of the four 'private' residences on the citadel. Rooms 51 and 52 were, according to Loud and Altman (1938: 68, pl. 71), part of the service area around the court.[6]

Seven large fragments of opaque, light blue glass (Khor2, Khor3, Khor4, Khor5, Khor6, Khor7, Khor8) can mostly be identified as large inlays serving as interior design features for the palace. The exact findspots are, however, unknown.[7]

3.1.3 Arslan Taş

Arslan Taş, ancient Ḫadatu, is situated in the Jezireh, 30 km east of the Euphrates, on the Seroudj plain near the modern city of Kobane in Syria. The site is located along an important east-west route connecting the ancient cities of Tell Halaf, Harran, Til Barsip and Carchemish (Albenda 1988: 5–6).

The glass objects from Arslan Taş comprise inlays that were formerly inlaid in ivories. In 1928, Thureau-Dangin (1931) unearthed the so-called 'Bâtiment aux ivoires'; the excavation report, however, largely lacks information concerning the small finds, including glass. This is the reason why in 2006, a joint mission of the General Directorate of Antiquities and Museums of Syria, the University of Bologna and the Musée du Louvre initiated a study and restoration program that focused on the ivories that were found in the 'Bâtiment aux ivoires' (Cecchini 2009; Cecchini *et al.* 2012).[8] Most of the ivories that were decorated with glass inlays were found in room 14 of this building, which was already recorded by Thureau-Dangin (1931: 89–92). Regarding the stratigraphy of the 'Bâtiment aux ivoires', no connection to the Assyrian Palace dominating could be identified in the recent excavations. According to Cecchini and Venturi (2012: 330), the erection of this building dates prior to the construction of the Assyrian Palace, probably in the first half of the 8th century.[9] Regarding the use of the ivories, Cecchini (2009: 94) defines them as inlays for couches and chairs. With regard to the classification of the building and its function – especially with regard to the analysis carried out in Chapter 5 – the categories of a 'palace' are proposed here. This results from the ground plan of the building (Thureau-Dangin 1931: 41–54) and the discovery of ivory plates with the Aramaic inscription of Hazael, king of Damascus.

Another chapter in the history of the Arslan Taş ivories occurred in the 1950s, when around 80 pieces said to come from the site emerged on the antiquities market and were acquired by the collector Borowski (the so-called 'Borowski Collection'). Some 65 glass rosettes of different types were acquired from this collection by different museums, such as the Badisches Landesmuseum (AM13, AM14, AM15, AM16, AM17, AM18, AM19, AM20, AM49), the Kunst- und Gewerbemuseum in Hamburg (AM36), the Metropolitan Museum of Art (AM21, AM22, AM23, AM24, AM25, AM26, AM27, AM31, AM32, AM33,

[5] An overview of the research of the Oriental Institute at Khorsabad is provided by Wilson 1995.

[6] My thanks goes to Kiersten Neumann, for allowance to study and publish the colourless bowl fragment from Khorsabad which is part of the Oriental Institute Museum in Chicago.

[7] I would like to thank Ariane Thomas for the granting me access to the inlays from Khorsabad, which are kept in the Louvre museum and allowing me to publish them.

[8] During the field-campaigns between 2007 and 2009, the 'Bâtiment aux ivoires' was relocated, allowing for the possibility of carrying out research on the missing data of the small finds.

[9] After a large non-Assyrian occupation phase, the site came under Assyrian control in the 9th century (phases III–II). It is very likely that, during this period, the city had far-reaching political autonomy. At the beginning of the 8th century, Arslan Taş became a provincial capital and was ruled by the turtānu Shamshi-ilu and his eunuch Ninurta-bel-usur (phase I); see Cecchini and Venturi 2010.

AM34, AM35, AM37, AM38, AM39, AM), and the British Museum (AM28, AM29, AM30). It will remain unknown whether the ivories were removed from Arslan Taş before or after the excavations of Thureau-Dangin. Their authenticity, as well as their provenience from Arslan Taş, is widely accepted, based on stylistic and typological considerations, but nonetheless this cannot be demonstrated with absolute certainty.

3.1.4 Nimrud

Nimrud, ancient Kalḫu, is located on the east bank of the Tigris, 30 km south-south-east of Mosul in northern Iraq.

The majority of glass objects from the Iron Age period from Mesopotamia that have so far been published come from Nimrud (Barag 1983; Brill 1978; Curtis 1999; Mallowan 1966; Orchard 1978; Saldern 1966a). Also, some archaeometrical work was carried out on some of the objects (Turner 1955; Turner and Rooksby 1959) (see Chapter 7.4.3). Many of the glass objects incorporated in this work were studied by the author in the British Museum and in the Metropolitan Museum.

Most of the glass finds from Nimrud come from palatial contexts (Northwest Palace, Burnt Palace, Fort Shalmaneser), and comprise mostly objects of the cast-and-cut type. Regarding a date for most of the glass objects, an attribution to a specific period is difficult since most of them were found in the destruction layer of the residential buildings. Throughout the history of the site, different kings built new residences and temples, and also rebuilt the existing structures. Ashurnasirpal II (883–859) turned Nimrud into a royal centre, and the activities were still carried out under his two sons until 614, when Nimrud was attacked for the first time. In 612 Nimrud was finally sacked, which is attested in the archaeological record by a vast destruction layer that can be observed in many rooms of the palaces and temples.[10] After the first attack, attempts for a resettlement were made, seen by the hoards that were gathered, for example in room AB in the Northwest Palace, or in room SW37 in Fort Shalmaneser. Curtis (2013: 2) distinguished three phases regarding the chronological attribution of the objects: 1) pre-614; 2) 614–612; and 3) post-612. Since the floors of the major buildings were kept intact and clean, objects found prior to the first destruction in 614 cannot be dated precisely. Therefore, in most cases, no distinction can be made between phases that go back to even the 9th century, when most of the buildings were founded, and the late 7th century, when they were destroyed (Curtis 2013: 2). Since most of the glass objects were found in the destruction layer, solely a *terminus ante quem* correlated with the fall of the city in 612 can be supposed.

3.1.4.1 Ninurta Temple

The Ninurta Temple was built under Ashurnasirpal II (883–859), and partly under Shalmaneser III (858–824); it was probably in use throughout the entire Neo-Assyrian period (Curtis 2013: 9). Nim6 was found in a layer with traces of fire (melted lead objects, burnt mud-brick and beams, ash layers) in the Ninurta Temple (Layard 1856: 357–358; Mallowan 1966: 92). Nim6 is a larger than life-size inlay that belonged to a composite statue, as do Nim1, Nim2 and Nim5, which can be identified as composite attachments of different beards that were also found in the same context.

3.1.4.2 Northwest Palace

The Northwest Palace was built by Ashurnasirpal II (883–859) as a royal residence in the new capital. It was repaired multiple times, most notably under Shalmaneser III (858–824). The palace lost some of its status after Sargon II (721–705) shifted the capital to Dūr Šarrukīn (Khorsabad). From that time, many of the rooms were moved over to storage rooms and therefore lost their original function. This circumstance becomes particularly important regarding the location of some of the glass objects discussed. In the south wing, four occupation phases were differentiated, two of which date prior to 612 (Curtis 2013: 2–3).

Nim27 ('Sargon Vase') was found in room 1, which is situated in the north-westernmost part of the palace. It was part of a set of rooms flanking the northern side of the throne room courtyard through which the palace was entered (Figure 3.3). The courtyard and adjacent rooms form the major part of the *bābānu*, serving as an administrative centre and easily accessible (Kertai 2015: 26–27). Besides Nim27, several pieces of alabaster vases, and armour, were found in room 1, suggesting its use as a magazine, until it was finally destroyed in 612 (Kertai 2015: 85).

The glass objects Nim72, Nim74, Nim75, Nim76, Nim77, Nim78, Nim120, Nim176, Nim177, Nim178, Nim179, Nim180, Nim181, Nim182, Nim184, Nim185, Nim186, Nim187, Nim188, Nim189 were uncovered by Mallowan (1952), in the so-called 'Layard's dump' in room V, a backfill created by Layard. Therefore the original position of this debris is uncertain, although Mallowan (1952: 45) attributed it to the original inventory of room V. This is because most of the ivories found in the 'dump' match the ivories that had been previously found in it. Room V is a small stone-paved room (*c.* 8 x 5 m), only accessible through room X, situated to the south of the Great Courtyard Y, an area identified as the private domestic quarters of the palace (*bītānu*) (Figure 3.3).

[10] The observation that Nimrud was sacked twice is mainly based on research carried out by Oates at Fort Shalmaneser.

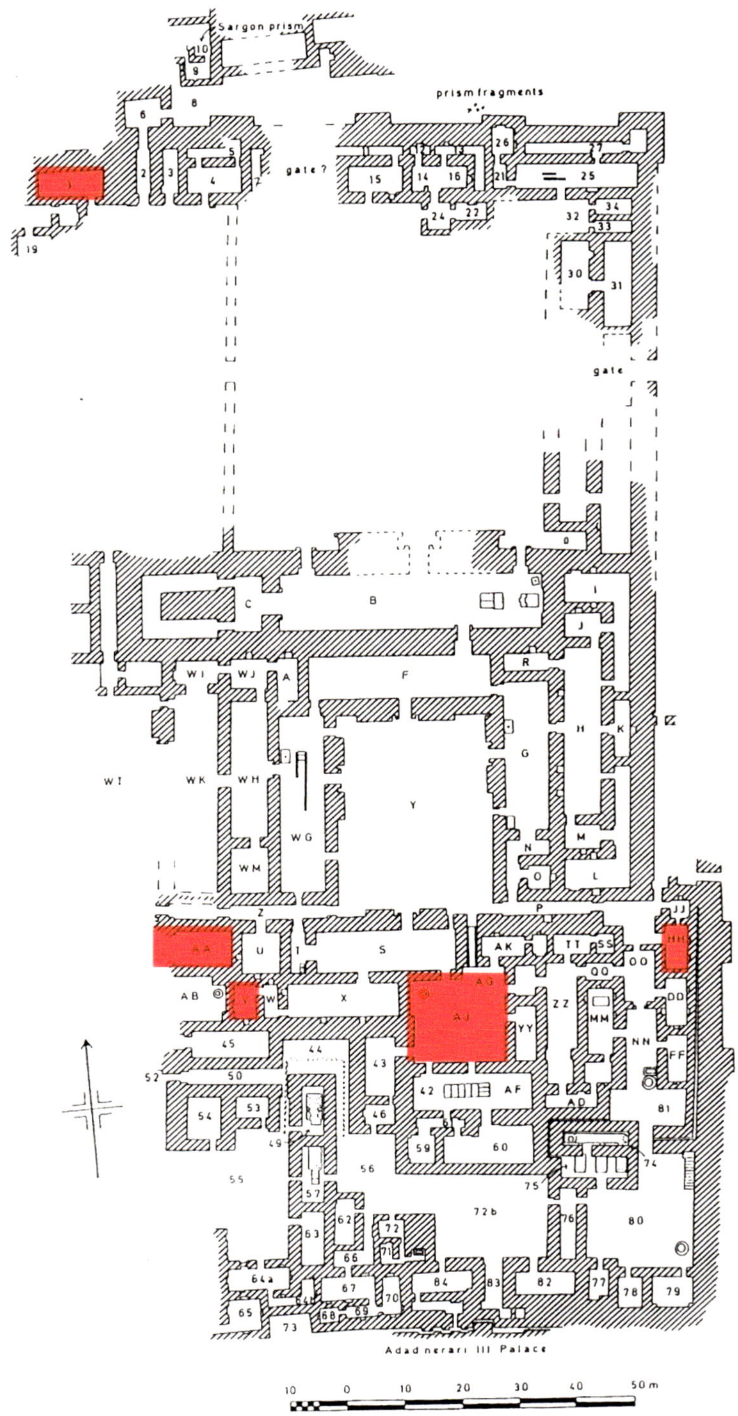

Figure 3.3: Plan of the Northwest Palace. Findspots of glass are indicated by red boxes (after Oates and Oates 2001: 60).

This, as well as the fact that its doors were narrowed, suggests a storage function in the later periods of the palace. Kertai (2015: 41) identifies the original function of the room as a bath, and the adjoining room X as the bedroom of the king. As the glass finds came from the filling, a precise date on the basis of stratigraphy cannot be made here.

Four hemispherical bowls Nim7, Nim8, Nim9 and Nim10 were found in the southern part of room AA (Figure 3.3). As AA was only partially dug, its reconstruction, as well as its function, remains largely unknown. As four pieces of the same vessel type had been stored in the room it makes its function as a magazine likely. The room was obviously not severely damaged by fire, because Nim7, Nim8, Nim9 and Nim10 were almost entirely preserved. Several pieces of vessels and transparent bowls have been found in Courtyard AJ (Royal Courtyard), which seems to represent the major residential area of the palace (Figure 3.3).

3.1.4.3 Burnt Palace

The building underwent various renovation phases since its foundation in the Middle-Assyrian period. The final phase of the palace was the work of Sargon II (721–705), therefore most of the finds can be attributed to the time between the reign of Sargon II and the destruction of Nimrud in 612 (Curtis 2013: 8). A large number of glass objects was found within the palace, among which are also sherds of transparent hemispherical bowls. Only a small number of them have been published in detail.

Three core-formed vessel fragments Nim137 come from throne room 8 (Figure 3.4), among other finds from the same context are a larger number of ivory fragments, as well as cuneiform tablets, and a bowl made of rock crystal. Another fragment (Nim138) of a core-formed vessel was uncovered in room 23, on the northern side of the court (Figure 3.4). As Nim137 and Nim138 were found within the destruction layer, a date prior to 612 is suggested. Core-formed vessels at Nimrud were only found in the Burnt Palace.

The findspot of the red glass ingot Nim198, within room 47, is of particular interest, as on its southern side two separate kilns with long, narrow rectangular flues were uncovered (Figure 3.4) (Mallowan 1954: 77). Another ingot was found in close association with one of the kilns, making it very likely that they were used for glassmaking. This is supported by the additional ingots, cullet, slags, as well as by the large amounts of ash, dispersed throughout the room. Regarding the stratigraphic relations of the kilns, Mallowan (1954: 77) states that they were built on top of the walls of the

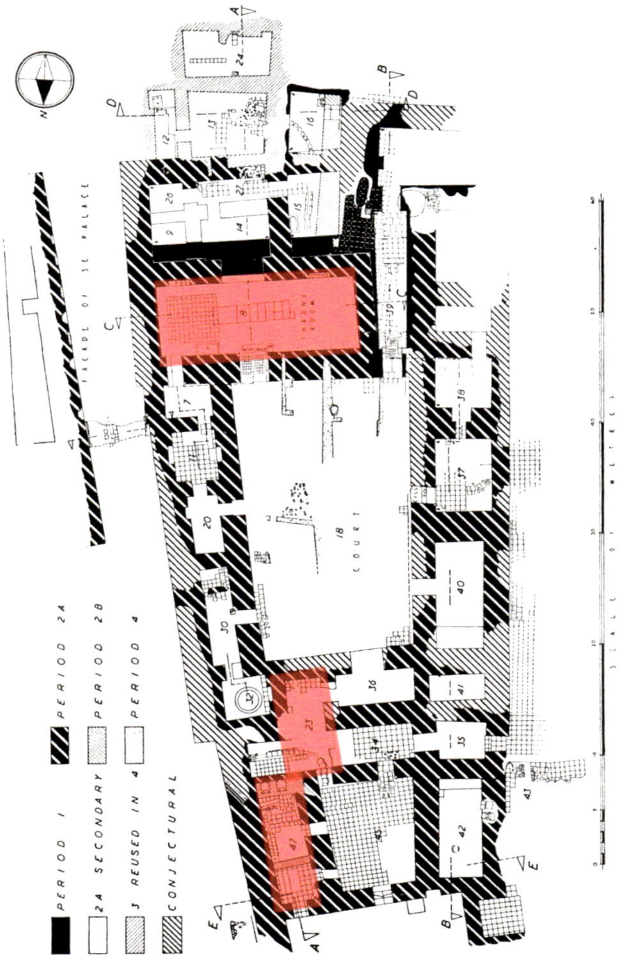

Figure 3.4: Part of the plan of the Burnt Palace. Glass findspots are indicated by red boxes (after Mallowan 1954: pl. XI).

latter squatter occupation, and therefore post-date the destruction level. He further remarks that the kilns were heavily damaged, so that it was not possible to trace parts of their outline. He nonetheless suggests a date in the early 6th century. In the same article, on the other hand, he mentions that 'the furnaces are probably to be correlated with period 4 [Hellenistic], although they were sunken considerably below the normal surface level of that particular period' (Mallowan 1954: 83). Even though a datable ingot would be desirable – since almost none of the primary glass objects from the Iron Age were found *in situ* – the find situation of Nim198 shows too many uncertainties to make any assumptions on its date. For further (archaeometrical) studies the object was, however, included in this monograph.

3.1.4.4 Fort Shalmaneser

Fort Shalmaneser is situated at the southern edge of the city. The building complex consisted of a large empty space, surrounded by an enclosure wall that functioned as a parade ground. The inner courtyards were flanked by workshops, treasuries, barracks, private quarters and state apartments (Oates 1963: 7). Shalmaneser III (858–824) initiated the construction of the Fort, which was completed around 844 (Yamada 2000: 35–40). In Shalmaneser's time the fort was used for military and administrative purposes and as a storehouse for booty and tribute.[11] Under succeeding kings the layout of Fort Shalmaneser underwent major changes due to renovation and restructuring works, especially during the time of Esarhaddon (680–669). The function of the building as an arsenal and storehouse remained until the time of Sennacherib (704–681). Esarhaddon, who introduced the name *ekal māšarti* 'Review Palace', restored Fort Shalmaneser, and it remained in use until its final destruction in 612, which is witnessed by destruction debris and burnt material throughout the complex. Shortly before its final destruction the building was repaired several times. It is probable that the blocking of some of the doorways described below was caused by this event as well. Most of the finds, and therefore also the glass objects, cannot be dated precisely as the floors were kept clean for two centuries (Curtis 2013: 9–12).

The inlays Nim109 and Nim65 were found in room T20, located in Area T of the southern wing of the palace, which was erected during the reign of Shalmaneser III, whose inscriptions were found on many of the door-sills in the area (Mallowan 1966: 452) (Figure 3.5). The glass finds come from a deposit that is sealed between two floors, the older one dating to the time of Shalmaneser III and the younger Esarhaddon (Curtis 1999: 59). Accompanying finds comprised horse harnesses, iron and bronze armour scales, a blade and an arrowhead (Curtis *et al.* 1993: 8–16). The floor was laid directly on top of these finds; the inlays therefore date between the construction of the building complex in 844 and the reign of Esarhaddon (681–669).

The vessel sherds Nim18 and Nim19 come from Gate Chamber SE13, which was destroyed by fire in 612 (Figure 3.5). A layer of burnt debris filled the room up to one metre. In addition to glass fragments, ivories were found in the gateway, leading Mallowan (1966: 417) to assume that the pieces were lost or thrown away by looters before burning the building. The objects found in this context were therefore thought to have been in a secondary context.

Area S was identified as the residential area of the palace with its main courtyard S6 and surrounding reception and storage rooms (Kertai 2015: 69). In room

[11] This becomes obvious regarding numbers of finds that clearly point to an origin outside of Mesopotamia. Examples from the western parts of the Empire are, for instance, an ivory plaque inscribed with the name of Hazael, king of Damascus, and shells inscribed with the name of Irhuleni, king of Hamath (western Syria).

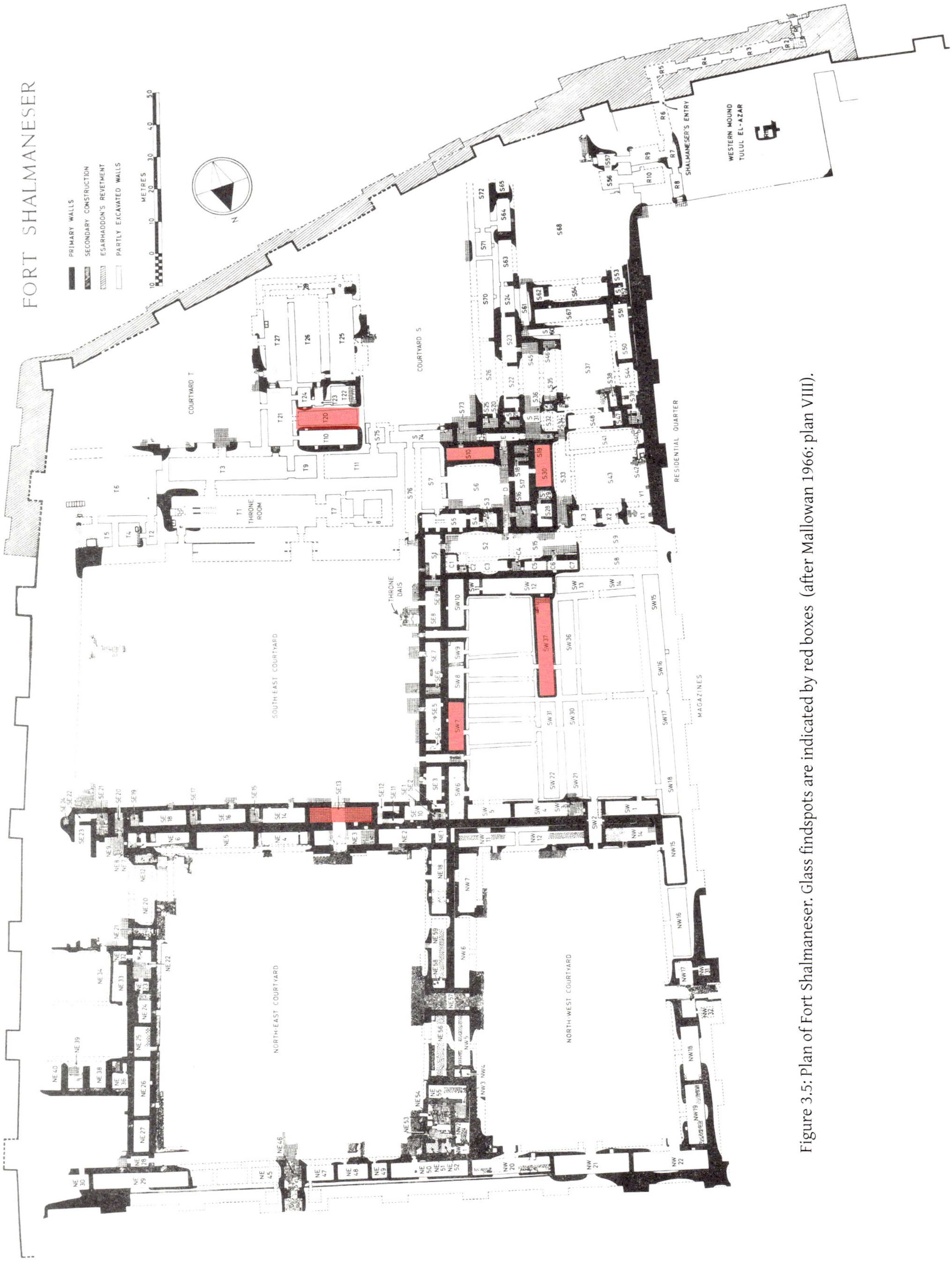

Figure 3.5: Plan of Fort Shalmaneser. Glass findspots are indicated by red boxes (after Mallowan 1966: plan VIII).

S10 huge amounts of glass items, mainly inlays, were found. The room was heavily damaged by fire, yielding burnt debris 1.50 m deep. In this filling vast numbers of burnt ivories was discovered, consisting of broken plaques that can be attributed to the Phoenician type. No ivories of the Assyrian type were retrieved (see Chapter 4.2.2.9). Among the finds were also cuneiform tablets that belonged to the archive of the *šakintu*, the female comptroller of the household (Ciafaloni 2009: 308). The doorway to S10 was blocked and only a narrow passage of 1.50 m served as an entrance, indicating its use as a storeroom or treasury, at least in its final stage (Herrmann 1992: 10; Oates and Oates 2001: 186–189, fig. 117).

Room S30 yielded glass inlays of rosettes of different types, although the exact number cannot be reconstructed. The inlays were found together with ivories that can mainly be attributed to the Assyrian type (Oates and Oates 2001: back cover). S30 is interpreted as a retiring room that surrounded the major courtyard S6 (Kertai 2015: 69). Hence the ivories could have belonged to beds or similar furniture.

The south-western quadrant (area SW) of *ekal māšarti* comprises large storage magazines. The area is subdivided into four large courtyards that enable the reception and dispersal of goods. A number of painted glass inlays (Nim112, Nim113, Nim114, Nim115, Nim118) were found in room SW7 in the south-eastern corner of one of the courtyards. This assemblage of inlays indicates that a specific type of furniture was kept here. The eastern wall of room SW7 yielded a ventilation shaft, as well as a storage jar at the northern end, and fragments of bronze and iron armour. At the south end great quantities of ivories were uncovered, packed in a layer of mud-bricks (*c.* 2 m) that had fallen from the roof and walls. The ivories belonged to veneers of chairs or couches, whose original position can be reconstructed as having been stacked one on top of the other (Oates 1959: 104–105). Mallowan (1966: 413) suggests that the stacking must have been carried out after the sack in 614, during the phase when the palace was rebuilt. In this case, the furniture and therefore the glass inlays were produced prior to 614. It is very likely that Nim112, Nim113, Nim114, Nim115 and Nim118 served as inlay plaques for ivory panels.

In SW37, 41 glass objects were unearthed. The majority of finds consists of fragments of transparent glass vessels. Furthermore, a small amount of rosette inlays (Nim61, Nim62, Nim63, Nim64), as well as three painted inlays (Nim116, Nim117, Nim119) come from here. Room SW37 is of considerable size (30.30 m x 4.30 m) and was filled with broken mud-brick debris. Fiorina (2009: 36–37) identifies two groups of objects: one incorporates shells, armour, and arrow heads, the other group incorporates objects made of more delicate materials, like glass and ivory.[12] More than 1800 ivory pieces were distributed among this debris. It was stressed by the excavators that no single piece of the same series was found next to each other. The pieces do not show any traces of fire. In addition, the gold foil that formerly covered parts of the pieces had been removed. It was supposed that SW37 was not burnt during the first sack in 614, but was destroyed later in 612 (Fiorina 2009: 36–38: Oates 1961: 3–4). Fiorina (2009: 38) suggests that craftsmen were still using the room when it was destroyed, therefore Phoenician ivories were still being produced until the end of the 7th century. It is interesting to note that among the ivories found none can be attributed to the Assyrian type. Furthermore, Fiorina (2009: 37) refers to a cuneiform tablet[13] from SW37 that lists Jewish and Phoenician names, and supports the idea that these names might refer to craftsmen.

3.1.4.5 Summary[14]

In summary, the majority of glass finds from Nimrud are inlays uncovered within the destruction layer of the year 612. The inlays were part of different objects, often furniture made of wood and ivory. Some of the objects in which the glass inlays were set show traces of intentional destruction, as observed in room SW37 of Fort Shalmaneser. Some seem to have been displaced in the course of looting (SE13, Fort Shalmaneser). The fact that the rooms of several buildings in Nimrud show traces of blockings that were probably erected after the first sack of the city in 614 indicate the value of the stored objects.

A number of buildings in Nimrud lost part of their status over time, which led to major changes in function. This becomes particularly obvious regarding the Northwest Palace after Sargon II (721–705) had moved to Khorsabad during his reign. The function of Fort Shalmaneser as an arsenal and storehouse, as well as a residence, especially for tribute, is well attested. Therefore objects in which glass inlays were often inserted could have been acquired as booty from different parts of the empire.

Concerning primary glass finds, there is an accumulation in room 47 of the Burnt Palace in which the two glass kilns were found. Another ingot (Nim190) comes from

[12] The results are based on the excavations of the Italian mission. For a plan showing the distribution of the finds, see Fiorina 2009: 41, fig. 2.

[13] No text number is given.

[14] I would like to thank Nigel Tallis, St. John Simpson and Alex Truscott (British Museum, London) for the opportunity to work on the artefacts from Nimrud which are kept in the British Museum collection. I also thank Joan Aruz, Kim Benzel and Tim Healing for granting me access to the collection of the Metropolitan Museum of Art, New York, and for their support in examining the glass objects from Nimrud.

the Northwest Palace. Further primary glass finds have been recorded from the Burnt Palace and the Northwest Palace that make a primary production workshop at Nimrud, probably within these buildings, likely.

Since the floors of the large buildings in Nimrud were kept clean, the finds cannot be retraced to an earlier phase of the Neo-Assyrian period of the building. Therefore most of the finds from Nimrud only show a *terminus ante quem* of 614. It can certainly be expected that objects found in the destruction layer date back to the 9th or 8th century (Curtis 2013: 2).

3.1.5 Nineveh

The city of Nineveh, which consists of the two modern tells of Kouyunjik and Nebī Yūnus, is situated on the east bank of the Tigris on the opposite side of Mosul.

There is only one (Nin2) out of three glass finds from Nineveh that can be attributed to secure find contexts. Mace-head Nin2 was found at the bottom of a well within the Nabû Temple. Numerous bricks inscribed with dedications for Sargon II (721–705) indicate that he had either constructed the well or carried out major renovations during his reign (Thompson and Hutchinson 1929: 106). According to Barag (1985: 74), the well probably remained in use until the destruction of Nineveh in 612. This renders a possible date for the deposition of Nin2 between 722 and 612, an older date cannot however be excluded.

Nin4, a fragment of raw glass, was unearthed from an unknown context during excavations carried out by Smith in 1873 and 1874.[15]

Ingot Nin3, as well as attachment Nin1, were purchased by Budge in 1889. Details of the find contexts are therefore not known.

3.1.6 Sultantepe

Sultantepe, modern Ḫuzirina, is situated on the River Balih, 15 km south of Sanlıurfa in Turkey.

There is only one glass object known from the site published by Barnett (1953: 50). The core-formed vessel Su1 was found in room M2, one of three rooms excavated there. As the rooms have only been partially preserved, little is known about the architectural structure in which they were incorporated. The associated finds include a great number of glass beads, coloured pebbles and sea shells, as well as miniature stone and frit vessels. Barnett (1953: 48) suggested that the latter served as containers for cosmetic products. Furthermore, remains of gold and lapis lazuli inlays, ivories, as well as stamp and cylinder seals were found, showing the richness of this room. In close proximity of the rooms, library tablets that belonged to a family of the priest were found. The finds, as well as the texts, make an Assyrian presence likely (Radner 2012: 287).

3.1.7 Til Barsip

Til Barsip, modern Tall Ahmar, lies on the east bank of the Euphrates, 20 km south of Jerablus, close to the Lake Assad in Syria.

Only a very small number of glass finds, comprising inlays, were unearthed at the site. A rosette inlay (TB1) as well as an undefined number of yellowish glass remains were found set into palmette petals of ivories, similar to the ones found at Nimrud (Bunnens 1990: 444, no. 7). TB1 comes from room1, Building C, phase B, where almost all the ivories were discovered. Building C, room 1 has three sub-phases (A, B, C) of which the earliest, phase C, came to an end by destruction. In the destruction layer, cuneiform tablets with Assyrian eponyms were uncovered, the latest one to be connected with the year 649, indicating that this phase came to an end in the middle of the 7th century. The building remained in use during phase B. Phase A can probably already be identified as a squatter occupation (Bunnens 1997: 436–438). Bunnens (1997: 438) gives the year 612 as a *terminus ante quem* for the entire collection of ivories, including the glass inlays, found in this context.

3.1.8 Ziyaret Tepe

The site of Ziyaret Tepe, ancient Tušḫan, is located 60 km east of Diyarbakir in the upper Tigris region in south-eastern Turkey.

Ziyaret Tepe was first excavated in the course of the construction works of the Ilısu Dam under Algaze (Algaze *et al.* 1990). From 2000–2013 Matney carried out fieldwork at the site.

There is only one known glass object from Ziyaret Tepe, uncovered during field-work directed by Matney between 2000 and 2013 ('Tušḫan Archaeological Project'). Bottle Ziy1 was found in a burial ('Brandgrubengrab' N-070), which can stratigraphically be attributed to the Neo-Assyrian period.[16] According to the context and finds a date in the 8th or 7th century is likely (Wicke, pers. comm.).

[15] According to his records, major field-works in these years were performed at the North Palace, Southwest Palace, Nabû Temple and the Ištar Temple (Smith 1876: 94, 139–142).

[16] I would like to thank Tim Matney for permission to publish the object, as well as Dirk Wicke for providing photographs and information on the bottle.

3.2. Babylonia

3.2.1 Babylon

Babylon is located on one of the branches of the Euphrates, 90 km south of Baghdad.

Only a small number of glass objects is published in different publications relating to their findspot (Reuther 1968).[17] In general, many of the glass objects from Babylon lack an object number ('Bab-number'), making the reconstruction of the specific context impossible. Those pieces which can be attributed to a specific locus were all uncovered in burials. The author was able to study the original objects in the Vorderasiatisches Museum and the British Museum; the pieces from the Istanbul Arkeoloji Müzeleri were not accessible.[18]

3.2.1.1 Graves

The majority of glass objects found in Babylon come from funerary contexts. The vessels Bab4 and Bab7 were both uncovered in grave 119 (the type of burial according to Reuther 1968 'Hockersarg'), situated in the area of the Merkes. Bab4 was placed in front of the knees of the deceased along with a number of miniature vessels, some of which were glazed. Bab7, in contrast, was positioned in front of the shins in the same burial (Reuther 1968: pl. 68). Regarding stratigraphy, the excavation report lacks significant information about the findspot. The plan of the trench section does not show any floor level that could be associated with burial 119 (Reuther 1968: p. 4). From a stratigraphic point of view, burial 119, therefore, cannot be dated precisely. However, regarding the type of burial and the accompanying finds, chronological observations can be made. Mofidi-Nasrabadi (1999: 117) dates this type of burial 'Hockersarkophag' from Babylon to the second half of the 1st millennium. Among the accompanying finds from burial 119 is a fibula whose bow is described as an arm. Regarding the date of this fibula type, Pedde (2000: 38) suggests the 7th to early 6th century. These suggestions imply a date for burial 119 that is not earlier than this time period.

Bab3, Bab5 and Bab8 were found together within burial 109 (the type of burial according to Reuther 1968 'Ovalsarg'). The plan of the section provides some indication of the stratigraphy of the burial (Reuther 1968: pl. 63, 109b). The coffin was sealed by an ash layer underneath the pavement of a house that was rebuilt in the Neo-Babylonian period. The pit of the grave is not drawn in the plan, but Reuther (1968: 209, pl. 63. 109b) reports that it cuts through parts of the walls of the Neo-Babylonian house. The pit of the grave was therefore dug after the erection of the Neo-Babylonian house, thus Bab3, Bab5 and Bab8 are therefore Neo-Babylonian or later in date.

Grave 109 is a burial of a child, which was equipped with a large number of grave goods (Figure 3.6). Bowl Bab3 was found covering the face of the individual, Bab8 was placed in front of the knees, and Bab5 was found between the elbows and knees. Baker (1995: 216) assigns grave 109 to the 'bowl-burial' (type 5), which was common among inhumations of children. Comparable graves from Nippur date to the early 7th century. Apart from the three glass vessels, two necklaces were also part of the grave goods (Figure 3.6) The links of the necklaces show close parallels to four chain links that were found in Ur, in a hoard underneath the Achaemenid phase of the E-nun-maḫ but above the pavement erected by Nebuchadnezzar (605–562) in room 5 (Woolley and Mallowan 1962: 30, pl. 21, U458). Musche (1992: 218), as well as Barnett (Barag 1985: 160, footnote 115), date this type of necklace to the Neo-Babylonian period. Since the ivory head (Figure 3.6) was not found within the burial, but next to it, its dating does not necessarily correlated with that of the burial (Reuther 1968: pl. 63, 109b).

3.2.1.2 Other contexts

Bab6 was, according to the excavation journal, found in the area of the Merkes 23l2 + 6,00, east of the residential quarter. Square grid 23l2 lies in between the residential quarter, east of the Ištar Temple (Reuther 1968: pl. 2, 29, 30). Unfortunately, no further information about either the context or accompanying finds has been published.

A considerable number of ingots and raw glass objects were found in Babylon (Bab9, Bab10, Bab11, Bab12, Bab13, Bab14). Unfortunately, most of the ingots cannot be associated with a specific context, as no information about the findspot is known.

Bab9 comes from a context underneath the Parthian layer in the house of the 'Perlenfabrikant' (Wullen and Marzahn 2008: 603). It is very likely that the ingot was part of the so-called 'Schatzfund' from room E of this building. This hoard was composed of objects made of different materials (lapis lazuli seals, and inlays, a mace-head made of onyx, and rock-crystal objects among other finds) that were placed in two baskets, impressions of which were preserved in the pavement (Wetzel and Schmid 1957: 32, 36, pl. 15). The oldest of these finds dates to the Kassite period, but also Neo-Assyrian, Neo-Babylonian and Hellenistic objects were

[17] The results were published in the monograph series 'Wissenschaftliche Veröffentlichungen der Deutschen Orient-Gesellschaft'. Here of particular interest for this study, see Reuther 1968.

[18] I would like to thank Markus Hilgert and Lutz Martin for granting me access to the collection of the 'Vorderasiatisches Museum'. I would like also thank Nigel Tallis for making the objects in the British Museum collection accessible to me.

Figure 3.6: Glass objects and accompanying finds within and around coffin 109 (Reuther 1968: pl. 65, 109).

recorded. It was suggested by Koldewey (1913: 215), and later by Wetzel and Schmid (1957: 32), that parts of this collection of finds formerly belonged to the treasure of the Esangila, which was in Parthian times used as a depot of raw materials for bead-making. Whether or not this specific use can be assumed, regarding a date for Bab9 ranging from the Kassite to the Parthian periods needs to be taken into consideration. As Bab11 exhibits very close parallels to Bab9, it is very likely that both objects belonged to the same ingot (see Chapter 4.5.1).

Finally, Bab1 and Bab2 need to be mentioned, which are identified as attachments for composite statues. A stratigraphic attribution is not possible. The objects were part of a consignment of objects from Babylon and Dilbat (Dul1, Dul2) that were excavated by Rassam in 1881.

3.2.1.3 Duleym

Tell ed-Duleym, ancient Dilbat, is situated on the eastern banks of the Euphrates, south-east of Babylon in modern Iraq.

The glass attachments (Dul1; Dul2) as well as the ingots and raw glass fragments (Dul3, Dul4, Dul5) have no findspot to which they can be traced. The pieces were acquired, most likely by Rassam, in 1897 and came this way to the British Museum. The ancient city was more or less continuously inhabited from the beginning of the 3rd to the middle of the 1st millennium. Texts from Dilbat date primarily to the Old Babylonian (first half of the 2nd millennium) (Charpin 2004 with references) and to the Neo-Babylonian, Assyrian, or Persian periods (Lilyquist 1994: 5).

3.2.2 Eridu

Eridu, modern Abu Shahrain, is located in southern Iraq on the west bank of the Euphrates, 24 km west of Ur.

Only a small number of glass objects from Eridu have been recorded. Three pieces of raw glass fragments (Er1, Er2, Er3) are said to have been found during the excavations of Tylor and Thompson in 1918 at either Eridu or Ibrahim el-Khalil, a smaller nearby site (Er2, Er3). No information about the findspot is known (Curtis *et al.* 2008: 5; Hall 1930: 187–189; Safar *et al.* 1981: 29). The finds are part of the British Museum collection.

3.2.3 Isin

Isin, modern Išān al-Bahrīyāt, is located about 200 km south-east of Baghdad, 35 km from Nippur.

A number of blue attachments and inlays have been found at Isin, of which 13 have been published (Hrouda 1987, 43 pl. 19, 32). All of these fragments have belonged to composite statues, which were found within a layer of asphalt that covered a brick-paving in Court B of the Gula Temple. The bricks were stamped with the name of Nebuchadnezzar II (605–562) (Hrouda 1987: 11–12). Among the finds in the asphalt layer were objects dating prior to the reign of Nebuchadnezzar II, which makes an earlier date for the attachments also possible (Hrouda 1987: 43).[19]

3.2.4 Kiš

Kiš, modern Tell Uhaimir, lies 15 km east of Babylon in modern Iraq.

Both of the core-formed vessels from Kish (Kis1, Kis2) incorporated in this study were published by Barag (1970: 159). Kis2 was studied by the author in the Field Museum (Chicago), where also a number of later glass finds are stored.[20]

The core-formed bottle Kiš1 was found in grave 23, and core-formed bottle Kiš 2 in grave 54 at the site of Ingharra, Mound W. The graves on Mound W, Tell Ingharra, were scattered, without any demonstrable concentrations, and with dates from the Neo-Assyrian to the post-Achaemenid period (Gibson 1972: 70). Moorey (1979: 52) attributed Kiš 1 and Kiš 2 to the 5th century, based on 'archaeological context'. As no stratigraphic assumptions can be made on the basis of the locations of the graves, a date for burials 23 and 54 can only be suggested regarding accompanying grave goods. Unfortunately, grave goods have not been published for each grave (Moorey 1979: fiche 4. A11. 12. 13), therefore the exact dates for Kiš 1 and 2 cannot be determined.

3.2.5 Nippur

Nippur, modern Nuffar, is situated in the southern part of the Mesopotamian floodplain, some 180 km south of Baghdad, and therefore forms the boundary between the ancient lands of Sumer and Akkad.

In 1840, Rawlinson was the first to visit the site, followed by Loftus in 1851. From 1888 to 1900, excavations were carried out by the Babylonian Expedition of the University of Pennsylvania. Not until 1948 were the excavations revived by the Oriental Institute and the University of Pennsylvania under the direction of McCown and Haines. After 1962 the American Schools of Oriental Research, under the direction of Knudstad, continued excavation. In 1972, Gibson was appointed director.[21]

Glass material from Nippur was published in the excavation report by McCown *et al.* (1967: 56, 8, pl. 148, 4) as well as by Barag (1970: 158, no. 1, 2). Two of the three core-formed vessels incorporated in this study where found in burial contexts. Bottle Nip1 comes from burial 4B 77 in locus NT 25 I 2, the youngest phase of the badly preserved residential quarter around the North Temple. The burial was uncovered in sounding SE, level II, which can be attributed to the Achaemenid period (McCown *et al.* 1978: 39–40, 62, pl. 35). The burial was comprised of an upright ring-based jar, a type particularly common in the Neo-Babylonian period but also – although to a lesser extent – in Achaemenid times (Baker 1995: 216).

Core-formed vessel Nip1 was associated with a Neo-Babylonian pottery jar. The stratigraphic affiliation, as well as the burial type itself, points towards a date prior or contemporary with the Achaemenid period. The associated pottery jar makes a Neo-Babylonian date for Nip1, however, likely.

Nip3 was found in locus TA 51 IV which belongs to the 'scribal quarter', dating to the Neo-Assyrian period (McCown *et al.* 1967: 70). No further finds have been recorded from this locus which would confirm this date.

3.2.6 Ur and Diqdiqqa

Ur, modern Tell el-Muqayyar, is situated in southern Iraq along a former branch of the Euphrates, about 17 km west of an-Nasiriya. The major site consists of a number of mounds (1200 m NW–SE x 800 m NE-SW). Further west of the Ziggurat, a number of tells can be identified, including the site of Diqdiqqa.

In Ur, a number of core-formed vessels, one cast-and-cut bowl and an attachment for a composite statue, are recorded that come from different find contexts.

Bottle Ur6 was found in a dwelling context that belongs to area XNCF, Level 2. The area overlays the architectural remains that can be attributed to the Kassite king Kurigalzu, and therefore dates to a period after his reign.

Ur2, a fragment of a transparent glass bowl, was found in room 5 of the E-nun-maḫ, where it was sunk into the surface of the 'Nebuchadnezzar pavement' (Woolley 1962: 25, 29–31, 110). A date contemporary with, or

[19] I would like to thank Kai Kaniut for the discussion on the find context of the glass objects from Isin.

[20] I thank Jamie Kelly for the opportunity to study the glass bottle at the Field Museum. Also in the Ashmolean Museum, there are several glass finds from Kiš, dating to the later period (pers. comm. Paul Collins).

[21] For further details on the history of research and further literature, see Gibson et al. 2001: 548–550.

later than the reign of Nebuchadnezzar II (605–562) is therefore likely.

Ur5 was found in burial 26, which is described as a 'bell-shaped urn, 1.50 m below the Persian pavement' (Woolley 1962: 60). Baker (1995: 212) dates this type of double-pot burials at Ur into the early 1st millennium. The finds recorded from this grave incorporate Pazuzu heads, scaraboids, copper bangles, bracelets and rings, an iron kohl stick and a silver disc. Regarding the Pazuzu heads, they begin at around the end of the 8th century (Heeßel 2002: 31–32).

Diq1 was found in Diqddiqa. As with most of the finds unearthed at the site, a date on the basis of stratigraphic considerations has to remain uncertain (Woolley 1976: 84).

3.2.7 Uruk

Uruk, ancient Warka, lies in southern Iraq, 15 km east of the provincial capital as-Samawa.

First excavations were carried out by Loftus from 1849 to 1853 (British Museum).[22] In 1912, Jordan was appointed director, and after 1928 Jordan, Nöldeke, Heinrich and Lenzen carried out excavations for the Deutsche Orient- Gesellschaft. After World War II, excavations were led by Lenzen (1953–1967), from 1968 until 1977 by Schmidt, and from 1980 to 1995 by Boehmer. Since 2001, van Ess has been director.[23]

Most of the glass finds are recorded from graves dug on top of the so-called 'Ringmauer', outside the city wall, as well as from the so-called 'Sassanian hills' in the south-eastern part of the site outside the city wall. The great majority of objects cannot be dated on the basis of stratigraphic considerations, as most of them are surface finds (van Ess and Pedde 1992: 159). An accumulation of finds in the south of the city area was observed by Rau (1991: 60) during survey. According to Rau, the existence of workshops in this area is very likely due to the large quantities of slag. However, the date for these installations remains unknown.

Urk2 and Urk3 were found in the double-pot burial 129 (W17961), that dates to the Neo-Babylonian period. The exact location on the city plan, however, cannot be identified precisely (Boehmer *et al.* 1995: 89). Regarding the type of double-pot burial, coffin 129 can be identified as type 1B, which was predominant in the early 1st millennium (Baker 1995: 212). Burial 129 dates to the Neo- and Late Babylonian domestic area west and south-west of the Eanna (Boehmer 1987: 14). Accompanying objects, present in burial 129, show a wide range. The oldest pieces date to the Kassite period, and the youngest pieces to the Late Babylonian (bronze rings, Lamaštu amulet, beads) (Boehmer *et al.* 1995: 89–90). Van Ess and Pedde (1992: 160), however, attribute the glass vessels to the 7th to 6th centuries.

Urk1 was found in a pit, dug by looters in the area of the Eanna, together with cuneiform texts that can be attributed to the time from Sargon II to Darius II (van Ess and Pedde 1992: 160). Regarding its stratigraphic context, Urk1 has therefore to be considered Neo-Assyrian, Neo-Babylonian or Achaemenid in date.

3.3. Levant

3.3.1 Amman

Amman, Ammonite Rabbath-Ammon, and Roman Philadelphia is the modern capital of the Hashemite Kingdom of Jordan. The city exhibits a long history, and the earliest phases date to the 6th millennium. As with many other cities that continue to the present day, only small parts of the ancient site are able to be explored archaeologically, and because of several later occupation phases (Roman, Byzantine, Umayyad), the Iron Age phase, of particular interest here, has not been excavated to a large extent. A coherent plan of its architectural structures is therefore not available. Only one inlaid glass bowl (A1) was found at the site, which is, however, outstanding as there is no comparable object known so far.

A1 was found on the citadel, the *Qal'a* of Amman, on the lower eastern terrace. The object was found within the earliest destruction phase of the so-called 'Ammonite Palace' (locus 110) (Humbert and Zayadine 1992: 257), that dates to around 700, giving a *terminus ante quem* (O´Hea 2011: 161; Zayadine *et al.* 1976: 362). In the same context, four double-faced Hathor stone heads were found that are attributed to the 7th and 6th centuries on the basis of stylistic considerations (Zayadine 1973: 27–28; Zayadine *et al.* 1976: 362). The best parallels for the double-faced heads can be identified on ivory plaques from Arslan Taş and Fort Shalmaneser (SW37), depicting the so-called 'Lady at the Window' (Dornemann 1983: 160–162; Zayadine 1973: 34).[24]

[22] Since this time, illegal digging activities have unearthed a great number of cuneiform tablets which today belong to the collections of the British Museum, Musée du Louvre, University of Pennsylvania, Princeton University and Yale University.

[23] See preliminary reports 'Vorläufiger Bericht über die von der Notgemeinschaft der deutschen Wissenschaft in Uruk unternommenen Ausgrabungen 1930–1934' (UVB 1–5), later 'Vorläufiger Bericht über die von dem Deutschen Archäologischen Institut aus Mitteln der Deutschen Forschungsgemeinschaft unternommenen Ausgrabungen in Uruk-Warka (1934–1979) (UVB 6–30)', and the final monograph series 'Ausgrabungen in Uruk - Warka. Endberichte (1985–2004)' (AUWE 1–25).

[24] A publication of the monumental stone objects from Iron Age II in Jordan is currently being prepared for publication by Häser and Schmidt. Furthermore, I would like to thank Jean-Babtiste Humbert, Barbara Porter, as well as the Department of Antiquities, Jordan for allowing me to study an publish the glass bowl from Amman.

3.3.2 Tel 'Aroer

Tel 'Aroer, modern Ararah, is situated on a natural hill, 22 km south of the modern city of Bersheba. The south-Arabian trade route connected Arabia and Edom via the prominent transversal route between the King's Highway on the Transjordan plateau and the coast, probably making Tel 'Aroer an important trading post.

Only two transparent glass fragments have been found at the site and which belong to the same vessel, Ar1. These are published by Barag (2011) in a chapter within the final publication of the site (Thareani *et al.* 2011).

Tel 'Aroer was founded in the Iron II period, with strata IV–III dated to the 8th to mid 7th century, and strata IIa–IIb (Iron III) dated to the late 7th to early 6th century. The site was destroyed at the beginning of the 6th century, either by the Babylonians or Edomites (Thareani *et al.* 2011: 305–307). Thareani *et al.* (2011: 305) refer to Tel 'Aroer as an important Assyrian 'caravan town' on the south-Arabian trade route. This 'cosmopolitan nature' of strata III/II is reflected by Edomite epigraphic material and pottery, Assyrian style weights and architectural features, Judahite pottery, pillar figurines, a shekel weight, and three *lmlk*-impressions (Thareani *et al.* 2011: 305).

Glass fragments Ar1 were found together with one of the *lmlk*-inscribed handles, in a context, sealed by a destruction layer, and can therefore be attributed to the period of king Hezekiah (*c.* 700).[25] The context is situated in a residential building outside the city wall, which is interpreted as a caravanserai (Thareani *et al.* 2011: 161–170). Ar1 shows close comparisons to Assyrian transparent glass vessels, and thus fits well into the 'cosmopolitan' range of objects found in strata III/II.

3.3.3 'Atlit

'Atlit is situated 30 km south of Haifa along the Mediterranean coast, and consists of a vast archaeological area, including the shipwreck 'Atlit Ram, the 'Pilgrims Castle' (Johns 1933), as well as the South-Eastern cemetery that is of interest here. Solely the bottom part of a transparent glass alabastron (At1), found in the South-Eastern cemetery, was explicitly mentioned by Saldern (1970: 227, no. 54). Further core-formed glass fragments are recorded in the excavation report (see, for example, Johns 1933: pl. XVIII).

The South-Eastern cemetery comprises burials dating from the Early Iron Age up to the Hellenistic period. In most of the cases, the burials cannot be attributed to a specific period, as they were very often disturbed and cut into each other. This is also true for tomb L21b, in which At1 was found (Johns 1933: 76–82). The burial was heavily disturbed, and contained several bodies of deceased, as well as a large number of funerary items. Among these were Iron Age I pottery sherds, a core-formed glass vessel, as well as a silver coin dated to the first half of the 4th century, which shows the wide range of objects and different periods. Johns (1933: 78) associates A1 with the singular burial 616 that could, however, also have been disturbed by neighbouring graves. Taking all stratigraphic difficulties into account, tomb L21b can only be considered as exhibiting a *terminus ante quem* for the 'late Achaemenid/early Hellenistic' periods (Saldern 1970: 227, no. 54).

3.3.4 Beth-Shean

Beth-Shean, or Tell el-Hosn, ancient Scythopolis, is located in the Beth-Shean valley on the southern bank of the River Harod. It is located at an important west-east (Jezreel to Transjordan), and north–south (along the Jordan valley) crossroads, which is also the reason for its historical importance (Mazar 2006: 6–9).

Whereas the glass and faience* objects of the Late Bronze Age have been published (McGovern *et al.* 1993), the Iron-Age glass finds have not been identified and published on a large scale so far. Since excavations at the site were carried out by the University of Pennsylvania Museum (1921–1933), and later by the Hebrew University of Jerusalem (1983; 1989–1996), the glass finds incorporated in this study are part of the Penn Museum collection (BS1; BS2), and the collection of the Israel Museum in Jerusalem (BS3). The three glass objects included in this study are, most likely, not exhaustive, but show the different find contexts and problems that are incorporated with Iron-Age glass finds. The mosaic glass fragment BS3 is a key object for the chronological attribution of the entire group of mosaic objects, therefore attention is devoted to it.

The largest number, and most important finds of the site, come from the Late Bronze Age and Iron Age I period (15th–10th centuries), when Beth-Shean served as a regional Egyptian administrative centre with temples, public structures, a 'Governor's Residency', and dwelling quarters (Mazar 2006: 61–172).

Ingot BS1 and the core-formed vessel sherds classified under BS2 were found in room 1028 of the South Temple, Level V, indicating an Iron Age IB/IIA date between 1150–925. James (1966: 39) remarked regarding this findspot the following: 'the room was in fact, almost entirely removed in the constructions of cisterns 10 a and b'. This shows that the finds, among them BS1 and BS2, could have belonged to different periods.

[25] The Paleo-Hebrew writing of lmlk (lamelech) on the handle is generally translated as 'belonging to the king'. These inscriptions were usually found on handles of large storage jars in the period of king Hezekiah (c. 700); for details see Grena 2004 and Kletter 1999.

The mosaic object BS3 is of special importance because it is typologically very similar to the mosaic bowl of Aššur, especially the bowl As2, making it the only object of its kind found outside Aššur. BS3 was discovered in the early Seti I temple (1313-1292), in room 1062, stratum VI (Rowe 1940: 8, 19, 20, 80, pl. 33, Nr. 46), which is situated outside the northern wall of the temple. The early part of the reign of Seti I corresponds to the Iron Age IA period (12th century to 1140), according to the division by the Hebrew University (Hebrew University Excavation, layer S-3) (Mullins 2012: 127, with references to earlier works). According to Mullins (2012: 142), layer VI incorporates the last Egyptian settlement in Beth-Shean, and in the later phases of layer VI the Canaanite city is revived (Mullins 2012: 144).

3.3.5 *Busayra*

Busayra is situated 45 km north of Petra, and 1 km south-west of the Iron Age copper-production area of Wadi Faynan in Jordan. This geographical situation is the basis for Busayra's political power, the control of the copper production, and Arabian trade to the Negev and Gaza, as well as control of contacts with Mesopotamian powers (Bienkowski 2002: 37–39, 480). Core-formed glass bottle Bus1 originates from locus B3.2.6, an ash layer within a rectangular room close to the postern gateway. The context is attributed to either Iron Age II or the Achaemenid period (Bienkowski 2002: 476).

3.3.6 *Tell Jemmeh*

Tell Jemmeh lies in the Wadi Nahal Besor, 10 km south of Gaza. The site was settled intensively from the 10th to the 8th centuries, and was conquered in 677 by Esarhaddon (680–669), who established a military base with a casemate wall.

The core-formed glass fragments of two different bottles, TJ1 and TJ2 are part of the collection of the Institute of Archaeology (London) (IoA), and were studied by the author. Information regarding their contexts can be obtained from the find boxes in the IoA.[26]

TJ1 was found in the 'Area under north wall CD' (entry on the find box in the IoA). The findspot can be identified as a room, probably of a larger building complex, situated in close proximity to the north-eastern wall which dates to the 23rd dynasty or the third intermediate period (1070–664). A large amount of Assyrian pottery, not specified, was unearthed in a grain pit in the southern part of building DZ. The remark 'under room CD' indicates that TJ1 comes from the preceding architectural structure that dates to the 22nd dynasty, which would still fall into Iron Age II. As during the Achaemenid occupation phase, granaries were dug through older layers, there remains the risk of younger finds having migrated into earlier layers (Petrie 1928: 8–9, pl. 13).

TJ2 comprises five partly joining pieces of a core-formed bottle with no information on the findspot and date.

Apart from the glass pieces discussed here, Petrie (1928: 24) refers to 30 glass fragments that were 'too small to be worth illustrating' that were found in different areas of the site. Therefore, the two glass vessels included in this monograph represent examples of a larger corpus of glass objects from Tell Jemmeh.

3.3.7 *Megiddo*

Megiddo, Tell el-Mutesellim, is situated on the western fringes of the Jezreel Valley at important crossroads connecting the coastal region (*via maris*) with the hinterland of the Acre plain (Kempinski 1989: 3).

A number of glass objects were studied by the author at the Oriental Institute, where information on the findspots came from consulting 'Fisher's field cards'.[27] For the reconstruction of the findspots, the concordance list published by Lamon and Shipton (1939: 216–232), as well as the master grid published by Braun (2013: 4, fig. 3) were consulted. Most glass finds date younger than the Iron Age period, only pallete Meg1 has turned out to be of an Iron Age date.

Meg1 was found in SQ Q-8, locus 1275, identified as a room of a house of residential area A (layer II). The buildings in this area are of the domestic type, with no outstanding characteristics.[28] Lamon and Shipton (1939: XXVii, 63, for a plan, see fig. 73) date stratum II to 650–600. The pottery from locus 1275 and its vicinity date to the late 8th or 7th century (Barag 1982: 13).

3.3.8 *Pella*

Pella, modern Tabaqat Fahl, is located in the eastern Jordan valley, 4 km east of the Jordan River. It lies near the junction of important trading routes connecting Egypt with Syria and Anatolia eastward along the Tigris

[26] I would like to thank Rachel Sparks for providing the glass objects at the IoA collection.

[27] In 1925 the Oriental Institute, under the supervision of Clarence Fisher, resumed work at the site. It was under his guidance that the 'locus number' was introduced, which represented a novelty for the archaeology of Palestine. Furthermore, Fisher also established the system of 'Record Cards' on which information about the context and date of every find was recorded. I would like to thank Kiersten Neumann, for allowance to study the glass palette from Megiddo, and for providing professional photographs of the objects for this book.

[28] In this regard, it has to be noted that only the stone foundations remained; these were almost completely buried under the occupation level. Actual floors could only be identified rarely; see Lamon and Shipton 1939: 64.

and Euphrates to Mesopotamia (Bourke 1997). The glass objects from Pella have not been published, but results on the chemical analysis have been presented by Reade *et al.* (2009), which are recognised in this study (see Chapter 7.4.4).

While the Bronze Age glass assemblage consists of vessels, ingots, pendants and beads of mainly blue glass,[29] the corpus of Iron Age glass objects incorporates solely beads. These are predominantly spherical in shape and greenish-black in colour. The Iron Age glass beads selected for chemical analysis came from tomb 89 (19 pieces), which is part of an Iron Age cemetery on the north-western slopes of Jebel Abu el-Khas, and dates to Iron Age IB/IIA (1050–850) (Bourke 1997: 112–113). A further four Iron Age glass beads come from the Iron Age temple area (Reade *et al.* 2009: 48).

3.3.9 Samaria

Samaria, modern Sebaste, lies 56 km north of Jerusalem/al Quds, overlooking the main route that connected Egypt with the Jezreel Valley and the northern routes to Damascus. It was the capital of the northern Kingdom of Israel until the Assyrian conquest in 722/21, and is therefore of great importance.

Glass finds from Samaria comprise inlays (Sam1–Sam22), which were found in association with carved ivories in the so-called 'Ivory House'. Some of the inlays were even found set in ivory panels, which were partly burnt. The stratigraphic attribution of these ivories is, however, problematic, as the finds were dug up in single small trenches that show no connection to the architecture of the complex. Crowfoot and Crowfoot (1938: 2–3, 23), and Tappy (1992: 163) interpreted the destruction layer as either caused by the invasion by Tiglath-pileser III (744–727) or the final Assyrian conquest in 711/21.

3.4. Related glass finds in other regions

Only glass objects that are typologically very similar to objects from Mesopotamia are incorporated in this study. These finds that were mainly found in the western and southern Mediterranean were most likely imported, either as raw material or finished objects, to the west. In addition to these finds from the Mediterranean basin, also glass objects from Hasanlu in north-western Iran and Susa in south-western Iran are incorporated into this study, as they stem from regions that interacted intensively with Mesopotamia.

3.4.1 Carthage

Carthage is located on the coast of Tunisia, and is part of the modern city of Tunis. Car1 was found in tomb 27 in Dermesch, a site that belongs to the ancient site of Carthage (Barag 1970: 167). The cemetery of Dermesch was excavated by Gauckler (1915), Car1 is not mentioned in the excavation records specifically. Therefore, the exact findspot, as well as the whereabouts of Car1 are unknown.

3.4.2 Fortetsa

Fortetsa lies close to the ancient site of Knossos on the island of Crete. The cemetery of Fortetsa is situated on the lower slope of the western face of the 'Acropolis' of Knossos, and was discovered in 1933 by Payne and Blakeway (Brock 1957: 1).

Vessel Fo1 was found in one of the burials, about which no further information is known, either on its stratigraphy or accompanying finds.

3.4.3 Gordion

Gordion, modern Yassıhüyük, is situated in central Turkey in the river plain of Sakarya. The city was the capital of ancient Phrygia, and is therefore of great importance. Most interesting here are the monumental buildings and tumuli that date to the 9th century, and which witness the large work force and great wealth of the city. Trade and gift exchange with regions to the east (ivories in the North Syrian style) are attested, as well as workshops discovered at the site that imported ivory as a raw material.[30]

The glass finds included in this study were first published by Jones (2005), and comprise two decolourised petalled bowls, Gor1 and Gor2. Gor1 was found almost intact inside a bronze bowl in Tumulus P, which was identified as the grave of a child. Tumulus P was richly endowed, containing furniture and other small wooden objects (bowls, animals, quadriga), 28 bronze bowls, indigenous and imported pottery, as well as faience* vessels, to mention only a few (Young 1981: 1–77). Young (1957: 330) assigns Tumulus P on the basis of its rich grave goods to a child of rank, probably belonging to the ruling family and suggests a date around 700 for its erection. This dating was recently challenged by radiocarbon dates from the Early Phrygian destruction level. The samples put the debris into the calibrated range of 827–803. Therefore, the erection of Tumulus P must date to after the destruction of the city. Sams (1994) in this regard assigns Tumulus P and K-III to the Middle Phrygian period, and to no more than one

[29] Some of the glass objects were found in burial contexts on Tell Husn (tomb 62); see Reade et al. 2009: 47 with further literature.

[30] For a comprehensive overview, see Graff 2014: 107.

generation after the destruction, which would point towards the beginning of the 8th century.

Gor2 is a fragment from a petalled bowl similar to Gor1, but much larger (d. 24 cm) (Jones 2009: 22). It was found in a building on the City Mound, which can be identified as a large domestic or palatial structure. This building can be attributed to the 8th century. Further information on the findspot has not been published.

Four other fragments of a translucent blue bowl are said to show close parallels to pieces from Nimrud (Jones 2009: 22).

In addition to the transparent and translucent bowls, core-formed glass was also found, which must be considered as Later Middle Phrygian in date (7th century). No further information was published on these pieces. In the Achaemenid period, the number of glass finds in Gordion increases significantly but is limited to core-formed glass (150 fragments).

3.4.4 Hasanlu

Hasanlu is situated in the Solduz valley, south of Lake Urmia in north-west Iran. Hasanlu was occupied from 1500 until about the Achaemenid period.

The glass finds from Hasanlu have been published in excerpts. Therefore, Marcus (1991) and de Schauensee (2001) focussed on the mosaic glass objects (see Chapter 4.1), whereas Stapleton (2003; 2011), and Stapleton and Swanson (2002; 2002a) referred to the chemical analytical data (see Chapter 7.4.2).

A number of Hasanlu glass finds in the Penn Museum collection are included in this monograph. The excavation records of the Hasanlu project, which were accessible to the author in the Penn Museum, enabled an almost complete reconstruction of the findspots. However, a detailed chronological assignment of the glass finds is not attempted as the allocation of the finds to the individual strata V, IV and III according to Zettler (pers. comm.) must be considered with caution.[31]

Between the oldest phase V (1450–1050) and the succeeding phase IVC, no major changes in the layout of the main buildings can be observed. Phase V came to an end by fire and the settlement was rebuilt in phase IVB with only slight modifications.

Hasanlu IVB exhibits large building structures: Burnt Building (BB) II, BB IVE and BBV, which were destroyed by fire as a result of military action. This is witnessed by the scattered weapons and bodies of armed warriors excavated at the site (Dyson and Voigt 2003: 219). According to radiocarbon dates, the fire and the destruction occurred at around 800, therefore the artefacts found in the demolition debris can be attributed to the timespan between 1050 and 800 (Dyson and Muscarella 1989: 3–4; Roaf 2012).[32]

All glass objects discussed here were found in the vicinity of buildings BBII, BBV and BBIV-V. Whereas BBII and BBV had already existed in phase IVC (Dyson 1989: 112, fig. 6a), and therefore previous to 1050, BBIV-V was added later in phase IVB (1050–800). The so-called 'Lower Court' and its surrounding building structures, including BBII, BBV, BBIE, BBIV, BBIVE, and BBIV-V, were erected in phase IVB (Figure 3.7) During this phase, access on the northwest side of the Lower Court was restricted to a narrow gate. Dyson (1989: 119) suggests that the building complex around the Lower Court was accessible to only a limited group of people. A second storey is proposed for all the buildings around the Lower Court, based on the existence of stairways, and the identification of two layers directly above each another (Dyson 1989: 120).

3.4.4.1 Burnt Building II (BBII)

BBII is the largest structure on the mound and is interpreted as a residential building or a temple. The fact that it existed already in phase IVC, as well as its size and structure, assigns an outstanding function to this complex. The entrance of BBII to its main room (5) led through two long narrow spaces (rooms 1 and 2), that exhibit single rabbets on both their interior and exterior. Room 1 is equipped with a portico, indicated by post-holes. The rooms are flanked with benches and platforms built of mud-brick and stone (Dyson and Voigt 2003: 224–226). Tube Has5 was discovered in room 2, which directly joins room 1 (Figure 3.7).

The centre of the building consists of a large columned hall (room 5) and on its eastern and western wall a row of wooden columns is reconstructed. Furthermore, benches on the east and west sides were uncovered, as well as a high offering table decorated with mud-brick pilasters on the north side. Within room 5, the alabaster cup inlaid with glass mosaic inlays (Has1, Has2, Has3) were found, as well as tube Has8.

On the southern side of room 5, the entrance, with a width of 60 cm, led to the small room 6, which was identified as the main room or 'cella'. The doorway to room 6 is elaborately decorated with a triple rabbet

[31] I would like to thank Richard Zettler for his information on the discovery of Hasanlu's glass finds and Katherine Blanchard for her support and assistance in making the objects available. Furthermore I thank Stephan Kroll, who provided me with all the data and photographs from Hasanlu.

[32] The succeeding squatter occupation (phase IVA) lasted only for a short period and was then burnt down once again. Hasanlu IIIB exhibits a fortified summit with open spaces and storage pits that are identified as Urartian in date.

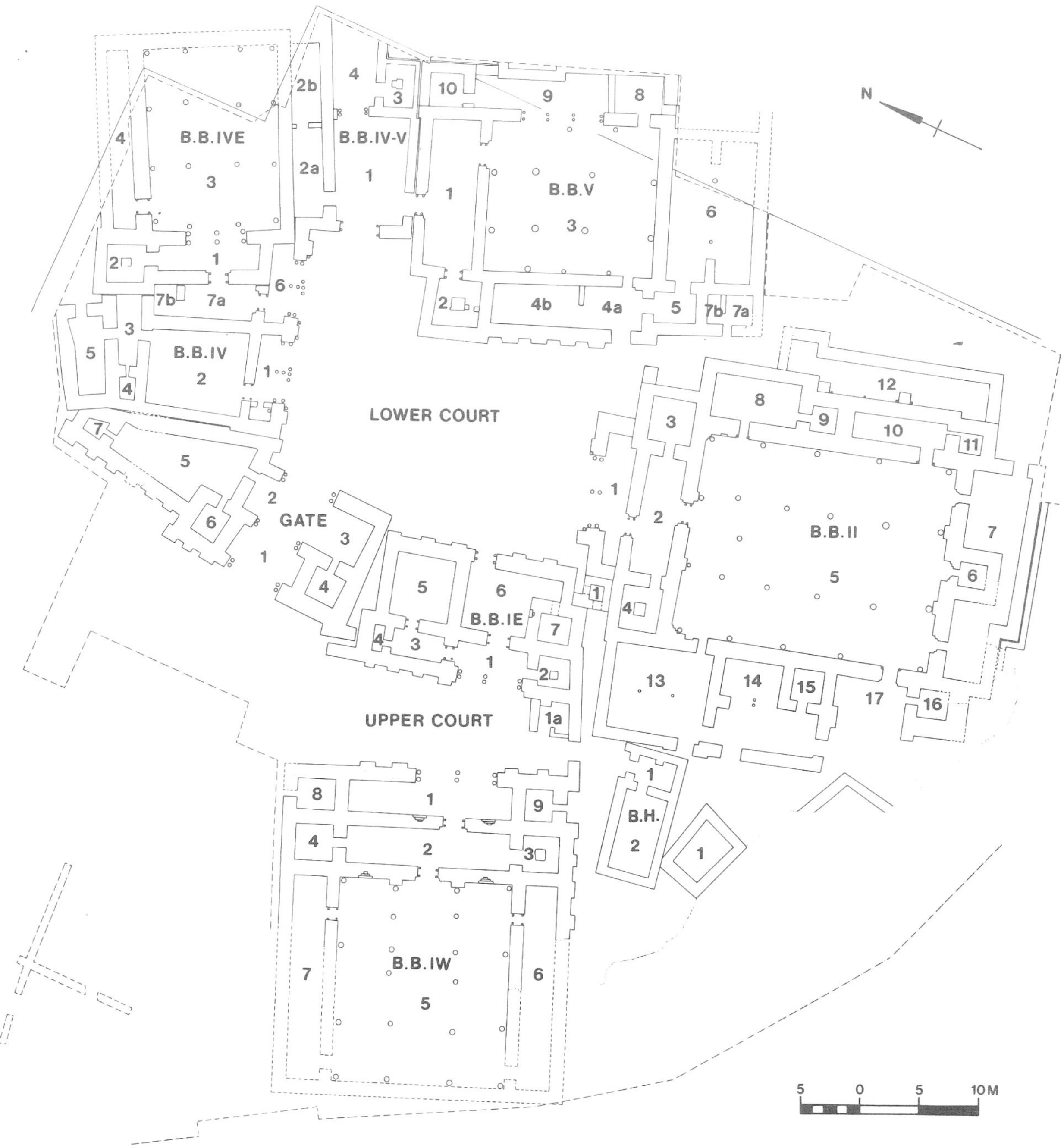

Figure 3.7: Plan of phase Hasanlu IVB, showing columns and post-holes (Muscarella 1980: 3, pl. 1).

structure facing room 5 (Figure 3.7). Furthermore, two entrances on the southern side of the hall, flanking the doorway to the 'cella', lead into room 7. By the time of the destruction of level IVB, only the entrance on the western side was in use (Roaf 1998: 64–65).

On the eastern side of room 7 another entrance leads to an outside space (Figure 3.7). Room 7a, i.e. the eastern part of room 7, was used for cooking, as suggested by the presence of pots, jars, pithoi and 30 pottery funnels. Additionally, eight glass tubes, two of which have been incorporated in this monograph (Has6, Has10), as well as mosaic glass fragments were found here (Dyson 1989: 122; Marcus 1991: 539). The objects, among which are also glass and faience* items, were spread widely throughout the building complex (for a detailed distribution map, see Dyson 1989: 122 fig. 19A, b). The distribution by Dyson (1989: 122) shows, that the glass and glazed containers (grouped together) were mainly kept on the 2nd floor, together with other 'precious materials' (glazed wall tiles, furniture attachments, shaped wood, ivory inlays and stone, copper, silver, gold containers). Marcus (1991: 539) identifies this conglomerate of finds on the upper storey as a treasury, in which precious goods were stored. Dyson (1989: 124) agrees on this point, but refers to the numbers of metal and stone vessels that would rather point towards an interpretation as a living space.

3.4.4.2 Burnt Building V (BBV)

In BBV, six glass objects were unearthed. Building V was erected in stratum IVC (see Chapter 3.4.4.3). Originally oriented to the north, BBV was then re-oriented to the south-west, joining the outline of BBIVE, in phase IVB. Glass finds uncovered in BBV were found in the main hall (room 3). The main hall (13 m x 15 m) has four wooden columns in its centre, which were already part of the preceding building (phase V). Tube Has7 can be attributed to room 3.

3.4.4.3 Burnt Building IV-V (BBIV-V)

Finally, the two raw glass fragments Has13 and Has14 were found in building BB IV-V, in room 4 which joins BBV on its north-western side. This complex was built between BBV and BBIVE to fill in the space between these two complexes. The glass finds from room 4 incorporate raw material of red and light blue glass, confirming the existence of primary and/or secondary production at the site. Other items found in this context were thought to have fallen from the upper storey. They incorporate a mass of smashed pithoi sherds, textile fragments, and bronze horse gear. It is very likely that the raw glass material was also stored on the first floor, as it is unlikely that the pieces were kept in a passageway, a function indicated by the presence of stairways in room 3.

3.4.4.4 Summary

As indicated previously, the glass finds from Hasanlu incorporated in this study come from contexts dating to the period between 1050 and 800; a more detailed chronology cannot be provided.

Regarding the glass tubes, it is interesting to note that they were solely found in rooms 2 (Has5), 5 (Has8), 7a (Has6, Has10, Has11) of BBII, as well as in room 3 of BBV (Has7). It is very likely that the tubes, as well as the raw glass fragments, were stored on the first floor of the relevant building, belonging to the treasuries, like the jewellery and ivory that were kept there. The existence of raw glass objects indicates that either primary or secondary production was carried out at Hasanlu.

3.4.5 *Idalion*

Idalion is situated beside the River Yialias at the foothills of the Troodos Mountains in Cyprus. The site is located at a crossing of important trade routes leading to the east and south of the island.

The vessel Id1 was uncovered during the excavations by Lang, when a temple area dedicated to 'Apollo Amyklos' was cleared (Lang 1875). As the exact location of the findspot is unknown, it is impossible to make further considerations about the context and date of this vessel based on stratigraphic considerations.

3.4.6 *Kameiros*

Kameiros is situated on the north-eastern coast of Rhodes. The vast necropolis of Kameiros was first excavated by Salzmann from 1858–1865 (Salzmann 1875). Salzmann published some of the finds, but neither referred to any of their findspots, nor included a plan in his publications. From 1868 until 1889 Biliotti continued excavating the site. Most of the finds from these excavations were brought to the British Museum. Kam1 was found in a tomb 'excavated by Salzmann and Biliotti' (Barag 1970: 165), no further information on the exact location is known.[33]

3.4.7 *Praeneste*

Praeneste is situated 37 km east of Rome. In 1876 the so-called 'Bernardini Tomb' was uncovered in Praeneste, which can be enqueued into a number of other rich tombs, such as the 'Barberini Tomb', found in close proximity. The richly furnished tomb of a single individual dates prior to 650 (Canciani and Hase 1979: 77–78). Accompanying finds represent gold and silver fibulae and vessels, partly decorated with pseudo-Egyptian motifs, inlaid ivories of Phoenician type, metal vessels and standards, to mention only some of the grave goods. Among the finds was also vessel Pr1, which was found together with three other hemispherical glass bowl fragments that remained unpublished (Canciani and Hase 1979: pl. 1–73).

3.4.8 *Susa*

Susa, the modern city of Shush, is located in south-western Iran, about 250 km east of the Tigris.

Only two core-formed glass vessels, Sus1 and Sus2, are published (de Mequenem 1931: 334). Their exact findspots are not known, however, de Mequenem (1931: 334) reports that the objects come from a cemetery on the acropolis, with Neo-Babylonian as well as Late Elamite tombs. The pieces were associated with 'good phials of moulded and coloured glass' and various copper vessels and other objects (de Mequenem 1931: 334).

[33] For an overview of the excavations and of the site, see a comparison by Launay 1895: 185–194.

4. The Glass Objects: Manufacturing Techniques, Typology, and Function

The following chapter is concerned with the typological classification of glass objects of primary and secondary production (Chapter 1.2). Objects produced in the primary glass production process comprise raw glass ingots and waste products. Most objects of primary glass production presented cannot be clearly dated, as a secured find context often does not exist. In order to include a discussion of the primary glass industry, some of the pieces will nevertheless be discussed below (Chapter 4.5). The process of primary glass production described in the Nineveh Glass Recipes thus also becomes accessible from an archaeological perspective (Chapter 6).

Glass objects produced in secondary production, on the other hand, are the finished glass objects, such as vessels, inlays, attachments and tiles. The glass objects can be further divided into typological groups, which depend directly on the manufacturing techniques, known as 'core-formed', 'mosaic', and 'cast-and-cut'. The three manufacturing techniques have developed in different periods but are partly interlinked and are presented in the following chapter. The shape of a glass object is closely linked to the manufacturing technique used. In fact the method used to process the glass of a specific objects leads directly to the typological group to which it can be assigned. This chapter therefore aims to identify the manufacturing technique used to produce a particular object. This requires a detailed description of each object, including irregularities, deformations, traces of tool marks, colour and discolouration, as well as the overall shape and distribution of the bubbles within the glass.[1]

The identification of a certain manufacturing technique is not always easy to determine. To formulate conclusions, however, this work draws heavily on the knowledge of modern glassmakers, whose practical experience makes an enormous contribution to the possible identification of a technology.[2] In this context it is important to note, however, that the use of a certain technique in modern times does not prove a technique to be used by ancient glassworkers. The examples given here are therefore to be understood as suggestions.

[1] In this regard the examination of the objects included in this study was either done by the author, or was based on published descriptions and photographs. If the object was not studied by the author then the appropriate literature is provided. A detailed description of corrosion is included in the catalogue section.

[2] I would like to thank Mark Taylor and David Hill, the Roman Glassmakers, Andover (UK), and William Gudenrath, Resident Advisor at the Corning Museum of Glass, for sharing their experiences about the production and working with glass. This book in its current form could only have been written with their support.

The processing of glass into an object requires high temperatures, but less than in glass production. It should be noted here that glass must generally undergo an annealing process*, which means that the hot object cools down slowly. This ensures that no compressive and tensile stresses are trapped in the glass during the cooling phase. In general, the larger and thicker the glass, the longer the annealing time must be. If glass is not annealed properly, it breaks.

Apart from technological aspects, the objects will be considered also in broader chronological and functional contexts. In this regard, objects of other materials, such as faience*, stone or ceramic will be cited for comparison. Together with contextual and chronological considerations, the overall function and broader meaning of the artefacts will be discussed.

4.1. Mosaic (glass) objects

4.1.1 Definition of the term 'mosaic'

The term 'mosaic' is commonly defined as 'decoration of a surface with designs made up of closely set, usually variously coloured, small pieces of material (...).'[3] The designation as a mosaic object thus directly depends on the manufacturing technique. Regarding glass or sintered materials, the mosaic technique allows the creation of decorative patterns. The major characteristic which distinguishes most objects made in the mosaic technique are determined by the fact that the resulting pattern is more or less the same on both the interior and exterior of an object. Furthermore, the mosaic technique allows the detailed depiction of figural scenes and geometric patterns to an extent which cannot be achieved by any other technique known in the period under consideration.

Mosaic glass objects already existed in the middle of the 2nd millennium (Haevernick 1968: 67; Saldern 2004: 20; Wartke 1982: 24). Therefore, this technique was established right from the outset of glass production itself. The earliest examples of mosaic glass objects from Mesopotamia come from Tell al-Rimah (Barag 1985: 40, no. 4), Dur-Kurigalzu (Saldern 1970: 213–214), and Nuzi (Barag 1970: 140, no.15), among other sites. In this early stage, the mosaic technique is used for open bowls, beakers and inlays. After the Iron Age, in the Hellenistic period, inlays with complex and often figural designs are common, as well as bowls with elaborate mosaic

[3] http://www.britannica.com/art/mosaic-art http://www.britannica.com/art/mosaic-art (accessed: 7.12.2015).

patterns (Wight 2011: 42).[4] This early mosaic glass is often incorrectly referred to as 'millefiori' ('thousand flowers'). 'Millefiori' originated in the late 15th century as a technique introduced by Venetian glassmakers for the production of flower designs in glass (Tait 1979: 94–95). Although 'millefiori' describes the same technique as mosaic glass, it relates to glass objects that specifically date to the Renaissance. The term 'millefiori' in connection with Late Bronze Age or Iron Age glass objects is therefore anachronistic and should be avoided in favour of the term 'mosaic'.

4.1.2 Manufacturing techniques

The principle of forming a mosaic glass object is to fuse individual glass segments together, and to shape them into an object by the use of a mould*. Therefore, the individual mosaic pieces – which later on formed the mosaic pattern – had to be produced first. Since the manufacturing techniques for each of the mosaic object groups in this monograph vary, the different processes will be discussed separately for bowls, inlays, and tiles. At this point, it should be noted that the description of production methods is always only one possibility of how a certain object could have been made. In many cases, the manufacturing process may have been different. Therefore the techniques described here should be considered as suggestions.

4.1.2.1 Bowls

The process described here refers to the bowls and bowl fragments As1, As2, As3, As4, and As5. Before discussing the steps of manufacturing, it is necessary to consider closely the material out of which the mosaic bowls were made: they were neither made of glass nor faience*. Rather they can be defined as a hybrid, '(...) eine Mischung aus den üblichen Ausgangsstoffen für Quarzkeramik mit Zugabe von Glas in Form feinen Pulvers' (a mixture of the usual basic materials for faience with addition of glass in the form of fine powder) (Wartke 2012: 406), which is called glassy faience* (Chapter 2.2.2).[5] To briefly return to the definition of the materials glass, glassy faience and faience at this point, all three are composed of soda, lime and silica (also Chapter 2). In glass, the quartz particles are completely fused, whereas in the glassy faience some quartz particles are still present in a vitreous phase. Faience consists of a sintered glass core with a glass layer on top (for details Chapter 2.2.1). Glass can be easily distinguished from glassy faience and faience with the eye. This is different with regard to glassy faience and faience which cannot be distinguished with the eye. Since there are no analyses available for the bowls discussed here, it is difficult to decide whether the material is truly glassy faience or faience. SEM analyses would shed light on this and are a future project. The mosaic bowls – even though not made of glass – are included in this monograph because of their relevance with regard to the history of Iron Age glassworking. However, it should be noted here that the production of these bowls was not made in the hot state, as with glass, but in the cold state, which is typical for glassy faience and faience. This difference in manufacturing is crucial and should be taken into account when reading the chapter.

In the following the manufacturing technique of bowl As2 will be described in detail. However, most of these descriptions also apply to the remaining bowls and discrepancies are indicated. To make As2, first of all, mosaic segments made of glassy faience had to be made, which were cut from polychromatic canes in cold state. The mosaic canes were modelled by coating a coil with a flattened layer of a second base material in another colour (Figure 4.1).[6] This was carried out in a workable plastic state. The diameter of the cane, and therefore of the mosaic segments, could be manipulated by stretching. By cutting the rods, small mosaic pieces were produced (Figure 4.1). The cutting* was done in a 'leather-hard' condition of the material.

In a second step, the mosaic segments were arranged next to each other to form a flat disc (Figure 4.2; Figure 4.3). During this process, the single mosaic pieces were dragged into a particular shape, creating the distinctive mosaic pattern of the latter bowl. For example, with regard to As2 the honeycombed pattern was a result of pressing the round pieces closely together. As1 and As3 were formed into a star and flower pattern. This process was carried out in the plastic state.

To bring the mosaic disc into the shape of a bowls, the wet and plastic disc was modelled over a convex mould* and pressed by hand to its surface. As a next step, the bottom-rim was added as a separate part on the bottom of the bowl. This was achieved by pressing a ring of mosaic coil against the surface. The coil was attached to

[4] For complex inlays dating to the Hellenistic period, see Tait 1991: 52–53. The manufacturing techniques of mosaic glass bowls have been intensively studied with regard to Hellenistic and Roman mosaic glass vessels by Grose 1989: 33–34 and Wight 2011: 39–43, among others. Experimental studies contributed largely to an understanding of the different procedural steps of the manufacturing process. A number of experiments were carried out in the Toledo Museum, see therefore Stern and Schlick-Nolte 1994: 84, fn. 275.

[5] This is the result of microscopic as well as experimental studies carried out on mosaic bowl As2 by Wartke 2012. Microscopic analysis revealed that quartz is present at a level of 50%–60%, as well as a large amorphous phase; see Busz and Gercke 1999: 335. Observations on the other mosaic bowls during their investigation imply that the entire group of objects was made of glassy faience. This can, however, only be fully verified by chemical analysis; a conclusive identification of the material is therefore left for further archaeometrical analysis.

[6] The processing of glassy faience is similar to faience. The basic materials (quartz, lime, and natron) are ground and mixed together with water to make a clay-like paste which was suitable for moulding and modelling; see Taj-Eddin 2014: 71. The blue colour of As2 is due to copper oxide, red to iron oxide and yellow to lead antimonite; white was not analysed; see Busz and Gercke 1999: 335; Wartke 2012: 407.

Figure 4.1: Cutting of mosaic rods, carried out in cold state while it is still plastic (after Wartke 2012: 409, fig. 5).

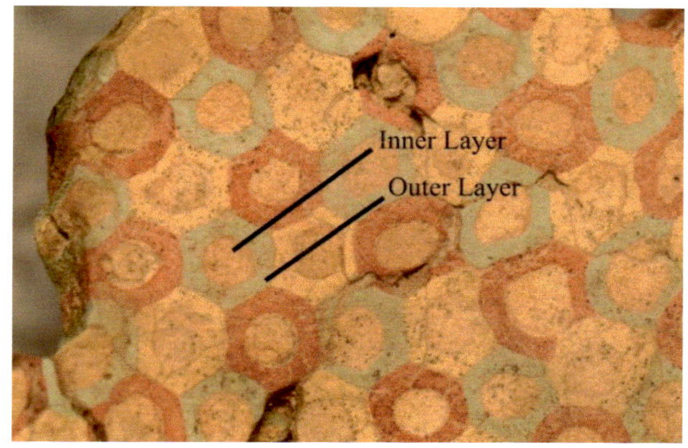

Figure 4.2: Detail of bowl As2, showing the inner and outer layer of mosaic pieces (after Staatliche Museen zu Berlin – Vorderasiatisches Museum; photo: Olaf Teßmer).

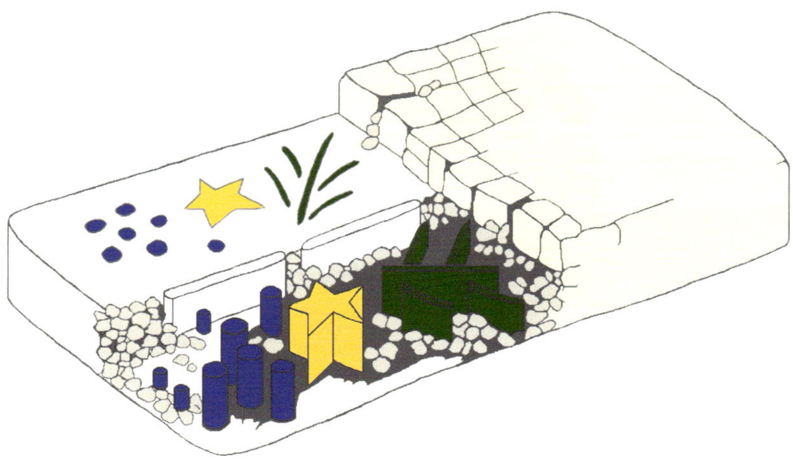

Figure 4.3: The principle of making mosaic glass inlays: mosaic glass segments of different sizes and shapes are fused together (after Stern and Schlick-Nolte 1994: 63, no. 96).

the surface by the use of an adhesive, which was most likely water. It is likely that the base-rim was attached separately, since the mosaic patterns on the walls of the bowl do not overlap with that on the rim-base.[7]

As a last step, the rim of the mosaic bowl was finished, either in the 'leather-hard' or cold state, most likely by the use of a blunt tool. A groove at the rim was applied by cutting into the cold vessel with a sharp tool before or after final heating (Busz and Gercke 1999: 336).

4.1.2.2 Inlays

To produce mosaic inlays singular glass segments were first cut from glass canes of different size and shape. Then, in order to make a larger and more complex mosaic pattern, these glass segments were fused together to create inlays with different patterns and colours (Figure 4.3).[8]

Among the mosaic inlays of this study, two different types can be distinguished which result from slightly different processes of manufacture. Patterns of inlays that belong to the first type can be identified by their visibility from the obverse and reverse side (Has3, As6, As8). In contrast, patterns of inlays of the second type are visible from one side only (Has1, Has2, Has3, As7).

The inlays of the first type consist solely of the mosaic layer (Figure 4.3). Inlays of the second type, however, were additionally fused onto a monochrome glass plate,

[7] With regard to its manufacturing, Wartke (2012: 410–411) suggested that it was made by moulding, i.e. pressing the disc into a former mould. For his full argumentation, see Wartke 2012.

[8] The different techniques of making the mosaic glass canes of various types are described in detail by Stern and Schlick-Nolte (1994: 54–61). For detailed descriptions of different kinds of mosaic glass, see Stern and Schlick-Nolte 1994: 61–63.

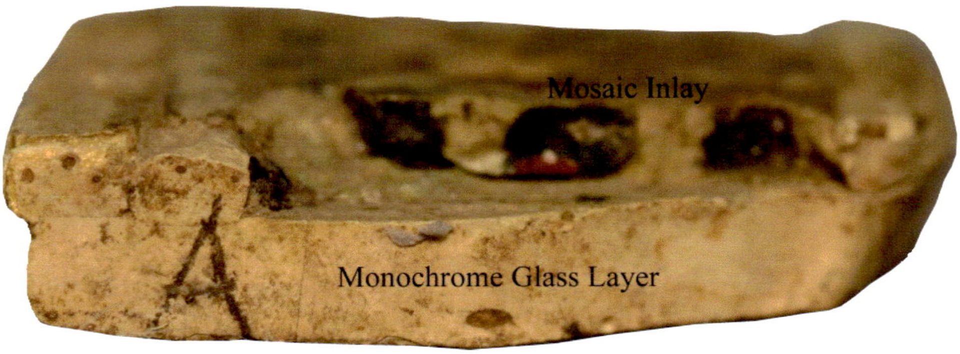

Figure 4.4: Mosaic inlays of the second type: a mosaic glass inlay is fused onto a monochrome glass layer (Near Eastern Collections, University of Pennsylvania Museum of Archaeology and Anthropology).

making the pattern visible from only one side. This is clearly indicated in section by a seam between the two layers (Figure 4.4).

4.1.2.3 Tiles

The major characteristic of glass tiles is their thickness - approximately 3 cm - and this is how they can be distinguished from inlays. Tiles were made by fusing monochrome mosaic segments and rosette mosaic pieces together, as described regarding the mosaic glass inlays (Chapter 4.1.2.2). Similar to inlays of the second type, the glass layer of the tiles was fused onto a thick, monochrome glass layer that formed the reverse of the object. Therefore the mosaic patterns of the tiles were only visible from one side. The thickness of the objects indicates that a large amount of glass was used for the production of tiles.

4.1.3 Description and discussion of objects

4.1.3.1 Bowls

Mosaic bowls relevant for this study were only found in Aššur (As1, As2, As3, As4, As5).[9] All these objects bear affinities in shape, surface, and colouration.

As2 shows a honeycomb pattern and belongs to the group of shallow bowls with a rounded wall and foot rim. The rim pulls slightly inwards and has a fine groove around the upper part of the vessel. The surfaces of As2, As3, A4, and As5 appear porous, dry, and finely grained in texture, no traces of glaze are present. The colour tones are similar to one another but vary from other colours that are attested in opaque glass of this period. The boundaries between the different colour fields are sharp, and sometimes fine cracks occur.

Fragments As3 with flower pattern and As4 with chevron pattern only preserve parts of the foot rim, but they can also be identified as shallow bowls. Regarding As2, As3, and As4, the foot rim was worked separately and added to the vessel after it was manufactured.

As1 (star pattern) also has the shape of a shallow bowl with rounded wall. The rim of this vessel is flattened and was, most likely, worked after its manufacturing. Since this vessel does not have a foot rim, the base was either straight or the ring is no longer preserved.

Stylistically, the star pattern of As1 shows close parallels to a glass fragment from Dur Kurigalzu which dates to the Kassite period (Marcus 1991: 52, fig. 24). The other bowls do not demonstrate any close parallels with comparable objects in terms of their decoration.

As2 was assigned to the post-Assyrian period by Haller (1954: 29), by Pedde (2015: 101) to the Middle-Assyrian period, and by Miglus (1996: 385) to a Neo-Assyrian date (see Chapter 3.1.1). None of the other bowls can be dated due to the find context. Only because of the close parallel between the decoration of BS3 and As2 can a date be proposed: BS3 was discovered in the early Seti I Temple in Beth-Shean, which dates from 1313-1292. The bead was discovered in a layer that, after Rowe (1940: 30), belongs to the early part of Seti I's reign. A *terminus ante quem* for the 12th century can therefore be assumed with regard to the production of BS3. Based on this, a date for the bowls from Aššur (As1, As2, As3, As4, As5) can be placed tentatively into the Middle-Assyrian period (see in detail Chapter 4.1.4). Since the mosaic bowls were all found in Aššur and form a narrow group in terms of material, form, style and applied manufacturing technique, an approximate contemporary date is likely for As2, As3 and As5.

[9] A mosaic glass bowl from Uruk, not included in this study, was previously dated either to the second half of the 2nd millennium (Busz and Gercke 1999: 336, with figure), or to the first half of the 1st millennium (Wartke 2012: 405, with figure). The vessel was discovered in a secondary context. With regard to the pattern of this bowl (spiral of two turnings), a date to the Hellenistic period is suggested here, as this pattern is regarded as a hallmark of this period (Grose 1989: 189, 178, fig. 184, 204).

The grave in which As1 was found contained no outstanding funerary items (Chapter 3.1.1). This might indicate that mosaic bowls were thus not restricted to usage in palaces or temples. The fact that the mosaic bowls were solely found in Aššur could indicate the existence of a workshop that produced this specific vessel type made of glassy faience.

4.1.3.2 Inlays

Inlays made in the mosaic technique have been found in Hasanlu and Aššur. The inlays Has1, Has2, and Has3 from Hasanlu were set into an alabaster cup (HAS64-127, UPM 65-31-23) (Figure 4.5).[10]

Has1 has an overall rectangular shape (1.5 x 1.4 cm) and a thickness of 0.5 cm. Three sides are preserved and have rounded edges, with one side broken off. Has1 belongs to the inlays of the first type (Chapter 4.1.2.2). The obverse is flat, while the reverse has an irregular surface and exhibits a horizontal crack (de Schauensee 2001: 100–102). The decorative motif is only visible on the obverse and is comprised of a slightly off-centre rosette. The round centre of the flower is weathered (previously probably white) and is outlined by opaque yellow glass. The petals are made in the same way, with the weathered inner colour outlined by opaque yellow glass.

Has2 exhibits an irregular rectangular shape (1.1 x 0.9 cm), 0.5 cm thick, with its edges rounded and partially broken off. This object also belongs to inlays of the first type (Chapter 4.1.2.2). The decorative motif is only visible on the obverse, of which the surface is smooth and flat. The reverse is plain and irregular. The nine-petal flower is set into an opaque dark-blue background inlaid into a background of opaque whitish glass, which is a result of fading. The rosette inlay is completely off-centre with the round centre of yellow colour, and the petals made of white colour.

Has3 differs largely from Has1 and Has2 in terms of motif, colouring, weathering, and manufacturing type. The overall shape of Has3 is an irregular rectangle with rounded edges (1.6 x 1.4 cm) and a thickness of 0.5 cm, and belongs to the second type. The pattern is therefore visible on both sides of the piece. The yellow parts of the pattern are set in an opaque, strong blue background that partly exhibits light greenish spots. The centre of Has3 has a rounded centre which was probably inlaid.

As6 is an irregular fragment which is broken on all sides (2 x 2.5 cm) and has a thickness of 0.7 – 0.8 cm. This piece belongs to the first type, therefore, two petals of the decoration are preserved and visible on the obverse and reverse of the object. The obverse is flat and smooth. The pattern is sharp, and the colours are clear. The petals are composed of two opaque white elongated triangles that are separated in the middle by a strip of greenish glass. The petals adjoin an originally rounded centre of opaque yellow glass. The flower is set into the background of dark greenish glass, with red streaks. On the reverse, this pattern is visible but unclear because it is covered by a thin layer of greenish glass, most likely due to the production process.

Figure 4.5: Alabaster vessel from Hasanlu with inlays made of mosaic glass, carnelian and Egyptian blue. The mosaic inlays are in secondary use. (Near Eastern Collections, University of Pennsylvania Museum of Archaeology and Anthropology).

As7 is an irregular piece of an overall rectangular shape (2.7 x 2.6 cm) and a thickness of 0.6 cm. The piece belongs to the first type and is bounded by two straight edges and is broken on the other two sides. The obverse side exhibits a ten-petal dark blue flower that was formerly translucent with an opaque yellow centre. The rosette is inlaid into an opaque whitish background framing the rosette on two sides by a strip of glass. The flower is therefore only visible on one side. The manner in which the petals are shaped has parallels with flowers that were set into a Late Bronze Age glass beaker found at Hasanlu.[11]

[10] The pieces could not be removed from the alabaster vessel during examination and thus the reverse could not be examined.

[11] Two glass beakers (UM 65-31-403, 404, UM 65-31-405) have been

As8 is comprised of six pieces of mosaic glass most likely belonging to the same object, as they are very similar in pattern, colour and material. The pattern on As8 exhibits a long, pointed leaf that is slightly bent, as well as volutes and a pomegranate-shaped inlay. Two round depressions were also most likely inlaid with glass pieces. The pieces belong to inlays of the first type. All fragments are flat and the mosaic pattern is clearly visible from both sides. As8 shows close parallels to a beaker made in the mosaic technique that was found at Aššur (Ass 19929), and two mosaic beakers from Hasanlu; all of these comparable finds can be dated to the late 2nd millennium (Chapter 4.1.4).[12] It is therefore likely that the manufacturing of As8 dates to the Late Bronze Age since the find context of As8 is unknown.

Regarding the three inlays from Hasanlu (Has1, Has2, Has3), it is most likely that the pieces reflect secondary use and were cut out from larger glass objects (de Schauensee 2001: 101). This is suggested since the rosettes Has1 and Has2 are off-centre. Furthermore, the irregular and rough edges indicate that the objects were cut out from a larger object. Some of them seem to have been smoothed even though no tool marks are visible.

Regarding the origin of the inlays from Hasanlu, it is worth considering the alabaster vessel in which Has1, Has2, and Has3 were set: the cup was found in the IVB destruction layer of Burnt Building II (room 5) (Chapter 3.4.4). The shape of the vessel is unique at Hasanlu and shows parallels with peg-based ceramic vessels from Mesopotamia, rather than from Iran (de Schauensee 2001: 103).[13] In addition, also the use of alabaster is rare in Iran, and without comparison at Hasanlu. Consequently, de Schauensee (2001: 103) suggested the alabaster cup to be an import from Mesopotamia. In contrast, with regard to the Egyptian blue, as well as to the carnelian beads decorating the alabaster cup, de Schauensee (2001: 103–104) indicated that they exhibit parallels to beads found in the cemetery at Hasanlu. The gold-banding techniques also appear among other objects from the site, which strongly indicates that the beads were produced locally. With regard to the glass inlays, Has1 and Has2 show close parallels with the inlays As6, As7 and As8 from Aššur, not only in their manufacturing technique but also in the colour tone of the glass itself.[14] This also applies to two mosaic glass beakers from Hasanlu (UM 65-31-403, 404, UM 65-31-405), dated to the late 2nd millennium on the basis of stylistic considerations, and to one mosaic glass vessel from Aššur (Ass 19929) found in the Ištar Temple in a phase attributed to the time of Tukultī-Ninurta I (1243–1207).[15] With regard to a date of production for the inlays Has1, Has2, As6, As7 and As8, this indicates that they were most likely produced at some point in the late 2nd millennium, probably in the 13th century. Since the origin of the mosaic vessels from Hasanlu and Aššur point towards Babylonia in the Kassite period, this can probably also be suggested for inlays Has1, Has2, As6, As7 and As8 (see also Chapter 4.1.1).

Returning to the alabaster cup, it is likely that the vessel was imported from Mesopotamia and inlaid at the site using locally worked Egyptian blue and carnelian beads, as well as older mosaic glass inlays that had been used in a different context before.

4.1.3.3 Tiles

Glass tiles, like the bowls, were found only at Aššur.

As13 is 5.0 x 4.0 cm wide and has a thickness of 2.7–2.9 cm. The piece has one straight-sided edge and is broken on the other sides. The reverse side is irregular with a beige corrosion layer. The obverse side is flat and shows mosaic decoration of long black and yellow stripes of irregular thickness. One black-yellow-black stripe runs along the straight edge, and two further bands branch out from there; the area in between the stripes has a light blue colour. A round rosette inlay is worked into the tile which has eight white leaves with a central red dot on a dark blue background. The upper mosaic layer is set on a monochrome, translucent dark blue glass layer.

As14 is an irregular fragment broken on all sides. The black-yellow-black stripes on the obverse side are arranged in an irregular square pattern. Two round rosette inlays, each with eight white leaves, a central red dot on a dark blue background, are also worked into the tile.

As15 is irregular and also broken on all sides. Traces of black-yellow-black stripes are present, as well as a round inlay with an eight-petal rosette of yellow colour on a red background.

studied intensively by Marcus 1991, who demonstrated convincingly that both objects date to the late 2nd millennium.

[12] The beaker from Aššur was found in a context that can be attributed to the time of Tukultī-Ninurta I (1243–1207). The ones from Hasanlu were convincingly attributed to the late 2nd millennium on the basis of close stylistic parallels to Kassite objects; see therefore Marcus 1991: 535. In this regard, it was left open whether the beakers reached Hasanlu from Babylonian via Assyria, or whether they were brought directly (Marcus 1991: 559). For a detailed discussion of the findspot of the beaker from Aššur, see Chapter 3.1.1; for Hasanlu, see Marcus 1991.

[13] For details on peg-based ceramic vessels, see de Schauensee 2001: 104, footnote 22.

[14] A parallel was also suggested by Marcus (1991: 546).

[15] Whereas the vessel from Aššur only demonstrates technical and stylistic parallels, the beakers from Hasanlu are also comparable regarding the consistency of the glass. With regard to stylistic considerations, Marcus (1991: 559) suggested a Kassite origin for the two mosaic glass beakers from Hasanlu. Recent analysis on lead, oxygen and strontium isotopes support this suggestion as they indicate a correlation between glass objects from Hasanlu and from Babylonia (Babylon, Nippur) and its adjacent territories (Susa, Persepolis) (Brill and Stapleton 2012: 219).

The irregular and damaged fragment As16 is 7.6 x 7.1 cm wide and 2.9 cm thick. The black-yellow-black stripes create hexagonal forms that are inlaid with rosettes which have eight yellow petals. The flower is set on a red ground.

As17 has an elongated irregular shape and is broken on each side. Black-yellow-black stripes form a zigzag band.

As18 has a straight-sided edge but the other is broken. A black-yellow-black stripe decoration can be reconstructed to a hexagonal form. Within the frames there is a round rosette inlay with a yellow, eight-petal rosette on a red ground.

The technique of fusing a layer of mosaic glass onto a second layer of glass only finds parallels among inlays from Hasanlu (Has1, Has2) and Aššur (As7), and similar inlays of the second type. The tiles were found out of context around the ziggurat and the Ištar Temple. The findspot does not allow for any conclusion regarding the date of the objects. No comparison between the tiles and glass objects from the Hellenistic and Roman periods can be established, albeit mosaic glass was widely spread in this time. Based on the close stylistic similarities of the tiles, it is likely that they are contemporaneous, either made in the late 2nd millennium or the Neo-Assyrian period, even giving slight preference to the Late Bronze Age, since the mosaic technique was then far more widespread.[16]

The tiles form a unique group of objects, making it likely to locate their origin at the same place, probably at Aššur. Their findspot in the former area of the Ištar Temple and ziggurat (Chapter 3.1.1) could indicate that they were originally used in connection with these building structures, probably as architectural decoration for walls or floors. The use of glazed coloured bricks integrated in the building structures of ziggurats, such as for instance at Nippur, Ur, Khorsabad and Borsippa was illustrated by Clayden and Schneider (2015: 361), James and van der Sluijs (2008), as well as by Nunn 1988. These monuments not only stood out of the urban landscape because of their size, but also because their colours appeared even more intensively with their glazed surfaces. The use of multi-coloured glass tiles within the building structures of the ziggurat in Aššur would therefore not be surprising. However, a secondary use of the tiles in later periods cannot be ruled out.[17]

4.1.4 Discussion: date of mosaic (glass) objects

With regard to the mosaic objects discussed in this chapter, it is imperative to distinguish primary and secondary use, since several of the Late Bronze Age objects were found in later early 1st millennium contexts. To do so, an overview of the development of mosaic glass technology from the Late Bronze Age to the Iron Age period is essential. The most comprehensive study on mosaic glass objects of this period has been published by Marcus (1991), which will serve as the major reference.

Mosaic objects from the Late Bronze Age period were found at sites in north-western Iran, such as Hasanlu and Marlik (Figure 4.6) (Saldern 1970: fig. 3, 4).[18] They were also unearthed in Babylonia at Dur Kurigalzu (Saldern 1970: fig. 2), or in northern Mesopotamia, for example at Tell al-Rimah (Figure 4.7) (Saldern 1970: fig. 1), Tell Brak (Oates *et al.* 1997: 84, fig. 122), and Aššur (Ass 19929). Some of these objects are very similar to one another. Considering technique (mosaic), shape (cylindrical, wide body), and pattern (round mosaic pieces), a connection between a mosaic glass beaker found at Marlik (Figure 4.6), and mosaic glass vessels from Tell al-Rimah (Figure 4.7), and Dur Kurigalzu was convincingly established by Marcus (1991: 553), who argues for a Mesopotamian origin of the so-called 'Marlik beaker', on the basis of technique, shape and style.[19]

Figure 4.6: Mosaic glass beaker from Marlik, ht. 17 cm (Marcus 1991: 545, no. 13).

[16] A Late Bronze Age date was suggested by Saldern (1970: 215, no. 7), and a Neo-Assyrian date by Busz and Gercke (1999: 339, no. 150–151).

[17] The recycling of glass is not attested in the Late Bronze Age and Iron Age periods and can therefore be neglected in this study.

[18] Despite the general existence of glass objects in Elam (Susa, Chogha Zanbil), none of these was made by the mosaic technique; see Marcus 1991: 553.

[19] The beaker was found together with a rectangular beaker and a chalice with pedestal base in tomb 25 of the cemetery of Marlik. The tomb was attributed to the late 2nd millennium; for a detailed argumentation and further literature, see Marcus 1991: 553.

Figure 4.7: Mosaic glass vessel fragments of an almost cylindrical beaker from Tell al-Rimah (7.7 x 6 cm) (Tait 1991: 23, fig. 17).

Similarly, regarding two mosaic glass beakers from Hasanlu (UM 65-31-403, 404, UM 65-31-405), and one from Aššur (Ass 19929),[20] an origin from Kassite Babylonia is suggested. This is based on close stylistic similarities to Kassite objects (Marcus 1991: 535), and is supported by isotope analysis on lead, oxygen and strontium, carried out on these glass vessels (Brill and Stapleton 2012: 219) (Chapter 7.4.2). Accordingly, the two beakers from Hasanlu found in the destruction debris IVB were most likely imports from Babylonia, either in the late 2nd or early 1st millennium and stored in Burnt Building II as heirlooms (Chapter 3.4.4).

With regard to the mosaic glass inlays from Hasanlu (Has1, Has2) and Aššur (As6, As7, As8) discussed in detail in Chapter 4.1.2.2, close similarities with the two mosaic glass beakers from Hasanlu, as well as with the vessel from Aššur (above), were established, indicating similar origin and date. It is therefore suggested here that also the inlays Has1, Has2, As6, As7, and As8 originate from Babylonia and were made in the late 2nd millennium.

To summarise, the weight of evidence for the origin of mosaic glass vessels from Marlik and mosaic vessels and inlays from Hasanlu (Has1, Has2) clearly points to a Babylonian production in the Kassite period. The reason for this wide distribution of similar mosaic objects can be explained by the cultural practices in the Kassite kingdom and its 'large-scale cultural interaction' with the regions to the north and east of Mesopotamia (Marcus 1991: 535). Also, the role of Aššur in the mosaic glass production needs to be outlined since there is a noticeable amount of mosaic glass objects of different techniques and styles found at this site. A number of mosaic glass fragments from this site show significant parallels with mosaic pieces from Hasanlu, 'in fact (...) the Hasanlu vessels compare best with fragments of vessels and inlays from the Ishtar Temple at Aššur (...)' (Marcus 1991: 546). This indicates a close alliance between these two sites, which could also be connected with the redistribution of Kassite goods through Assyria before they reached north-western Iran, and notably Hasanlu.[21] At Aššur unique mosaic objects were also found, such as mosaic glass tiles (As13, As14, As15, As16, As17, As18) and mosaic bowls made of glassy faience* (As1, As2, As3, As4, As5). The latter only show close connections to a flat mosaic bead from Beth-Shean (BS3), which has almost the same mosaic pattern. Whether this bead (BS3) was the object of secondary use cannot be decided here.[22]

The question now arises how to date the glassy faience bowls? Unfortunately, due to their find context they cannot be dated directly (Chapter 3.1.1).[23] Only through the close parallel with bead BS3 can a date be cautiously suggested: BS3 was discovered in the Early Seti I Temple in Beth-Shean, which dates from 1313–1292. The bead was discovered in a stratum that, according to Rowe (1940: 30), belonged to the early part of the reign of Seti I. A *terminus ante quem* for the 12th century can therefore be supposed with regard to the manufacturing of BS3. Based on this find, the time of origin of the bowls from Aššur (As1, As2, As3, As4, As5) can be dated carefully into the Middle-Assyrian period. It cannot be ruled

[20] This vessel was found in the Ištar Temple, dating to the time of Tukultī-Ninurta I (1243–1207).

[21] For a detailed elaboration of the connection between Babylonia and Iran, and the flow of different goods in the Kassite period, see Marcus 1991: 558. For objects made in Kassite style discovered in Aššur and Nineveh, see Brinkman 1972: 277. For historical information on the presence of the Kassites and Babylonians in western Iran, see Fuchs 2017: 127–154; for the distribution of Kassite pottery, also in western Iran, see Armstrong 2017: 430–435.

[22] A closer study of the perforation could indicate whether the hole was made before or drilled after drying. But even this would not finally answer the question of secondary use.

[23] Haller (1954: 29) dated As2 post-Assyrian, Miglus (1996: 385) Neo-Assyrian, and Pedde (2015: 101) Middle-Assyrian.

out that the pieces – similar to those from Hasanlu – remained in use during the following period. With regard to the mosaic glass tiles, a Late Bronze Age date can be suggested (Chapter 4.1.3.3).

Even though the production of mosaic glass was absent, or at least much more reduced, in the early 1st millennium, the appreciation for this type of glass, however, did not come to a complete halt, as mosaic glass was reused or kept as heirlooms. It is nonetheless interesting to note that there is a general decline of polychromatic glass in the early 1st millennium in favour of monochrome glass, majorly driven by a new production technique established in this period known as the 'cast-and-cut' technique (Chapter 4.2).

4.2. 'Cast-and-cut' glass

4.2.1 Manufacturing techniques

4.2.1.1 Principles of 'cast-and-cut' glass

The 'cast-and-cut' technique requires two steps of manufacture: casting and cutting. The principle of casting* describes filling a void within a mould* with glass (Cummings 1997: 81). The glass can be poured into the mould in viscous form, or it can be inserted as crushed powder or chunks and heated afterwards. When removed from heat, the viscous glass solidifies in the mould. The casting process can be implemented in open or multi-part moulds, and the latter can also include the technique of lost-wax* (see Chapter 4.2.1.2, 4.2.1.3). Another form of casting is the principle of slumping and sagging. Slumping is the process of forming a hot glass disc over a mould (Figure 4.12), while sagging is the process of forming a hot glass disc into a mould (Figure 4.10) (Taylor and Hill, pers. comm.).

Casting in moulds*, as well as slumping and sagging, are methods carried out under heat exposure. The manufacturing of the specific shapes in a hot environment was thereby the first step in the production of cast-and-cut objects. In contrast, the second step of manufacturing is carried out in cold state and includes cutting, grinding, polishing, shaping, and even painting as finishing techniques. It is important that both hot- and cold-working stages have to be clearly distinguished from one another, as they were most likely carried out in different workshops by different craftsmen (see Chapter 8.2.2).

In this introductory part on the manufacturing techniques of cast-and-cut glass, the general techniques are discussed, with the details on individual objects described in the following sub-chapters.

4.2.1.2 Casting in open moulds

In general, casting* glass in open, or one-part moulds* can be considered a fast and easy method to form specific types of glass objects. In this study, only the attachments and inlays for composite statues (Chapter 4.2.2.11) and the palettes (Chapter 4.2.2.1) can be confidently identified as being made in open moulds. To make objects in open moulds, glass was introduced into the cast, either in liquid form, as chunks, or crushed glass (Figure 4.8) (Stern and Schlick-Nolte 1994: 48; Wight 2011: 16).[24] During this process the temperature of the kiln, and thus the viscosity of the glass, played a decisive role: if the glass was too hot and liquid it would adhere to the mould and damage it; if it was too cold, the detail of the mould would not be reproduced (Taylor, pers. comm.). In any event the glass was in contact with the mould for as short a time as possible. The glass was pressed into the mould from above, or the mould could be pressed onto the glass below, similar to a stamp (Ertman 2013: 13-19). The glass was removed from the mould directly and was thereafter annealed* outside the mould.

To remove the cold glass object from the mould*, separators were mandatory: a light coating of carbon or soot (Taylor and Hill, pers. comm.), talc, plaster (Cummings 1997: 147) or lime (Goldstein 1979: 28) could be used. These substances formed a barrier between the glass and its mould to allow the glass to be removed easily. The use of separators was by no means unproblematic: in most cases the separator would stick to the glass, which would require cold-working to remove it. In addition, separators could hide some details of the mould if they remained in place during heating (Taylor, pers. comm.).

A smooth surface of the finished glass was created by the use of cold-working tools (Chapter 4.2.1.6). By so doing the remains of the separator and any unevenness in the glass could be removed, and also the design could be reworked if required.

Objects cast in open moulds show an obverse which is shaped by the mould* and a flat or irregular reverse. Traces on the reverse sides of the different cast-and-cut objects, therefore, provide valuable insights into the casting process. Irregular surfaces and traces of tool marks can occur on the reverses of glass objects, which result from pressing semi-viscous glass into the

[24] Cummings (1997: 84) points out that melting a single ingot would be easier than melting finely crushed glass into a mould. Taylor (pers. comm.) indicates that when using chunks or crushed glass the mould and the glass had to be heated and melted in the furnace. This would require time since the mould and glass would have to be brought to high temperatures, which would cause the glass to adhere to the mould even when using a separator. He therefore thinks that crushed glass or chunks can be rejected in favour of hot glass that would be poured into the mould directly.

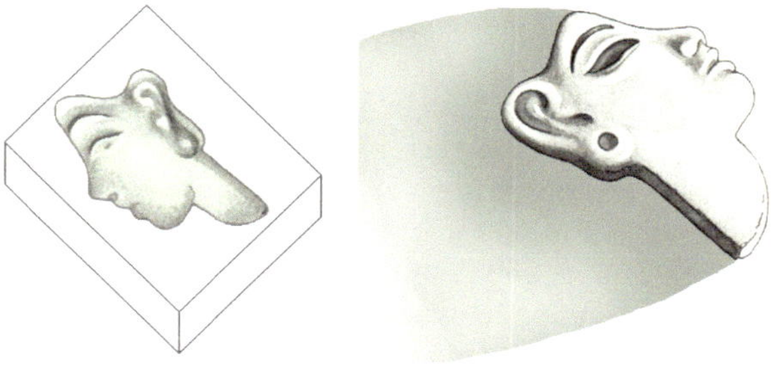

Figure 4.8: Casting glass in an open mould. On the left an open or one-part mould, on the right the cast object (after Wight 2011: 17, 1a, 1c).

Figure 4.9: Nude female glass figurine (remaining ht. 4.9 cm) from Tall Zirā'a, Jordan from the 13th century made in an open mould (BAI/GPIA, photo: Johannes Kramer).

mould. If the mould during this process is too cold, fine veins, so-called cooling marks,* occur on the reverse side (Taylor, pers. comm.; Stern and Schlick-Nolte 1994: 49) (see therefore Khor4 reverse). Apart from cooling marks, also holes from popped bubbles are sometimes visible on the reverses, which indicate that the glass was kept melted in the mould for a long period of time, allowing bubbles to rise to the surface (see therefore Ur1).

Moulds* specifically used to form glass objects have yet to be identified in the archaeological context at any site in Mesopotamia. It is likely that open moulds did not have to be destroyed necessarily. This stands in contrast to complex or multi-part moulds, which were most likely destroyed after use. To resist the great heat, the moulds were likely made of silica and plaster. These materials were capable of withstanding the heat and to cope with the shrinking and expanding rates, which would avoid cracking (Taj-Eddin, pers. comm.). The use of other stone moulds would have also been possible. Experimental studies using tempered pottery moulds, however, also yielded good results (Taylor and Hill, pers. comm.). For the Mycenaean period in Greece, Stern and Schlick-Nolte (1994: 49) refer to moulds made of soapstone.

The technique of casting* glass objects in open moulds was widespread already in the late 2nd millennium* to produce plain or star disc pendants, e.g. from Nuzi (Starr 1939: pl. 120 NN, OO, XX), nude female plaques, e.g. from Tell al-Rimah (Barag 1970: fig. 26; Oates 1965: 74;), and Tall Zirā'a (Figure 4.9), or spacer beads, e.g. from Megiddo (Loud 1948: pl. 209, no. 29. 210, no. 39).

4.2.1.3 Casting in multi-part moulds and the lost-wax technique

In addition to the objects made in open moulds* (Chapter 4.2.1.2), there are glass objects with more complex shapes that have also to be made in multi-part moulds. Complex glass objects in this monograph include mace-head Nin2, the group of jars and alabastra, and one of the ribbed and petalled bowls. All these objects have in common that they had to be formed from all sides.

Casting* in multi-part moulds includes the use of at least two moulds: Similar to casting in single moulds, the glass was introduced between the moulds in hot state or as cold powdered or crushed glass. During heating in the kiln, the glass would flow and the object could be shaped between the moulds, creating a moulded surface on both sides. After annealing*, the moulds were destroyed or carefully removed one after the other, which was very difficult to carry out. The problem was the so-called 'undercut'*. While with open moulds the object could be removed through the opening, with complex shapes the glass object was trapped between the different parts of the mould. Therefore the different parts had to be removed one after the other. On some glass objects this technique can be recognised by the existence of seams, which formed

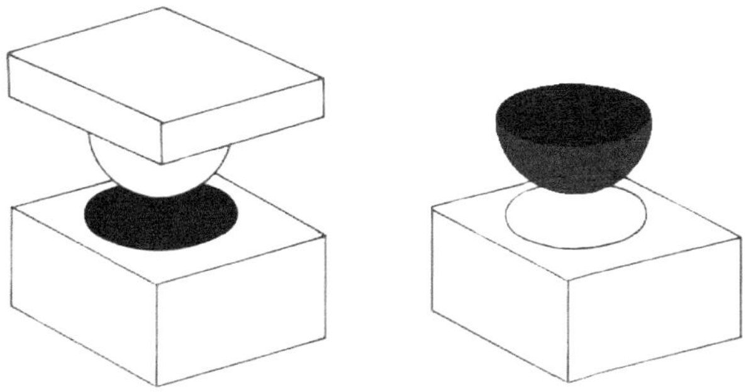

Figure 4.10: Principle of casting in multi-part moulds by the use of a second positive mould that is pressed down to form a monochrome glass bowl (after Goldstein 1979: 31).

exactly where the moulds met. These were, however, in most cases removed by grinding and polishing. A good example for a complex glass object is mace-head Nin2, which was made, most likely, with a two-part mould.

Another type of multi-part mould was probably used for the petalled bowl Gor1, which shows a petalled-shaped design, both on the obverse and reverse sides. Both sides are likely to have been shaped by the use of a negative open mould, in combination with a second positive mould that was pressed down when the glass was in a viscous state (Figure 4.10).[25]

The lost-wax process could also have been used to produce complex shapes that would enable the production of solid glass objects.[26] In the lost-wax process a positive is produced from wax that was coated with the mould material. Both mould and wax are thereafter heated to melt out the wax, which was then replaced by glass. The lost-wax technique could have been used to make solid glass objects. Not every kind of thin-walled glass object could be produced with this technique, as the glass would have adhered to the mould. The slumping technique would be preferred for the production of such objects (e.g. hemispherical bowls).

4.2.1.4 Slumping and sagging

Slumping is understood as shaping a hot glass disc over a form, whereas sagging means the shaping of a hot glass disc into a form (Figure 4.10, 4.12) (Taylor and Hill, pers. comm.).[27] It is supposed here, that the technique of slumping and sagging was applied for a majority of objects listed in this chapter, such as the hemispherical bowls (Chapter 4.2.2.4), the shallow undecorated, ribbed, and petalled bowls (Chapter 4.2.2.5), and the cut-and-inlaid vessels (Chapter 4.2.2.6). In fact, Tylor and Hill (pers. comm.) believe that every non-blown bowl form was made using slumping as the final hot glass technique, which gives this production technique a particular importance in the production of cast-and-cut glass objects. All these objects have in common that they do not have undercuts*, which means that they could be released out of a single mould* without being trapped.

The initial use of the technique of slumping and sagging had previously been dated to the Hellenistic period, where it was used for the production of mosaic and monochrome glass bowls (Barag 1985: 31). It is, however, supposed here that the technique of slumping and sagging was introduced already in the early 1st millennium. The group of hemispherical bowls can therefore be seen as the first prototype for this technique. The principle, however, draws upon the mosaic glass vessels that were already produced in the late 2nd millennium (Chapter 4). The slumping method can therefore be understood as a technical recourse to earlier techniques.

To make a hemispherical bowl by slumping, a disc of glass had to be made in the first place, in the same way as with the manufacturing of a mosaic glass bowl (Chapter 4.1.2). A glass disc could have been made by pouring out a crucible of glass taken from a furnace or by gathering a quantity of glass on a gathering iron* and allowing it to pool, cool and stiffen on a hard surface, such as a stone (Figure 4.11) (Stern and Schlick-Nolte 1994: 67). In order to achieve the appropriate thickness and width, the glass disc could be pressed by, e.g., a large block of damp wood with a flat surface, and by exposing it to heat several times (Taylor and Hill, pers. comm.).[28]

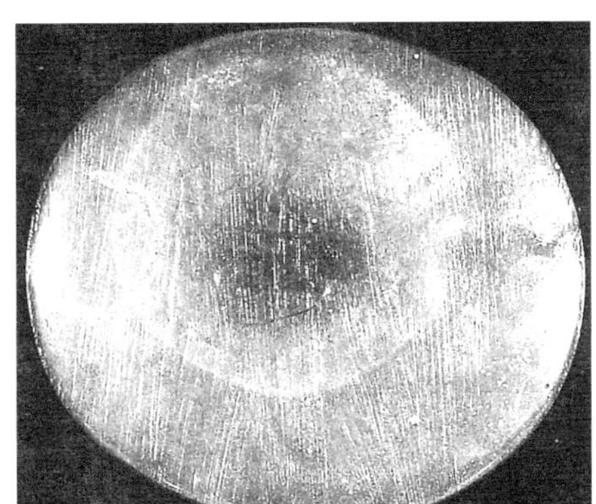

Figure 4.11: Monochrome flat glass disc before being slumped (Stern and Schlick-Nolte 1994: 67, no. 112).

[25] This principle was suggested with regard to the manufacturing of mosaic bowls by Grose (1989: 31–32).

[26] For a general explanation of the lost-wax technique, see Roaf 1990: 126.

[27] Taylor and Hill (pers. comm.) believe that there is no sagging involved in any of the ancient open bowls, cup shaped objects and shallow lens-shaped dishes.

[28] Taylor and Hill investigated this technique by producing a cast monochrome patella with an integral foot-ring in fired clay moulds.

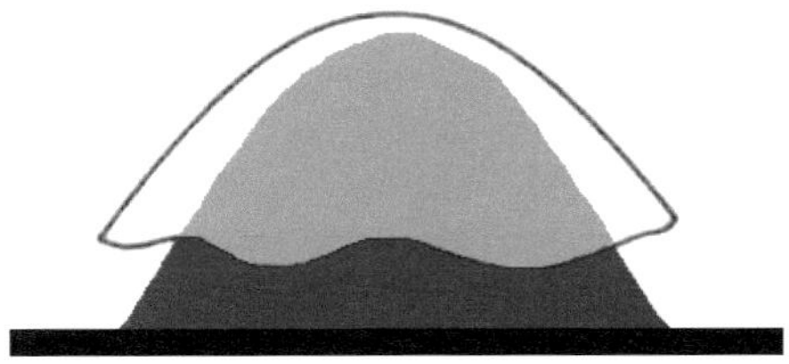

Figure 4.12: The principle of slumping a glass blank over a dome-shaped mould. The heat causes the glass disc to slump down over the form (after Grose 1989: 245, fig. 118).

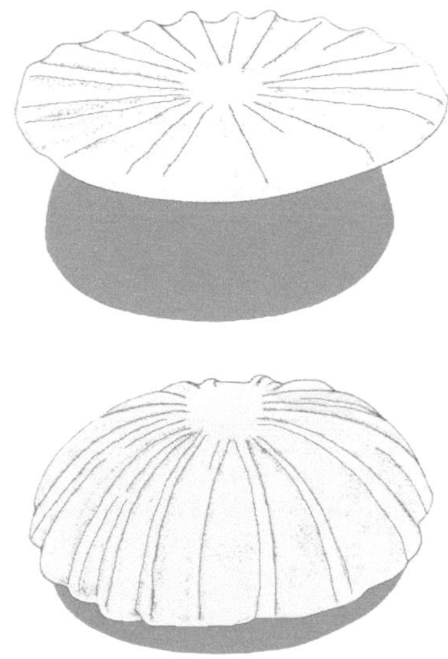

Figure 4.13: The principle of slumping a glass blank – here with ribs – over a dome-shaped mould. The heat in the kiln causes the glass disc to slump down over the form (after Grose 1989: 245, fig. 118).

The surface needed to be coated with a separator and the disc was then placed over a dome-shaped form and heated in the furnace mouth to be able to manipulate the slump using metal tools for control (Figure 4.12). The temperature of *c.* 1050°C would then have caused the disc to slump down over the form (Taylor, pers. comm.) (Figure 4.12).

In the same way, to form a ribbed bowl the ribs on the disc would be shaped by pinching, which creates characteristic tool marks sometimes visible on the surface (Grose 1989: 246; Taylor, pers. comm.) (Figure 4.13). By doing so, the ribbed pattern would be formed on the exterior of the vessel, as can be observed for Nim15, Nim16, and Nim17. The interior of the vessel would stay smooth.

Irregularities in shape and diameter could be the result of removing the glass vessel from the mould*. This process was facilitated by pressing the slightly hardened but still hot glass against the walls of the mould, which would result in an irregular vessel shape.[29] The vessel and form were annealed together. The strain crack on Nim8 most likely resulted from inadequate annealing* at this point. After annealing the rim was ground and shaped by cold-working techniques and the surface was smoothed. For the form, a tempered clay vessel coated with a separation layer, for example, could have been used (Taylor and Hill, pers. comm.).

4.2.1.5 Significance of bubbles in the manufacturing process

Whether the study of bubbles contributes to determining manufacturing techniques is a matter of debate. Generally, the amount of bubbles in the glass can be decreased by allowing the glass in the mould* to be melted for a longer period of time. By doing so the bubbles migrate to the top of the glass and pop (Taylor and Hill, pers. comm.).[30] Gudenrath (pers. comm.) indicated that if air gets trapped in glass it behaves in accordance with gravity. If viscous glass is distorted, the bubbles are distorted in the same way. Consequently, elongated bubbles can indicate the torque and movement of the glass, and thus help to identify the manufacturing process applied. Taylor and Hill (pers. comm.), however, suggested that bubbles only represent and record aspects of the state of the glass at the final stage of glassmaking, just before the glass is annealed*. Therefore, in their eyes, elongated bubbles do not necessarily prove or disprove a certain technique.

4.2.1.6 Cold-working techniques

Almost all objects made by the cast-and-cut technique were completed in a final stage when the glass was already hard. Therefore, edges, rims, bases and handles were formed and finished by grinding, smoothing and polishing. Also the surface of most objects was

[29] This was shown by Stern and Schlick-Nolte (1994: 70–71) on the basis of experimental studies.

[30] The bubbles could have also been absorbed into the melt. However, allowing the bubbles in the glass in a mould to lessen in number could have taken a long time (Taylor, pers. comm.).

smoothed and polished. For this the use of stone-working techniques, such as engraving tools or turning wheels were necessary.

Stone-working techniques were applied for almost all cast-and-cut glass objects; the group of cut vessels that are decorated with cut decorations are particularly characteristic in this regard. Many of the decorations were cut into the surface by the use of engraving tools and turning wheels. In particular the group of cut-and-inlaid vessels (Chapter 4.2.2.6) and inlays (Chapter 4.2.2.7, 4.2.2.8, 4.2.2.9) are important, since the recesses, grooves, lines, friezes and cut designs among them were all made by wheel-cutting. The cut decorations occur only on the exterior of a vessel. Also, with regard to the inside, particularly of vessels that belong to the group of jars and alabastra, stone-working techniques were applied to shape the inside surface of the vessel (Chapter 4.2.2.3). This is attested by typical grinding marks on the inside of most of these vessels.

With regard to the inlays, techniques were applied which make use of the specific material property of glass, and which is called 'grozing'. Therefore, in order to form flat inlays, as for example the rosette inlays (Chapter 4.2.2.8), the small monochrome inlays (Chapter 4.2.2.9), or the large monochrome inlays (Chapter 4.2.2.10), glass was poured onto a flat surface when still hot. The hot glass was formed to a flat glass sheet with the thickness of the later inlays. Still in hot state, squares would be slightly incised on the surface of the viscous glass sheet, which would later result into the shape of the inlays. The glass was annealed and resulted in a hard glass sheet with a checked pattern, similar to the surface of a chocolate bar. Exactly underneath one incision, a stick would be placed, and by pressing down the stiff glass sheet, the individual segments were cracked out, which is also called grozing (Taj-Eddin, pers. comm.) (Figure 4.14).

Grozing was relatively easy to carry out and therefore not very time consuming. Since the inlays are all slightly different in shape and size, this shows that there was no need to produce exactly the same objects. Instead, and this is also supported by the great number of inlays existing, a quick and easy method of manufacturing was needed.[31] According to the needs, the objects could be ground and polished afterwards. The irregular reverses of all the inlays indicate that they were intended to be visible from only one side. The irregular reverse of the inlays would facilitate the adhesive to stick to the inlay.

Inlays with bent cross-sections, such as the painted inlays (Chapter 4.2.2.7), or the inlays on A1, were probably made by 'chipping them off' a hemispherical

Figure 4.14: Break-lines on the edge of a monochrome inlay, caused by 'grozing' (© The Metropolitan Museum of Art).

bowl (O'Hea 2011: 161). If glassworkers wanted to 'chip off' glass at a specific part of an object, they probably used a sharp tool to scratch a fine line into the surface of the glass object. The piece of glass was then cracked off by slightly knocking against it.[32] This chipping or cracking off has the advantage that it was a very quick method to shape an object, and that no specific skill of the craftsmen was required. The sharp surfaces were later smoothed by grinding and polishing.

Taylor (pers. comm.), however, suggested another technique of manufacture for the inlaid bowl A1. A thick opaque-green flat glass disc would be slumped, and in cold state ground to create recesses for inlays. The bowl would then be mounted again on its form, reheated, and the pre-heated inlays would be attached to the surface. This would solve the problem of curving the inlays – rosettes as well as the rectangular inlays – to match the bowl, as they would be soft when applied to the bowl and would bend as necessary. The edges of the inlays would also be fire-rounded and polished during this process. Another positive effect would be that only a little waste glass is produced.

[31] Barag (1991: 2) suggested that the inlays were cast in moulds, which can be completely discarded here.

[32] For the method of cracking off, see Gudenrath in https://www.youtube.com/watch?v=E4yovEi7j7E (accessed: 17.02.2018).

4.2.2 Description and discussion of objects

4.2.2.1 Palettes

Description

The palette Meg1 is made of translucent glass that has a strong light-green tone. It has a diameter of 10 cm at the surface and a diameter of 7 cm at its bottom. The walls of the object are rounded and interrupted by two sharp demarcation lines. At the top of the object a round cavity is cut slightly off centre. It is most likely that palette Meg1 was cast in an open mould*, as it is regular in shape and exhibits sharp edges. The surface of the palette would have been ground and polished with tools.

Only two further glass palettes, although of unknown provenience, are known: AM2 is made of transparent colourless glass, as is AM3, which also shows a greenish tinge (Barag 1982: 11).[33] Both objects are very similar in shape and diameter and exhibit walls with grooved lines, rounded cavities on the top and flat surfaces. The cavities differ only regarding their depth.

Discussion

As only three glass palettes are known, it is worthwhile to consider palettes made of other materials to find out more about their dating and function. The closest parallels to the glass are those stone palettes which were usually made of limestone or marble.[34] These palettes are either undecorated or incised with dots (sometimes even inlaid) and lines to form symmetrical patterns. Palettes in general were found at sites in the hinterland of the Levantine coast, for instance at Beth-Shean, Jericho, Beth Shemesh, Gezer, Lachish, Beersheba, and Tell Abu al-Kharaz, and date from the 8th to the mid or late 6th century.[35] In contrast, very view finds were made at coastal sites. Parallels have furthermore been drawn to singular pieces from Nineveh, Nimrud, and Sippar, but which either differ in shape or in material from the pieces found in the Levant (Barag 1982: 15–16).

Palettes were used as vessels to grind and mix pigments for cosmetics or powder for medicine (Crowfoot *et al.* 1967: 464). The identification for cosmetic use is based on some pieces showing traces of pigment on the rim (Lamon and Shipton 1930: pl. 110–111; Fischer 2014: 495, fig. 453, 8, 9). A pestle found in close proximity to one of the stone palettes from Hazor reinforces this argument (Yadin 1960: 61). The palette from Sippar, however, carries the inscription 'Property of Marduk' (NÍG.GA ^{d}*Marduk*), indicating its use in a temple complex (Meyer 1980: 99, no. 30, pl. 27, 30).[36] With regard to the palette from Tell Abu al-Kharaz, Fischer (2014: 43) suggested their use as lids for containers when turned upside-down.

Meg1 was found in a domestic context dating to the second half of the 7th century, not showing any other outstanding finds. However, the amount of glass, as well as its rarity in comparison with other materials, indicates a certain degree of value despite its findspot. The fact that palettes can largely be attributed to the southern Levant, and are almost absent in Mesopotamia and northern Syria, makes it likely that Meg1 was made in this region (Figure 5.3).

4.2.2.2 Mace-heads

Description

Nin2 is made of strong blue translucent glass and has a piriform shape with a flat bottom (6.6 x 5.7 cm).[37] A large quantity of pin-prick bubbles is evenly spread throughout the glass (Barag 1985: 74, no. 60). A band with a central ridge and cut grooves decorates the lower part. A rectangular socket is inserted into the middle of the base. The object is very regular in shape and exhibits sharp edges. It is likely that the object was made by the use of a two-part mould* (Chapter 4.2.1.3). The shaft-hole in the middle could have been created by filling the mould only two-thirds with glass and inserting a stick to create a void. The hole could have also been cut afterwards.[38] The socket has traces of horizontal grinding marks, indicating that the object was cold-worked after annealing*. The decorative band was most likely finished by grinding and polishing.

AM1 is dark blue[39] and has an overall irregular, spherical shape (4 x 4.2–4.5 cm), and exhibits a cylindrical shaft-hole which becomes wider on both sides. Because of its irregular shape, AM1 was likely rod-formed; the shaft hole is wider on both ends, which could be the result of this forming process. However, it cannot be ruled out that the hole was drilled through the object

[33] The assignment of Fig. 1 in Barag 1982: 12 has to be exchanged with Fig. 5.

[34] Pallets made of ivory are not discussed in this study. Stone palettes were primarily uncovered at Hazor, Samaria, and Megiddo. At Hazor, stone palettes can largely be attributed to funerary contexts (Yadin 1960: 32, 34, 62, pl. LXXVIII.7, 8; CV, 24, 25; CVII, 21), as can be observed regarding three palettes from Samaria (Crowfoot et al. 1967: 463-464). At Megiddo palettes made of faience have also been recorded, most of them coming from domestic contexts; see Lamon and Shipton 1939: pl. 108–111.

[35] For literature on particular finds from these sites, see Barag 1982: 15; for Tell Abu al-Kharaz, see Fischer 2014: 39.

[36] For another palette involved in ritual use, see Searight et al. 2012: 82, pl. 63, no. 618.

[37] Barag (1985: 74 no. 60) describes the colour of Nin2 as 'almost opaque, but translucent along the edges'. This indicates that the object was made of translucent dark blue glass.

[38] Cummings (1980: 13) suggested that a solid cast lump was lathe-turned into shape, similar to the method of working stone. Due to the large material loss and the high amount of work involved, this method is not considered here.

[39] Barag (1985: 75) describes the colour as similar to the beard inlays from Nimrud.

with a driller.[40] In this case, the mace-head could have been made by applying a mould, or, similar to Nin2, by lathe-turning. This is the reason why AM1 is included under cast-and-cut glass objects. The findspot of AM1 is unknown. Reade (2002: 155) proposes that AM1 came from Rassam's excavation of the Kidmuri Temple in Nimrud (Chapter 3.1.5).

Discussion

Generally, piriform and spherical mace-heads made of different materials occur from the 4th to the early 1st millennium with only minor changes in shape. Mace-heads are found predominantly in temples and were made of stone, metal, ceramic, faience and glass (Muhle 2008: 147–149). In addition to the two objects discussed, examples of glass mace-heads are only known from the late 2nd millennium from Chogha Zanbil and Tell al-Rimah (Oates 1970: 3c). These objects, however, vary greatly from their Iron Age successors, as they were made by applying the core-forming technique, showing the characteristic feather design. In the 1st millennium, piriform and spherical mace-heads are only found sporadically, for instance, at Tell Halaf, Aššur, Nimrud, and Uruk, and they are made solely of stone and faience*, in addition to the two examples made of glass (Moorey 1994: 180; Muhle 2008: 51–52). A good parallel for Nin2, made of lapis lazuli, comes from the Ninurta Temple at Nimrud.[41] The inscription is most likely by Ashurnasirpal II (883–859), referring to this mace-head as a cult object (Grayson 1991: 353–354). AM2 finds close parallels with two mace-heads made of faience and stone that were found in the Kidmuri Temple at Nimrud (Curtis and Reade 1995: 177, no. 182).[42]

Even though the evidence is scanty, it can nonetheless be stated that blue glass was used simultaneously with lapis lazuli and also with faience* for the same types of object. As with the blue inlays and attachments for composite statues (Chapter 4.2.2.11), it could therefore be supposed that the colour rather than the material was of foremost importance in this particular context. Mace-heads were generally used as votives, ceremonial objects or royal insignia (Braun-Holzinger 1991: 40). Because of this, and their material, it is likely that Nin2 and AM1 served ritual or ceremonial purposes, a suggestion supported by the findspot of Nin2 in the Nabû Temple.

[40] Unfortunately, this detail can neither be seen on the drawing nor in the photograph.

[41] See for coloured photographs the British Museum Database: http://www.britishmuseum.org/research/collectiononline/collectionobjectdetails/collectionimagegallery.aspx?assetId=1306085001&objectId=365131&partId=1 (accessed: 11.3.2016).

[42] See for coloured photographs the British Museum Database: http://www.britishmuseum.org/research/collectiononline/collectionobjectdetails.aspx?objectId=369290&partId=1&searchText=mace+head&page=2 (accessed: 18.12.2015).

4.2.2.3 Jars and 'alabastra'

Origin and definition of the terms 'jar' and 'alabastron'

Jars and alabastra fall within this chapter as it is likely that all the vessels discussed here were made using largely the same techniques. Strictly speaking, Nim27, Ur2 and AM4 are defined as jars, whereas At1, Id1, AM6, AM7, AM8, AM9, AM10, AM11, and AM12 belong to the group of alabastra. Jars are characterised as compact containers with rounded walls and broad neck and rim. Alabastra, in contrast, have long, elongated bodies, narrow necks and flared rims.

The term 'alabastron' is used here, however, with great caution and requires further explanation. The term derives from the Greek world referring to vases with an elongated body, a narrow neck and a rounded base. Whereas alabastra from Egypt were made of alabaster, the alabastra found in the Greek area of influence refer to painted ceramic vessels depicting funerary rituals (Caubet 2014: 168–169). The function of this vessel type was usually to hold ointment or oil, as the narrow neck allowed the contents to be carefully poured. Often a wide rim is attested, from which the liquid could be dispensed (Yon 1981: 16). The word alabastron therefore implies a specific function. The term is used in this study as it has become a common designation for the glass vessels of this type discussed here. This chapter includes a number of unprovenanced vessels that have been attributed to the early 1st millennium from the eastern Mediterranean, and are therefore included in this work despite their problematic origin.

Manufacturing technique

Both jars and alabastra show sharp angles at the rim, shoulders, and handles. The handles are always shaped out of the body material; they were not attached separately as with core-formed* vessels (Chapter 4.3). The inner walls of some of the vessels exhibit concentric grinding marks. The jars and alabastra have in common that they had to be formed from all sides. It is therefore likely that the pieces were made in at least two moulds*.[43] The core of the vessels could be formed by filling the mould only half way with glass and pushing a core – similar to those used for core-formed vessels – into the hot glass, creating a void. Traces of the core could have been removed after annealing* by scratching. As all the jars and alabastra are either transparent or translucent, it was imperative to remove the core complete, leaving no residues on the inside walls. Therefore, further shaping of the inside of the vessel could have been achieved by the use of a drill,

[43] It has been repeatedly suggested that these vessels were made by the lost-wax technique and finished by cold-working; see Barag 1985 and Saldern 1970. Regarding Nim27. It has also been supposed that the entire vessel was cut out of a large block of glass and then drilled, like a stone vessel; see most recently Miho 2013: 371, no. 37.

since rounded concentric grinding marks are preserved (Figure 4.15). The drills could have been made of copper, bronze or iron and could have been bow driven.[44] Drilling would always involve the use of an abrasive: sand, crushed quartz, or emery (Gorelick and Gwinnett 1986: 16). At this point, it is very interesting to note that the shaping of the jars and alabastra much depended on the use of stone tools, which even made a decisive contribution to the shaping of these vessels.

Figure 4.15: Concentric grinding marks on the inside of the 'Sargon Vase' (Nim27), caused by the use of a drill (Stern and Schlick-Nolte 1994: 47, no. 22).

Description and discussion of jars

One of the most renowned glass objects from Mesopotamia is the so-called 'Sargon Vase' (Nim27) from Nimrud. Nim27 is made of colourless glass that exhibits a light green tinge. Small spherical bubbles are spread throughout the vessel with a high concentration in the solid rectangular knob handles. There are no traces of corrosion on the exterior, as they were probably removed either after excavation or later in the 19nth century.[45] The vessel has a compact ovoid body and a short, broad neck with a flaring, flat rim (8.6 x 6.2 cm). There is a sharp angle between the shoulder and body, and two vertical knob handles (2 x 0.8 cm) on the shoulders. The bottom is round. The side of the vessel is incised with a lion and an inscription. The lion, facing towards the right, has a wide, open jaw. The paws and mane are finely detailed. The lion is followed by an inscription that continues on the other side of the vessel:

É.GAL ^{M}MAN.DU
MAN KUR AŠ

Palace of Sargon
King of Assyria[46]

The interior of the neck is covered by a thick, whitish corrosion layer. There are also regular concentric grinding marks running throughout the interior wall of the vessel. The grinding marks, the sharp angle between shoulder and body, the flat and sharp edges of the rim, the handles, as well as the inscription, all indicate that grinding and cutting techniques were employed after shaping the vessel. The knob handles are part of the body and were not applied separately. The walls of Nim27 are thick and vary from 0.7 to 1.7 cm.

AM4 is made of transparent, colourless glass exhibiting a slightly greenish tinge. The inner surface and small sections at the bottom were unaffected by deterioration, but the exterior is covered with a thick weathering* layer. Small, spherical bubbles are regularly spread throughout the glass. AM4 has a short, broad neck. The rim has a flat surface and exhibits sharp edges. It is pointed towards the interior. The shoulders, duck-head handles, and base of the vessel exhibit sharp angles. AM4 has an overall rounded shape and a height of 7.4–7.6 cm. The maximum thickness is 8 cm. The walls on which the duck-head handles are attached are straight, resulting in a blocky shape. The vessel exhibits a low disc-base bottom. On the exterior, close to the duck-head handles, there are thin vertical grinding marks. The interior has regular concentric grinding marks from the wall to the bottom. The interior bottom is slightly convex. The wall thickness varies from 0.7 to 1.4 cm.

Ur2 is almost transparent, showing only a slightly greenish tinge and only few bubbles. The surface is heavily corroded, exhibiting severe depressions. Ur2 has a thick and pointed rim that strongly pulls towards the exterior (9 cm). There is a strong incision below the rim. Ur2 has a height of 9 cm with a rounded body shape. The thickness of the wall decreases towards the bottom of the vessel (0.7 – 1.4 cm). The base of the vessel is not preserved. The height, and the overall vessel shape, with its rounded walls and curved neck, similar to AM4, makes it likely, however, that the item had a flat bottom, probably even a disc-based bottom similar to AM4.

A very close comparison for jar Nim27, with regard to shape and similar colourless, greenish appearance of the material, is an example made of rock crystal bearing a cartouche of Roud Imen (750–700) a pharaoh of the 23rd dynasty. The vessel has no provenience, thus a reference to an Egyptian origin cannot, apart from the inscription, be confirmed with certainty (Figure 4.16).[47]

[44] For experiments, see Gorelick and Gwinnett 1986.
[45] See the curatorial comments at: http://www.britishmuseum.org/research/collectiononline/collectionobjectdetails.aspx?objectId=369247&partId=1&searchText=sargon+vase&page=1 (accessed: 20.12.2015).

[46] See Barag 1985: 60, no. 26.
[47] This parallel has previously been suggested by Lehrer (1974: 13).

Figure 4.16: Jar made of rock crystal with similar shape to Nim27 (Lehrer 1974: 13).

Figure 4.17: Faience jar from Sultantepe (Lloyd and Gökçe 1953: pl. VIIb, c).

Further well-comparable objects are three calcite jars from room I of the Northwest Palace in Nimrud. Even though these vessels vary in shape, they carry the same inscription as Nim27.[48] The calcite vessels are considered as booty probably brought to Nimrud from Samaria by Sargon II (721–705) (Searight 2008: 16).[49] Regarding the appearance of the glass itself, there are close parallels between Nim27 and Id1, not only in its colour, but also in the distribution of bubbles. In addition to the material, Nim27 and Id1 both exhibit sharp transitions between neck and body. Whether both vessels were manufactured by the use of the same raw materials cannot be decided here. It is however likely that they were made by applying very similar manufacturing processes. Typological comparisons seem to link Nim27 to objects that originate from the southern Levant (stone jars), and which were brought to Assyria as booty. Chemical analyses show similarities with the hemispherical bowls from Nimrud (Chapter 7.4.3). It could be possible that Nim27 was made in Assyria, probably in Nimrud, and modelled after vessels that were brought from the Levant.

Jar AM4 finds its closest parallel with regard to its shape in small faience* jars, of which examples were uncovered in Sultantepe (Figure 4.17), Tille Höyük (Blaylock 1999: fig. 11.4) and Tell Halaf (Sievertsen 2012: fig. 154, no. 2).

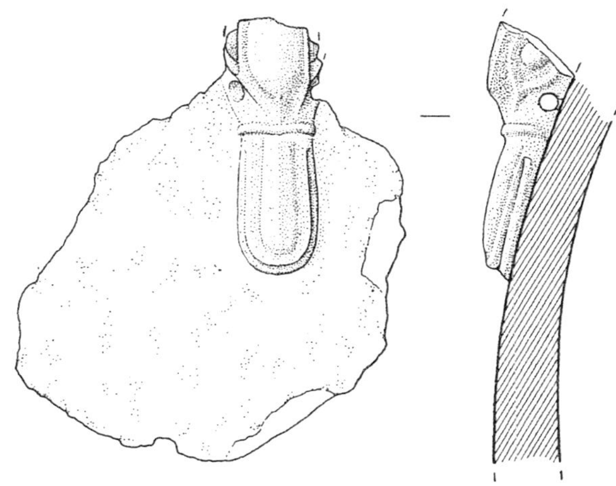

Figure 4.18: Neo-Assyrian duck-head handle on a stone vessels from Nineveh (Searight *et al.* 2008: pl. 13, no. 104).

The stylised duck-head handles find their closest parallels among stone vessels from both Assyria and the Levant. It is very interesting to point out here that stylised duck-head handles occur for the first time on stone vessels found in the southern Levant (e.g. at Lachish, Tel Dan, Kamid el Loz, Ugarit) (Bevan 2007: 212), and continued to be made in the early 1st millennium (Searight *et al.* 2008: pl. 13, no. 105–107). Examples of duck-head handles that were shaped in detail were found, for example, at Nineveh (Searight *et al.* 2008: pl. 13, no. 104).

Description and Discussion of the alabastra

AM9 is an outstanding vessel because of its high degree of transparency. The vessel is large with a height of 18.1 cm. The body is slightly convex, and the bottom is rounded. There are many small, spherical bubbles spread throughout the glass. A large amount of bubbles

The vessel is part of the collection in the Musée du Louvre (Inv. No. 23385).

[48] For coloured photographs, bibliography and the inscription, see (http://www.britishmuseum.org/research/collectiononline/collectionobjectdetails.aspx?objectId=368984&partId=1&searchText=sargon+vase&page=1 (accessed: 30.12. 2015).

[49] Inscriptions on stone slabs of Sargon II refer to booty brought back from his campaigns, which he stored in the Northwest Palace; see Luckenbill 1927: 72–73.

is present in the knob handles; the rim and shoulders have sharp edges, and the shoulders differ slightly in shape. The knob handles are rounded and shaped out of the body; they were not attached separately. The walls are thin and exhibit, as far as can be judged, a regular thickness.

AM7 (ht. 21.3 cm) is made of transparent glass with a green to light blue tinge. The colour differs greatly from the other vessels. There are small, hemispherical bubbles evenly spread throughout the glass. On the exterior surface, vertical thin streaks run down from the rim to the bottom, similar to AM6. AM7 has an elongated body with straight walls and slightly widens towards the bottom. Rim and shoulders have reasonably sharp edges, making a rework in modern times likely. Below the shoulders there are two duck-head handles with a sharp medial ridge. The walls are regular in thickness.

AM8 (ht. 22.5 cm) is made of transparent, colourless glass with a light blue tinge. Only few bubbles occur in the middle of the body, most of them elongated. Strain cracks are present on several parts of the vessel, and grinding marks appear around the duck-head handle. The vessel is slightly taller than AM7 but very similar in shape: the body is long with straight walls and is slightly wider around the bottom. Rim, shoulders, and duck-head handles exhibit sharp edges. The colour of the vessel has a very unusual bluish tinge compared to the other alabastra, and also the rim has relatively sharp edges.

AM6 (ht. 14 cm) is made of colourless, transparent glass. There are thin streaks running from the rim towards the bottom of the vessel, which has a slightly convex shape. The rim shows sharp edges, and the shoulders are not very sharply detached from the body. The knob handles are rectangular.

AM11 is 14.8 cm tall, and is made of colourless glass with a slightly greenish tinge (Saldern 1970: 227, no. 53). There is a thick corrosion layer on the surface, which is heavily pitted. The vessel has an elongated body and is slightly broader towards the bottom. Rim and shoulders have sharp edges, and the handles are small and shaped out of the body.

AM12 is about the same size as AM6 and AM11, and is made of colourless glass with a greenish tinge (Saldern 1970: 226, no. 49). The vessel has a slightly convex shape; the flaring rim has sharp edges, as does the shoulder. The knob handles are small and are shaped out of the body and were not attached separately.

AM10 is considerably smaller (ht. 9 cm) and is made of colourless, transparent glass (Arveiller and Nenna 2000: 167, no. 195). The vessel has a slightly convex shape. The rim and shoulders exhibit sharp edges. There are simple vertically rounded knob handles placed directly under the shoulders.

The fragment At1 is made of transparent, colourless glass with a slight greenish tinge (Saldern 1970: 227, no. 54). The piece can be identified as the bottom part of an alabastron.

Id1 is made of transparent, colourless glass with a slightly greenish tinge.[50] There are many spherical and elongated bubbles of different size (up to 1 cm long) which are unevenly spread throughout the glass. The bubbles in the upper part of the vessel are elongated, while those in the lower part are spherical. On one side there is a large strain crack surrounded by an accumulation of bubbles. The top of the vessel is broken off. The body is ovoid and widens slightly towards the base. The shoulders exhibit sharp edges. The vessel is regular and straight and has horizontal grinding marks. The thickness of the walls is therefore regular. The colour and nature of the glass can well be compared to Nim27.

The closest similarities for Ur2 in shape and in the character of the material can be drawn to rock-crystal vessels. One good parallel is a rock-crystal bowl from tomb II of the royal princesses at Nimrud (Figure 4.19).

Figure 4.19: Bowl made of a transparent rock-crystal from the royal graves at Nimrud (Damerji 1999: 46, fig. 24).

The alabastron AM9 varies from the other alabastra discussed here, as its body is wider and has a much rounder shape. Comparable vessels, made of stone can be quoted as comparisons. A number of similar vessels

[50] Id1 is incorporated in this study, although it falls outside the geographical scope, because it is one of the rare examples of this group with a known findspot.

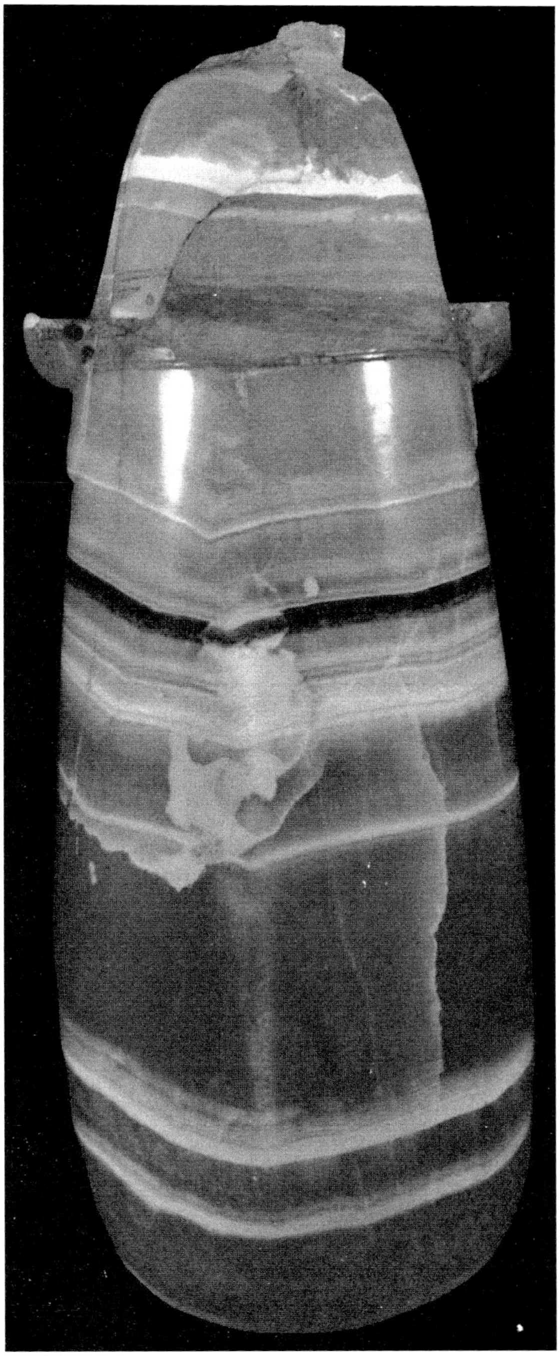

Figure 4.20: Stone alabastron from room 25 of the Northwest Palace, Nimrud (ht. 47 cm) (Mallowan 1966: 169).

- albeit larger in size - come from room 25 of the Northwest Palace at Nimrud. One of the vessels is made of calcite with black-and-white veins and a pseudo-hieroglyphic inscription (Figure 4.20).

Further similar vessels were found in area ZT of the Northwest Palace. All of these alabastra were most likely brought from the southern Levant to Assyria by Esarhaddon (680–669), or came to Nimrud as part of Sargon's (721–705) booty, in a similar way to the inscribed stone vessel mentioned above (Mallowan 1966: 169; Searight *et al.* 2008: 41).[51] Another comparable group of alabaster vessels was found in the Old Palace ('Fliesenzimmer') in Aššur, and is largely attributed to the third intermediate period (1075–652). Some objects of this group carry inscriptions of Esarhaddon (680–669) (Ass22813, Ass185a, Ass153) and were therefore most likely made during the time of his reign (Onasch 2010: 10–11, 66–70).[52] Preusser (1955: 20–23) refers to them as oil jars and suggests they were originally made in Egypt, being transported as booty to Assyria via Sidon and Tyre. A Levantine origin for one of the alabaster fragments (Ass153) was supposed by Onasch (2010: 69, no. 219a-12), which originates from the palace of Abdi-Milkutti from Sidon. Comparable stone alabastra were found, furthermore, in Babylonia. Examples from Babylon come from grave no. 148 (two pieces), and from grave no. 140. All three vessels are Neo-Babylonian in date (Preußer 1968: pl. 76, 148, pl. 77, 140).

To summarise, the shape of AM9 finds close comparisons in stone (calcite) vessels from Babylonia and Assyria, which were, most likely, brought to Mesopotamia as booty from different cities in the southern Levant, either by Sargon II (721–705) or Esarhaddon (680–669). The high degree of transparency of AM9 is exceptional, especially in terms of the absence of corrosion, which was most likely achieved by modern rework. It should be mentioned here that the hemispherical bowls from Nimrud (Chapter 4.2.2.4) were made of a similar type of glass. The combination of shape, transparency and size, as well as the absence of any comparable object, makes it, however, questionable whether the piece can be attributed to a Neo-Assyrian or Neo-Babylonian date. A chronological and geographical attribution of AM9 is therefore difficult to suggest.

The rest of the alabastra discussed in this chapter can be grouped together because of their similar shape and size. AM7 and AM8 form a group of large vessels with a height of around 18–22.5 cm. Id1, AM6, AM10, AM11 and AM12, in contrast, are smaller with a height between *c.* 9–15 cm. AM6, AM11, and AM12 exhibit very close parallels to alabastra made of stone, which are, however, rarely attested in Mesopotamia before the Hellenistic period; a good comparison comes from Tharros, Sardinia (5th–4th century) (Searight *et al.* 2008: 37, no. 286).

[51] Another calcite vessel, indicating an Egyptian origin (ht. 17.1 cm) from Nimrud can also well be compared to AM9, see Searight et al. 2008: pl. 20, no. 304. Similar to the stone vessels bearing the inscription of Sargon, mentioned above, this vessel was also considered to be part of Sargon's booty from the Levant. For the history of the object, see Searight et al. 2008: 41.

[52] The shape of this group of vessels ('Typus 219a') has been considered Egyptian in origin, as Onasch assumes that calcite was quarried in antiquity solely in Egypt (Onasch 2010: 10–11, 66–70).

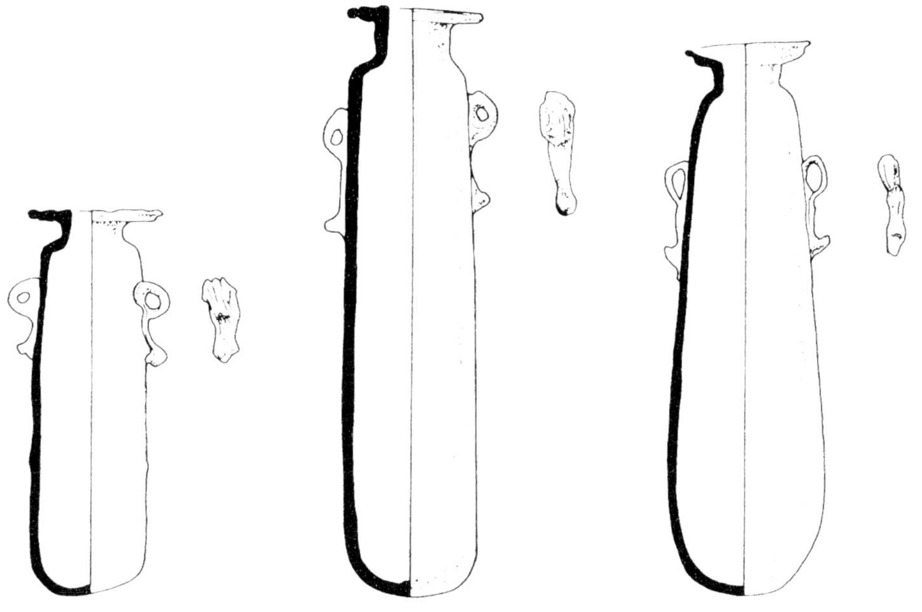

Figure 4.21: Glass alabastra of the 'Mediterranean Group 1' from the period after the 6th century (Grose 1989: 130).

AM10 and Id1 are smaller and more compact, finding good parallels with an alabastron from Aššur (Onasch 2010: 14, 227-21). This piece has a height of 10 cm and similarly shaped knob handles. The piece cannot be dated by stratigraphy, but the alabastra with which it has been grouped dates between the 26th dynasty (post 664) and the Roman period.

Regarding the two large vessels AM7 and AM8, comparisons come from the cemetery in Deve Höyük. Here, ceramic alabastra exhibit comparable sizes and shapes and have been dated between 540–360 (Lehmann: 1996: pl. 38; Moorey 1980: fig. 4, no. 32, 51).

Apart from the examples listed, the closest comparisons for all elongated alabastra with straight walls can be drawn from the core-formed vessels of the 'Mediterranean Group 1' (late 6th – early 4th centuries), and the 'Mediterranean Group 2' (mid 4th – late 3rd century). Core-formed glass vessels of an earlier period (Chapter 4.3) tend to have more rounded than elongated shapes. Therefore, elongated alabastra made by applying the core-form technique are not attested before the late 6th century (Figure 4.21).

In summary, the alabastra presented in this study were previously dated between the 8th and 4th centuries (Barag 1985; Saldern 1970). Due to the typology presented here, and the objects mentioned for comparison, a later date after 600 seems, however, more probable for this group. Since most of the alabastra contained in this study have no provenance, this statement should be viewed with caution. The best comparisons with the cast-and-cut glass alabastra can be drawn to pieces made of stone, but also to glazed ceramics and core-formed glass vessels.

Discussion

The function of Nim27 ('Sargon Vase') should be considered in more detail. The interior of the neck is covered by a thick, whitish corrosion layer, with only a thin residue of weathering* present. Based on this observation, it was suggested that only a limited amount of water entered the vessel after it had been deposited. Therefore, it is likely that the vessel was sealed by an organic stopper to keep the contents, most likely a 'semi-solid' substance, such as oil or cream, inside the vessel.[53] The wide opening of the vessel also supports an identification of the content as semi-liquid material, such as a cream, or probably powder, rather than a liquid, because it could be removed from the vessel with fingers or a wider tool. Residue analysis would give rewarding insights regarding this question. The inscription identifies Nim27 as an object that belonged to the palace of Sargon II in Nimrud. Nim27 and further inscribed stone vessels were found in room I of the Northwest Palace. In addition to these vessels, the room yielded a number of glazed ceramic vessels (Layard 1849: 341). Stone and glass/ glaze are considered suitable materials for liquids and oils. An accumulation of vessels that show these characteristics in one findspot, as well as the find context itself – room I exhibits drainage arrangements –

[53] See the curatorial comment at http://www.britishmuseum.org/research/collectiononline/collectionobjectdetails.aspx?objectId=369247&partId=1 (accessed: 21.09.2017).

suggests a use in connection with bathing, ablution, or anointment (Searight *et al.* 2008: 162). Chemical analysis of the glass (x-ray fluorescence) revealed that Nim27 contains significant amounts of magnesium (3.44%) but a lower concentration of potash (1.37%) (Miho 2013: 365). This could indicate that both plant ash and mineral natron* were used in the composition. The composition is similar to that of the colourless hemispherical bowls from Nimrud (Chapter 7.4.3). This could point towards a similar place of origin for the primary glass. The inscription indicates royal use, probably even restricted to Sargon II himself. Similar to the shape of AM9, Nim27 could also have recalled shapes that only existed in stone.

With regard to AM4, the wide mouth of AM4 could point towards a similar use as described regarding Nim27, containing semi-liquid substances. The short but sharply cut neck of AM4 makes it plausible that the vessel was sealed, for instance, by a cloth and the help of a cord.

The elongated alabastra, with their slim shape and wide, flat rims, are usually considered as vessels to hold ointment or oil, as the narrow neck allowed the content to be carefully poured and the wide rim allowed the liquid to be dispensed (Yon 1981: 16). Glass is furthermore a particularly favourable material for containing oily substances, since its vitrified surface is impermeable.

4.2.2.4 Hemispherical bowls

Description

The characteristic feature of Nim7, Nim8, Nim9, Nim10, Nim11, Nim12, Nim13, Khor1, Fo1 and Pr1 is, in addition to the shape, the transparent or translucent nature of the glass. All of the objects exhibit a brownish weathering* layer that covers the interior and exterior of the bowls. Where the weathering layer has flaked off, heavy pitting* and iridescence* can be observed, which makes, in most of the cases, the observation of tool marks impossible. In some areas, the distribution of bubbles within the glass can be studied, which cautiously allows for conclusions on the manufacturing techniques.[54]

Nim7 is made of transparent glass with a slight greenish tinge. Small, spherical bubbles are regularly spread throughout the glass. Nim7 has a flat base with a slight depression. The walls draw slightly inwards. The base is off-centre, and the vessel therefore stands askew. The height of the bowl varies from 5.4–7.2 cm. The walls are slightly thicker towards the bottom (0.5 cm) of the object, decreasing towards the rim (0.3 cm). The rim is rounded with a sharp edge on the interior, which was most likely cut after annealing*. The diameter of the bowl is irregular (9.6–10.3 cm).

Nim8 is colourless and transparent with a slightly greenish tinge. Small, spherical bubbles can, however, be observed in some areas of the wall and bottom. They are small in number but regularly spread throughout the glass. There is a strain crack on one side, which is surrounded by large, round and elongated bubbles. The bubbles indicate that the crack occurred during the manufacturing process, most probably during annealing. Nim8 exhibits a rounded bottom which is slightly flattened. The thickness of the wall is irregular and the middle part of the walls is the thickest (0.8 cm). The rim is regularly rounded and was most likely worked after annealing*. The diameter of the bowl is regular and slightly wider than Nim7 (12.3 cm). The height varies around 6.2 and 7.4 cm.

Nim9 is also made of transparent, colourless glass. Unlike Nim6 and Nim7, the glass shows no greenish tinge. Small, mostly spherical bubbles are distorted regularly throughout the glass. The thickness of the walls varies from 0.2 to 0.5 cm (Barag 1985: 63, no. 30). The shape of the bowl is almost regular, and the wall turns out slightly. Nim9 exhibits a regular diameter of 14.6 cm. The rim is rounded and was most likely finished after annealing*.

Nim10 is made of colourless and transparent glass which has a slightly greenish tinge. Small pin-prick-shaped bubbles occur throughout the entire glass. The vessel is slightly taller than hemispherical, with a diameter of 13.8–14.7 cm. The walls are thicker towards the rim and thinnest towards the body (0.1–0.4 cm) (Barag 1985: 63, no. 31).

Nim12 is a rim fragment of a hemispherical bowl. The diameter of the bowl can be reconstructed to about 14 cm. The glass is almost colourless and shows only a slight greenish tinge. Small spherical bubbles are spread throughout the glass. The rim exhibits sharp edges and was most likely cut after annealing* (Barag 1985: 63, no. 32).

There are also two small fragments from Khorsabad that formerly belonged to one vessel (Khor1). One fragment can be identified as part of the base, but the other is too small to identify the part of the vessel from which it came. The fragments are thinner than the bowls from Nimrud (0.1 cm), which could also be due to corrosion. Khor1 is made of transparent, colourless glass that does not show any traces of a greenish tinge. Small, spherical bubbles are regularly spread throughout the glass.

Besides the colourless, transparent specimens discussed above, translucent violet, turquoise, and dark blue hemispherical bowls are also attested among the finds from Nimrud. In this regard, Nim11 can be identified as the lower part of a hemispherical bowl of translucent violet colour (Barag 1985: 64, no. 36). Fragment Nim13

[54] The distribution of the bubbles is recorded on the drawings included in this study. For a discussion of the significance of bubbles in glass, see Chapter 4.2.1.5.

is made of translucent turquoise glass (Barag 1985: 64, no. 35) and can be reconstructed to a hemispherical bowl with walls strongly drawing to the exterior. Eleven additional translucent turquoise fragments and five translucent dark blue fragments of hemispherical bowls have been reported from the Burnt Palace at Nimrud (Barag 1985: 63–64, no. 33).

Fo1 exhibits a pale green colour and is made of transparent glass. The shape of the vessel is slightly irregular.

Pr1 is made of translucent blue glass, exhibiting many hemispherical bubbles that are spread throughout the glass. Pr1 is asymmetrical and slightly taller than hemispherical. The walls draw slightly inwards, making the overall appearance of the vessel different to the other hemispherical bowls. The rim is very sharp and regular and was most likely re-cut in modern times.[55]

Discussion

The hemispherical bowls form a typologically very tight group, which makes it likely that they were made at the same place, probably even in the same workshop. Parallels for the hemispherical shape of the bowls are limited. The closest similarities in shape and character of the material can be drawn to rock-crystal vessels. Of note in this regard is a fragment of a rounded rock-crystal bowl from the throne room of the Burnt Palace in Nimrud (Figure 4.22).

Figure 4.22: Fragment of transparent rock-crystal bowl from the Burnt Palace, Nimrud (Oates and Oates 2001: 239, fig. 151).

Only two additional objects, however slightly shallower than the hemispherical bowls (ht. 6 cm; d. 14.1 cm), can be usefully compared. They come from the Great Tumulus in Gordion, dated to the late 8th century (Toker and Öztürk 1992: 103, no. 81; Young 1981: pl. 73, C, D). These two examples can be joined with the group of shallow bronze bowls found in great quantities in the Great Tumulus.

Hemispherical bowls are depicted on different reliefs at Khorsabad (Figure 4.23) (court VII, façade N, room 6, 10, 11) showing foreign tributaries. The tributaries are similar in their overall representation and their headdress, but differ regarding their garments and footwear. On slabs 22–20 (façade N), the garment is long and has a scalloped rim; the persons here wear high boots (Albenda 1986: 66, pl. 24). On slabs 1–4 (room 10), the tributaries wear simple knee-length dresses, banded with a belt and a winged-shaped band with fringes. These men either wear strapped sandals, or no shoes (Figure 4.23b). In room 6, the tributaries wear short-sleeved coats with straight corners; no shoes are depicted (Figure 4.23a), and they either lead horses or carry different objects, such as large sacks, models of cities, or hemispherical bowls. Regarding the identification of the tributaries, at least from rooms 6 and 11, a singular relief found in room 6 sheds light on this question (Figure 4.23a left). Even though the slab was not found *in situ*, it depicts similar bare-footed men wearing the same type of garment. A fibula identifies the person in the middle, and therefore probably also the other tributaries from room 6, as Phrygian (Figure 4.23a; Bär 1996: 199).

Hemispherical bowls, made of metal were found in the tumuli in Gordion, as discussed above, and therefore form a good parallel for their depiction on the relief slabs. Regarding the tributaries from room 10 and court VII, no identification on the basis of their garments or accessories can be made. It has generally been suggested that the specific headdress is the standard type ('Standardtyp'), of foreign tributaries (Bär 1996: 210).[56] Regarding the hemispherical bowls depicted on the reliefs of Sargon II, it is therefore suggested here that the bowls depict metal rather than glass objects, which could have been brought as tribute to Khorsabad. As glass bowls from Gordion never occur in hemispherical shape, glass as the material of the bowls depicted is unlikely.

The number of hemispherical bowls that existed in Nimrud was much larger than it would appear from the number of objects discussed here. In addition to the seven bowls included in this chapter, over 250 bowl fragments are recorded from the Burnt Palace (Barag 1985: 64, no. 35).[57] Bowls Nim7, Nim8, Nim9, and Nim10 were found in room AA of the Northwest Palace. Further fragments can be attributed to room SW37 in Fort Shalmaneser

[55] Another hemispherical bowl, not published with image and without provenience is reported from the Museo Nazionale in Rome; see Helbig and Reisch 1969: 490.

[56] Wäfler (1975: 19) also identified this headdress as 'sargonidische Standardkopfbedeckung für Fremde'.

[57] Barag (1985: 64 no. 35) estimated that the fragments belonged to two different bowls, but the number of sherds seems too large for only two specimens.

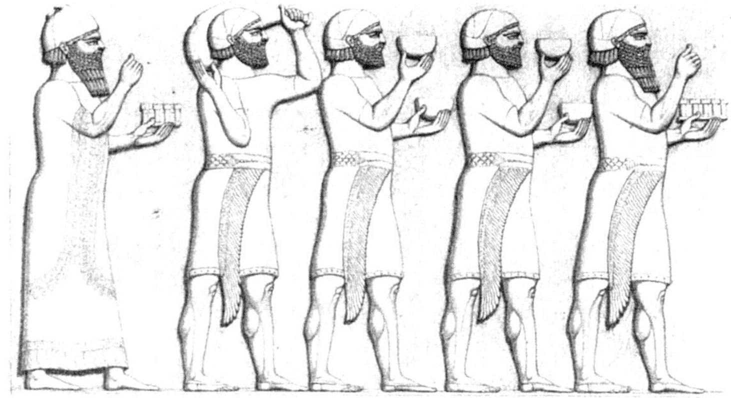

Figure 4.23a Left: Foreign tributary carrying hemispherical bowls; the central figure with typical Phrygian fibula (Albenda 1986: pl. 67, 68).

Figure 4.23b: Right: Foreign tributary with hemispherical bowls (room 10) (Albenda 1986: pl. 27).

(Nim11, Nim12, Nim14) (Chapter 3.1.4). It is therefore likely that the vessels formed sets stored and used for the same purpose. Bowl fragments Khor1 can also be attributed to a palatial context in Residence K, room 51/52 in Khorsabad. One further piece that belonged to a hemispherical vessel is reported from Nineveh, but no photograph has been published (Barag 1985: 66, no. 41).

The hemispherical bowls can with great certainty be identified as drinking vessels: drinking vessels 'must hold liquid and are generally portable but occur archaeologically in a variety of shapes and sizes.' (Hunt 2015: 183). The shape and size of the hemispherical bowls, as well as the rounded and thin rim, make this function likely. The bowls exhibit a maximum height of 10 cm and a maximum rim diameter of 15 cm, which cautiously suggests that they were used as containers for individuals rather than for communal drinking. Another factor is the material itself. The smooth surface of glass not only enhances the enjoyment of drinking compared to, for instance, ceramic, but is particularly useful for keeping liquids, as its vitrified surface is impermeable. It is in this regard important to note that Neo-Assyrian Palace-Ware specimens, which were made of very fine clay and also served as drinking vessels, sometimes have vitrified surfaces.[58] The findspots of the hemispherical bowls indicate that the pieces were used in the environment of the palace, most likely in connection with occasions such as royal banquets and feasts. Royal banqueting was an important event at the Assyrian court and often served as reaffirmation of loyalty between the king and the governors and vassal rulers (Radner 2001: 22). Of great importance, therefore, was the drinking of wine (Stronach 2000). In this regard it is probable that the transparent nature of the hemispherical bowls would allow the red colour of the liquid to shimmer through the wall, resulting in a particular visual effect.

It is interesting to note that both of the hemispherical bowls found outside Assyria (Fo1, Pr1) were found in burials. Similar to the situation regarding the colourless petalled bowl from Gordion (Gor1), obviously a change in function had occurred. Whether the bowls were used as drinking vessels before they became part of the burial equipment has to remain uncertain. It is likely that the bowls from Fortetsa and Praenestre were imported and were most likely manufactured at the same place as the rest of the hemispherical bowls. An import becomes even more likely if one considers the accompanying finds from the 'Barberini Tomb', incorporating inlaid ivories of the Phoenician type, and vessels decorated with pseudo-Egyptian motifs, pointing towards an origin in the eastern Mediterranean. Without chemical analysis, this must remain, however, uncertain.

Because of their findspots (exclusively in palaces), and the specific characteristics of their material and shape, hemispherical bowls must have served as drinking vessels, most likely in connection with royal banquets. Comparable objects of hemispherical shape are almost non-existent, and only the rare example of a rock-crystal sherd, with similar characteristics, allows a parallel to be drawn. This indicates that the significance in shape did not recall examples in another material but stood in close connection to the glass material itself. It is therefore suggested here that the hemispherical shape is a result of the technological process of slumping, rather than a reference to a specific shape in other media.

[58] In this regard, the hemispherical bowls exhibit typological and functional close parallels to the Palace-Ware bowls, which were widely spread throughout the Neo-Assyrian Empire, and which also served as drinking bowls. For an extensive study on Palace-Ware, see Hunt 2015; for the Palace-Ware bowls, see Hunt 2015: 48–49, and for considerations about their function, see Hunt 2015: 183–187.

4.2.2.5 Shallow undecorated bowls, ribbed bowls and petalled bowls

Origin and definition of the terms 'omphalos' and 'phiale'

'Shallow undecorated, ribbed and petalled bowls' discussed in this chapter are included within the same group here because of their overall similar body shape. Shallow bowls with curved walls are often referred to as an omphalos bowls or phiale. 'Omphalos' derives from Greek (ὀμφαλός), and means 'navel', literally, or metaphorically the central point of an object. The convex bulge on the central point of these bowls is eponymous for omphalos bowls, and describes the bulge attached to the shield of Omphalos (Luschey 1939: 10; Schütte-Maischatz 2001: 1). Both terms can be used interchangeably but are avoided in this study in favour of the neutral designation 'bowl'. This vessel type emerges in the first half of the 1st millennium in Assyria, northern Syria and Egypt, and reached Greece around 700. In the Achaemenid period, this shape for bowls becomes dominant. Therefore the pieces discussed in this study can be identified as precursors to the Achaemenid form, even though significant differences remain.

Manufacturing technique

Most of the bowls presented in this chapter were made by the use of two- or multi-part moulds*, apart from ribbed bowls, which were sagged (Chapter 4.2.1.4). Gor1, Gor2, and AM5 show petalled-shaped obverse and reverse sides. It is therefore likely that both sides were shaped by the use of a negative open mould in combination with a second positive mould pressed down when the glass was in viscous state, as has been similarly assumed by Saldern (1959: 24) (Figure 4.10). Any details of the decorations were probably re-cut and ground by applying cold-working techniques.

The shallow bowl Bab3 shows a very complex way of manufacturing that was also carried out in multi-part moulds.

Description and discussion of the shallow undecorated bowls

Nim14 is made of transparent, colourless glass with a greenish tinge. The rim diameter of Nim14 is 15 cm (Barag 1985: 64, no. 37). The walls draw slightly towards the exterior, and the base is straight. The thickness of the walls is irregular (0.2 x 0.4 cm) and decreases towards the rim, probably occurring because of the use of cold-working techniques when finishing the rim. The rim is sharp and there is a sharp angle between the sides and the base of Nim14.

Bab3 is made of transparent, colourless glass with a slightly greenish tinge. The bowl is shallow (4 cm) with curved walls and a flaring rim with a diameter of 15.2 cm. The thickness of the walls varies from 0.4 cm at the bottom to 0.2 cm at the rim (Reuther 1968: 210). It is likely that the piece was made using multi-part moulds*, most likely in three pieces. This is because the shoulder part of the object was difficult to slide out of the mould. Therefore, the more moulds used, the easier it was for the bowl to be removed from them. The bowl has very thin walls, making its production extremely difficult. Any kind of cold-working had to be taken out very carefully.

The shallow, undecorated bowl Nim14 was found in SW37 in Fort Shalmaneser and therefore cannot be dated precisely (Chapter 3.1.4). Similar plain bowls with slightly more rounded walls are depicted on relief scenes from the Northwest Palace at Nimrud, showing the banqueting of Ashurnasirpal II (883–859) (e.g. Cohen and Kangas 2010: 187, fig. 7.5). It is, however, unclear whether the bowl here represents glass or metal, as it is with any depiction of vessels seen on reliefs. Similar shaped metal bowls were found in Aššur, and only vary in terms of their straight walls, central rosettes or bulges (Curtis 2013: pl. 37, 514, 510).

The shallow, curved bowl Bab3 can be attributed to a burial context, dating to the Neo-Babylonian or Achaemenid periods (Chapter 3.2.1). The piece exhibits

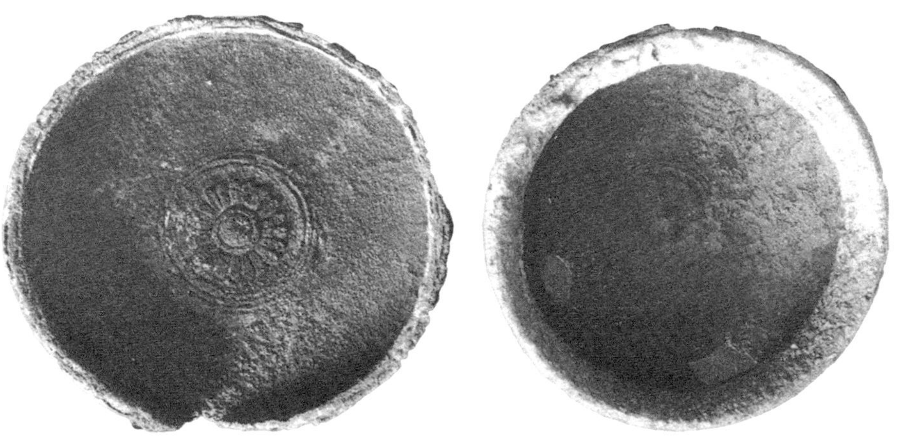

Figure 4.24: Two bronze bowls from grave 38 at Aššur (left: Ass14180; right: VA8354) (Haller 1954: pl. 25, d, e).

parallels to metal bowls from Aššur, also found in a burial, either dating to the Middle- or Neo-Assyrian period (Figure 4.24).[59] Further similar objects were found in room AB in the Northwest Palace at Nimrud. They were regarded as Assyrian in manufacture, dating to the Neo-Assyrian period. Further examples have also been uncovered at Ur, unfortunately with no indication as to date (Woolley 1962: pl. 32, no. 17).

Description and discussion of the ribbed bowls

Nim15 and Nim16 were found together in room SW37 in Fort Shalmaneser. A third fragment of the same type of vessel is reported from the same findspot (Saldern 1970: 222, no. 33). The fragments are made of colourless glass, and exhibit a greenish tinge (Saldern 1970: 222, no. 33). The fragments are curved, and there are regularly shaped ribs on their exterior. The ribs on Nim16 are curved, and the distance between them becomes narrower towards one side of the fragment. The different thicknesses of the walls (0.2–0.3 cm and 0.3–0.4 cm), as well as the different radiation of the ribs, indicate that the pieces either belonged to two different vessels (Saldern 1970: 222 no. 33), or to different parts of one vessel. The deterioration makes it impossible to observe tool marks.

Nim17 is a rim fragment that can be reconstructed to a shallow bowl with rounded ribs. Nim17 is made of colourless glass with a slight greenish tinge. Small spherical and pin-prick bubbles are spread throughout the glass (Barag 1985: 65, no. 38). There are broad, rounded ribs on the exterior, with a shallow groove on the upper end of the ribs. The interior of the vessel is smoothed. The ribs as well as the rim were shaped only on the exterior of the vessel. The ribs are parallel, broad and rounded at the end.

The find context of the shallow ribbed bowl Nim17 cannot be dated (Chapter 3.1.4). Stylistic comparisons for the shape and the petal design can, however, be found on different reliefs scenes of Ashurnasirpal II (883–859) in the Northwest Palace at Nimrud (Figure 4.25). The ribbed bowl fragments Nim15 and Nim16 form a typologically tight group exhibiting simple thin and sharp ribs. The former shapes of the vessels cannot be distinguished, making typological comparisons difficult.

Gor1 is made of colourless glass with a strong greenish tinge.[60] Only a few spherical and pin-prick bubbles are spread throughout the glass. Gor1 exhibits the shape of a shallow bowl (3.5–3.8 cm) with a plain, slightly flaring and rounded rim of *c.* 15 cm. The walls are in some areas more than 1 cm thick. There is a concave depression in

Figure 4.25: Detail of Ashurbanipal banqueting with his queen holding a petalled bowl (Cohen and Kangas 2010: 207, fig. 8, 9).

the centre of the bowl, corresponding to the rounded bulge on the interior. The exterior exhibits 32 radially arranged concave petals with slightly varying widths. The relief of these petals increases towards the curve of the bottom of the bowl and disappears as they approach the rim at about 1.5 cm below it. On the interior, the petals are concave in shape and start from a point further away from the centre and finish 1 cm below the rim. It needs to be emphasised that the petal motif occurs on both the exterior and the interior surface.

Gor2 can be identified as a fragment of a vessel similar to Gor1, but with a larger diameter of *c.* 24 cm (Jones 2009: 22). The petal motif of Gor2 also appears on the interior and exterior of the fragment, indicating similar manufacturing techniques as Gor1.

AM5 can well be compared to Gor1 and Gor2. It is made of colourless glass with a slight greenish tinge; pin-prick bubbles are regularly spread throughout the glass (Barag 1985: 66, no. 42). Five radial petals are preserved, separated by grooves on both the exterior and interior. The petals of AM5 have sharper edges however. It can be stated that the pattern was cut on both sides of the vessel, similar to Gor1 and Gor2.

14 ribbed glass sherds are catalogued as Has12. The pieces are almost entirely corroded to white, only partially revealing a translucent, dark blue core. The

[59] The piece was found in 'Gruft' 38 (Ass14180). The grave was dated by Haller (1954: 116) to the Neo-Assyrian period; Miglus (1996: 227) cautiously attributes the layer in which grave 38 was found to the Middle-Assyrian period.

[60] The description of Gor1 is based on Jones 2005: 106 and Saldern 1959: 24.

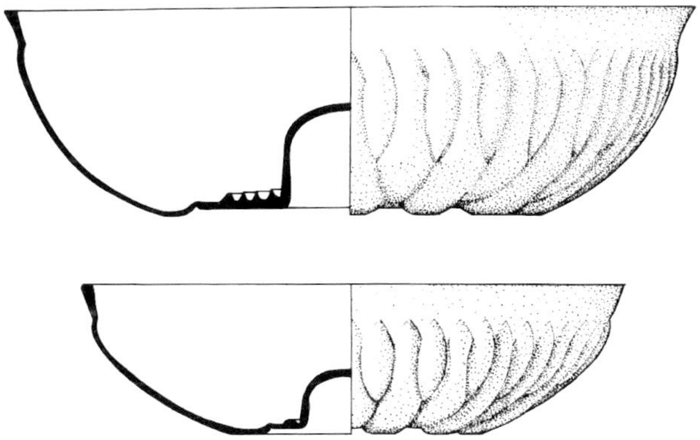

Figure 4.26: Bronze bowls from Gordion with the same decoration as found on the glass bowls (top: dm. 22.3 cm; bottom: dm. 17 cm) (Young 1981: 135, A, B).

surface of the pieces is covered with a dark brown corrosion layer. It is unclear whether the pieces belong to one or more vessels, as the fragments do not join. It is furthermore difficult to identify the vessel type. The shape of one fragment indicates, however, that this piece belonged to the shoulder of a bottle. The ribs are narrower at the top part and become wider towards the bottom. Another fragment is a thinner walled fragment with wide and regular ribs. Finally, another fragment, thick and misshapen with wide ribs, probably belonged to the bottom of a vessel. It is interesting to note that all of the fragments show traces of the rib impression on the interior and exterior of the vessel. On the exterior these impressions are usually sharper. Only the shoulder fragment exhibits the ribs on the exterior only, which is probably due to the position of the fragment on the upper part of the vessel. Has12 can therefore be identified as a vessel with a narrow neck and ribbed surface, with no close parallels to other vessels known to the author.

Gor1, Gor2, and AM5 also form a particularly tight group and find their closest parallels among mid-8th-century bronze bowls from Gordion (Figure 4.26). All of the Gordion bronze bowls have a diameter of 17–22 cm, as do the glass bowls. Similar to the glass bowls, the design is present on the interior and exterior of the vessel, as is the concave bulge. The glass bowls from Gordion, with their elaborate petal design, can be attributed, based on their findspots, to the first half of the 8th century (Chapter 3.4.3). The characteristic shape of both the glass and metal bowls indicate that both vessels were most likely manufactured at Gordion. Based on the typological coincidences between metal and glass bowls, it has been speculated whether Gordion had its own glass production (Jones 2005: 108). Chemical analysis indicates that glass from Gordion, including Gor1 and Gor2, does not exhibit a distinctive chemical signature. Rather, they can be grouped with glass from Nimrud (Chapter 7.4.3, 7.4.5). This indicates that the same type of raw glass was used at both sites. The close parallels with the bronze bowls from Gordion, however, make it highly likely that Gor1 and Gor2 were shaped at Gordion, and therefore confirm that secondary glass production occurred at this site. It is furthermore possible that transparent raw glass was imported from Nimrud to the site.

Summary

In summary, the Neo-Assyrian relief scenes are a particularly good source of comparison, and therefore of interest, as they depict different types of drinking bowls which vary in decoration according to the different kings depicted. Reliefs that can be attributed to the time of Ashurnasirpal II (883–859) show bowls with straight, rounded petals, similar to Nim17, or plain bowls of rounded shape similar to Nim14. Comparisons for the flaring shape of Bab3 can, in contrast, be found on reliefs dating to the reign of Ashurbanipal (669–627) (Figure 4.25). In addition to the flaring rim there is now an elaborate petal design which will become dominant in the Achaemenid period.[61] Based on these comparisons, it could therefore be suggested that Nim14 and Nim17 date earlier, probably already in the 9th century, as Bab3. As the materials used for the bowls on the reliefs cannot be identified, conclusions regarding the glass bowls have to be considered with caution.

Nim14, Nim15 and Nim16 were found in SW37 in Fort Shalmaneser. It is therefore likely that these objects were used in connection with royal occasions. The find contexts and shape of the shallow undecorated and ribbed and petalled bowls indicate their use as drinking vessels, as suggested for the hemispherical bowls (Chapter 4.2.2.4). Regarding the central bulge that occurs on some of the bowls (Gor1, Gor2, AM5, and probably Nim 17), Luschey (1939: 7) indicated its use as a finger hold to keep the vessel steady in the hand while pouring. This would support a drinking or libation function. Furthermore, the indentation in the side wall of Bab3 and Gor1 should not only be identified as a typological development but as a practical device to catch sediments present in the liquid (Stronach 2000: 188). In addition to drinking, the scenes on the reliefs of Ashurnasirpal II (883–859) suggest that both the petalled and the plain vessels were used for royal libations (Paley 1976: 38).

Regarding the glass bowls from Gordion, more detailed suggestions can be made. First of all, Gor1 and Gor2 (and also AM5) were decorated on the exterior and interior surfaces, indicating that special attention was drawn to their optical effect (Jones 2005: 106). This was certainly enhanced by the transparent character of the glass. As Gor1 (tumulus P) and Gor2 (city mound) occur simultaneously in funerary as well as palatial contexts, it could be concluded that this vessel type was not

[61] Bowls with flaring rims are depicted on reliefs showing the lion hunt; see Barnett 1976: pl. LIX.

primarily used as a funerary item but was also used for other purposes, probably for banqueting or libation.[62] Gor1 was found in a richly furnished tumulus for a child, inside a bronze bowl of similar type (Chapter 3.4.3). This establishes a close connection between the bronze and glass bowls, which is supported by their coinciding shape. In this context, it could be suggested that both metal and glass bowls served not only a similar purpose but were also valued in similar way.

It is striking that Bab3 was also found in a child's burial (Chapter 3.2.1. Here, the bowl was placed covering the face of the individual. This burial was equipped also with a large number of grave goods, even though it does not match the opulence of tumulus P at Gordion. Indeed, evidence is too scarce to suggest regularities between the occurrence of glass bowls and inhumations of children, but it should be considered if more data becomes available.

In summary, shallow, undecorated and ribbed and petalled bowls most likely served a function of holding liquids, either for drinking or libations. In Nimrud, this type of vessel was used in the environment of the palace, while at Gordion and Babylon the bowls also occur in inhumations for children.

Excursus: Achaemenid glass bowls

An extensive typological study on Achaemenid glass bowls would go beyond the scope of this study. Some remarks can be made, however, to point out typological differences concerning Achaemenid glass bowls. Datable Achaemenid decorated bowls were found at Persepolis (520–330) (Schmidt 1957: 92–93), Aslaia in Libya (430–425) (Vickers 1972), Sairkhe in Georgia (5th–4th century) (Makharadze and Saginashvili 1999), Gordion (mid 4th century) (Jones 2005: 108–113; Jones 2006: 22–25), and Ephesos (400–350) (Oliver 1970: 11). One of the earliest datable Achaemenid glass bowls comes from Ihringen in southern Germany. This plain bowl was found in a tomb (Tumulus I) dated to around 500 (Dehm 1997; Kistler 2010: 63).[63]

Even though there is only a relatively small number of bowls available for comparison, some major typological differences between Achaemenid and earlier bowls can be made:

- Achaemenid bowls exhibit a greater variety in shapes. In addition, deeper bowls with fluted rims occur (Figure 4.27), while earlier bowls are generally shallower.
- Achaemenid petal designs exhibit central ridges and triangular elements that fill the spaces between the tips of the petals (Figure 4.27). The earlier examples are more stylised and do not exhibit a central ridge (Nim17, Gor1, Gor2). Almond-shaped bosses are considered a hallmark of Achaemenid glass vessels and never occur on earlier bowls (Figure 4.27).
- Achaemenid patterns only occur on the exterior of bowls (Grose 1989: 75). This is not the rule in terms of earlier bowls. The bowls from Gordion, for example, as discussed above, are shaped on both the interior and exterior.

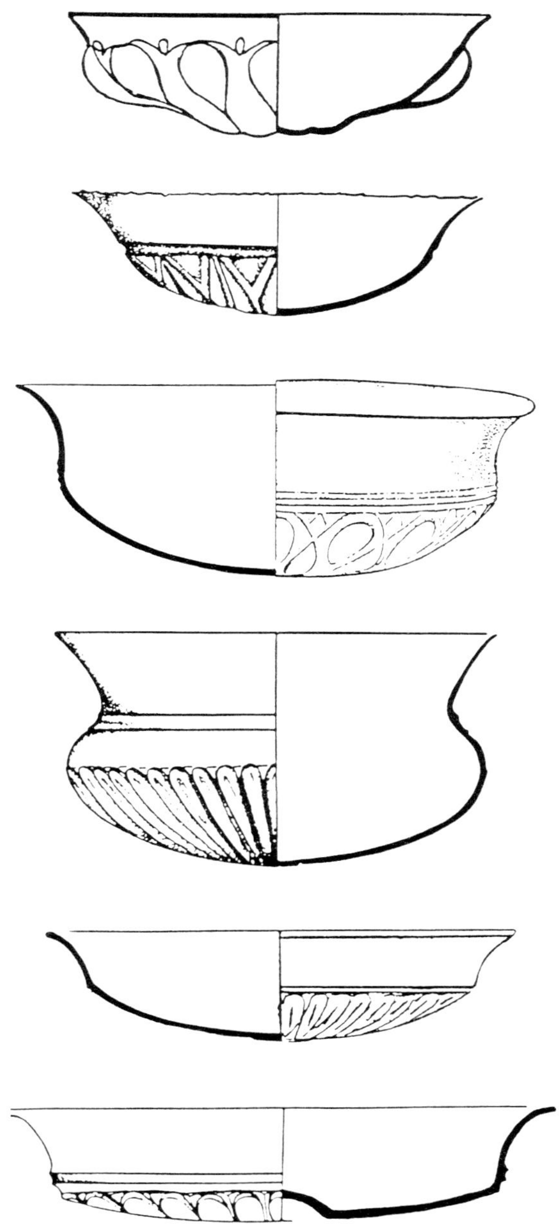

Figure 4.27: Typical forms of Achaemenid bowls (5th–4th century) (Grose 1989: 80, fig. 48).

[62] The metal bowls from Gordion were, however, considered as drinking vessels that were filled with liquids from cauldrons, using ladles or jugs; see Young 1981: 233.

[63] It is interesting to note that this bowl was not decorated. This is, however, unique, as all the other glass bowls dating to the Achaemenid period exhibit cut decorations. It could be speculated whether plain glass bowls were distributed and cut in secondary workshops at another site.

4.2.2.6 Cut-and-inlaid vessels

Description

Nim18 incorporates two fragments of the upper part of a beaker with a diameter of around 7 cm, a height of 3.5 cm, and slightly convex sides. The glass was originally colourless but is now heavily corroded with a thick, white corrosion layer. The vessel exhibits a cut groove (0.5 cm) below the rim, into which two small rosette inlays (group 3) (Chapter 4.2.2.8) are set. There is another groove below (1.6 cm) in which four vertical lines occur; these are either traces of an adhesive or deliberately painted. It is interesting that similar vertical but cut lines also appear on fragments Nim24 and Nim25, giving a similar impression. A third groove of the same size as the upper one (0.5 cm) occurs below, with traces of an adhesive. It is likely that rosette inlays were also inserted here. The rim turns slightly inwards and was most likely cut. The walls exhibit a regular thickness (0.25 and 0.4 cm).

Fragment Nim19 has a groove of roughly the same size as Nim18 and was originally part of a colourless transparent vessel that has been heavily corroded (Saldern 1970: 220, no. 28a). Nim19 most likely belonged to an inlaid colourless vessel. On the basis of its close similarity to Nim18, it could even be possible that the two fragments belonged to the same bowl, as they were both found in the Gate Chamber (Room SE13) of Fort Shalmaneser (Chapter 3.1.4).

Nim20 is a rim fragment of a transparent, colourless bowl with a diameter of 11 cm (Saldern 1970: 220, no. 27). The arrangements of the grooves, as well as their sizes, can well be compared to Nim18 and Nim19. The groove below the rim was probably inlaid. The rim is cut and turns slightly inwards.

Nim21 is a rim fragment of a colourless, hemispherical bowl (Saldern 1970: 221, no. 31). The piece is similar to Nim18, Nim19 and Nim20 regarding the structuring of the surface, but exhibits grooves that are wider and shallower. The piece has a groove below the rim, which is followed by a frieze of elevated panels of rhomboid shape. Each panel is *c.* 2 x 2.5 cm wide. It could be possible that the grooves were inlaid, however, no traces of an adhesive are present.

Nim22 is a rim fragment of a deep bowl with straight walls and a diameter of 11–12 cm.[64] The size is therefore similar to Nim20. The glass is colourless with a slightly greenish tinge, and many small spherical and oval bubbles are spread throughout. On the exterior, there is a groove 1 cm below the rim, followed by a diamond patterned frieze (1.5 cm wide), cut by horizontal and diagonal lines. The horizontal lines were clearly cut after the vessel was formed, most likely using a wheel (Barag 1985: 65, no. 38). The diamonds are ground flat at the top and are slightly irregular.

Nim23 shows the exact same pattern and was made of the same type of colourless glass. Since Nim22 and Nim23 were found together in room SW37 of Fort Shalmaneser, it is likely that they belonged originally to the same vessel.

Ar1 is a body fragment of convex shape made of colourless, transparent glass, now covered with a thick, white corrosion layer. The fragment belonged to a bowl with rounded shape, with a maximum diameter of *c.* 13.5 cm. Ar1 has a frieze on the exterior (more than 3.3 cm wide) that is decorated by a diamond pattern with diagonally cut lines, similar to Nim22 and Nim23. This pattern could have been cut also with a wheel, the tops of the diamonds patterns are ground.

Fragments Nim24, Nim25 and Nim26 must have belonged to the same vessel, probably to a bowl with a diameter of *c.* 10 cm, since they are described by Saldern (1970: 222, 33) 'as fragment with curvature'. The glass is transparent with a faint green tinge (Mallowan 1966: 416). Nim24 depicts a winged sphinx or griffin turning to the left, wearing a pointed crown with a volute at the front, a headcloth or wig, and an apron before the chest.[65] To the right of the sphinx is a horseman wearing the same long, pointed headdress. The horseman carries a whip; only the back of the horse is visible. Fragment Nim25 shows a kneeling male figure, facing towards the left. The figure raises its arms towards a plant with two volutes. The figurative band is bordered by a cut frieze of vertical lines. The arrangement of these lines, in particular the vertical ones, coincide with the lines on Nim18. Nim26 was identified as a base fragment exhibiting three rounded petals of a rosette (Mallowan 1966: 416).

Discussion

Nim18, Nim19, Nim20 and Nim21 form a tight group of small beakers exhibiting grooves for inlays. The inlays preserved among Nim18 indicate that the grooves on these vessels were most likely inlaid with rosettes that belong to group 3, as discussed in Chapter 4.2.2.8. It is interesting to note that bowl A1 was subdivided in a similar way as the glass bowls, alternating between narrow and wider grooves. Like Nim18, the narrow grooves were also inlaid with rosettes and the wider grooves were equipped with glass frames. Regarding the wider grooves on Nim18, Nim19, Nim20 and Nim21, it could therefore be possible that glass inlays were also set into them.

[64] Reconstructed in Saldern 1966a: 630, and Barag 1985: 65, no. 38.

[65] An identification as griffin (bird-head) or sphinx (human-headed) cannot be made here.

Figure 4.28a/b: Ivories found in SW37, depicting a sphinx (left: ht. 5.8 cm) and a griffin wearing the Egyptian double crown (right: ht. 5.9 cm) (Herrmann 1986: pl. 26, no. 110, pl. 28, no. 120).

Nim22, Nim23 and Ar1 form a tight group of cut vessels which are decorated with a diamond pattern. As parallels for this decoration in other media are absent, it could be possible that this feature is unique to glass.

Nim24, Nim25 and Nim26 most likely all belonged to the same vessel. The winged creature on Nim24 finds close analogies with the winged sphinxes on the painted glass inlays Nim112, Nim113 and Nim115 (Chapter 4.2.2.7). Further parallels can also be drawn to carved ivories depicting winged sphinxes and griffins (Figure 4.28a/b) (Herrmann 1986: pl. 26, no. 110, pl. 28, no. 120).[66] In accordance with these depictions, the headdress on Nim24 can be identified as the Egyptian double crown with double volute.

The headdress of the horseman on Nim24 is similar to that of the winged creature, and can also be identified as the Egyptian double crown with double volute. The kneeling person on Nim25 finds counterparts also on ivories (Figure 4.29) (Herrmann 1986: 73–74, pl. 1–2).

The flower pattern on Nim26, as well as its position on the bottom of the vessel, recalls stone and faience lids, palettes, and also mortars with similar designs.[67] In particular the radial and vertical decorations coincide with those observed on Nim24, Nim25 and Nim26. The overall composition of motifs, as well as the style, on this bowl can well be compared with a Phoenician bronze bowl found at Nimrud (room AB, Northwest Palace) showing antithetical groups of winged griffins

Figure 4.29: Kneeling person with raised arms on an ivory panel found in SW37 (Herrmann 1986: pl. 2, no. 24).

and standards.[68] This type of bowl was generally widespread in the Mediterranean.[69]

Similar to the painted inlays (Chapter 4.2.2.7), also here the differentiation between primary and secondary production plays an important role (Chapter1.2). As

[66] For a discussion of the ivories, see Chapter 4.2.2.9.

[67] For comparisons, see Searight et al. 2008: pl. 53, 54. The lids date between the 9th and the 2nd centuries, the pallets between the 8th and 6th centuries.

[68] Similar bronze bowls in Phoenician style were found on Crete and at Lefkandi; see Aruz 2014: 117, fig. 3.4, 287.

[69] A large quantity of these bowls found outside Assyria come from graves; see for an overview Markoe 1985: 75–86. A detailed analysis would, however, go beyond the scope of this study.

the type of decoration can best be compared with Phoenician objects, it is likely that the incisions were made by craftsmen trained in this style, similar to what has been observed with regard to the painted Phoenician-style designs. It is likely that the colourless bowl was manufactured in Assyria, probably at Nimrud (Chapter 4.2.2.4). It is therefore likely that the incisions were also made this site by craftsmen trained in the Phoenician style.

Regarding Ar1, which was found in Tel 'Aroer, Barag (2011: 260) stated: 'How such an exquisite vessel reached 'Aroer in the late 8th century BCE will, however, remain an enigma'. It is indeed surprising that Ar1 was found here, as no other specimen of this vessel type - and very few pieces of transparent glass in particular - were found outside of Assyria. The presence of Ar1 at the site could be explained by its findspot, which was interpreted as a caravanserai (Thareani *et al.* 2011: 161–170). Tel 'Aroer served as an important trading post along the south-Arabian trade route that connected Arabia and Edom, via the Beersheba valley, which is visible in the architecture, as well as in the archaeological record (Thareani *et al.* 2011: 305). The findspot, and the rarity of transparent glass outside of Assyria, makes it likely that Ar1 was imported from Assyria. Since the destruction of the stratum III caravanserai, in which Ar1 was found, occurred violently, it is possible that Ar1 was only stored there temporarily on its way from Assyria to the south (Arabia, Egypt).

Nim20, Nim21, Nim22, Nim23, Nim24, Nim25 and Nim26 were all found in SW37, and Nim18, Nim19 and Nim20 in the Gate Chamber SE13 in Fort Shalmaneser. It is therefore likely that all of the cut-and-inlaid glass vessels belonged to the palace inventory. With regard to Nim18 and Nim19, it could be possible that these pieces were lost or thrown away while the building was looted (Chapter 3.1.4).

The function of the cut-and-inlaid vessels is likely to be similar to that of the hemispherical bowls and the shallow undecorated and ribbed and petalled bowls. The size of the cut-and-inlaid vessels and the thin rim indicate that the pieces must have been used as drinking vessels. The incised diamond pattern on Nim22, Nim23 and Ar1 in this regard could have facilitated holding, and therefore drinking. Also a special visual effect, in particular regarding the consumption of wine, could be imagined.

4.2.2.7 Painted inlays

Manufacturing Technique: paint and colouration

The inlays on glass bowl A1, as well as the painted inlays, were created by cutting them out of hemispherical colourless bowls (Chapter 4.2.1.6). Regarding the painted decoration, Orchard (1978) presented a detailed study which is complemented by the following discussion. The paint was applied onto the inlays in cold state. The black lines of the drawings are irregular in thickness, which makes it likely that they were applied with a fine brush.[70] The black colour on Nim119, Nim112 and Nim113 was identified as a bituminous substance.[71] The blue and red pigments on Nim112, Nim114, and Nim115 were identified by Brill (1978: 28) as accidental contamination by Egyptian Blue fragments that were found in association with the inlays. Orchard (1978: 7), however, defined this as intentional decoration because the colour seems to be precisely bounded by the black lines, as can be seen in his reconstruction; Figure 4.30 (Orchard 1978: 9–10, 14). Apart from pigments, he also reconstructs a gold-leaf decoration on the disc, *uraeus*, and on parts of the sphinx.[72] It is remarkable that the overall combination of colour on the painted inlays suggestedcorresponds well with inlaid and ornate ivories found at Fort Shalmaneser (Figure 4.31).

Figure 4.30: Reconstruction of Nim115 by Orchard (1978). (after Orchard 1978: 8, fig. 1d).

Description

Eight painted inlays were found in Fort Shalmaneser at Nimrud, and can be linked to a typologically tight group. The inlays were originally transparent but are now heavily weathered with a thick, white corrosion layer with traces of iridescence*. This was also confirmed by chemical analysis of Nim119, which revealed elevated levels of antimonite (0.52%), thus indicating a deliberate decolouration (Brill 1978: 29) (Chapter 7.1.7, 7.4.3). The inlays are rectangular in shape, and almost

[70] Orchard (1978: 6) suggested that the black colour was applied by the use of a pen.

[71] X-ray fluorescence was carried out regarding Nim119. Nim112 and Nim113 were also examined under the microscope and strongly confirm this; see Brill 1978: 34–36.

[72] Glass inlays covered with gold-leaf were also found in situ at Samaria; see Crowfoot and Crowfoot 1938: 45.

Figure 4.31: Ivory panel with gold-leaf and blue inlay (Herrmann and Laidlaw 2013: colour pl. XVI).

all are slightly bent. They show similar motifs and are all painted in black. A winged, human-headed sphinx, a winged, anthropomorphic figure, or a floral motif occurs among them. The reverse is always plain.

Inlays Nim112, Nim113 and Nim115 show nearly identical motifs, arranged in linear frames. A winged, human-headed sphinx advancing towards the left is depicted on Nim112 and Nim113. On Nim115 this creature is facing towards the right. The sphinxes all wear wigs and a disc with an *uraeus*. In front of the chest on Nim112 there is another *uraeus* with disc. The sphinx on this inlay also wears a pectoral and apron with horizontal bands in front of the chest. The tails of all the sphinxes are upright, and the genitals of the creatures on Nim112 are also articulated, as are the ribs on Nim115. On Nim112 and Nim115, the sphinx is approaching a long-stemmed lily; on Nim113 a long-stemmed papyrus can be identified. Nim112 exhibits traces of blue paint on the lower snake and apron, and also traces of red are reported.[73]

Nim114, Nim116 and Nim118 have a tall, rectangular shape; Nim116 and Nim114, furthermore, show concave obverses. Nim114, Nim116 and Nim118 show a standing anthropomorphic, winged figure, framed by linear lines, facing either towards the left (Nim116) or the right (Nim118, Nim114). The figures each exhibit four wings, one advancing leg, and both arms raised. The figures are beardless and all wear wigs. On Nim118 the head of a papyrus flower can be identified.

Nim117 varies from the others because it has a flat cross-section. The motif also differs from the other inlays, showing antithetical floral motifs with two papyrus flowers and volute branches descending from a stem.

Nim119 is heavily corroded and no traces of decoration can be identified.

Discussion

Regarding their style, the painted motifs on the glass inlays all show close stylistic parallels with sphinxes depicted on ivories of the Phoenician style (Figure 4.32) (Herrmann and Laidlaw 2013: 62).

Regarding their shape, the painted inlays show similar measurements and thickness, but vary with regard to their cross section: Nim116, Nim117, Nim11 are straight, and Nim112, Nim113, Nim114, Nim115, Nim118 are bent. The distribution of the painted inlays within the different rooms of Fort Shalmaneser shows interesting patterns according to this feature: the glass inlays found in room SW37 show all straight cross-sections, whereas those found in room SW7 are curved. It is reasonable to assume that they were, therefore, used in slightly different contexts. Generally, a great number of ivories of North Syrian and Phoenician style were retrieved from both rooms SW37 and SW7.[74] Because of their thin section, it is likely that the painted glass inlays served as inlays, either for these ivories or for wooden furniture stored in the same rooms (Mallowan and Herrmann 1974: 6–8).[75] The curved glass inlays from room SW7 were found in association with curved ivory inlays of the North Syrian style that served as panels of wooden chairs with curved backrests (Herrmann and Laidlaw 2013: 102–104; Winter 2010: 227).[76] Accordingly, the

[73] For a detailed description of the coloured parts of the inlay, see: http://www.britishmuseum.org/research/collectiononline/collectionobjectdetails.aspx?objectId=1596744&partId=1&searchText=134900&page=1 (accessed: 23.09.2017).

[74] The ivories are discussed in greater detail in Chapter 4.2.2.9.

[75] In this case it could have been possible that the glass inlays were mounted into metal frames, similar to the rosettes, in bronze frames (Chapter 4.2.2.8) and secured directly onto the wooden surfaces of furniture.

[76] This was convincingly argued by Mallowan and Herrmann (1974: 6–8) on the basis of the diameter of the complete panels and the distances between them. The ivories were mounted with dowels on the backing of a material, which did not survive but which can most likely be identified as wood. The screens were found on top of each other, leaning against the

Figure 4.32: Trapezoidal ivory inlay in the Phoenician style from room SW37 of Fort Shalmaneser, ht. 5.9 cm (Herrmann 1986: 28, no. 103).

curved painted glass inlays could have served a similar function, and could have also been used as inlays for curved furniture panels.

Apart from curved ivories, sphinxes, as depicted on the painted glass inlays, regularly occur in rectangular frames between the legs of the chairs on ivories of the North Syrian style (Figure 4.33). On some of the ivories there are rectangular recesses instead at the same position, into which the painted inlays could have been set (Figure 4.34).

The find contexts of the painted glass inlays (curved and straight) decorated with sphinxes of the Phoenician style show that they were most probably set into ivories of the North Syrian style.[77] It is likely that the Phoenician motifs on the glass inlays were drawn by Pa group of craftsmen, trained in the Phoenician style, probably from the northern or southern Levant, specialised in inlay painting. However, the ivories of the North Syrian style were likely cut by ivory cutters , trained in the North Syrian style, who originate probably from northern Syria. It is therefore plausible that the glass inlays and ivories were decorated by different groups of craftsmen, working at the same location, to produce glass inlaid ivories at this site. As there is no other archaeological context or site that has revealed ivories with painted glass inlays, neither in Mesopotamia, or in northern and southern Levant, of this kind, this assumption is the most convincing.[78]

As outlined in Chapter 1.2, the distinction between primary and secondary production plays a decisive role, as the same craftsmen were not necessarily involved on both processes. This becomes particularly obvious in connection with painted glass plaques, which were processes by different groups of craftsmen trained in hot- or cold-working techniques.

The close link between painted inlays and inlaid ivories rests on the visual appearance of both media. It was shown (Figure 4.30, 4.31) that the ivories were decorated with blue and red coloured inlays and gold foil, a colour combination that also determines the painted inlays. The transparent colour of the glass itself coincides with the colour of ivory, and also traces of blue, red and gold pigments occur that coincide with the colours used on the ivories. Therefore, the painted glass inlays recall the colour combinations of the ivory panels and *vice versa*. The colours of both media did not, therefore, create a sharp contrast.

4.2.2.8 Rosette inlays

Manufacturing techniques of groups 1 and 2 inlays

How the inlays themselves were produced is described in detail above (see Chapter 4.2.1.6). At this point the different techniques for the production of the rosette decoration can be described. With regard to the rosette decoration on inlays, most likely two different types of production can be observed. Therefore all the petal depressions on inlays of group 1, and some of the decorations of group 2 (Nim45; Nim61; Nim64, and most likely Nim72; Nim73; Nim74; Nim75; Nim76; Nim77; Nim78; Nim109; TB1), were made in hot state. All these petals have in common that they are almond-shaped, and that their shape and size only varies slightly. It is therefore likely that this shape was achieved by stamping an almond-shaped stamp onto the viscous glass. Depending on how deep it was stamped, a wider or narrower pattern emerged. Some of the petals even overlap slightly, which is caused by placing the impressions inaccurately. This theory is supported by the absence of any grinding marks around the petals, as is the case with all rosettes of group 2. Furthermore, it is almost impossible to achieve almond-shaped incisions by the use of grinding tools.

With regard to the petals of almost all group 2 inlays (Chapter 4.2.2.8), it is likely that they were cut after annealing*. The petals of this group have an irregular oval and round (central part), which could be easily produced by a cutting tool. The most obvious evidence for cutting,

south wall. For a plan of the distribution of the panels and a sketch of the stacking, see Mallowan and Herrmann 1974: 4–5, 7.

[77] This does not rule out that the painted inlays could also have been inlaid into ivories of the Phoenician style.

[78] It could also be possible that the ivories and glass paintings were manufactured at different places far away from each other, and came to Nimrud as booty. This is, however, the less conclusive assumption from the existing data.

Figure 4.33: Sphinx carved into the space underneath a chair, room SW7, Fort Shalmaneser (ht. 24.5 cm).

Figure 4.34: Ivory panel with recession below a chair, room SW7, Fort Shalmaneser (ht. 17.5 cm) (Mallowan and Herrmann 1974, pl. LIX, LXIII).

however, are the grinding tool marks seen around the petals (Figure 4.36). Regarding the filling material, a white paste was filled into the recesses in cold state (see rosettes group 3). In contrast, the light blue glass filling was applied in hot state, with the base glass also heated, to avoid cracking (Taylor and Hill, pers. comm.).

Manufacturing technique of group 3 inlays

Group 3 inlays were not made by the mosaic technique, as already indicated by Spear (2005: 30) and Barag (1991: 8), who proposed that the 'cane technique' was instead applied. An experiment carried out by Taylor and Hill for the purpose of this monograph partially confirms this.[79]

The principle of manufacturing group 3 inlays appears to be based on creating hollows in the glass and stretching it, see in order Figures 4.37, 4.38, 4.39, 4.40, 4.41, 4.42, 4.43:

The single inlays could have been cut from the rod after annealing by applying cold-working techniques. Regarding the application of the white fillings, Taylor and Hill (pers. comm.) suggested that a white paste had been inserted into the hollow spaces. This is likely, as it can be observed that the fillings were not fused to the glass. Analysis of the fillings showed that the colourant was calcium antimonite (Spaer 2003: 30). The almost complete absence of silica indicates that the substance cannot be identified as either glass or faience*. All of the rosette inlays are irregular in shape and often exhibit tool marks at the edges. This indicates that the pieces were individually shaped to set them into specific recesses.

Description

Rosette inlays were found at Nimrud (78), Arslan Taş (27), Samaria (5), and Til Barsip (1). The inlays are made of either dark blue translucent glass or blue to light blue opaque glass. Based on typology and manufacturing technique, the rosette inlays are divided below into three major groups.[80]

[79] The experiment was undertaken (4 April, 2015) at the Glass Studio of the Roman Glassmakers, Mark Taylor and David Hill (Project Workshops, Lains Farm, Quarley, Andover, Hampshire SP11 8PX, UK). The photographs taken by the author are published with permission of Mark Taylor and David Hill.

[80] Curtis 1999 differentiated only between two major groups.

Figure 4.35: Detail of petal decoration with almond-shaped petals (© Trustees of the British Museum).

Figure 4.36: Detail of petal decoration with irregular oval petals and a round central part cut into the surface and showing tool marks of preliminary sketches (© Trustees of the British Museum).

Figure 4.37: Viscous glass is gathered* on an iron and marvered* on a stone surface (published with the permission of the 'Roman Glassmakers').

Figure 4.38: Next, a central hole is made by inserting a sharp tool into the gather, which will later become the central hole of the rosette (published with the permission of the 'Roman Glassmakers'). The glass is allowed to cool and stiffen before gathering a second layer of melted glass over it.

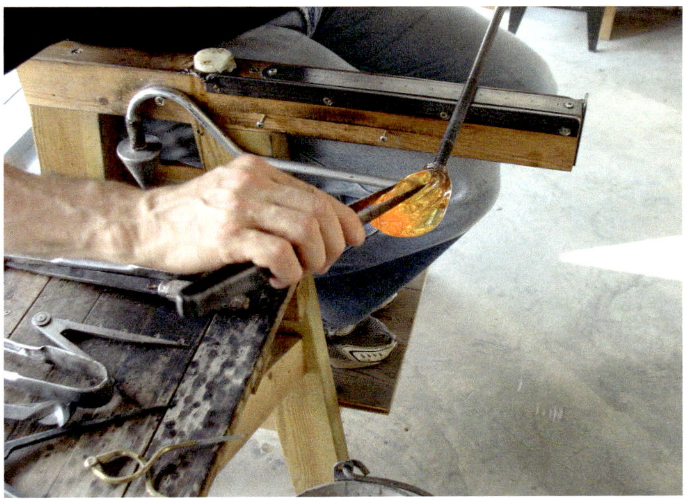

Figures 4.39, 4.40, 4.41: Using a pair of pincers, six slots are created in the hot glass. These slots will form the petals of the rosettes. The glass is allowed to cool and harden (whilst slowly turning the iron) (published with the permission of the 'Roman Glassmakers'). A third layer of hot glass is added, trapping air in the slots and creating hollow spaces. Using the marver, the glass is carefully shaped to achieve a square cross-section.

Figures 4.42, 4.43: A rod of the desired thickness is formed by reheating and stretching the glass (published with the permission of the 'Roman Glassmakers').

Group 1: Pieces belonging to group 1 were found at Nimrud and Samaria (Nim28, Nim29, Nim30, Nim31, Nim32, Nim33, Nim34, Nim35, Nim36, Nim52, Nim53, Nim54, Nim55, Nim56, Nim57, Nim58, Nim59, Nim60, Nim62, Nim63, Nim65, Nim70, Nim71, Nim111, Sam1, Sam2, Sam4, Sam5). The two main features of group 1 rosettes are the straight-sided edges and the six-petalled rosette design without a central depression. Often the rosettes are not placed in the centre of the inlay, and the petals have an irregular shape that more or less describes the shape of an almond. The inlays are mostly rectangular in shape, and the size varies at *c.* 1.2 x 1.8 cm. The obverse side is ground and polished, and the reverse side has an irregular surface which was not smoothed by cold-working. The recesses were originally filled with light blue glass (Nim55, Nim66), or, more often, with a white paste (Nim32, Nim62, Nim63, Sam1, Sam2, Sam4, Sam5).[81] The major criteria for inlays that belong to group 1 are the straight-sided edges of the inlay and the six-petal rosettes without a central point. Nim34, Nim35, Nim36, Nim58 and Nim59 are thinner than the other inlays of this group (0.1 cm or less), which could have been occurred accidentally.

Group 2: Inlays that belong to group 2 were found at Nimrud and Til Barsip (TB1, Nim37, Nim38, Nim39, Nim40, Nim41, Nim42, Nim43, Nim44, Nim45, Nim46, Nim47, Nim49, Nim50, Nim51, Nim61, Nim64, Nim72, Nim73, Nim74, Nim75, Nim76, Nim77, Nim78, Nim109). The main features of group 1 inlays are the bevelled side edges and the six-petalled rosette design with a central point. The side edges of group 2 inlays clearly differ from group 1 inlays as they exhibit bevelled side edges. The position of the rosettes is often off centre. The petals have an irregular shape, and on the surface tool marks are also preserved repeatedly. Inlays of this type are made of dark blue, translucent glass. The size and shape of the pieces vary and also larger inlays, around twice the size (*c.* 3.2 cm wide), occur (Nim43, Nim44, Nim46, Nim49, Nim50). The reverse side is irregular. The recesses were formerly filled with white paste (Nim37, Nim41, Nim43, Nim47, Nim49, Nim51, TB1).[82] Fine cuts around the petals are most likely the result of the process of cutting the petals into the glass (Figure 4.36). Tool marks of this kind only occur among inlays that belong to group 2. Therefore, the cutting of the rosettes must have been carried out after the surface was ground and polished.

Group 3: This group incorporates rosette inlays that were repeatedly referred to as 'mosaic' inlays (A1, AT1, AT2, AT3, AT4, AT5, Nim60, Nim66, Nim67, Nim68, Nim79, Nim80, Nim81, Nim82, Nim83, Nim84, Nim85, Nim86, Nim87, Nim88, Nim89, Nim90, Nim91, Nim92, Nim93, Nim94, Nim95, Nim96, Nim97, Nim98, Nim99, Nim100, Nim101, Nim102, Nim103, Nim104, Nim105, Nim106, Nim107, Nim108, Nim110, Nim134, Sam3, AM13, AM14, AM15, AM16, AM17, AM18, AM19, AM20, AM21, AM22, AM23, AM24, AM25, AM26, AM27, AM28, AM29, AM30, AM37, AM38, AM49) (Barag 1985: 52, 71–2; Curtis 1999: 60; 1989: 76; Moorey 1994: 200; Saldern 1966a: 630). This term is misleading, and therefore avoided in this study, as the mosaic technique was not applied make them. The major characteristic of the rosette inlays of group 3 is their visibility from both sides. The glass is dark blue and translucent. Inlays that belong to this group were found at Nimrud, Samaria and Arslan Taş. The shapes that occur are squares (Nim60, Nim82, Nim84, Nim85, Nim87, Nim88, Nim79, Nim83, Nim86, Nim80; Nim81, Nim110), circles (Nim90–Nim108, Sam3), and rectangles (Nim66–Nim69) and small in size (*c.* 0.6 x 0.8 cm). The shape and thickness of the inlays vary, as do the rosettes. There are traces of white paste in almost all recesses.

Group 3 inlays have also been preserved mounted into glass frames. Frame AM37 is, however, unique as it consists of 16 rectangular group 3 inlays of irregular shape, which are inserted into individual frames. The pieces were placed into the frame, most likely by the use of an adhesive, and were visible from both sides. It is likely that the inlays were shaped by cutting and grinding to fit into the spaces of the frame.

Group 3 rosettes were also mounted into single monochrome glass frames. 14 of the 15 examples (AM13, AM14, AM15, AM16, AM17, AM18, AM19, AM20, AM28, AM29, AM30, AM49, AT1, AT5, Nim89) of glass frames in this study can be attributed to Arslan Taş; the final example is from Nimrud (Nim89).[83]

Inlays from Arslan Taş were additionally mounted into bronze frames (AT2, AT3, AT4, AM21, AM22, AM23, AM24, AM25, AM26, AM27, AM38). Some empty red glass frames – now entirely corroded to green – are also recorded from Samaria (Crowfoot and Crowfoot 1938: 44). Glass frames, and also group 3 inlays, were set into glass bowl A1 from Amman.

Nim57 and Nim199 differ from the groups discussed above. Nim199 is rectangular in shape and exhibits a recessed rosette with six petals and a centre. Traces of blue are still visible on the back, as well as diagonal

[81] It is very likely that the white filling coincides with the white filling of the rosettes of group 3. Chemical analysis indicated that the paste used for inlays of group 3 contained high levels of calcium antimonite and low levels of silica. See the discussion for rosettes groups 3.

[82] For a discussion of the white paste, see rosettes group 3.

[83] For the attribution of many of the framed inlays from the Art Market to Arslan Taş Chapter 3.1.3.

Figure 4.44: Bronze cases in which rosette inlays may have been inserted and could thus be attached to furniture (Curtis 1999: 65, fig. 13).

tool marks. Nim57 is heavily corroded to green, but it is possible that the colour was originally red.

Discussion

Rosette inlays have rarely survived in position, but often occur in close connection to decorated ivories. The unworked reverse of group 1 and group 2 rosette inlays indicates that they had been inlaid into another material. Even though these arguments are true for all of the rosette inlays, it can, however, be supposed that the precise function of the different groups of rosette inlays varied. It is possible that group 1 inlays, which were featured by straight sides, were mounted into bronze frames, which have, for example, survived at Nimrud. The bronze frames exhibit on their reverse a sharp spike that most likely served for affixing the piece onto another material (Figure 4.44).

Regarding group 2 rosette inlays, it could be possible that the inlays were directly inlaid into another material, facilitated by their sloping side edges. In addition to their use in connection with ivories, the rosettes of groups 1 and 2 could have also been used as decoration for composite statutes. For example, on a statue from Toprakkale ivory rosettes were inlaid together with rounded glass inlays in similar metal frames.[84]

Incised decorations showing rosettes in frames are also present on Neo-Assyrian reliefs or ivory pyxides showing garments. In particular, the seams of garments or belts of kings, genies, attendants, and soldiers are often decorated with framed rosettes.[85] Furthermore, rosette inlays of group 3 were found *in situ* in red glass frames that could be mounted into bronze frames. It is possible that this entire set was then fixed on wooden furniture attributed to room 14 at Arslan Taş. The style of the ivories that decorated furniture in this room was identified as part of the 'Intermediate Group', and therefore with a Syrian origin. Cecchini (2009: 95) draws close parallels to ivories that can be attributed to the reign of Barrakib, king of Sam'al, who was contemporary with Tiglath-pileser III (744–727) and Sargon II (721–705). This could indicate a date for the glass frames with inlays of group 3 in the second half of the 8th century.

The distribution of the inlays has to be considered in connection with chemical analysis carried out on a small number of inlays. In this regard, 13 samples of blue glass from 'small inlays, frequently squared inlays with central white rosettes' (Reade *et al.* 2005: 23) from Nimrud were analysed (Chapter 7.4.3). The exact affiliation of these samples unfortunately cannot be determined. Analysis indicates that the pieces were coloured using cobalt as colourant, which explains the good state of preservation of the inlays, and which, for this reason, could lead to the assumption that all of the rosette and small monochrome inlays were coloured using cobalt. This has to be confirmed, however, by chemical analysis.

4.2.2.9 Small monochrome inlays

Manufacturing technique

The monochrome inlays were inserted into prepared sockets of the ivory panels, also in cold state.[86] Chemical and x-ray diffraction analyses identified residues of Egyptian blue powder and hematite-powder, which are present together with the blue and red glass inlays in the recesses of the ivories. The powder was most likely mixed with an adhesive to fix the inlays into the sockets. The unworked, irregular reverse sides of the inlays enabled the adhesive to stick to the glass. The coloured powders served to form smoother edges in areas where the glass did not join the ivory precisely (Spaer 2003: 28).[87] A smoother transition from glass to ivory was also achieved by the bevelled side edges of the inlays.

[84] For a colour photograph and detailed description, see: http://www.britishmuseum.org/research/collectiononline/collectionobjectdetails/collectionimagegallery.aspx?partid=1&assetid=44067001&objectid=369425 (accessed: 19.4.2016).

[85] For an example of this pyxis, see Wicke 2008: pl. 77, 80, no. KH 35. Very often these rosettes are only incised in very fine lines, see, for instance, Paley 1976: 95 and Cohen and Kangas 2010: 77, pl. 4.2.

[86] This was already described by Barnett (1957: 156).

[87] With regard to the ivories of the Syrian type, Barnett (1957: 156) indicated that the inlays could have been inserted by drilling one or two circular depressions which can still be seen in the recesses. Since no matching counterparts for the circular impressions were found on the glass inlays, gluing was likely.

Wing-shaped inlays

A large group of inlays are in the shape of wings. 18 pieces of this kind are included in this Catalogue, which were found at Nimrud (Nim121, Nim122, Nim130, Nim147, Nim148, Nim149, Nim150, Nim151, Nim152, Nim153, Nim154, Nim174), Samaria (Sam12, Sam14), and most likely Arslan Taş (AM32, AM34, AM35, AM39).[88] All of the inlays are made of blue glass, some of them are opaque* light blue, and others are dark translucent blue. The size of the inlays varies. On average the pieces that probably come from Arslan Taş (AM32, AM34, AM35, AM39) are larger (max. 6 cm length) than those from Nimrud and Samaria. All of the intact objects have a slanted, pointed end, except for Nim121, which has a rounded end. The obverse sides are smoothed and the reverses are flat and have an irregular surface.[89] Some of the pieces show a convex obverse (Nim122, Nim130, AM35, AM39, Nim147, Nim148, Nim149, Nim150, Nim151, Nim152, Nim153) and a smaller number exhibit a straight surface (Nim121, Nim154, AM32, AM34). The side edges of most of the inlays are bevelled, resulting in an almost v-shaped form (Nim147, Nim148, Nim149, Nim152, Nim153). In contrast, the side edges of inlays from Arslan Taş are straight (AM32, AM34, AM35, AM39).

The wing-shaped inlays served most probably as inlays to decorate wings, wigs (Herrmann 2013: 29, no. 39), floral designs (Herrmann 2013: pl. 53, no. 249) or boats. The larger winged-shaped inlays (Nim121, Nim153) must have decorated larger ivory inlays with the same motifs (Herrmann and Laidlaw 2013: pl. 87, no. 374, 375, 389).[90] Ivories which regularly exhibit wing-shaped recesses belong to the Phoenician style. In contrast, wings on other types of ivories (e.g. the Assyrian style) were often incised and further articulated by painting.

The wing-shaped glass inlays from Nimrud (Nim147, Nim148, Nim149, Nim150, Nim151, Nim152, Nim153, Nim154, Nim174) were found in room S10 in Fort Shalmaneser. Those from Samaria can most likely be attributed to the 'Bâtiment aux Ivories' (Crowfoot and Crowfoot 1938: pl. I–IV). As far as can be judged from the published material, only a few ivories found at Arslan Taş exhibit recesses into which inlays could have been set (Cecchini 2009; Thimme 1973; Thureau-Dangin 1931). The only type of ivory that exhibits spaces for inlays that were found at the site is the so-called 'Lady at the Window' type. The straight-sided pieces AM32 and AM34 could therefore have been set into these ivories.

[88] For the problem of the provenience of the piece from Arslan Taş, see Chapter 3.1.3.

[89] The grinding marks on the obverse of Nim153 are most likely modern.

[90] A great number of ivories exhibiting wings that had formerly been inlayed were found in rooms SW11 and SW12 of Fort Shalmaneser; see Herrmann and Laidlaw 2013: pl. 86–90, 156.

Scale-shaped inlays

11 scale-shaped inlays come from Nimrud (Nim123, Nim126, Nim146, Nim155, Nim156, Nim157, Nim158, Nim159, Nim160) and Samaria (Sam11, Sam22). Nine of these inlays are dark blue translucent, and two are made of yellow glass (Nim156, Sam22). The inlays exhibit three pointed tips and measure about 1 x 1 cm. The obverse sides are smoothed, and the reverses are irregular and have not been smoothed. The side edges are bevelled, creating a v-shaped profile. Irregular side edges and traces of tool marks can be observed among many of the inlays. This indicates that the pieces were shaped individually to fit into a specific recess.

Scale-shaped inlays were used to decorate the interiors of wings, dresses, and mantles, or as the interior designs of throne pedestals. Ivory inlays that reserve recesses for these types of inlays belong to the Phoenician style. Similar to the wing-shaped inlays, the scale-shaped inlays all came from room S10 of Fort Shalmaneser. Apart from blue, scale-shaped inlays occur also in yellow glass. Scale-shaped inlays decorated with gold-leaf and made of lapis lazuli were found at Samaria and form a good parallel to the blue and yellow glass inlays shown here (Crowfoot and Crowfoot 1938: 9, 44, pl. XXIV, 1).

Triangle-shaped inlays

Six triangle-shaped inlays can be attributed to Nimrud (Nim124, Nim125, Nim127, Nim173) and Samaria (Sam6, Sam13). Sam6, Sam13, Nim124, Nim125, Nim127 and Nim173 form a typologically tight group, not only because of their shape but because they always occur in the same colour. The pieces are corroded to light green. Most likely, the inlays originally had been red. The triangles are irregular in shape and have rounded ends. Nim127 and Nim173 show a flat and smoothed surface. Sam13 exhibits an elevated central part, most probably to facilitate the attachment of the inlay to a base. Because of their irregular shape, it is likely that they were cut or grozed to fit in specific recesses and therefore served as inlays.

Double-triangle-shaped inlays

16 double-triangle shaped inlays come from Nimrud (Nim129, Nim143, Nim161, Nim162, Nim163, Nim164, Nim165, Nim166, Nim167, Nim168 and Nim190). They are made of dark blue glass which exhibits a great number of bubbles. Nim128, Nim169, Nim170, Nim171 and Nim172 are made of light blue glass, which is strongly affected by corrosion. The objects have two pointed or rounded ends opposite from one another. On the side edges, there are each two curves. All of the objects are irregular in shape, and there are traces of tool marks, indicating that the pieces were shaped by cold-working techniques to fit into specific recesses. At

Figure 4.45: Diagonal grinding marks at the side edges of an inlay resulting from the smoothing of the surface in the process of cold-working (Image © The Metropolitan Museum of Art).

some edges, diagonal grinding marks are still visible. The pieces are all *c.* 2 x 1 cm. The turquoise pieces are slightly thinner (0.15 cm) than the dark blue inlays (0.2 cm), which is probably due to corrosion. The side edges are bevelled. The reverses have an irregular surface and show traces of tool marks.

12 of the 16 inlays can be attributed to room S10 of Fort Shalmaneser (Nim161, Nim162, Nim163, Nim164, Nim165, Nim166, Nim167, Nim169, Nim170, Nim171, Nim190). Recesses that match the glass inlays can be found among the ivories that are shaped in the champlevé technique. One ivory panel in this style still exhibits glass inlays *in situ*.

Floral inlays

Nim133, Nim140, Nim142, Nim144, Nim145, Nim175, Nim184, Nim185, Nim186, Nim187, Nim188 and Sam21 come from Nimrud and Samaria. They are heavily corroded and exhibit severe traces of pitting*. The colour is faded, but in some areas traces of light blue are still visible. The inlays have a slightly curled shape, terminating in a pointed tip. The obverse side is slightly convex and the reverse is flat.

Nim140 and Nim175 can be attributed to room S10 in Fort Shalmaneser. Ivories that exhibit spaces for inlays of this type show floral motifs with curving volutes. Nim133 and Nim142 are more strongly curled at their ends. Nim133 therefore can be identified as volutes of plants which occur frequently on different types of ivories. Nim142 is heavily corroded and exhibits a yellow surface. Since the piece does not exhibit a flat base, whether it served as inlay or whether it was used differently has to remain an open question. It is likely that Nim142 was either used as inlay for floral designs or served as a curled lock that can be observed among depictions of lions (Herrmann 1986: pl. 312 no. 1194).

Nim144 and Nim145 are in the shape of tulip heads. They exhibit a rounded base and two leaves that terminate in points. The pieces are made of light blue glass, which is heavily corroded. In addition to the strong corrosion, the side-edges of Nim145 are sharp. The objects are furthermore irregular in shape, which indicates the use of cold-working techniques. The side edges of Nim144 are bevelled. Comparable ivories that exhibit recesses for this type of inlay are of the champlevé technique (Herrmann and Laidlaw 2013: 40). Ivories made in this technique were primarily found in Rooms SW37, SW11 and SW12 of Forth Shalmaneser.

Nim184, Nim185, Nim186, Nim187 and Nim188 are also tulip-shaped, but differ greatly from Nim144 and Nim145, as the leaves are straight. Recesses for this type of glass inlay occur on ivories with papyrus flowers and buds. In contrast to the tulip-shape inlays of Nim144 and Nim145, none of these ivories was made in the champlevé technique.

Simple geometric inlays

There are 16 inlays within this group (Nim120, Nim131, Nim136, Nim139, NIm141, Nim176, Nim177, Nim178, Nim179, Nim180, Nim181, Nim182, Nim183, Sam8, Sam15, Sam19). Nim120, Nim176, Nim177, Nim178, Nim179, Nim180, Nim181, Nim182, Nim183, Nim139 and Sam19 can be identified as irregular rectangular-shaped inlays. Nim120, Nim139 and Sam19 exhibit bevelled edges, whereas the edges of the other pieces are straight. Rectangular recesses appear frequently on ivories that belong to the Ornate Group[91] and form parts of lotus fields, rods of standards (Herrmann 1986: 1006, 1009), or pegged wigs. It could therefore be possible that Nim176, Nim177, Nim178, Nim179, Nim180, Nim181, Nim182 and Nim183 served as inlays for these motifs.

Nim136, Nim141 and Sam15 are leaf-shaped and most likely decorated parts of necklaces and ornaments on garments of different figures.

[91] The Ornate Group is a sub-group of ivories of the Phoenician style; for details of ivories in this style, see Herrmann and Mallowan 1992: 35.

Nim131 and Sam8 are eye-shaped; Sam8 exhibits a white central space which was probably inlaid with a white paste similar to that found on the rosette-inlays.

Inlays for figurative designs

Nim191, Nim192, Nim193, Nim194, Sam7, Sam9, Sam10, Sam16, Sam17, Sam18 and Sam20 belong to the group of inlays for figurative designs. Sam16, Sam17 and Sam10 are slightly s-shaped, and were most likely used for decorating different parts of anthropomorphic and zoomorphic figures. They could have therefore been fitted into spaces of side-locks of infant Horuses, a scene repeatedly found on ivories of the Phoenician style (Herrmann and Laidlaw 2013: 27–30) (see *Discussion* below). Nim191 and Nim192 are inlays of entire wigs, which find corresponding recesses on ivories from room SW37 in Fort Shalmaneser. Ivories of the Phoenician style always exhibit the hair as one piece of inlay, whereas hair of the Ornate Group consists of multiple rectangular inlays (Herrmann and Mallowan 1992: 36).

Sam9, Sam18 and Nim194 can be identified as legs of humans, lions or griffins. Inlays of this type were most likely set into ivories that belong to inlays of the champlevé technique.[92] Legs of humans always appear in red colour on ivories (Herrmann and Laidlaw 2013: 40), an observation which is supported by the red-coloured glass of Sam9.

On Sam20 and Nim193 there are three parallel incisions on the obverse. Parallels can be found on ivory panels showing falcons whose feet exhibit similar incisions.

Inlays of the 'Lady at the Window' type

AM31, AM33 and AM36 are colourless glass inlays found among ivories with the 'Lady at the Window' motif.[93] The pieces are all *c.* 0.5 cm wide but vary in length. The inlays occur only below the head and can therefore be reconstructed as window ledges.

The 'Lady at the Window' motif is depicted on ivories of the Syrian type,[94] of which ivories were found in room 14 at Arslan Taş, but also at Samaria, Nimrud (Fort Shalmaneser) and Khorsabad (Nabû Temple). Because of the history of many ivories that occurred on the art market (Chapter 3.1.3), and because of the great number of ivories of the 'Lady at the Window' type found in room 14 which reserve recesses for inlays, it is likely that also AM31, AM33 and AM36 came from Arslan Taş. Regarding the use of the ivories found in room 14, Cecchini (2009: 94) reconstructs them as parts of couches and chairs.[95] Close parallels can furthermore be drawn to ivories that can be attributed to the reign of Barrakib, king of Sam'al, who is contemporary with Tiglath-pileser III (744–727) and Sargon II (721–705) (Cecchini 2009: 95).

Discussion

The ivories in which partly also glass inlays occur were mainly studied by Poulsen (1912), Barnett (1975), Winter (1973; 1976),[96] Herrmann (1986), and Herrmann and Laidlaw (2008; 2013),[97] who divided them into different regional groups based on stylistic classification, such as Phoenician, North Syrian, South Syrian, Intermediate, and Assyrian.[98] Different sub-groups within this main body have since been proposed and reconsidered by different scholars and have usually been associated with workshops linked to different geographical regions. In this regard, the groupings by Hermann, based on stylistic and technical criteria, are of particular importance (Herrmann and Laidlaw 2008: 56–57).[99]

The majority of the glass inlays discussed in this chapter can be attributed to ivories of the Phoenician group and more specifically to the 'Phoenician Group I (Classic)', and a smaller number to the ivories of the Syrian style, which has already been suggested by Barag (1983: 165).[100] This is because only this type of ivories shows recesses for small monochrome inlays. In contrast, other types of ivories, e.g. the Assyrian style, are often incised and further articulated by painting. Ivories of the Phoenician style show broad technical variances, and often draw on Egyptian motifs (Herrmann and Laidlaw 2013: 26). In this regard, it was proposed that Phoenician craftsmen borrowed Egyptian motifs and styles because of the close contact of the coastal region to Egypt (Herrmann and Laidlaw 2013: 26). However, the ivories of the Phoenician and Syrian style were predominantly found at Nimrud and Samaria and served as panels for furniture.

Based on their shape, some of the glass inlays can be attributed to specific sub-groups of the Phoenician style. Thus ivories of the champlevé technique have specific designs that were hollowed out and filled

[92] This group belongs to the ivories of the Phoenician tradition I.

[93] For a detailed study of this motif and its interpretation, see Suter 1992.

[94] See, therefore, Winter 1981 or Herrmann and Laidlaw 2013: 81.

[95] For a reconstruction of the numbers of furniture according to the different ivories, see Cecchini 2009: 93–95.

[96] Winter identified a South Syrian Group and proposed the existence of 'Halaf' and 'Zincirli' schools.

[97] Herrmann contributed a number of in-depth studies on the Nimrud ivories and proposed different styles, such as the 'Flame and Frond' group; see Herrmann 1989, among others.

[98] For a summary on the history of research of the ivories, see Feldman 2014: 13–21. A detailed study of the different styles of ivories would exceed the scope of this work, therefore only a short overview is provided here.

[99] The most recent summarising volumes on the topic are the outcome of a workshop held in Fribourg in 2001 that resulted in Crafts and Images in Contact edited by Suter 2005, and that of a workshop held in Pisa in 2004, resulting in the publication Syrian and Phoenician Ivories; see, therefore, Cecchini et al. 2009.

[100] The terminology follows Herrmann and Laidlaw 2013.

with coloured inlays (Herrmann and Laidlaw 2013: 40). Glass inlays that can be associated with these ivories incorporate double-triangles, flaring tulip-shaped inlays, and human legs. Another subgroup is the Ornate Group, among which the rectangular glass inlays can be counted (Herrmann and Laidlaw 2013: 35–38).

Ivories associated with glass inlays generally are dated earlier then the 8th century and continue into the 7th century.

The different shapes of glass inlays mostly appear in specific colours. Blue is predominant, but also yellow, red and colourless glass occurs. The colours of the glass inlays coincide with the colours of other materials, like Egyptian blue, gold leaf, and lapis lazuli, which were used for inlays.[101] For instance, blue and yellow glass scales occur simultaneously with scales made of lapis lazuli and gold foil. Hair, for instance, is often held in blue, whereas wings occur in blue, red, and gold. This could point towards the existence of a certain colour code regarding specific shapes and motifs on the ivories. Therefore, the glass inlays did not aim to achieve an eye-catching contrast to the other inlays but rather were integrated into the colour spectrum of the ivories (see in detail Chapter). This is particularly obvious with regard to the colourless inlays in the 'Lady at the Window' ivories. Here also the transparent glass inlays aim to achieve an integrating effect, suggesting that the combination of ivory and glass materials bore a certain significance. This leads to the conclusion that, with regard to glass inlaid in ivories, it can be observed that glass was used simultaneously with genuine materials, such as stone, as well as with artificial materials, e.g. Egyptian blue.

4.2.2.10 Large monochrome inlays

Description

Khor2 is made of opaque light blue glass. The inlay is broken in the middle and can be reconstructed to an irregular rectangular shape with a round hole (dm. 0.4 cm) in the middle, which is pierced through the entire object. The surface of Khor2 is convex and smoothed. The reverse is also slightly convex but has an irregular surface. The side edges are cut and bevelled, sloping towards the reverse. Deep cuts on the reverse side, towards the side edges, are up to 2 cm long and indicate that this side was not intended to be visible (Chapter 4.2.1.6). The hole, as well as the shaping of the side edges, was applied after the object had been annealed*.

Inlay Khor3 is an irregular fragment, of which one side forms an edge, and has an opaque* light blue colour. Both the obverse and reverse are convex and smoothed. The side edges slope towards the reverse.

Khor4 is also opaque* and light blue and has an irregular shape. The surface of the object is slightly convex and smoothed. On the reverse there is a slight depression in the middle and irregular incisions caused by the manufacturing process cover the entire surface. The side edges slope towards the reverse and were smoothed.

Khor5 is an irregular small fragment made of opaque* light blue glass. The fragment is broken off on all sides.

Khor6 is made of an opaque* glass with a formerly light blue colour that had partly turned into light green. The irregular small fragment is broken off on all sides and has a round hole which is cut.

Two irregular fragments are subsumed under Khor7, which are made of opaque, light blue glass. The bubbles present at the recent break are only small in number, spherical in shape, and large in size (0.2 cm). The surface of Khor7 is smoothed, and the reverse, in contrast, irregular. The side edges slope towards the reverse.

Khor8 is opaque* and made of formerly light blue glass, partly corroded on the reverse to a greenish colour. The obverse is smoothed, and the reverse side has not been flattened.

Discussion

The bevelled edges and unsmoothed reverses, which partly show traces of tool marks, are similar to those observed with regard to the rosette inlays of group 2 (Chapter 4.2.2.8). Accordingly, the unworked reverses of the large monochrome inlays indicate that they had been inlaid into another material. This would have been facilitated by their sloping side edges, and makes it plausible that Khor2, Khor3, Khor4, Khor5, Khor6, Khor7 and Khor8 served as inlays (Chapter 4.2.1.6). Because of their size and thickness, it can cautiously be suggested that they served as inlays for larger objects, like walls or interior installations. The perforations on Khor2 and Khor6 could indicate that the pieces were mounted on a base using a dowel, as observed similarly regarding the inlays for composite statues (Chapter 4.2.2.11). The Residences, as well as the Nabû Temple in Khorsabad, were extensively decorated with colourful elements, i.e. with painted plaster (room 12, Residence K) (Loud and Altman 1936: pl. 88. 89) or glazed bricks (Residence K, Nabû Temple). Here, a free-standing installation, probably an altar, in the Nabû Temple was covered with glazed bricks (Loud and Altman 1936: 14. 15. 42).

Both groups of objects Khor2, Khor3, Khor4, Khor5 and Khor7 on the one hand, and Khor6 and Khor8 on the other, were made of very similar types of glass, which could originate from the same ingot. From the similar appearance and nature of the glass, as well as their coinciding find spot, it is plausible that these two

[101] Faience was hardly ever used as an inlay medium for ivories; see Spear 2005: 28.

groups of large monochrome inlays were manufactured at the same place, probably even at Khorsabad.

4.2.2.11 Attachments and inlays for composite statues

Glass, among other materials, was used for attachments and inlays to decorate composite statues. Composite statues occur in Mesopotamia as anthropomorphic or zoomorphic statues, in miniature or life-size. Attachments and inlays were used in this context to replace different body features.[102] Attachments were made in the shape of head-dresses, beards, arms, or legs, and were usually mounted onto the core, probably made of wood or bitumen, with a dowel (Berlejung 1998: 100; Dick 2005: 50). Attachments made of glass were mounted in a similar way as attachments made of stone, faience or Egyptian blue, indicated by the positions of the dowel-holes, which is the same for different media. Inlays, in contrast, are usually smaller pieces, for example eyes, eye-brows, or parts of the garment that were inlaid into cut recesses, often by the use of an adhesive.

The find spots for the attachments and inlays discussed in this chapter all come from temple contexts. It is therefore likely that the inlays belonged to statues of gods, but an affiliation with statues of kings or other humans, however, cannot be ruled out.[103]

Description

Bab2 is made of opaque* dark blue glass (Barag 1985: 76 no. 68). The piece is irregular and broken on all sides, and the overall shape is convex (4.3 x 5.2 cm). The original diameter can be reconstructed to approximately 7 cm. The obverse of Bab2 is decorated with five wavy strands which are cut in the opposite direction by seven incised wavy lines. The reverse of Bab2 is uneven. The thickness of the fragment is irregular, which is due to its manufacturing process (*c.* 1.7 cm) (Chapter 4.2.1.2). There are two dowel holes, one of which is visible in cross-section, indicating that the piece was fixed onto the statue by a thin nail or dowel.

Ur1 has a beige corrosion layer, and the glass has faded to light blue. On the obverse of the fragment there are two rows with a total of eight curls coiling to the left. The lower and left side edges are original, allowing the piece to be identified as the lower part of a beard (Barag 1985: 76, no. 66). The reverse of Ur1 is slightly convex and projects the lower part of the beard 0.4 cm forward (Barag 1985: 65, no. 66). On the surface of the reverse, holes are visible that occur from popped bubbles.

Dul1 is made of opaque* light blue glass (Barag 1985: 76, no. 65) (3.1 x 2.9 cm). The reverse has an irregular surface and there are three curls which coil to the left on the obverse.

Nim2 is made of a translucent dark blue glass (Barag 1985: 75, no. 62). The piece shows 12 plaits which are rolled from the top right downwards, and on the left side from the opposite direction. On the right side the plaits curl to the right, and on the left side the plaits curl to the left (5.1 x 3.6 cm). The reverse of the inlay is slightly uneven and smoothed, exhibiting a thicker (1 cm) and a thinner part (Barag 1985: 75, no. 62).

Nim1 is made of dark blue glass (Barag 1985: 75, no. 64). The piece is broken on every side (5.1 x 5.5 cm), but the remaining surface exhibits five plaits with diagonal lines. The reverse of the object is slightly concave, similar to Ur1 (Barag 1985: 75, no. 64). It is likely that the finely drawn and deeply cut lines decorating the plaits were cut after annealing*. Unlike most of the other objects of this group, Nim1 was found together with Nim2 in the Ninurta Temple.

Nim5 is made of translucent dark blue glass (Barag 1985: 75, no. 62). The piece is broken on all sides (6.4 x 4.8 cm) and is covered on its obverse with seven diagonal rows of curls coiling to the left. The obverse is slightly convex (Barag 1985: 75).

Nin1 is large (6.5 x 6.2 cm) and covered with a thick white corrosion layer, but originally of blue glass. The object shows six diagonal lines, each with five curls coiling to the left side, and above them single wavy vertical lines separated by horizontal lines are represented. The reverse side has an irregular surface.

Bab1 is made of dark blue opaque* glass (Barag 1985: 76, no. 67). The shape of the fragment is curved, and the end is rolled up to a curl (3 x 6 cm). The obverse is decorated with two plaits of seven curls; the reverse is not decorated and exhibits an irregular surface.

Nim3 is made of opaque* blue glass (Barag 1985: 77, no. 70). The object has a sharp edge at the top, and the left side is original and forms a straight edge (3.9 x 5 cm) with a thickness of 0.6–1.1 cm. The obverse is decorated with a convex curved band with an incised fish-bone pattern. This is followed by another pattern of wavy grooves. The fish-bone decoration of the plait is irregular and the lines are fine and deeply cut. It is therefore likely that the decoration was cut after annealing*, rather than being made in a mould*. There

[102] For a detailed study on composite statues and the different materials used for decoration, see Schmidt forthcoming. For a general overview of composite artefacts, see Composite artefacts in the Ancient Near East edited by Di Paolo 2018.

[103] Cuneiform texts are solely concerned with the production of statues of gods; see Berlejung 1998. However, anthropomorphic statues of gods and humans are known that can only be distinguished on the basis of headdress, garment, and attribute(s). For a summary of statues of kings erected in temples, see Magen 1986: 137–169; for a detailed study of statues in general, see Strommenger 1970 and Spycket 1981. For statues of deities, see Seidl 1983: 314–319 and Spycket 1968.

is a slightly convex band on the reverse, corresponding to the convex band on the front. The overall surface of the reverse is mostly flat and covered with fine veins. This is a result of surface tension that arises while the glass cools (Chapter 4.2.1.2) (Stern and Schlick-Nolte 1994: 49). The band was cut after annealing, together with the dowel-hole.

The original colour of Dul2 was dark blue, but most parts of the objects are now corroded to turquoise green (Barag 1985: 77, no. 69). Dul2 has an overall convex shape (2.7 x 3 cm). On the obverse, two slightly bent plaits with incised diagonal decorations are visible. Similar to Nim3, the decoration of the plaits was most likely incised after annealing* since the incisions are irregular, deep and very fine. Dul2 has a dowel-hole, in which are preserved parts of a bronze dowel (Barag 1985: 77, no. 69).

The small fragment Nim4 is made of dark blue glass (Barag 1985: 77, no. 71). The object has a wavy side and is broken on the others (1.9 x 3.4 cm). The obverse is wavy and grooved. The grooves were incised after annealing*, as with Nim3.

Is2 consists of five fragments that can be most likely assigned to the same inlay on the basis of their shapes. The fragments are all covered with a yellowish-brown corrosion layer, which overlays the formerly blue glass. Fragments 1, 2, 3 and 4 all show rounded ends, and the obverse sides are incised with wavy or zigzag bands (No. 4). Fragment 2 has a round dowel-hole.

Is1 incorporates five fragments that can be grouped together on basis of their decoration. It is likely that they have served as fragments for a single attachment. The fragments have a yellowish-brown corrosion layer, similar to Is2, but the original blue colour of the glass is still preserved in some parts. On the surfaces of the pieces, rounded convex curls are present. Fragment 1 exhibits a hole that was cut after annealing.

Nim6 is a complete eye-inlay preserving also the dark blue glass pupil. The inlay has an overall size of 4.4 x 2 cm and a thickness of 1.2 cm. The outer black part of Nim6 was made of black stone. Various traces of tool marks are still visible on the reverse. The reverse of the black frame is open and preserves the white stone around the eyes, indicating the skin. The pupil, made of dark blue glass, does not sit directly in the white stone but rather in a paste that is corroded to light brown.

Discussion

Attachments for male statues

Ur1, Nim1, Nim2, Nim5, Dul1 and Nin1 are attachments for beards, comprised of curls and wavy strands; Bab2 clearly served as wig. The design featured on these beards varies considerably and follows the beard-designs of specific Neo-Assyrian kings and their entourages, which date to a certain period. It is important to note here that the glass attachments, however, do not necessarily appear to have belonged to composite statues depicting kings.

The single thin strands on the attachment Nin1 find parallels among depictions of Ashurnasirpal II (883–859).[104] The beard-design of Nin1 coincides with both 'Stilstufe 1' and 'Stilstufe 2', determined by Strommenger (1970), which incorporates objects dating from the early 10th century to the reign of Tiglath-pileser III (744–727).[105] In contrast, Nim1 and Nim2 show twisted elements between the curls which are similar to the ones found on the reliefs of Sargon II (722–705),[106] Sennacherib (704–681),[107] and Ashurbanipal II (668–669).[108] The manner in which Nim1 and Nim2 are designed coincides with Strommenger's 'Stilstufe 3', which she attributes largely to the time of Sargon II (Strommenger 1970: 32).

Bab2 served as a wig of a male statue. The wavy-hair pattern finds close parallels with reliefs and statues of gods and kings from the Neo-Babylonian and Neo-Assyrian periods. An earlier date however cannot be ruled out. Notable examples include the relief block of Šamaš-rēša-uṣur,[109] on which gods as well as the king with comparable hairstyles are shown (Figure 4.46).[110]

Attachments for female statues

The attachments Bab1, Nim3, Nim4, Dul2 and Is1 feature either braids or forelocks, thus identifying a typical female hairstyle. Braids, as well as forelocks, are a characteristic part of female hairstyles and can be found among representations of females from the 3rd millennium onwards. A particularly good comparison is the group of 11 faience* inlays that were found at Uruk (van Ess and Pedde 1992: pl. 90, 1091–1122), which can be reconstructed to a life-sized head that belonged to a statue of the Middle- or Neo-Babylonian period.

[104] For further examples, see the reliefs from the Northwest Palace at Nimrud, e.g. Cohen and Kangas 2010: 78, pl. 5, 80, pl. 6, or in the Northern Palace in Nineveh, e.g. Matthiae 1998: 198.

[105] The 'Stilstufen' are based on Neo-Assyrian statues; see Strommenger 1970: 31–32.

[106] For example on the relief from Khorsabad; see Orthmann 1975: no. 220 and Matthiae 1998: 61–62.

[107] For example on the reliefs from Nineveh; see Barnett 1998: pl. 478, 496.

[108] See, for instance, the depiction of the king in the banquet scene; Cohen and Kangas 2010: 195, fig. 7, 14.

[109] The relief block is Assyrian in style and was found in the region of the Middle Euphrates; it was dated to the second third of the 8th century. The gods can be identified on the basis of inscription as Ištar and Rammānu.

[110] For example Ashurnasirpal I (1114–1076); Strommenger 1970: pl. 1a, or the Kudurru of Marduk-apla-iddina II (late 8th century); see Strommenger 1962: no. 274.

Figure 4.46: Relief block of Šamaš-rēša-uṣur, showing comparable hairstyles with the glass attachments (Koldewey and Wetzel 1932: pl. 20).

Bab1, Nim3 and Dul2 feature lateral braids that could have been worn either behind the ear or next to the cheek (Figure 4.47). Bab1 belongs to the lower right end of a braid, and Nim3 and Dul2 belong to the upper end. Braids are common among female statues already in the Early Dynastic period (Frankfort 1939: pl. 74, 76, 82), but the details of the incised patterns changed over time.[111] On the relief of Šamaš-rēša-uṣur, Ištar is shown with a braid, serving here as an example for the early 1st millennium (Figure 4.46).

Another typical feature of female headdress is the forelocks that are worn on the front part of the forehead, and among which Nim4 and Is2 (No. 3 and No. 5) can be included (Figure 4.47). The forelocks display a wavy shape, and also exhibit wavy incision.

The Significance of Glass and other Materials in the Context of Composite Statues

As of today (2018), the earliest Mesopotamian glass attachments date to the 11th century and were found in the E-Babbar in Larsa.[112] With the currently available data, the majority of glass attachments and inlays can be attributed to the Neo-Assyrian and Neo-Babylonian period, based on stratigraphic and typological considerations.[113] This shows a distinct increase in the

[111] During the Kassite Period, braids begin to appear regularly among depictions of goddesses; see for example Orthmann 1975: no. 189. This can also be observed among the goddesses on the kudurrus No. 48, No. 49, and No. 50 which belong to the 'Vierte Gruppe', determined by Seidl 1989, and attributed to the reign of Marduk-apla-iddina I (1173–1161).

[112] Six glass fragments, one of a beard, and five belonging to a wig of a female statue, were found in the E-Babbar of Larsa; see Huot 2014: 169–174. One of the inlays was attributed by Huot (2014: fig. 105 L, 74, 76) to the 11th century.

[113] The number of glass attachments and inlays included in this study shows the present state of published research. This becomes clear by referring, for instance, to the site of Isin, where altogether 34 glass fragments for composite attachments inlays were found in the Gula Temple, but out of which only 11 were published; see Hrouda

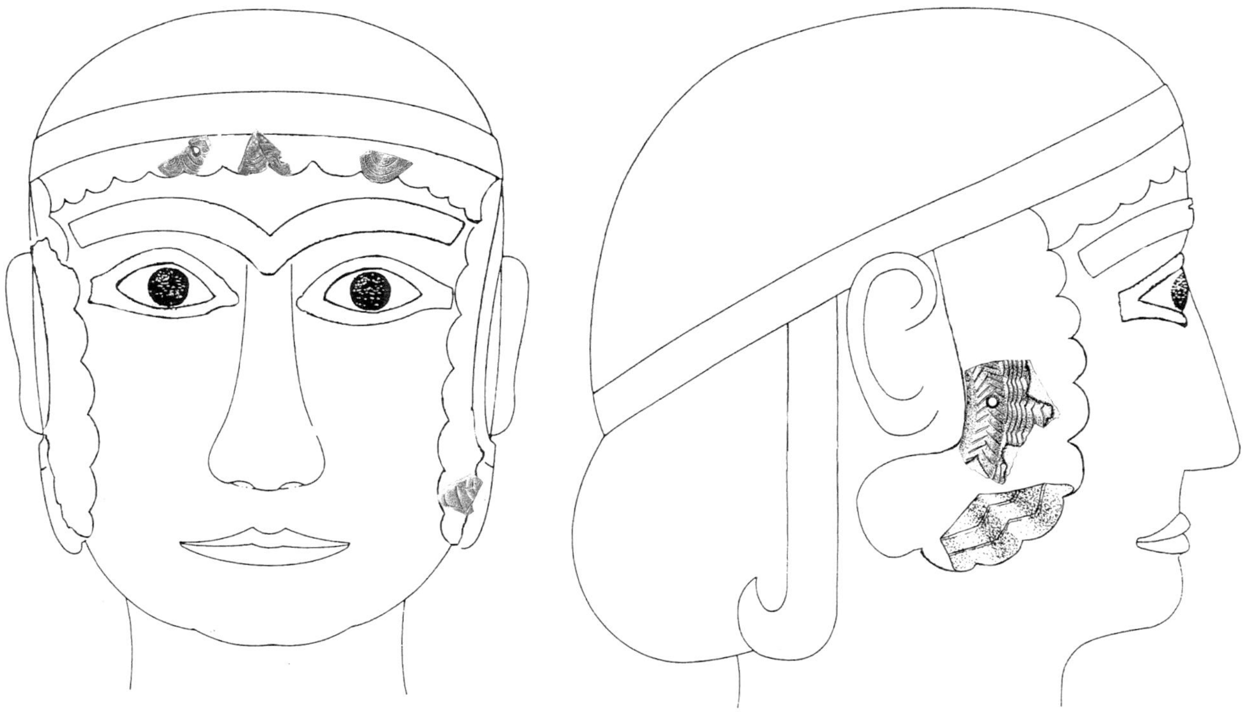

Figure 4.47a-b: Possible reconstruction of the mounting of the different attachments Nim3 and Nim4 (left) and Is2 (right). The attachments are not supposed to have decorated the same composite statue (after a drawing by van Ess and Pedde 1992: pl. 90, 1091–1122 showing the mounting of inlays from Uruk, attachments are not to scale).

use of glass in the context of composite statues and a new development in the production of composite statues, since at the major glassworking sites of the mid to late 2nd millennium, such as Nuzi or Alalakh, glass attachments do not yet appear in the repertoire.[114]

Apart from glass, attachments and inlays have been preserved in various materials such as different types of stones, sintered materials (e.g. faience*), Egyptian blue and glaze, metal, shell, and bitumen. Lapis lazuli, and also other types of dark-coloured stones, were used throughout all periods, whereas sintered materials, predominantly coloured in blue, like faience and Egyptian blue, are only attested from the mid to late 2nd millennium onwards.[115]

Regarding the 1st millennium, in addition to glass, a large number of composite attachments and inlays were made of sintered materials, as well as lapis lazuli, carnelian, and turquoise, which can stylistically well be compared to the ones made of glass. A number of attachments and inlays (faience, Egyptian blue, blue-glazed ceramic, lapis lazuli, carnelian, turquoise) are part of the so-called 'Schatzfund' from Babylon, providing good sources for comparison (Wetzel *et al.* 1957: 36, pl. 45 a, b; Wullen and Marzahn 2008: 202 fig. 133). Furthermore, there are faience attachments from Uruk which belonged, probably, to the same life-sized female head dated by van Ess and Pedde (1992: pl. 90, 1091–1122) to the 10th or 9th century. From Assyria, attachments and inlays are, for example, known from the Aššur Temple in Aššur (Gries, pers. comm.) and from the Ninurta Temple in Nimrud (Egyptian blue, glass) (Curtis and Reade 1995: 102, no. 49).

This overview is not exhaustive but shows a common thread: a wide range of materials was used for the production of inlays and attachments of composite statues. In this regard, not only blue glass – probably not earlier than the 11th century – but also blue faience*, Egyptian blue and lapis lazuli are commonly used and play a particular role. The different materials possess

1987: 42–43. Information about a large number of glass attachments and inlays for composite statues is also reported for the Ninurta Temple in Nimrud; see Layard 1856: 357–358. The actual number of glass attachments and inlays for composite statues can therefore be considered as larger, as this study represents.

[114] No glass attachments and inlays are present at sites with a generally large corpus of glass finds, or, at least, have not been published. Therefore composite inlays are not attested at Tell Brak (Oates et al. 1997: 81–89), Nuzi (Stark 1939: 459–460), or Alalakh (Moorey 1994: 193).

[115] For an overview of the different materials used for composite attachments and inlays in different periods, see Schmidt forthcoming.

specific material properties which can be of a chemical, mechanical, and optical nature.

On the one hand, the use of these different materials could have simply aimed at a divergence of substance within these statues, creating a strong contrast (Roßberger 2018). This would emphasise that each material was appreciated for its individual characteristic, and was considered an important element of the statue as a whole. The diversity of materials created a certain effect on the observer that could not have been achieved by painted or incised surfaces. This effect is enhanced by the intensity of colour and the shining surfaces. Therefore glass plays a particularly important role, as it not only has a colourful and shiny surface, but it is also, in most cases, translucent, creating a particularly intensive and deep effect.

On the other hand, it is vital to note that all these materials were used simultaneously in the context of composite sculpturing and served the same purpose. It seems therefore likely that not the choice for a specific material but rather the material characteristic, e.g. colour and shine, were important. It can thus be concluded that artificial and genuine materials were appreciated similarly in this particular context.[116] This reasoning is supported by the *mīs pî* ritual, the so-called mouth-opening ritual, which gives us an idea about the process of inducing a cult statue. Specialised artisans who worked in temple workshops (*bīt mummi*) were entrusted with the production of these statues in composite technique (Walker 2001: 8; CAD M: 198). The different materials needed for their production are mentioned in detail and are attributed to specialists, such as stone-cutters (*qurqurru*), stone-carvers (*purkullu*), or lapidaries (*zadimmu*).[117] However the most important stage in the production of the statue is the final ritual, in which the statue is transformed from these various composite materials into a divine being: the artisans have to leave the city and the statue is 'discovered' near a river, staged as a birth-like appearance of the statue. On the level of materiality, the ritual unifies the different parts into a uniform statue.[118] The various materials out of which the statue was composed can now be considered as secondary. Far more important is the appearance and effect of the divine being enhanced by the richness of colour and shimmering appearance.

Summary

Glass attachments and inlays can be attributed to both male and female composite statues of different sizes. Their find spots all come from temple contexts. It is very likely that the attachments and inlays therefore belonged to statues of gods. However, an affiliation with statues of kings or humans cannot be ruled out.

In the context of composite statues, a large variety of different materials for attachments and inlays was used throughout different periods. In the first half of the 1st millennium there was a distinct increase in the use of glass for attachments and inlays, in particular to make headdresses and beards. With the data available today, this was a new development in the production of statues, since at the major glassworking sites of the mid to late 2nd millennium, such as Nuzi or Alalakh, glass attachments do not yet appear in the repertoire. As of today (2018), the earliest Mesopotamian glass attachments or inlays date to the 11th century.

The important factor of composite sculpture is, however, the presence of various materials being used together to create a single statue. Moreover, the intensive visual effect created on the observer by the use of different materials is crucial, since it distinguishes composite from painted or incised statues. Glass was, in addition to blue faience*, Egyptian blue and lapis lazuli, also used as material for attachments, and therefore served the same purpose as the other materials. Their parallel use, however, indicates that all these materials were appreciated similarly. It can thus be concluded that artificial and genuine materials were also valued similarly in this particular context. This is supported by the *mīs pî* ritual performed for the 'mouth-opening' of a statue. Here, in the course of ritual action the material statue is transformed into a divine entity. Therefore the origin of materials, whether artificial or genuine, no longer plays a role. The paramount feature is the visual qualities.

4.3. Core- and rod-formed glass

4.3.1 Previous studies on core- and rod-formed glass

Core- and rod-formed glass vessels are already produced at the beginning of regular glass production in the middle of the 2nd millennium, and continue to be produced in the 1st millennium. Both groups use the same principles of core-forming*, but result in vessels with different shapes (Chapter 4.3.2): core-formed vessels therefore comprise rounded vessels shapes, rod-formed vessels result in more elongated, tubular forms.

In the late 2nd millennium, core-formed vessels are known from Mesopotamia (Barag 1970: 135–154; Barag 1985: 35–49; Grose 1989: 45–48; Stern and Schlick-Nolte

[116] This was also noticed by Schuster-Brandis (2008: 7) in respect to amulets made of stone, other sintered materials, and glass.

[117] For further text examples, see Berlejung 1998: 120, footnote 658, 124–125.

[118] This transformation was achieved by the mouth-washing and mouth-opening rituals; for summary and literature, see Hurrowitz 2003: 157.

1994: 119–129), as well as from Egypt (Grose 1989: 45–56; Lilyquist and Brill 1993; Stern and Schlick-Nolte 1994: 130–149). In the 1st millennium, core-formed vessels continue to be produced. Research, however, was focused on the time period between the 6th and 1st centuries BCE (Fossing 1940; Barag 1970: 154–180; Harden 1981: 51–57; Grose 1989: 60–61, 109–174; McClellan 1984).[119] The first comprehensive research on core-formed vessels, however only including the objects from the 1st millennium BCE, was carried out by Fossing (1940), who established four major chronological groups of this vessel type:

Mesopotamian Group	7th – early 6th century
Mediterranean Group I	mid 6th – early 4th century
Mediterranean Group II	4th – early 3rd century
Mediterranean Group III	3rd – 1st century

This classification was widely accepted and only partially supplemented by Harden (1981) and McClellan (1984). Harden (1981: 51) therefore added the so-called 'Italian Group' to the existing scheme, which he attributed to the time period from the late 7th to the 4th/early 3rd century. McClellan (1984) further redefined this scheme by adding objects from the western Mediterranean, i.e. North Africa, Spain and Italy.[120]

The Mesopotamian core-formed vessels were intensively studied by Barag (1970). He extended this typology to include core-formed vessels from the 2nd millennium BCE, and therefore established an 'early group' (15th–13th century) and a 'late group' (8th–6th century) of Mesopotamian core-formed vessels. In his *Catalogue of West Asiatic Glass in the British Museum* Barag (1985) collected about 100 core-formed vessels from the British Museum, most of which originate from northern Mesopotamia. Two-thirds of these vessels date to the Late Bronze Age period (Barag 1985: 135).

The technique of rod-forming is closely related to that of core-forming* (Chapter 4.3.2). Apart from elongated vessels, rod-formed objects also comprise beads and head-pendants (Chapter 4.3.5), as well as the 'tubes' from Hasanlu (Chapter 4.3.4). The most common vessel shape of this technique is the kohl tube, which is common in the shape of palm columns among Egyptian glass vessels from the 14th and 13th century (for example Grose 1989: 62, no. 10). Kohl tubes occur again in large number in Iran in the middle of the 1st millennium (Barag 1975). Most of these vessels, however, lack archaeological context and therefore a detailed typological study has not yet been carried out.

4.3.2 Manufacturing process

4.3.2.1 Core-forming

Core-formed glass occurs already in the mid 2nd millennium and was one of the most common glass forming methods until the invention of glass-blowing. The principal of core- and rod-forming incorporates the manipulation of hot glass around a core, which is removed after the glass has annealed*. The description of the process of core-forming that follows is largely based on experimental studies carried out by Taylor and Hill (2005)[121] and Gudenrath (1991). To begin with the core, experimental studies focused primarily on the requirements a core demands to allow the formation of a glass body in hot state. First, the shrinking and expansion of glass during annealing must be taken into consideration. Therefore the core would have needed to be strong enough to withstand the glassworking stage but weak enough to crush during the shrinking process of the glass while cooling. Second, the core must be easily removable after annealing* (Taylor and Hill 2005).

Regarding the ingredients used to form the core, Grose (1989: 31) cites clay, mud, sand and an organic binder as ingredients on basis of experiments. The experiments conducted by Taylor and Hill (2005), however, yielded the best results with cores made of clay, vegetal remains, and sand: vegetal remains were necessary to bind the mixture and reduce the shrinkage while drying. Gudenrath (1991: 214) and Giberson (2004) suggested the use of horse dung and clay as a basis. Traces of the former core, in the form of small grains, have been preserved on the insides of the core-formed vessels; sometimes even holes that are left by vegetal remains occur. A large-scale examination of the core-remains of Mesopotamian core-formed vessels would be rewarding regarding the composition of the core itself, as well as the vessel content.[122] The core material was mixed with water to form a mass, formed into a particular vessel shape and attached to the end of a rod, probably made of metal. The core was then either fired or air-dried, and probably built up in stages by applying thin layers that were dried before the next layer was added (Taylor and Hill 2005).

The glass was applied onto the core using different techniques, such as coiling*, dipping*, and the addition of crushed glass pieces. Coiling involves the trailing of hot glass around the core. This could be achieved through viscous glass canes coiled around the core (Cummings 1997: 29), or hot glass directly applied from the gathering rod* (Figure 4.48) (Grose 1989: 31; Stern

[119] For a list of important catalogues of core-formed vessels, see Grose 1989: 109 footnote 1.

[120] Chemical analysis of core-formed vessels from Italy largely confirmed these typological groups; see Panighello et al. 2012.

[121] See also http://www.romanglassmakers.co.uk/nl7text.htm (accessed: 25.04.2016).

[122] Bimson and Werner (1969) analysed 62 samples of 14th-century Egyptian vessels. They proved that the cores consisted of organic components, fine silt or clay, limestone, and clay.

and Schlick-Nolte 1994: 40; Taylor and Hill 2005). In a successive step, the surface was flattened and merged by rolling it on a marver* or by using a hand-held flat tool (for the illustration of marvering see Figure 4.37).

The technique of dipping incorporates the coating of the core by dipping it into melted hot glass (Gudenrath 1991: 214). By turning the rod, and therefore the vessel, the viscous glass was evenly spread on the core, creating a uniform glass thickness.

As a third technique, the core could have been also covered by applying crushed glass pieces in cold state onto the core, that would then be melted to form a glass skin by heating (Cummings 1997: 29; Giberson 2004). Schuler (1962) proposed that the core could be dipped into a suspension of water and powdered glass.

The vessels were decorated by winding glass rods of contrasting colour(s) onto the viscous base glass. This was carried out in direct heat to control the thickness of the threads (Figure 4.48). By dragging a pointed instrument up and down across the threads, the characteristic 'feather patterns' were achieved (Figure 4.49). The trails were thereafter pressed into the surface of the vessel by marvering* (Taylor and Hill 2005; Stern and Schlick-Nolte 1994: 40). Some of the core-formed vessels discussed in this study show a fluted and irregular surface, indicating that the process of marvering was not carried out carefully.

Figure 4.48: Winding hot glass threads around the glass core (Tait 1991: 215, fig. 7).

Rim, shoulder, handle, and base needed to be formed in viscous state by the use of pincers (Figure 4.50, 4.51). The handles were made by adding a pre-cut rod to the vessel, subsequently softened and looped, or by applying a gather of glass to the vessel, which was looped using pincers (Gudenrath 1991: 215; Taylor and Hill 2005). Both techniques resulted in the so-called 'duck-head' handles, the characteristic handle shape of core-formed vessels. Regarding the smaller knob handles, the glass was most likely drawn out of the vessel (Grose 1989: 31).

Figure 4.49: Making the 'feather' decoration by dragging a pointed instrument (Tait 1991: 215, fig. 8).

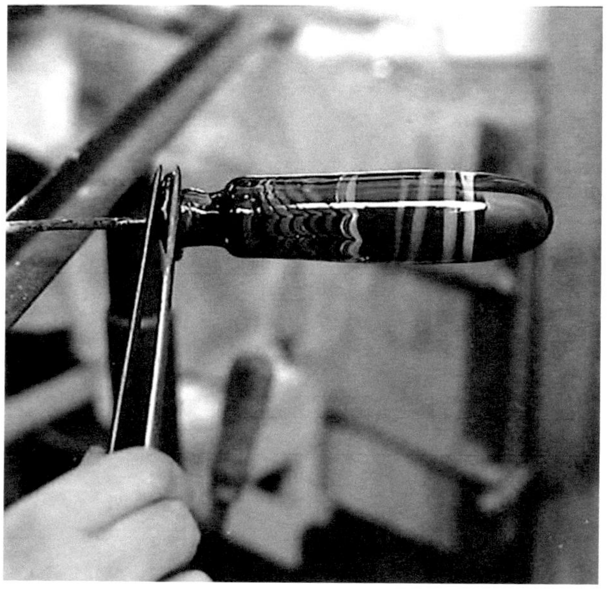

Figure 4.50: Forming the neck using pincers (Tait 1991: 215, fig. 10).

The vessel was annealed while still on its rod and twisted free in cold state. The core was removed by scratching, leaving a characteristic sandy and rough surface on the inside. Different-sized holes can be repeatedly observed on the surface of core-formed vessels, which result from gases released from the core.

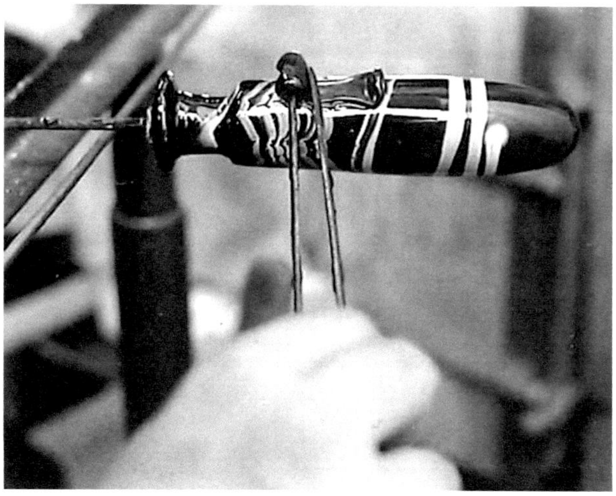

Figure 4.51: Forming 'duck-head' handles using pincers (Tait 1991: 215, fig. 13).

The proposal of Stern and Schlick-Nolte (1994: 31, 40) that the different methods of covering a core were used at different chronological periods cannot be proven, according to the present author. Too few core-formed vessels of different periods have been examined for such a statement. Furthermore, recognition of the different manufacturing techniques of the core-shaped forms is hardly feasible in any event with the naked eye. It should also be generally noted that the use of one technique does not automatically lead to the absence of another. Rather, techniques may have been used simultaneously, for example by different workshops or craftsmen.

4.3.2.2 Rod-forming

The technique of rod-forming differs only little from core-forming*, but was implemented for long and narrow containers, such as kohl tubes, and also for beads, head pendants, brooch runners and bracelets. Regarding the manufacturing of rod-formed vessels, a long, slim core was formed around a rod and the glass was manipulated in hot state around it. The rod was removed from the core before annealing (Grose 1989: 31). Kohl tubes occur frequently only after the 6th century and are therefore omitted from this monograph.[123]

Head pendants were made by winding hot glass around a rod on which features of the faces were created by adding threads or blobs of different coloured glass. Most of the heads have rings on their tops for suspension (Grose 1989: 82). Some of the pieces differ in terms of morphology and therefore in the process of manufacturing technique employed. Here the head was also formed on a rod, but the details of the face were partly produced by the use of moulds* (Grose 1989: 83; Gudenrath 1991: 216) (Chapter 4.3.5).

4.3.3 Core-formed vessels

4.3.3.1 Description of core-formed vessels

All core-formed vessels have closed vessel shapes and are divided into the two groups, 'bottles' and 'jars'. The group of bottles comprise vessels with a narrow neck opening which is often elongated. 'Jars' have on average a lower height than bottles, but still possess a broad vessel body and thus appear more compact. The neck opening of the jars is wider than that of the 'bottles', whereby the neck diameter is still smaller than the maximum extension of the vessel body. The term 'alabastron' is avoided due to terminological difficulties in this study (Chapter 4.2.2.3).

The typology presented here corresponds in major parts with the one suggested by Barag (1970: 174–180). However, the main difference lies in the fact that the groups are less strongly divided, but are based on the more general shapes of the vessels. Altogether seven different types of core-formed vessels are distinguished: 'ovoid bottles with pointed base', 'ovoid bottles with rounded base', 'piriform bottles', 'globular bottles', 'bottles with disc-base', 'large cylindrical bottles', and 'small jars'. The main characteristics of this classification is the superordinate vessel shape – which, as mentioned above, generally distinguishes between 'bottle' and 'jar' – and the shape of the base. The group of 'large cylindrical bottles' is in part defined by the size of the individual pieces, which are significantly above the average of the other core-formed vessels. Each typological group consists of at least of two specimens. Objects that cannot be assigned to one specific type are listed as individual pieces. All groups are based on objects with a secure provenance, apart from the group 'bottles with disc-base'.[124] Fragments of vessels were also included in this monograph to show as extensively as possible the presence of this vessel type at different locations.

Ovoid bottles with pointed base

The group of ovoid bottles corresponds to group 3 of Barag (1970), and includes Ur3, As9, Bab4, AM40, AM41, AM42, AM43 and AM50. Vessels of this type have an ovoid body shape with a pointed base. The neck is either straight or slightly flaring, and there are duck-head handles on the

[123] This was also indicated by Grose (1989: 79). It cannot be ruled out that individual objects may date from older times. However, according to the current state of research this cannot be finally assessed. Kohl tubes in form of a palm column are a common shape among Egyptian glass vessels dating to the 14th and 13th centuries, see for example Grose 1989: 62, no. 10. They occur again in large number in Iran in the middle of the 1st millennium, and contain no information on their origin and date; see Schlick-Nolte 1968: 39.

[124] Groups 6 and 14, established by Barag (1970: 174–180), were solely based on unprovenanced vessels.

shoulders (the handles of AM41 and Bab4 are broken off). The base colour of this bottle type is dark green to brown; the threads are white and yellow. The patterns consist of wavy lines (Bab4, AM42, AM50) and feather patterns (Ur3, AM40, AM43). The pattern on As9 is made of a combination of horizontal and wavy bands of at least two colours, the bands on the neck are drawn to feathers. AM40, AM42 and AM50 show the greatest correlation. AM43 has a long and wide neck, and AS9 stands out for its size.

AM50 features traces of dark blue colour. Thin threads of yellow and white glass are combed to a wavy pattern. The handles are made of dark blue glass with a yellow thread; rim and tip are yellow. The small damage on the surface in the upper part occurred most likely during the manufacturing process while the surface was still viscous.

AM40 has a dark brown, greenish base colour which was formerly dark blue. The threads of the irregular wavy pattern are white, the rim, tip, and handles are yellow. There is one short yellow thread covering the neck and upper body, which was most likely added to the surface accidentally. The two larger holes on the surface are due to gases set free from the core.

Bab4 was found with the transparent glass bowl Bab7 in grave 119, which is located in the area of the Merkes in Babylon. The burial dates from the late 7th/early 6th century (Chapter 3.2.1).

As9 was found in tomb 691 in Aššur, together with six pottery bottles and a beaker, two small glazed bottles, a terracotta head, a copper bowl with traces of food sacrifices, a handle, a bone needle, two silver rings, and an agate bead set in gold. The finds indicate a Neo-Assyrian date (Chapter 3.1.1).

The find context of Ur3 is not known.

Pointed bottles without handles are well attested among Neo-Assyrian pottery vessels of the 8th and 7th centuries (Hausleiter 2010: pl. 93, particularly Fl 3.1. Fl 5.3), and also occur among Neo-Babylonian funerary contexts (Reuther 1968: 223, no. 125, 224, no. 128, pl. 73) (Figure 4.52). Close parallels can furthermore be drawn to glazed ceramic vessels, in particular from northern Syria and the Aegean (Peltenburg 1969: 85, fig. 3). A well comparable bottle made of faience* comes from Sultantepe, found in the same building structure as glass vessel Su1 (Lloyd and Gökçe 1953: pl. VIIIdI) (Figure 4.46).

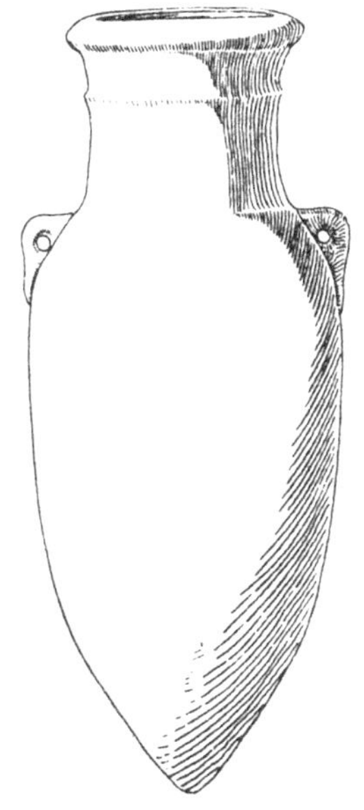

Figure 4.52: Pottery vessel from Babylon, ht. 30.6 cm (Reuther 1968: pl. 73, 125).

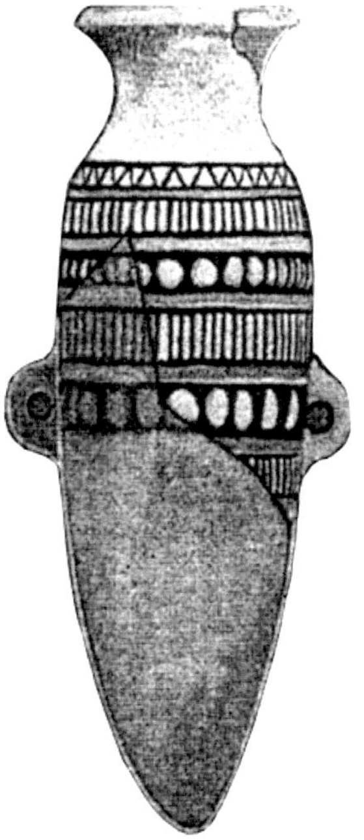

Figure 4.53: Faience vessel from Sultantepe, ht. 8 cm (Lloyd and Gökçe 1953: pl. VIIIdI).

Ovoid bottles with rounded base

Characteristic for Kiš2, Ur5 and AM46 are their ovoid bodies and a rounded vessel base. Kiš2 and AM46 have a rather elongated body, while that of Ur5 is rounder.

Kiš2 has an ovoid elongated body (ht. 9.2 cm). The neck is narrow, the base is rounded and slightly tapering, duck-head handles are placed on the upper part of the vessel.[125] The basic colour of the bottle is strongly corroded to beige-brown. One thread is white and a second one was formerly yellow. On the neck and close

[125] The piece was attributed by Barag (1970: 176) to group 4.

to the base, the threads are wound horizontally and on the middle part of the body, the threads are combed to form a festoon pattern. Kiš2 was found in a grave that cannot be precisely dated. The overall rounded shape of the body and the rounded base are similar to AM9.

AM46 has an elongated, straight body and a rounded base (8.6 cm). The long and narrow neck, as well as the shoulder section, is slightly miss-shaped. The colour of the vessel is completely corroded to beige, but part of the dark blue base colour is preserved. The ribbons on the body are combed into a festoon pattern. Deep, irregular vertical incisions remain and which are not marvered*.

Ur5 has an ovoid body, a rounded base and duck-head handles on the upper part of the vessel (ht. 8.7 cm).[126] The neck is not preserved. The basic colour of Ur5 is described as dark green and the thread decoration as white (Barag 1985: 157–158). The threads are wound horizontally around the neck and combed from the bottom towards the neck, creating a festoon pattern.

Ur5 from Ur was found in a tomb, dated to the time after the late 8th century and prior to the Achaemenid period (Chapter 3.2.7). The burial further contained an iron kohl stick and a silver disc, both of which are unpublished. Since no other vessel is reported to be found in this burial, it is possible that the kohl stick belonged to the glass vessel Ur5, which then could have been used to remove a liquid or balm (Chapter 4.3.3.1).

Kiš2 was found in a burial context (grave 54) in Kiš, dated by Gibson (1972: 70) from the Neo-Assyrian and the post-Achaemenid period, and by Moorey (1978: 52) to the 5th century.

Grose (1989: 85) attributed AM46 – under the label 'Mesopotamian core-formed vessel' – to the late 8th or 7th century.

Unlike the ovoid bottles with pointed bottom, the ovoid bottles with rounded base are very similar to the group of small alabastra made of the cast-and-cut technique, and their well comparable specimens, made of stone, for example from Aššur (Onasch 2010: 14, 227-21) (Chapter 3.1.1). Also the cast-and-cut alabastra show an overall similar vessel shape and a rounded bottom, as for example AM10 (Chapter 4.2.2.3).

Another parallel to the shape can be found in the bronze bottle discovered on Tall Siran in the modern city area of Amman in Jordan (Figure 4.54). The bottle is dated

Figure 4.54: Bronze bottle from Tell Siran in Jordan (ht. 10.5 cm) (Amiet and Mittmann 1987: 149).

to around 600, because of its Ammonite inscription mentioning king 'Amminadab'.[127]

Piriform bottles

The group of piriform bottles corresponds to Barag's second group, and incorporates Ur4, Bus1, Su1, Sus1, Sus2, Sus3, Sus4, Kiš1, Nip1 and AM44. Characteristic for this type is the piriform body-shape with wide shoulders and a wall that tapers sharply towards the base, and a pointed base. Another characteristic of this bottle type is the wide neck. Some examples show duck-head handles, some do occur without (Kiš1, Nip1). The heights vary from 10 to 15 cm.

AM44 is corroded to an amber to green colour, and was formerly most likely of blue colour. The colours of

[126] The bottle was attributed as single piece to group 7; see Barag 1985: 177.

[127] The bottle was closed by a lid when found and contained charred wheat and barley grains, and is interpreted as a gift; for literature on the Siran bottle, see the extensive bibliography in Amiet and Mittmann 1987: 149.

the threads are white and dark green. The handles are made of the same brown green colour as the base glass and are additionally decorated with four thin vertical white threads. The surface of the piece is fluted.

The pieces from Susa (Sus1, Sus2, Sus3, Sus4) have a greenish grey corroded surface and a yellow-white thread decoration. The basic colour of the bottles used to be blue, as traces on Sus1 show. Sus1 is decorated with a pointed feather ornament, which also contained a thick white thread. Sus3 and Sus4 show wavy lines on the neck and upward directed festoons on the body. Sus2 has white threads wrapped horizontally around the neck and shoulder. The surfaces of Sus1 and Sus2 were strongly ribbed.

Ur4 is made of blue glass and the threads have a yellow-white colour which is largely corroded. The threads are wound horizontally around the neck and are pulled to feather patterns on the vessel body.

Su1 consists of opaque* blue glass. The white threads are wound horizontally around a ribbed neck. A feather pattern can be seen on the fluted body of the vessel.

Bus1 is dark greenish-brownish corroded and was probably originally dark blue. The white threads are drawn into a horizontal, slightly wavy pattern. Bus1 is larger than the other vessels of this group, but its overall shape can be assigned to the piriform bottles.

The exact find contexts of Sus1 and Sus2 are not known; it is likely that the objects come from a cemetery on the acropolis of Susa, with Neo-Babylonian as well as Late Elamite tombs. An assignment to the Neo-Elamite period, and a date after the reign of Ashurbanipal II, is likely (Chapter 3.4.8). The vessels from Susa (Sus1, Sus2, Sus3 and Sus4), together with AM44, form a tight typological group that is characterised by knob-shaped bases and long necks. It is remarkable that the festoon patterns only appear on the specimen from Susa, indicating that this pattern is specific for core-formed glass vessels from that region. Typologically well-comparable faience* vessels come from grave contexts in Susa, dating to the 7th century (Mecquenem 1922: 131–132) (Figure 4.55). The faience bottles strongly resemble the glass vessels, as they show the same pointed, almost knob-shaped form as Sus1, Sus2, Sus3, Sus4 and AM44, and are also similar in size. This correspondence is another indication that the group of bottles from Susa is very likely a regional glass vessel type. Su1, Ur4 and Bus1 were all found in domestic contexts (Chapters 3.2.7, 3.3.5, 3.4.8).

Su1 draws close parallels, both with regard to form and design of the thread decoration to vessel Ur4. Typological comparisons for Su1 and Ur4, the vessel type with handles, can be found among glazed pottery vessels, for example from Babylon (Figure 4.56). With regard to Kiš1 and Nip1, the vessels without handles, parallels can also be found among glazed pottery vessel from Kameiros on Rhodes, a glazed pottery type that was common in the northern Levant during this period (Peltenburg 1969: 73) (Figure 4.57).

Figure 4.55 Faience vessel from Susa, ht.13.8 cm (Amiet 1966: 503).

Globular bottles

Bab6 and Bab7 from Babylon have almost identical shapes, decorations and dimensions (9.2 cm) and belong to the group of globular bottles.[128] To be added to this vessel type is a group of sherds TJ1 and TJ2 which can be joined together to form a globular vessel. The bodies are spherical, but show irregularities in the walls.

Bab6 and Bab7 have long necks that are slightly narrower at the rim. The rims are thickened and have yellow bands on their top surface. On the shoulders are duck-head handles which are made in the same blue colour as the base glass and have yellow threads. The base of Bab7 is rounded. The basic colour of Bab7 is dark blue and that of Bab6 light blue corroded. Thin white threads are wound around the vessel and combed at the neck to form feather patterns. On the body of Bab7 there are horizontal lines and decorations combed from the shoulder to the bottom of the vessel, alternating with thick yellow bands.

[128] Bab5 was probably wrongly assigned by Barag (1970: 178) to the group of vessels with tubular shape.

Figure 4.56: Glazed Pottery vessel from Babylon, ht. 15 cm (Reuther 1968: pl. 76, 133).

Figure 4.57: Glazed pottery vessel from Kameiros (Rhodes), ht. 23 cm (Peltenburg 1969: pl. XXI. A).

The neck and body of Bab7 are strongly fluted. Bab6 has a wavy pattern; the yellow stripes are thicker than the white stripes, but not as thick as those of Bab7. In the production of Bab6 and Bab7 it can be clearly observed that the yellow stripe was added last.

Bab7 was found together with Bab4 in a burial of the so-called 'Hockersarkophag'. The opening of Bab7 was covered with a piece of linen tied around the vessel and crossed at the bottom of the vessel (Reuther 1968: 221). This could indicate that the contents of this vessel were probably a fragrant substance that could have been released by the cloth. If the substance were to remain in the vessel, the use of a stopper would have been more likely.

Bab6 was found in the area of the Merkes in Babylon and can therefore be attributed to either a domestic or funerary context.

Because of their close similarity in shape, size and decoration, that differs significantly from any of the other core-formed glass vessels, it is very likely, that Bab6 and Bab7 were manufactured at the same site, probably at the same workshop, most likely in Babylon, as both specimens were found there. It is also striking that both vessels obviously come from similar types of contexts. Neither of the two pieces was found in a prestigious environment, for example a palace or temple, but it seems that this type of vessel was used in graves or environments of the living around the Merkes. As a result, globular vessels appear to be accessible to a broad group of the population of Babylon.

Comparable glazed pottery vessels with globular bodies were found in a 7th-century burial in Aššur (Hausleiter 2010: 36, 40) (Figure 4.58), as well as in burials from Babylon, such as 127 (Reuther 1968: pl. 76, no. 127 f), 138 (Reuther 1968: pl. 76, no. 138a), and 125 (Reuther 1968: pl. 73, no. 125c). It is interesting that the glazed globular vessels were found in the same type of burial ('Hockersarkophag') as Bab7. Because of the strong typological coincidence of the glass and pottery bottles from Babylon, it could be possible that Bab6 and Bab7 represent a typical Mesopotamian core-formed vessel shape of Babylon.

TJ1 and TJ2 were found in different contexts at Tell Jemmeh. TJ1 comes from a room of a larger dwelling context, dating to the 3rd intermediate period (1070–664) (Chapter 3.36). Because of the similar decoration of TJ1 and TJ2, it is likely that both vessels date to the same period.

Figure 4.58: Glazed pottery vessel with encircled dots, found in a grave in Aššur, ht. 6.6 cm (Andrae 1925: fig. 17).

Small jars

Bab5, Ur6, AM51, AM55 belong to the group of small jars. The vessels show variations, but their spherical overall shape and wide neck unite them into a broad group.

Bab5 has a spherical body, a disc-shaped base and a short, broad neck with a sharp edge on its shoulder. The vessel is 6.7 cm high. There are two duck-head handles that sit on the shoulder and are made of white glass with a blue thread above. Bab5 is of particular interest because the design is very different from all other core-formed vessels. The basic colour is monochrome; red dots with blue circles, called 'eyes' (Barag 1970: 160), decorate the neck and body. The rim has a blue band.

Ur6 is globular with a slightly flaring neck and rounded rim (ht. 8 cm). The neck is broad, similar to AM51; the basic colour of the vessel is corroded to dark green and was probably originally dark blue. The irregular zigzag decoration consists of white threads. As the vessel has no base and no handles, it may have been placed on a stand.

AM55 is 6 cm high, has a spherical body and a wide neck with a thickened rim and a disc-base. The thread decoration alternates between horizontal lines and a feather pattern.

AM51 is a miniature vessel only 4.4 cm high. The vessel has a rounded body and a short neck that tapers towards the rim. The margin is large compared to the total proportions of the vessel. The piece also has a flat base and two duck-head handles on the shoulder. The decoration is very similar to the other pieces, with simple horizontal threads on the neck and bottom and a wide feather pattern on the body. The surface of the body is grooved. The basic colour is greyish greenish and was probably originally dark blue.

Ur6 is assigned to a 'post-Kassite' dwelling context, the exact find spot of AM55, in contrast, remains unknown (Chapter 3.2.7).

Bab5 was found in a burial of an infant, dated to the Neo-Babylonian period or later.

The group of small jars does not form a particularly tight typological group, as is the case with other groups dealt with in this chapter above. However, its wide opening and the wide, spherical body are particularly characteristic, which could also indicate a similar function. In contrast to the closed vessels, the wide mouth could facilitate easier access to the contents of the vessel. This could mean that rather a semi-liquid or powdery substance was kept in them.

Comparable jars and bottles with spherical bodies and wide necks are be found among glazed Neo-Assyrian pottery vessels, for example, in graves in Aššur (Andrae 1925: fig. 14; Hausleiter 2010: pl. 45i) (Figure 4.59), which date to the 7th century. In ceramics, handles mostly occur on vessels with rounded bottoms.[129] The glazed vessels are also similar in size. Also among the grave inventories at Babylon (Reuther 1968: pl. 74, 135, 139) and Uruk (Boehmer *et al.* 1995: pl. 101e. 132b, 145a), glazed jars of the same type occur.[130] Another parallel with regard to Bab5 can be drawn to small faience* vessels found at Sultantepe (Lloyd and Gökçe 1953: pl. VIIb, c), Tille Höyük (Blaylock 1999: fig. 11.4) and Tell Halaf (Sievertsen 2012: fig. 154, no. 2).

Figure 4.59 Glazed Neo-Assyrian pottery vessel from Aššur (left: Andrae 1925: fig. 14, ht. 5.7 cm).

[129] For examples with handles, see Andrae 1925: fig. 12, 14, 23, and for examples without handles but bases, see Andrae 1925: fig. 11, 13 and Hausleiter 2010: pl. 35, 40. For the chronological attribution, see Hausleiter 2010.

[130] The type of jar that occurs regularly in graves in Uruk is identified as 'Typen B 4/4a'; see Boehmer et al. 1995: 42.

The dot-decoration on Bab5 is unusual among glass vessels of the early 1st millennium but it is found on vessels of the Late Bronze Age, which, however, show different vessel shapes. Therefore, Bab5 has characteristic features of both the 1st millennium (handle and shape) and the 2nd-millennium (dot-decoration) core-formed vessels. There is a similar glazed ceramic bottle from Aššur, dating to the 7th century, with similar decorations of encircled dots, although its shape more resembles spherical bottles (Figure 4.58). The dot-decoration could probably also be interpreted as 'eye decoration'. The eye motif often appears among objects in burials, mostly as part of eye-pendants or jewellery (e.g. Haller 1954: pl. 14, 28).

Bottles with disc-base

The characteristic feature of this group (Nip2, AM52, AM53, AM54) is the disc-base. In addition, the bottles have an ovoid body, a short narrow neck, as well as duck-head handles (except for Nip2). The colour of the base glass of Nip2, AM52 and AM53 is dark blue, and that of AM54 light blue. Nip2, AM52 and AM53 are decorated with festoon patterns, threads are made of white and yellow glass (Barag 1970: 158 170). AM54 is monochrome with horizontal bands on the rim and bottom. All vessels are well preserved, showing only slight traces of weathering* and flaking. AM52 and AM53 are particularly well preserved, and even show shiny surfaces.[131]

Nip2, AM52, AM53[132] and AM54 have very similar characteristics and correspond to Barag's Group 6 (Barag 1970: 177). None of the vessels can be assigned to a certain context, and three of the pieces were even acquired from the art market. The best typological comparisons can be drawn to vessels of the 'Mediterranean Group I' that date from the middle of the 6th to the early 5th century (Chapter 4.3.1).[133] This corresponds well with the suggestion made by Barag (1970: 177) to date them 'not before the Neobabylonian Period'.

Large cylindrical bottles

As10, Kam1, Kam2, Car1, AM45, AM47, AM48, as well as the fragments Nim137, Nim138 and Nip3, belong to the group of large cylindrical bottles.[134] Vessels of this type have a long cylindrical body that is slightly wider towards the base. The neck is long and straight and there are duck-head handles underneath the shoulder. The shoulder is very characteristic for this group of vessels since it is very sharp. The size of the vessels – averaging 15 cm in height – is remarkable, as it is far above the average size of the other core-formed vessels of this period.

Kam1, Kam2, Car1 and AM45 form a typologically tight group in terms of shape, size (15.6 cm; 15.6 cm; 15.5 cm; 17 cm), colour, decoration and traces of tool marks. Their basic colour is corroded to beige. On Kam1, Car1 and AM45 traces of the former dark blue basic colour are still visible. Car1 preserves traces of yellow and white threads, which were wrapped around the neck horizontally to form a wavy pattern; the pattern on the body was combed from the rim towards the bottom. The upper sides of the handles have a thin, white thread decoration, and above the base there is the same white thread but much wider in diameter.

As10, AM47 and AM48 are similar in shape, but differ slightly with regard to their decoration. AM47 has a narrow white feather pattern on the body. The dark blue basic colour of this piece is well preserved, as is the white of the threads. The feather decoration on AM48 is wide and consists of thick yellow threads. With As10 the pattern cannot be identified, the surface of the body shows, however, deep fluting.

Nim137, Nim138 and Nip3 are fragments which, due to their shape and size, most likely formerly belonged to large cylindrical bottles. Nim138 can be identified as shoulder fragment and Nip3 as a base fragment. Barag (1970: 155, 156, 158) and Turner (1955: 59), who studied the pieces in detail, described Nim137 as a fragment that derived from a cylindrical body. Nip3 has close parallels to the decoration of AM48 and traces of the feather decoration are still visible on Nim138.

The group of large cylindrical bottles is very distinct, and examples of this type were found both in the Aegean and Mesopotamia, which could point to some connection between these two regions. Due to the early dating of the fragments from Nimrud and Nippur, Barag (1970: 194–195) suggested that the group of large cylindrical bottles must have originated in Mesopotamia and not in the Aegean. He therefore concluded that Mesopotamian glassworkers founded a glass industry on Rhodes in the first half of the 7th century where this specific kind of core-formed vessels could have been produced (Barag 1970: 194–195). A closer look at the pieces shows that Kam1, Kam2, Car1 and AM45 (as well as AM48) can be grouped even more closely together. Apart from their shape and size, all these vessels have short vertical, irregular incisions (0.2 cm) on their necks (Figure 4.61). It is probable that these impressions were caused by combing the narrow feather decoration. The end of the carving tool would accidentally come into contact with the neck when pulling the glass threads

[131] Since the parts do not come from regular excavations, it may be possible that the surfaces have been reworked by polishing and grinding in modern times.

[132] It is recorded that AM53 was confiscated in Nasiriyah; see Barag 1970: 170, 14.

[133] For examples of this type, see Harden 1981: pl. X, no. 141, 143, 146; for monochrome pieces, see no. 140 and 149.

[134] Barag 1970: 177 included them in his group 8, in addition to further pieces from the art market.

(Taylor and Hill, pers. comm.). If one assumes these tool marks as accidents – this is likely because they cannot be observed among other core-formed vessels of this type – then it is plausible that these vessels were made in the same workshop, probably by the same craftsman. As these specific tool marks are absent on bottle As10, which was found far away from Rhodes in the city of Aššur, it is likely that this bottle was made by another glassworker, or even in another workshop. This could indicate the existence of local secondary glass production in the Aegean, most likely on Rhodes, rather than the import of Mesopotamian vessels. This is particularly evident against the background of the fact that Rhodes seems to have been hosting a distinguished glazing industry in the 7th century, with vessels traded in various regions. These include glazed pottery vessels that were found at various sites in the Aegean, in Asia Minor, northern Syria and Mesopotamia (Barag 1970: 195; Peltenburg 1969: 78) (Figure 4.58). It is impossible to outline the wide-reaching trade relations during this period here, but the examples mentioned above show that the spread of a particular type of core-formed glass vessel could well be imagined.

As10 was found in a burial dating to the late 8th or early 7th century (Chapter 3.1.1). The fragments Nim137 and Nim138 come from the Burnt Palace in Nimrud and date between 722 and 612 (Chapter 3.1.4.3).

The characteristic feature of the large cylindrical bottles is their large size, ranging between 15.6 cm and 17 cm, sharp edges above the bottom, and sharp shoulders. These features are also characteristic for storage jars made of ceramic or stone, found at sites such as Beth-Shean (Mazar 2006: 457, pl. 37, no. 4), Hazor (Ben-Tor *et al.* 2012: fig. 4,13; Yadin 1961: pl. CCXXIX) (Figure 4.60), Lachish (Tufnell 1953: pl. 95, no. 498), or Sarepta (Pritchard 1975: fig. 23, 20).Ceramic storage jars are, of course, much larger than the large cylindrical glass vessels and range from 45 cm to 100 cm. This vessel shape obviously indicates the largest vessel size in both glass and ceramics. Even if the material to be stored certainly differed in glass and clay, the function of this vessel type as a storage vessel and container for refilling could have been the equivalent.

Vessel fragments

Fragments of core-formed vessels that cannot be assigned to a particular type are incorporated in this chapter. It has to be pointed out that the pieces presented here are probably only a small part of the existing fragments of core-formed vessels. However, to ensure a somewhat balanced picture of the distribution of the core-formed vessel type in general, the fragments that were available to the author personally or in publication – with particular attention to Mesopotamian sites – are included here. Furthermore, the fragments help to obtain information on the function of core-formed vessels in general. The fragments therefore contribute significantly to this subject.

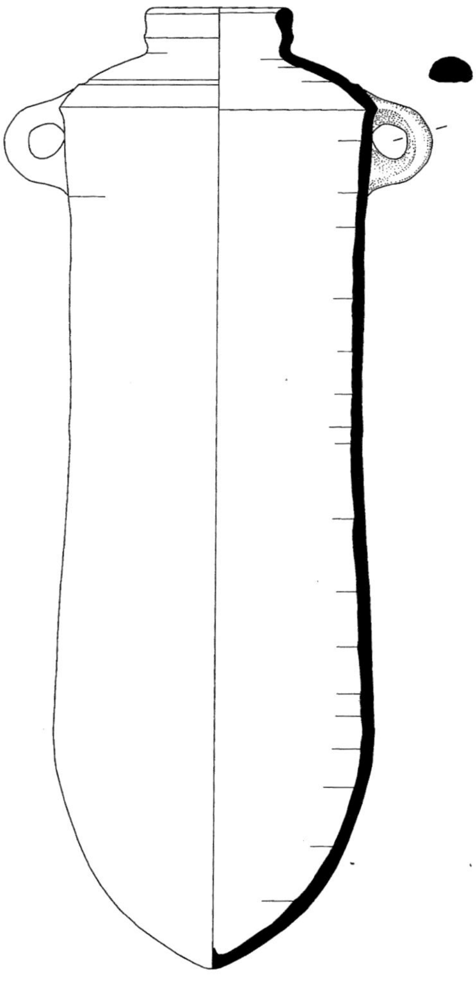

Figure 4.60: Ceramic storage jar from Hazor, stratum V, ht. c. 65c m (8th century) (Ben-Tor et al. 2012: 4.13, 6).

Ziy1 consists of 37 fragments, one of which represents the edge fragment and the other shows a rudimentary handle; no base fragment is preserved. The former base colour was most likely dark blue and occurs today as greenish blue; the thread decoration is made of white and yellow glass. The threads are wound around the entire body and are combed to a feather pattern on the body of the vessel. The rim is *c.* 3 cm wide and slightly flaring, with the body sherds indicating that the vessel had a rather elongated or slightly ovoid shape; the handle was formerly made of yellow glass.

Urk1 is the fragment of a shoulder, which cannot be further specified.

Urk2 represents a base, probably a ring or disc-base. It is impossible to provide any further information on the basis of the available photograph. The bottom may have belonged to a cast-and-cut vessel, similar to AM4, or a core-formed vessel, as for instance the 'bottles with disc-base'.

Urk3 is described as a small make-up vessel with pierced rim (van Ess and Pedde 1992: 160, no. 1190). On the basis of the published photograph, however, this cannot be confirmed.

Ur7 is the fragment of a body sherd with feather decoration, which cannot be further specified.

Bab8 is the fragment of an elongated neck of a formerly translucent blue glass vessel.

BS2 is a group of six body fragments of a core-formed vessel. The base colour was formerly most likely dark blue and is corroded to light blue. A feather decoration, made of yellow and white threads can be identified. Since the sherd has no signs of curvature, it is likely that the body of the vessels was elongated or ovoid.

Meg2 is a small body sherd with feather decoration.

There is only one known find spot, that of Ziy1 from the excavations of the 'Tušḫan Archaeological Project' at Ziyaret Tepe (Chapter 3.1.8). Ziy1 was found in a burial dated to the Neo-Assyrian period. According to the context and finds a date in the 8th or 7th century is likely.

Urk1 was found in a looted pit and has to be considered as Neo-Assyrian, Neo-Babylonian or Achaemenid in date.

Urk2 and Urk3 were found in the double-pot burial no. 129 (W17961) that dates to the Neo-Babylonian period (Chapter 3.2.8).

BS2 was found – together with ingot BS1 – in room 1028 of the South Temple, Level V, indicating an Iron Age IB/IIA date (1150–925) (Chapter 3.3.4). The context was, however, disturbed by pits, which makes a later date possible for BS2.

Meg2 was found in a dwelling context that can be attributed to the time period from 1150 to 925.

4.3.3.2 Discussion

Typological characteristics

Core-formed vessels dating to the early 1st millennium can be divided into six typological groups: 'ovoid bottles with pointed base', 'ovoid bottles with rounded base', 'piriform bottles', 'globular bottles', 'small jars', 'bottles with disc-base' and 'large cylindrical bottles'. From the attributes of these typological groups, superordinate features in form, decoration and colouring can be identified which are characteristic of vessels of the first half of the 1st millennium and distinguish them from their Late Bronze Age predecessors, as well as their successors of the 'Mediterranean groups'.[135]

The bodies of the core-formed vessels of the first half of the 1st millennium tend to have a rounded or oval shape compared to the vessels of the later 'Mediterranean groups', which mainly comprise 'alabastra', 'oinochoai', 'aryballoi', and 'amphoriskoi' (Grose 1989: 126–127). In comparison to the Late Bronze Age core-formed vessels that comprise various forms of cups (Barag 1979: fig. 14B, fig. 18; Barag 1985: fig. 1, 4; Marcus 1991: fig. 3a–d, 13) and bottles with long or short necks and strongly varying body shapes (Barag 1970: fig. 20, 23), the variety of shapes of core-formed vessels of the early 1st millennium is far less pronounced. Furthermore, the vessels of this period are much smaller, with a maximum height of 17 cm, than in the Late Bronze Age period,[136] which is particularly evident in the glass vessels from Aššur that have a minimum height of 13 cm (Barag 1970: 143–145; for the forms, 175–180).

Early 1st-millennium core-formed vessels have a rounded or pointed base. Only the vessels of the group of 'bottles with disc-base' have an additional base in the form of a disc. Any other types of bases are rare.

Core-formed vessels of the early 1st millennium occur either without handles or with so-called 'duck-head' handles. The shape of a duck-head on the core-formed glass vessels results directly from its manufacturing technique, in which a hot glass thread is applied to the vessel (Chapter 4.3.2). Duck-head handles appear not only on core-formed vessels, but also on cast-and-cut vessels (e.g. Chapter 4.2.2.3). They take on different shapes due to the different manufacturing techniques, however, both forms have their respective versions of duck-heads in common. The duck-head handles on core-formed glass vessels are most probably a development of the early 1st millennium, since they do not appear on the Mesopotamian core-formed vessels of the Late Bronze Age, as far as this can be decided on the basis of published material available. From the Iron Age onwards, duck-head handles form an integral part of the core-formed vessels, since they were always present on vessels of the later 'Mediterranean groups' (Grose 1989: 400–401).

[135] Here the Mediterranean group refers to Mediterranean Group I (mid 6th – early 4th century), Mediterranean Group II (4th – early 3rd century), and Mediterranean Group III (3rd – 1st century), which were discussed in detail in Chapter 4.3.3.

[136] This is particularly evident in the glass vessels from Aššur that have a minimum height of 13 cm; see Barag 1970: 143–145.

Ovoid bottles, pointed	Ovoid Bottles, rounded	Piriform Bottles	Globuar Bottles	Small Jars	Bottles Disc-base	Large Cylindrical Bottles

Table 4.1: The different types of core-formed glass with corresponding examples (not to scale).

The most common original basic colour among most core-formed vessels of the early 1st millennium was dark blue. A maximum of two other colours, mostly yellow and white, were used for the thread decoration (Barag 1970: 174–175). Late Bronze Age core-formed vessels, in contrast, show a wider colour spectrum, usually using more than two colours for decoration.

The usual decoration of Iron Age core-formed vessels consists of horizontal bands and feather patterns, which alternate. Horizontal bands were usually spirally wrapped around the neck and base, and the threads were slightly combed on the body to create a wavy or feathered pattern. Common patterns on core-formed vessels are feather patterns and horizontal lines, wavy lines (e.g. Bab6), zigzag lines (e.g. Ur6), festoons or garlands (e.g. Sus3). Very rarely, dotted decorations occur (only on Bab5). The transitions from wavy lines, zigzag and meander to feather patterns are often blurred and cannot always be precisely determined.[137] The decoration of the core-formed vessels of the Late Bronze Age was more varied. There were more designs, which combined with each other, resulting in a range of different patterns (e.g. Barag 1970: figs. 20–23). On the later 'Mediterranean vessels' the threads were usually drawn into a feather decoration only on the central part of the vessel body. The feather decoration here often appears as a pointed zigzag line (e.g. Grose 1989: 96–108).

Often vertical depressions occur on the surface of early 1st-millennium core-formed vessels, creating a fluted body (Figure 4.61). Sometimes the threads of the decorations are also still visible on the surface, creating irregularities (Kam1) (Figure 4.61). Both, depressions and threads on the vessel surface, are due to insufficient smoothing, which could indicate a

Feather pattern	
Horizontal lines	
Wavy lines	
Zigzag lines	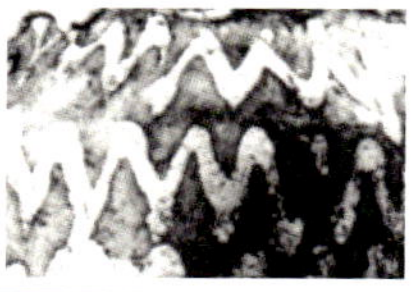
Festoons/ Garlands	
Dots	

Table 4.2: Various decorative elements on core-formed vessels of the early 1st millennium (not to scale).

[137] For a list of the designs, see also Barag 1970: 175.

Figure 4.61: Vertical depressions on the surface of a core-formed glass vessels creating a fluted body. The depressions are due to insufficient smoothing of the surface during the finishing process (courtesy of The Corning Museum of Glass).

certain degree of carelessness in the production of the vessels, or the inability to bring the glasses to a certain temperature for fire-polishing. In contrast to the early 1st millennium, both the vessels of the Late Bronze Age and those of the later 'Mediterranean group' were usually carefully smoothed. It is therefore also possible that a smooth surface at the beginning of the 1st millennium was not regarded as important and the fluting now somehow corresponded to the taste.

Nowadays, the basic colour of the Iron Age core-formed vessels is usually heavily corroded. Typical corrosion colours are dark green or dark brown. The poor conservation status of Mesopotamian glasses in general is due to the soil composition in this region. Humidity and high salt contents attack glass finds more strongly than is the case with Egyptian glasses (Bouquillon *et al.* 2008: 93) (Chapter 2.1.3). However, the well-preserved colours of Egyptian glasses and those of the later 'Mediterranean group' give a good impression of the original appearance of the Mesopotamian pieces. This must be taken into account when looking at core-formed vessels presented in this study. The state of preservation of the 'Mediterranean vessels' is usually better and the pieces show bright colours. Since many of them do not come from regular excavations, it could be possible that some treatment (placing of handles and edges, smoothing and polishing of surfaces) was carried out in modern times so as to increase prices on the art market. There is also a large number of fake vessels among the vessels of the 'Mesopotamian group', that are either partly or completely forged.

Chronological classification

Ovoid bottles with pointed base, as well as piriform bottles, have close parallels to glazed and unglazed ceramics from Assyria and Babylon from the 8th to the 7th century. Very close parallels occur among glazed pottery vessels from northern Syria and the Aegean of the same period (Figure 5.3). In the case of the glass bottles from Susa (Sus1, Sus2, Sus3, Sus4), a close connection to faience* vessels found in burial contexts at the same site (late 7th to 6th century) can be established, which could mean that both groups of vessels were produced locally. The find contexts of the glass vessels themselves, as well as their comparative pieces, allow a dating of these two types into the 6th century.

'Ovoid bottles with rounded bases' are not very typical, neither for the glazed pottery vessels, nor for the faience* vessels of that time. Comparisons for this vessel type are more likely to be found among the smaller cast-and-cut alabastra, e.g. AM10, and their counterparts made of stone. One rare comparison made of bronze comes from Jordan, which dates to around 600. Kiš2 and Ur5 cannot be attributed to a specific period. Therefore, the groups of 'ovoid bottles' and 'piriform bottles' roughly date to the time period between the 7th and the middle of the 6th century (Figure 5.3).

The 'globular bottle' shape is rare among early 1st-millennium vessels, and only Bab6 and Bab7 fall into this group. Globular bottles find parallels only among glazed pottery vessels from Aššur and Babylon. Bab6 and Bab7 both date to the late 7th or early 6th century (Figure 5.3).

Bab5 belongs to the group of small jars and dates to the Neo-Babylonian period or later. Close comparisons can be found among glazed pottery finds of the 7th century from Aššur (Andrae 1925: fig. 14; Hausleiter 2010: pl. 40d, 45i), Babylon (Reuther 1968: pl. 74, 135, 139) and Uruk (Boehmer *et al.* 1995: pl. 101e, 132b, 145a).

None of the 'bottles with disc-base' was found in a datable context. Because of their good comparison with core-formed vessels of the 'Mediterranean Group I' (mid 6th – early 5th century) it is likely that this group dates later, as do the other vessel types presented in this study (Figure 5.3).

The large cylindrical bottles form a particularly tight group, out of which three pieces were found in the Aegean (Kameiros, Carthage), and one in Mesopotamia (Aššur). All of the vessels, apart from the one from Aššur, have vertical, irregular tool marks on their necks, which could indicate that the pieces were made in the same workshop. The best comparisons for this type are among large storage jars (45–100 cm) made of ceramic, which were found at sites such as Beth-Shean (Mazar 2006: 457 pl. 37, no. 4), Hazor (Yadin 1961: pl. CCXXIX; Ben-Tor *et al.* 2012: Figure 4.60), Lachish (Tufnell 1953: pl. 95, no. 498), or Sarepta (Pritchard 1975: fig. 23, 20) and date to the 8th century (Figure 5.3).

Reflections on materials and function

Even if the overall appearance of the core-shaped vessels, i.e. the multi-coloured nature of the different glass threads drawn into different patterns, is directly related to the manufacturing technique of core-forming, the shapes of the core-formed vessels of the Iron Age were not material-oriented. In fact, core-formed glass vessels find parallels in various other materials, such as faience*, glazed ceramic, stone, and even unglazed ceramic and metal. However, particularly close parallels can be found in glazed ceramic vessels. Peltenburg (1987: 20–21) pointed to the close relationship between glazed ceramic vessels and glass vessels and their strong typological correlation: 'With regard to typology it is remarkable that the earliest glazed pottery vessels [he refers to Late Bronze Age objects] are primarily handleless flasks, identical with the most characteristic early glass vessel shape. (...) They occur together at the same time and on the same sites (...).' He even goes as far as to assume the same craftsmen for both the glazed pottery and glass vessels (Peltenburg 1987: 21). The material characteristic of glassy materials plays an important role when it comes to the contents the vessels might have kept: glass and glazed surfaces are particularly suitable for oily substances, such as oils, perfumes or ointments. Glassy surfaces form a coating on the surface which does not allow the often expensive oily substances to be absorbed by the material or to penetrate outside, in contrast to ceramic, for example, which has a porous surface. Therefore, the typological similarity of core-formed glass vessels to the glazed vessels may well be due to their function, which included the storage of oil-containing substances.

The function becomes even clearer when taking a closer look at the vessel shapes. Almost all of the core-formed glass vessels exhibit narrow necks, which must stand in close connection to their use. It is likely that the vessels contained a liquid that could have been carefully poured out through the narrow neck. Bab7 was covered with a piece of cloth, indicating that the content of this vessel was potentially a fragrant substance that could be released through the cloth. The contexts in which these vessels were used mainly comprise burials. It is interesting to note that typologically comparable glazed pottery vessels also occur primarily in funerary contexts. This function does not exclude the use of core-formed vessels in different contexts, but indicates a certain tendency towards their use as funerary gifts containing oily liquids and balms.

4.3.4 Tubes

4.3.4.1 Manufacturing process

All the tubes are irregular, but mostly rectangular in shape; only Has4 is octagonal. Because of this irregular shape, the process of casting can be entirely ruled out. It is likely that the tubes were made by rod-forming (Chapter 4.3.2). The large rod would leave a central, rounded hole through the entire object. The irregularly flaring holes, the rounded ends, and the bulges at the end of the tubes are a result of the rod being turned during the manufacturing process. This, as well as the overall irregular shape of the tubes, indicates that the core-forming technique, rather than casting, was applied.

4.3.4.2 Description

A total of 17 fragments of differently sized tubes were found at Hasanlu, out of which eight are included in this study. The tubes are made of both opaque* and translucent blue glass. These objects were only found at Hasanlu.

Has4 (15 x 4.3 cm) is one of the largest tubes included in this study. It has an octagonal shape with a central hole that is slightly off-centre and rounded. The colour is corroded and shows traces of opaque* blue, which has partly turned into green. The outer surface is polished and the inner surface is uneven and creased.

Has10 (7.2 x 11 *c.* 4 cm) is broken on one side; the other side is slightly narrowing towards the end. The tube has an irregular rectangular shape. Two opposite sides are slightly longer (4.2 cm) than the other sides (3.6 cm). The hole is off-centre. One side forms a rounded edge (dm. 1.5 cm), which is the original end of the tube; the other side is broken off (2.8 x 2.2 cm). The colour is corroded and the original opaque* light blue colour is only partly preserved. The outer surface is smoothed; the inside is smoothed towards the end, but in the middle it is irregular and creased.

Has11 represents the short end of a tube (4.1 x 3.3 cm), which is broken on one side. The overall shape is rectangular; the edges are rounded and the surface is smoothed. The hole (1.8 x 1.3 cm) is off-centre and irregularly ovoid in shape, forming a bulge at one end. Both the tube and hole widen slightly towards the end.

The colour is corroded, exhibiting traces of the original light blue opaque colour.

Has9 has an irregular rectangular shape, and narrows slightly towards the end of the tube (l. 5.6 cm, 3.8 x 3.1 cm). One side forms the rounded end with a bulge, the other side is broken off. The hole has an irregular oval shape (1.9 x 1.5 cm) and widens at the end of the tube (dm. 2.3 cm). The colour is corroded but shows partly the original translucent blue base colour. Has9 therefore differs from the other tubes with regard to its colour. There are bubbles present in the glass, which are sometimes heavily distorted and elongated.

Has5 is smaller in size than the other tubes and has a rectangular shape (2.3 x 2.3 cm). One side of the tube has rounded ends and the other side is broken off. The hole is irregularly round (dm. 1 cm). The surface of Has5 is smoothed. The colour is heavily corroded, showing no traces of the original colour. There is a great number of rounded bubbles of different sizes present in the glass.

Has8 forms the end of a long tube (6.9 x 1.6–1.9 cm). It has a rounded end, the other side is broken off. The overall shape is rectangular and broken off on one side. The outer surface is smoothed and the interior surface is irregular and shows bulges at the end. The colour is completely faded. A great number of rounded bubbles of different sizes are present.

Has6 is a long (15.2 cm), but slim (3.1 x 3.1 cm) fragment of a rectangular tube that is slightly misshapen and broken off at both ends. The hole is round and slightly off-centre. The surface is smoothed but badly weathered. Only in some areas are traces of the original opaque* blue base colour visible. There are many irregular bubbles, which are heavily distorted and elongated.

Has7 is a fragment of a tube. The piece has three irregular sides, which are smoothed. On one edge, traces of a black substance, probably bitumen, are preserved. The surface of the interior is irregular and exhibits irregular folds. The glass is badly corroded and only partially preserves traces of the original opaque blue colour.

4.3.4.3 Discussion

All tubes were found in Burnt Building II in rooms 2, 5, 7a, and in Burnt Building V in room 3. This could indicate that the objects fulfilled a certain function related to these buildings.

The holes in the middle of the objects suggest that they may have been attached to other objects or vice versa. The fact that all sides of the surface were well smoothed results from the manufacturing technique on the one hand, and on the other makes it possible that they were to be viewed from all sides. It could therefore be possible that the objects were used as shafts or handles, for example for standards. Another possibility would be that they belonged to furniture. Thus the parts could have been used as legs or horizontal struts. Also imaginable would be their use as decorative elements to highlight specific parts of the rooms. This could be possible as the rooms in which the tubes were found mainly served as entrances or main spaces and exhibited prominent architectural features, such as rabbets or wooden columns. The tubes could probably have served as architectural elements, used in connection with these installations. However it cannot be ruled out that some of the tubes were stored on the upper floor level, together with other 'precious materials' (glazed wall tiles, furniture attachments, shaped wood, ivory inlays and stone, copper, silver, gold containers) that fell down when the building collapsed (Chapter 3.4.4).

Even though the function of the tubes cannot be identified with certainty, it is nevertheless striking that the objects are attested only at Hasanlu, and that they were made of glass that is very much likely to have been produced at the site. Chemical analysis indicates that copper was used as a colourant*, which derived from alloyed metals (Stapleton 2011: 89) (Chapter 7.4.2). Due to their specific shape and chemical composition, these objects show how independent the glass production in Hasanlu was.

4.3.5 Head pendants

Small heads made of glass, with a loop at the top, are called 'head pendants'. Head pendants are either human, animal or grotesque in form, and were made by the technique of rod-forming (Chapter 4.3.2). Their function was hypothesised as apotropaic, magical and decorative (Arveiller-Dulong and Nenna 2011: 21). Head pendants were distributed widely along the Levantine coast, the western Mediterranean, the Greek mainland and islands, as well as Cyprus, Egypt, southern Russia, Bulgaria, Yugoslavia, Switzerland and France (Grose 1989: 82). Most of the head pendants, however, do not come from archaeological contexts, as, for example, the pieces from the Borowski Collection (Bianchi *et al.* 2002: 177–213), the Musée du Louvre (Arveiller-Dulong and Nenna 2011), and largely also the British Museum (Harden 1981), which is why their chronological and geographical distribution is based on typological considerations. Studies have been carried out by Arveiller-Dulong and Nenna (2011); Grose (1989: 82–83), Haevernick (1977) and Tatton-Brown (1981 and 1985) in attempts to achieve a typological and chronological division. Seefried (1982) presented the most extensive study and divided the objects into six main groups (with subgroups) on the basis of morphological similarities.[138] Regarding their

[138] The typological groups comprise: 1) demonic heads; 2) male heads with straight hair; 3) male heads with curly hair; 4) feminine heads; 5) animals; and 6) miscellaneous; see Seefried 1982: 5–11. With regard

chronological attribution, Seefried (1982: 25–26) and Arveiller-Dulong and Nenna (2011: 21) attributed the earliest of the head pendants to the early 7th century on the basis of find contexts of head pendants from Egypt. An increase in the production of head pendants on the Levantine coast can be observed during the 5th century.[139] Tatton-Brown (1985: 115) and Barag (1985: 57) suggested that the origin of the rod-formed heads could be located on Rhodes at the time of the core-formed vessels of the 'Mediterranean 1 group' (mid 6th century) (Chapter 4.3.3). Also Mazar and Dunayevsky (2007: 269) attribute most of the head pendants to the 'Achaemenid period'. Further consideration regarding the Levantine rod-formed head pendants would only be rewarding in the context of a large-scale study incorporating new material from Egypt and the Aegean. As this would exceed the scope of this study, the Levantine head pendants are not considered further in this study. Head pendants found in Mesopotamia are rare. One piece that could stylistically be attributed to Seefried's 'Type A' was found in a grave in Aššur, but cannot be dated precisely (Werner 2009: pl. 18, no. 168).

4.4. Summary on different manufacturing techniques

The manufacturing technique is decisive for the typological classification of glass objects, since a certain manufacturing technique always results in a certain type of object and lends this a characteristic appearance. The detailed description of the manufacturing techniques in this work shows how diverse, and at times complex, glass processing was at the beginning of the 1st millennium. It becomes clear how imaginative and inventive the glassmakers of that time were. They not only looked for new forms, but also drove the development of the specific material properties of glass to its zenith.

The two manufacturing techniques applied during the investigation are core-forming (Chapter 4.3) and cast-and-cut (Chapter 4.2). The mosaic technique (Chapter 4.1) seems to have fallen out of use at the beginning of the 1st millennium and only occurs again in the Hellenistic period (Chapter 4.1.1). While the core-forming and mosaic techniques were already in use in the middle of the 2nd millennium, with the start of regular glass production, the cast-and-cut technique was only introduced at the beginning of the 1st millennium. Objects produced by the cast-and-cut technique are therefore an innovation of this period.

Core-formed vessels have a closed vessel shape and are always polychrome decorated (glass threads are wrapped around the object in different colours and drawn into a pattern) (Chapter 4.3.3.2 *Typological Characteristics*, Table 4.2). It should be noted that the glass of core-formed vessels now found heavily corroded was brightly coloured in ancient times. The core-formed vessels of the early 1st millennium differ typologically from those of the late 2nd millennium. The later vessels are smaller in size, have a reduced variability of form and decoration, and the colour spectrum is also clearly reduced. This is similar to the later vessels of the 'Mediterranean groups', which have a completely different form and colour spectrum (Chapter 4.3.3.2 *Typological characteristics*, Table 4.1).

Objects produced in the mosaic technique are also polychrome decorated, the geometric and often also figurative patterns to be achieved by this technique are very elaborate and realised by no other technique (Chapter 4.1.3). With the decline of this technique at the beginning of the 1st millennium, no glass objects are produced which show such a variety in the combination of colour and design.

Due to their production technique, cast-and-cut objects are monochrome, unless they are inlaid or painted with inlays (Chapter 4.2.1.1). Of particular importance in connection with this technique is the post-processing of the glass object in cold state, which also owes its name to this technique (Chapter 4.2.1.6). A number of tools that were also common in stone processing were used, showing the close connection of these production spheres. In terms of the production process of cast-and-cut glass, hot processing must be strictly separated from cold processing, since the equipment of the workshops differed greatly and the craftsmen also had to possess different skills (pyrotechnological knowledge versus stone processing and inlay work). A separation of the two sectors becomes particularly clear with regard to the group of cut vessels and painted inlays (Chapters 4.2.2.6, 4.2.2.7). With regard to these groups, the figurative decorations, if present, are kept exclusively in the so-called 'Phoenician style'. This could mean that craftsmen who were trained in this style – possibly also decorating the ivories in the 'Phoenician style' – were responsible for the cold processing, respectively for the decoration of these glass objects. If one separates the two production areas of hot and cold processing, this explains why objects with Phoenician style decorations do not necessarily have to be made in that region. It is much more likely that various groups of craftsmen originating from different regions worked at the Assyrian court (Chapers 4.2.2.6 *Discussion*, 4.2.2.7 *Discussion*, 4.2.2.9 *Discussion*, Chapter 8.2.2)

to the date of the different morphological groups, Arveiller-Dulong and Nenna (2011: 21), on basis of the typology of Seefried (1982), incorporate demonic heads, male heads with dark skin, and ram heads to the earliest groups. Grose (1989: 82–83), however, divides the objects into two broad chronological groups on basis of their manufacturing techniques.

[139] Seefried (1982) sets the start of the production in Carthage and Rhodes at approximately the same time as the increased manufacturing in the Levant, although they considerably outlasted Levantine production.

The term 'cast-and-cut technology' refers to a number of different manufacturing processes, e.g. the slumping and sagging technique* (Chapter 4.2.1.4), or the casting in complex moulds* technique (Chapter 4.2.1.3), which were either complex and time-consuming to perform (casting in complex moulds), or guaranteed the rapid production of specific shapes (sagging). The introduction of this new manufacturing technique shows that new possibilities of glassworking were explored within glass technology to produce objects with different and also more complex shapes than those previously known (Chapter 4.2.2). Another decisive feature that occurs with the introduction of the cast-and-cut technique is the occurrence of translucent glass, which allows the light to shine through. This effect was only surpassed by another invention of the early 1st millennium, namely the invention of transparent glass, which could only happen in connection with the cast-and-cut technique (Chapter 7.1.7). Transparent glass made it possible to see through glass for the first time. This effect represents a completely new material property for this time, which could previously only be achieved with rock crystal, a material that was rare and difficult to process. Thus, at the beginning of the 1st millennium a significant technological development took place which resulted in a completely new way of producing glass objects. The development from a preference for polychrome to monochrome glass is particularly noticeable in this respect (Chapter 8.2.1).

4.5. Primary products: ingots, raw glass fragments and waste material

Ingots are produced during the primary production process, in contrast to accidental waste products which are mainly generated during the primary, but especially in the secondary production process.

In this study 'ingots with rounded shape' and 'ingots with rectangular shape' are distinguished from one another. Ingots with rounded shape have a round shape, a low height and a convex bottom and were made in crucibles. How ingots with rectangular shape were produced is difficult to determine as no crucibles of this form have survived. Generally, raw glass was traded in the form of ingots. This is well illustrated by the shipwreck of Uluburun, in which a large consignment of rounded glass ingots was transported (Chapter 7.4.7). The raw glass ingots were crushed at their destination for further processing into objects (secondary production). These pieces of glass have an irregular shape, broken edges and are referred to as 'raw glass fragments' in this study.

Waste products are produced during the processing of glass, both in primary and, especially, secondary production. If waste products are found at a site this is a good indicator that glass was manufactured or processed at that same place. It is often difficult to identify waste products because their shape is irregular and they have no specific characteristics. This often leads to the fact that such waste products are rarely included in publications.

4.5.1 Description and discussion of the ingots

Ingots with rounded shape

Nim198 is an almost complete part of a round ingot originally 16.4 cm in diameter. The ingot is 3.6 cm thick and weighs 1273 gr (Barag 1985: 108, no. 166). The obverse is flat and the reverse is slightly convex. The surface of the object exhibits a thick brown corrosion layer covered with a black layer, which Turner (1955: 62–63) identified as charcoal adhered to the surface. Many strain cracks are also present in the glass (Barag 1985: 108).

Nim197 is a fragment of a formerly round ingot, originally 22 cm in diameter (7.6 x 1.9 cm; 53.5 gr). The glass is opaque red; the surface has weathered to green and shows a thick, white-beige corrosion layer (Barag 1985: 109, no. 167).

The two ingots of opaque red glass Nim132 and Nim195 were published under the same find number (British Museum 1992,0701.1). Nim132 and Nim195 comprise 18 fragments that can most likely be joined into two different ingots, both of red colour. Nim132 and Nim195 weigh together *c.* 2254.5 gr; further details about their measurements are unknown. The ingots have a rounded shape; the obverse is flat and the reverse is slightly convex. Nim195 is slightly larger and thinner in width and can well be compared to Nim198. Nim132 is slightly thicker and exhibits an irregular shape. The ingots are covered with a thick white corrosion layer. At some edges opaque* red glass is visible.

Nim196 contains seven fragments of an opaque* light blue glass that can be joined to an ingot. Chemical analysis shows that the glass was coloured with copper (Turner 1955: 68) (Chapter 7.4.3). At least three of the fragments have slightly bevelled edges, indicating a convex shape for the ingot. An overall diameter cannot be determined. The total weight of the piece is 588.5 gr. On the surface there is a thick, beige-coloured corrosion layer.

Nim190 is a broken fragment that was most certainly part of a red ingot. The surface is covered with a thick, whitish beige layer showing traces of green corrosion.

Dul3 is a large fragment which can be reconstructed to a round ingot with an original diameter of 38 cm and a height of 3.5–4.7 cm. The piece weighs 2949 gr. The ingot has a flat obverse and reverse; the sides are straight.

Besides a thick, beige corrosion layer, a number of holes occur on the surface (Barag 1985: 109, no. 168). The ingot is coloured in opaque* red with traces of orange.

Dul4 is a fragment of a rounded ingot with a slight curving edge. The obverse and reverse are flat. The obverse exhibits a number of holes similar to Dul3. Dul4 is corroded to green, and the original colour is opaque* dark red to orange (Barag 1985: 109, no. 169).

The rounded fragment Dul5 exhibits a flat bottom and flat sides and can therefore be reconstructed to an overall rounded ingot. The colour of the glass is dark red and orange; the surface shows a green-whitish corrosion layer (Barag 1985: 109, no. 170).

Ingot Bab14 has remains of a quarter of a formerly round crucible (13 x 9 cm) which can be reconstructed to a diameter of *c.* 15 cm. The crucible is made of clay which contains fine pores and small amount of chaff. The sides of the crucible are not preserved but a slight depression along the rounded edge can be observed. The crucible is curved and slightly thicker towards the centre. The glass has a dark green colour and red stripes, similar to Bab12, and round and elongated bubbles of different sizes. The surface of the glass is flat and shows only a thin layer of iridescence. The original thickness of the glass is a maximum of *c.* 2.3 cm.

BS1 has a flat reverse and a curved side edge, which indicates that the fragment belonged originally to a rounded ingot which can be reconstructed to around 20 cm in diameter. The side edge is straight (2.2 cm high) and not bevelled. The piece is covered by a greenish corrosion layer; the red core is visible at the edges.

Ingots with rounded shape are the most common shape among the ingots included in this study. Ingots with rounded shape were common in the Late Bronze Age period in Egypt and Mesopotamia, but were only rarely found at Mesopotamian sites such as Tell Brak (Oates *et al.* 1997: 85–86). Here, ingots with a diameter of *c.* 15 cm come from the Mitanni Palace. Also from Egypt similar rounded convex ingots are known, such as from Qantir-Pi-Ramesse for example (Rehren and Pusch 2007). The red ingots Nim132, Nim195, Nim197, Nim198 and the blue ingot Nim196 form a typologically tight group because of their rounded shape and colour (except for Nim196), as do Dul3, Dul4 and Dul5 because of their colouration. The diameters among the red ingots vary (16.4–22 cm), and the maximum height is *c.* 3.6 cm. Apart from Nim198, none of the ingots has been analysed so far. The red colouration of ingot Nim198 is due to copper oxide (9.6%), but also high levels of lead (25%) are present, which identifies the glass as high-lead/high-copper glass (Bimson and Freestone 1985: 122) (Chapters 7.2.3.2, 7.4.3.3). The production of lead-containing high copper glass required a high technical performance, which mainly consists in maintaining a reduced furnace atmosphere. Maintaining a reduced kiln atmosphere was probably facilitated by the use of 'reductants' which promoted a reduction from the cupric to the cuprous state (Freestone 1987: 183). Possible reductants could have been carbon (charcoal), antimony, iron and lead, which were added to the melt. Turner (1955: 62–63) argues that the charcoal present on the surface of Nim198 was probably thrown on it to maintain a reduced atmosphere* for achieving the red colouration (Chapter 7.2.3.2).

Ingot Nim198 was found in a room in the Burnt Palace which contained furnace remains and other raw glass finds. The stratigraphic situation is ambiguous and probably points to a date towards the end of the late 6th century or even the Hellenistic period (Chapter 3.1.4). Chemical analysis of the charcoal found on the surface of Nim198, however, shows that the piece can be attributed to the period of 860–740 with 59% probability (rather than to 440–380 with 16% probability). Nim198 is thus most likely Neo-Assyrian in date.

Fragment Nim190 also comes from a palatial context (South East Palace), however the exact find context and date are not known.

The fragment of ingot BS1 originates from room 1028 in the South Temple. Unfortunately the room was disturbed by a later cistern, so that a dating of the object into Iron Age IB/IIA cannot be assumed without doubt.

Ingot Nim132, ingot Nim195, the light blue ingot Nim196 and fragment Nim197 have no find contexts. This also applies to ingot Dul3, ingot Dul4, fragment Dul5 and ingot Bab14.

None of the round ingots can be clearly assigned to a period. With regard to the find context, however, it can be stated that the ingots from Nimrud originate from palace contexts. Only BS1 from Beth-Shean comes from a temple context. This allows the conclusion that raw glass was at least stored in the palace.

Ingots with rectangular shape

Bab9 is a large fragment (25 x 7.5 cm; 5.5 cm thick) of an ingot, of which the original dimensions cannot be estimated. The obverse and reverse of the piece are straight; the reverse has a coarse surface. The obverse side is smooth with a thin white corrosion layer. The glass is translucent dark blue with no traces of corrosion. A great number of rounded bubbles of different sizes are spread throughout the glass.

Bab11 can be identified as a large fragment (12 x 6.6 cm; 5.8 cm thick) of an ingot, of which the original size cannot be fully reconstructed. Only a small spot of the coarse surface is preserved; the obverse is smooth and covered by a thin layer of white corrosion. The glass is translucent dark blue and shows many rounded irregular bubbles. Bab9 and Bab11 are well comparable regarding the nature of the glass and its thickness. It is therefore highly likely that Bab9 and Bab11 belonged to the same ingot. If this is true, an ingot of at least 25 x 13.5 cm could be reconstructed.

Bab9 was part of a hoard that contained objects from the Kassite, Neo-Assyrian, Neo-Babylonian and Hellenistic periods (Chapter 3.2.1). The hoard itself was placed underneath a dwelling context and was interpreted as the raw material deposit of a craftsman.

Bab11 does not have a find spot; a datable comparable rectangular ingot was found in Persepolis and is Achaemenid in date. The colour of this ingot, however, is dark red with traces of orange. Bab9 and Bab11, however, are thicker (5.8 cm) than the piece from Persepolis (2.1 cm).

4.5.2 Description of the raw glass fragments

Raw glass fragments are irregularly broken on all sides and cannot be reconstructed to an ingot of known shape or size. Raw glass fragments are parts of an ingot broken off for further processing.

Has13 incorporates four fragments of irregular size and shape. The opaque* light blue colour is still preserved in the core. There is a thick whitish corrosion layer around it and the surfaces of the fragments are completely corroded to white.

Has14 contains four irregular fragments of different sizes that most likely belonged to the same ingot. The four irregular fragments are of different size and shape. Fragment C exhibits one straight side, and the rest of the fragments are broken on all sides. Fragments A and D are made of red glass that has corroded to orange and light green. Regarding fragment B, only a small nucleus of red is preserved and it is surrounded by a thick turquoise layer of glass. The surface is white. Fragment C has fully corroded to white and has only a slight tinge of green on its surface.

As11 is an opaque*, light blue glass fragment, broken on all sides (5.2 x 3.1 cm, thickness 1.2–1.4 cm). Two sides are flat, making it possible to identify them as front and back. The surface of As11 is coated with a thick, beige-brown corrosion layer. The opaque light blue core is only visible in the fractures. The glass contains rounded bubbles.

As12 is an irregular fragment broken on all sides (4.1 x 5.6 cm). The surface is irregular and the colour is faded entirely. The previous colour is not preserved. Characteristic conchoidal breaks and the chipped surface make the identification of glass, however, possible (Chapter 2.1.1).

Bab12 is an irregular fragment broken on all sides, but which exhibit smooth surfaces. One side is flat and the other is convex due to a break. The piece has an overall dark colour and exhibits a slight greenish tinge and streaks of red.

Bab10 is an irregular raw glass fragment, broken on all sides. The piece has a light blue colour and whitish corrosion layer.

Bab13 is an irregular fragment (8.8 x 8.4 cm) broken on all sides. The surface is partially coated with a whitish-beige corrosion layer, but in some places the glass is visible. The glass has a translucent light green colour, which is typical for iron impurities. It is therefore possible that the piece could have belonged to a large, almost colourless glass ingot. There are many small round bubbles.

Nin4 is an irregular small fragment of an opaque* red glass ingot (1.8 x 2.8 cm) and Er1 is an irregular fragment of translucent, dark blue glass (3.1 x 1.7 cm) that contains a number of spherical bubbles.

Er2 is a lump of opaque brownish-red glass (2.8 x 1.8 cm). Er3 is an irregular piece of glass, totally faded to a whitish-beige colour.

The raw glass fragments Has13 and Has14 were found *in situ* in Burnt Building IV-V, room 4, at Hasanlu, and therefore date to the Hasanlu IV debris (1000–800). The raw glass pieces had been stored together in the same place on the first floor, which collapsed during destruction (Chapter 3.4.4). All other finds (As11, As12, Bab10, Bab12, Bab13, Nin4, Er1, Er2 and Er3) were found out of any archaeological context.

4.5.3 Description of the waste material

Nim189 (2 x 1.3 cm) and Nim135 (1.2 x 1.1 cm) have an irregular, deeply grooved surface with rounded, drop-shaped parts and a flat back. From their shape they can be identified as waste material. The glass of both of the objects has a very similar translucent dark blue colour. The colour and characteristics of the pieces, however, coincide with the monochrome blue inlays and rosettes found in great numbers at Nimrud. It is therefore possible to connect Nim189 and Nim135 with these inlays and, therefore by association, with a Neo-Assyrian date.

Nin3 has an irregular surface which has become a brownish-grey colour (Barag 1970: 113, no. 187). The object can most likely be identified as a droplet that occurred during the secondary manufacturing process.

4.5.4 Summary

Most of the raw glass and waste material from primary production presented here comes from palatial or elite residential contexts. Only one object can be attributed to a temple. Two of the opaque red ingots from Nimrud come from the Burnt Palace (Nim198), and the South East Palace (Nim190) (Chapters 3.1.4.3, 3.1.4.5).

The raw glass fragments Has13 and Has14 from Hasanlu were found *in situ* in Burnt Building IV-V, room 4, where they were most likely stored on the first floor, which collapsed during destruction (Chapter 3.4.4.3). The large, rectangular translucent blue ingot Bab9 – and probably also Bab11 – were part of a hoard that was carefully placed or even hidden in a basket underneath a dwelling context. Fragment BS1 of a red opaque ingot from Beth-Shean originates, however, from a temple context. Nim198 was found in the vicinity of a furnace and can therefore be associated with workshop activities. However, this has to be taken with caution, as the date of the context and the date of Nim198 do not seem to coincide (Chapter 3.1.4.3).

Most of the objects of primary glass production are difficult to date as their find contexts are not clear. Typological comparisons are of limited use as the forms often extend over several periods (rounded ingots) or are not significant (fragments and waste products). A possible dating attempt would be the comparison of the chemical composition of the glass ingots and raw material with datable glass objects from corresponding sites, which has yet to be carried out.

The find situation of the primary glass products at Hasanlu (Chapter 3.4.4.3) and Babylon (Chapter 3.2.1.2) suggests, however, that it was not stored in, or in the immediate vicinity of the workshop, but in another location – partly with precious materials, as the find situation at Babylon shows – that was not directly accessible. This could indicate that raw glass had a certain level of value.

Looking at the colours of the raw glass fragments, the disproportionately high presence of red glass is obvious (Table 4.3). This becomes particularly clear with regard to the ingots from Nimrud, where only one out of six rounded glass ingots is made of blue glass. This stands in contrast with the frequency of colours used for glass objects in secondary production, where blue is clearly over-represented.

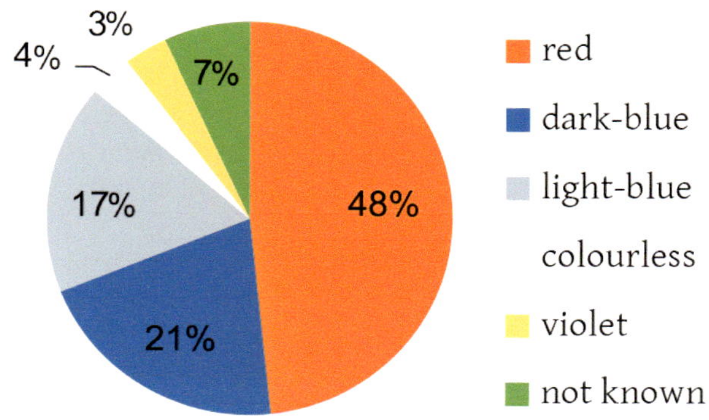

Table 4.3: The distribution of colours in percentages, with respect to ingots, raw glass fragments and waste products.

5. Discussion of the Archaeological Data

5.1. Remarks on the archaeological dataset

The following section aims to evaluate the geographical and contextual distribution as well as the chronological classification of the different glass objects. There are, however, potential biases that could influence the distribution patterns: the archaeological exploration of the area under study is generally irregular. This is particularly true for Neo-Assyrian sites such as Aššur, Nimrud, Khorsabad or Nineveh, which were extensively excavated in the 19th and early 20th centuries, with the aim of exploring monumental architecture such as palaces and temples. For this reason, finds have mainly been made from these structures and not from residential houses (see Chapter 3.1.1, 3.1.2, 3.1.4, 3.1.5).

With regard to the sites in the Levant, there is also the problem that many of the Iron Age strata cannot be excavated due to modern overbuilding (see Chapter 3.3).

Another point is the poor preservation of glass, which affects this material particularly severely. In this regard, glasses from Mesopotamia are much more affected by weathering* due to the soils in that region than, for example, Egypt (see Chapter 2.1.3).

However, the first point mentioned above can be resolved by negative evidence: in relation to the distribution of the cast-and-cut glass objects, it is shown that these objects do not occur in the well-documented dwelling houses at Aššur (Chapters 3.1.1), nor at all at Tall Halaf, Sincirli and Ziyaret Tepe (only one example is known) (the excavation databases were consulted), but only in Assyria in palatial contexts (Chapter 8.3). This shows that the contexts in which the finds were made have to be recognised, but not over-emphasised.

5.2. Distribution according to the different types of glass objects

A total of 389 objects are included in this study, of which 334 originate from regular excavations and 55 acquired via the art market and then entered various collections (Table 5.1).

The objects of secondary production are 362 in total and can be divided into three general object groups according to the production technique: 'mosaic (glass)'; 'cast-and-cut glass'; and 'core- and rod-formed glass' (Chapters 4.1, 4.2, 4.3; Table 5.1). 285 pieces belong to the group of cast-and-cut glass, which thus forms by far the largest group of objects (Table 5.1). Core- and rod-formed objects represent 58 pieces of the total quantity and form the second largest group (Table 5.1). Objects produced using the mosaic process comprise only 18 pieces and are thus the smallest group of secondary products (Table 5.1).

In addition, ingots, raw glass and waste material form a separate group, as they are considered primary production products and comprise 27 objects (Table 5.1).

Site	Mosaic	Cast-and-Cut	Core- and Rod-formed	Primary Products	Total
Amman		1			1
Tel 'Aroer		1			1
Arslan Taş		5			5
Aššur	14		2	2	18
'Atlit		1			1
Babylon		3	5	6	14
Til Barsip		1			1
Beth-Shean	1		1	1	3
Busayra			1		1
Carthage			1		1
Tell ed-Duleym		2		3	5
Eridu				3	3
Fortetsa		1			1
Gordion		2			2
Hasanlu	3	1	8	2	14
Idalion		1			1
Isin		2			2
Tell Jemmeh			2		2
Kameiros			2		2
Khorsabad		8			8

Site	Mosaic	Cast-and-Cut	Core- and Rod-formed	Primary Products	Total
Kiš			2		2
Megiddo		1	1		2
Nimrud		188	2	8	198
Nineveh		2		2	4
Nippur			3		3
Praenestre		1			1
Samaria		22			22
Sultantepe			1		1
Susa			4		4
Ur		2	5		7
Uruk			3		3
Ziyaret Tepe			1		1
Art Market		40	15		55
Total	18	285	59	27	389

Table 5.1: Assignment of typological groups and sites of discovery.

The large group of cast-and-cut glass objects can further be subdivided into groups: pallets; mace-heads; vessels; and inlays. Inlays form the largest group of these categories and comprise 'rosette inlays' (117), 'small monochrome inlays' (93), 'painted inlays' (eight), 'large monochrome inlays' (seven) and 'attachments and inlays for composite statues' (14) (Table 5.2). The second largest group (42 pieces) consists of vessels again comprising of 'jars and alabastra' (12), 'hemispherical bowls' (10), 'shallow undecorated bowls, ribbed bowls and petalled bowls' (9), and 'cut-and-inlaid vessels' (10). The groups of 'palletes' and 'mace-heads' contain (with three and two objects respectively) the smallest number of objects (Table 5.2).

Type of cast-and-cut object	Amount
Rosette inlays	116
Small monochrome inlays	93
Painted inlays	8
Large monochrome inlays	7
Attachments and inlays	14
Jars and 'alabastra'	12
Hemispherical bowls	10
Shallow undecorated bowls, ribbed bowls and petalled bowls	9
Cut-and-inlaid vessels	11
Palletes	3
Mace-heads	2
Total	285

Table 5.2: Total number of cast-and-cut glass objects and their respective groups.

The group of core- and rod-formed glasses can be divided into core- and rod-formed vessels (51) and tubes (eight). Core-formed vessels are subdivided in the seven groups 'ovoid bottles with pointed base' (eight), 'ovoid bottles with rounded base' (three), 'piriform bottles' (10), 'globular bottles' (four), 'small jars' (four), 'bottles with disc-base' (four), 'large cylindrical bottles' (10), 'vessel fragments' (eight) (Table 5.3).

Type of core- and rod-formed object	Amount
Ovoid bottles with pointed base	8
Ovoid bottles with rounded base	3
Piriform bottles	10
Globular bottles	4
Small jars	4
Bottles with disc-base	4
Large cylindrical bottles	10
Vessel fragments	8
Tubes	8
Total	59

Table 5.3: Total number of core- and rod-formed glass objects and their respective groups.

Mosaic glass is the rarest of these three different types of glass. Only 18 pieces are included in this study, consisting of 'tiles' (six), 'inlays' (six), 'bowls' (five), and one bead, which was most likely used as a secondary element (Chapter 4.1).

Type of mosaic object	Amount
Tiles	6
Inlays	6
Bowls	5
Bead (re-used)	1
Total	18

Table 5.4: Total number of mosaic objects and their respective groups.

Altogether 27 ingots, raw glass objects and waste material account for glass production processes. 12 pieces can be identified as ingots, 12 as raw glass fragments, and three as waste material. The 'ingots with rounded shape' include 10 objects, the 'ingots with rectangular shape' two (Table 5.3).

Type of primary object	Amount
Ingots with rounded Shape	10
Ingots with rectangular Shape	2
Raw glass	12
Waste material	3
Total	27

Table 5.5: Total number of primary objects and their respective groups.

5.3. Distribution according to sites and regions

5.3.1 Cast-and-cut glass

5.3.1.1 Vessels

Nimrud plays a key role with regard to cast-and-cut vessels, as this is where by far the most specimens of this vessel type were found (Table 5.6). This applies in particular to the groups of hemispherical bowls, the cut-and-inlaid vessels, and the shallow undecorated bowls, ribbed bowls and petalled bowls which occur frequently in Nimrud compared to the other sites mentioned below. Jars and alabastra come mostly from the art market without any indication as to their provenance and cannot be included in the distribution map. At almost all other sites listed below cast-and-cut vessels only appear as single pieces (Table 5.6).

In addition to the seven hemispherical bowls from Nimrud included in this catalogue, more than 250 fragments of this type of vessel were found at the site, so that a much larger number of objects must be assumed there (Barag 1985: 64, no. 35; Saldern 1970: 220, no. 25, 26). Only from Khorsabad is one further fragment of a hemispherical bowl known.[1] Further hemispherical bowls were found outside Assyria only on Crete (Idalion) and from Italy (Praenestre). The group of hemispherical bowls form a narrow typological group, in particular the pieces from Nimrud. This, combined with the large presence of hemispherical bowls from Nimrud and the simultaneous absence, or limited presence, of pieces in other (well-excavated) Assyrian, Babylonian, and Levantine sites, makes it likely that secondary production of the objects from Nimrud was carried out there. The wide geographical distribution to areas outside Assyria makes it clear that hemispherical bowls belonged to a group of objects that were also desirable beyond the borders of Assyria. Typologically, Fo1 fits well into the group of bowls made at Nimrud. With Pr1, the edge of the bowl is turned slightly inwards and thus differs typologically from all other hemispherical bowls. This variation could indicate an independent secondary production in Praenestre, but this does not necessarily have to be the case.[2] On the other hand, there is the disproportionately high presence of hemispherical bowls from Nimrud, which date earlier than the pieces from the Mediterranean (Chapter 1.2.2.4). If an import was assumed, this would therefore be more likely from Assyria to the western Mediterranean region than the other way around. The context in which Pr1 was found should also be mentioned here, since the so-called 'Barberini Tomb' contained inlaid ivories of the Phoenician type, and vessels decorated with pseudo-Egyptian motifs pointing towards an origin in the eastern Mediterranean. Due to the arguments given here, it remains, however, difficult to decide where the secondary production of the hemispherical bowls took place. With regard to primary glass production, reference is made at the end of this chapter.

A similar situation can be observed in the distribution of the cut-and-inlaid vessels, which almost exclusively originate from Nimrud. Two other vessels were found on the citadel of Amman and Tel 'Aroer (Table 5.6). This means that, except those from Nimrud, almost no vessels of this kind can be observed in any other place. Similar to the previous paragraph on hemispherical bowls, it seems tempting to relocate the origin of the vessel from Amman and Tel 'Aroer to Nimrud. Regarding the location of the vessel from Tel 'Aroer, this piece was found in the ruins of a building interpreted as a caravanserai (Chapter 3.3.2) and it is not certain whether the piece was intended to reach Tel 'Aroer or whether it was on its way along the southern Arabian trade route to an unknown destination in the southern Levant, Egypt, or southern Arabia.

In contrast to the hemispherical bowls and the cut-and-inlaid vessels, the geographical distribution of shallow undecorated and ribbed and petalled bowls is considerably wide, with objects occurring at Nimrud, Babylon, Hasanlu and Gordion (Table 5.6). The wide distribution is also reflected in the typological diversity of this group of vessels, which could favour local secondary production at the various sites (Chapter 4.2.2.5). With regard to Gordion, reference is made to the discussion of primary production at the end of this chapter, which illustrates a link between Gordion and Nimrud.

The group of jars and alabastra consists, to a large extent, of unprovenanced pieces (Table 5.6). Whereas the jars from Ur, Nimrud and Hasanlu can clearly be attributed to the Mesopotamian form spectrum, alabastra are difficult to provenience, as the only two known pieces from 'Atlit and Idalion cannot be attributed to a specific find context.

[1] There is one further fragment of a hemispherical bowl from Nineveh. This piece (N.1464) is not illustrated and therefore not included in the catalogue of this study; see Barag 1985: 66, no. 41.

[2] An additional complication of the typological comparison is that the two bowls Fo1 and Pr1 could not be examined personally by the author.

In summary, cast-and-cut vessels occur predominantly in Assyria and are very rare in other regions. This allows the assumption that at least the vessels from Nimrud were most likely produced at the site. Cast-and-cut vessels are closely connected to transparent glass since most of these vessels were made from it. The primary production of transparent glass and the secondary production of cast-and-cut glass must therefore be considered together (Chapters 4.4, 8.1).

With regard to the primary production of transparent glass, chemical analyses show that three chemically distinguishable glass groups existed in Nimrud. Samples of the colourless hemispherical bowls were also taken to identify these composition groups, i.e. these objects are also relevant in this regard. The large quantity of transparent glass in Nimrud and the high variance in chemical composition of the groups cannot be determined at any other site of the investigated time period, which gives Nimrud a special status in relation to this type of glass (primary and secondary). The transparent bowls from Gordion coincide chemically with one of the transparent glass groups from Nimrud, and it is therefore possible that colourless glass was imported to Gordion in the form of raw glass from Nimrud or a common source (Chapters 7.4.3, 7.4.5).

5.3.1.2 Inlays

The group of inlays can be divided into attachments and inlays for composite statues (Table 5.7), painted inlays and rosette (groups 1, 2, 3), small monochrome inlays and large monochrome inlays (Table 5.8).

Attachments and inlays for composite statues (14) were used as decorations for statues of gods and are found at Assyrian and Babylonian sites (Table 5.7). The majority of inlays (six) comes from Nimrud. In addition to one further piece that was found in Nineveh, the inlays are solely attested at Babylonian sites, such as Ur (one), Babylon (two), Duleym (two), and Isin (two) (Table 5.7).

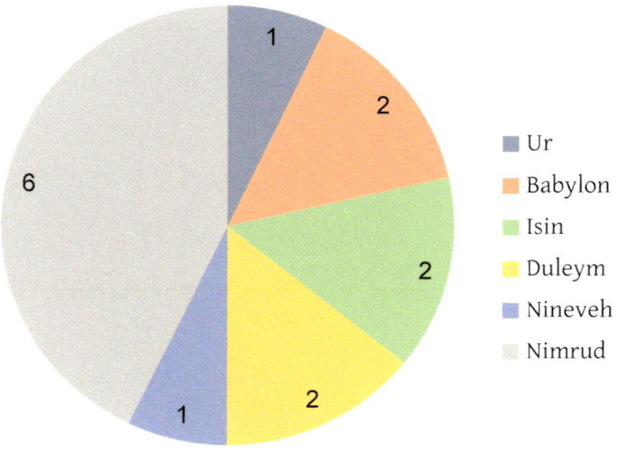

Table 5.7: Total number of inlays for composite statues and their distribution at the different sites.

With the exception of attachments and inlays for composite statues, the other inlays listed in this study were used in conjunction with ivories and furniture, or they were used as inserts for glass vessels. The largest number of inlays was found at Nimrud, where also the largest range of different inlay types is found (Table 5.8).

By far the most rosette inlays were found at Nimrud (105), with only five from Arslan Taş and Samaria and one from Til Barsip. All rosettes inlays found in Arslan Taş belong to group 3 (five) and were found inserted in a red glass frame, which was partly held by a bronze frame. The inlays of group 3 (21), which originate from the art trade, also had a red frame, with the exception of one piece. In Nimrud and Samaria (one) the rosettes of group 3 were only found without frames. Group 1 and group 2 inlays were spread over the sites of Samaria, Til Barsip and Nimrud.

A similar picture emerges for the group of small monochrome inlays: the largest number of these inlays comes from Nimrud (68), the second largest from Samaria (18), and eight pieces from the art market. Painted inlays were found only in Nimrud and large monochrome inlays only in Khorsabad. It is in this regard interesting

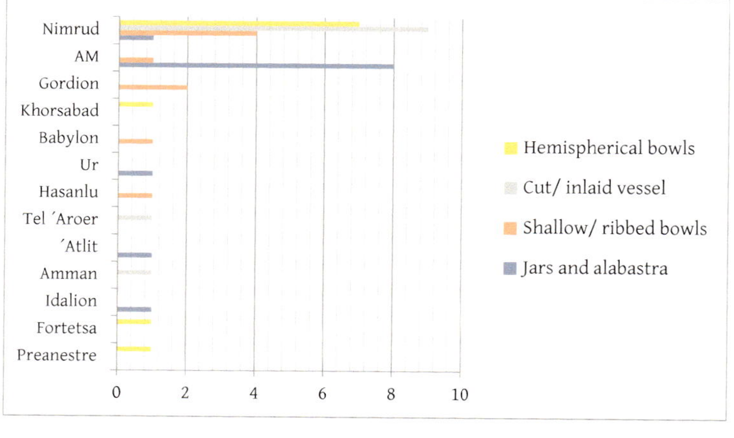

Table 5.6: Total number of cast-and-cut vessels and their distribution among different sites.

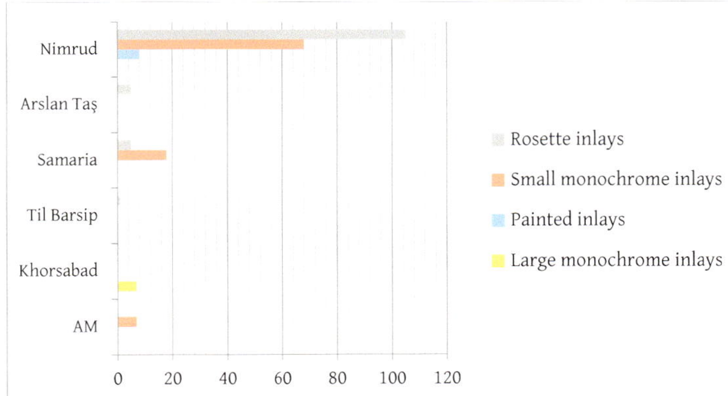

Table 5.8: Total number of inlays for ivories, furniture and vessels and their distribution among different sites.

to note that other types of inlays connected to ivories or furniture are entirely absent at Khorsabad, even though ivory plaques were found there.

With regard to the primary production of inlays, 13 samples from Nimrud of blue glass from small inlay plaques, frequently squared plaques with central white rosettes (Reade *et al.* 2005: 23), were analysed, which most likely correspond to group 3 rosettes. The results indicate that the pieces show similar composition using cobalt as colourant (Chaper 7.4.3). The cobalt used for colouring does not however match the cobalt alum used for colouring Egyptian New Kingdom glass (Reade *et al.* 2005: 26, fig. 3). Since no comparable analytical results are currently available for the Nimrud cobalt-blue glasses from Mesopotamia, the Levant or Egypt, it remains questionable where the cobalt was taken from. However, it can be stated that the inlays were obviously made from the same raw material. Future chemical analyses of the glass would be worthwhile with regard to the various inlays and the objects in which they were found, which are very similar in typology and appearance.

If one compares the distribution of all attachments and inlays for composite statues with the rest of the inlays, it becomes obvious that inlays for composite statues were spread widely in Babylonia and occur only to a lesser extent in Assyria (Table 5.7). In contrast, the remaining inlays were distributed only in Assyria and also occur largely among sites in the conquered territories of the Assyrian Empire in northern Syria and along the Levantine coast (Table 5.8). This indicates a strong connection of the production of these inlays with the Neo-Assyrian Empire. Against this background, the close connection of the inlays with ivory and other furniture should also be emphasised and thus the functional differentiation of the attachments and inlays for composite statues.

5.3.2 Core- and rod-formed objects

Core- and rod-formed glass comprises tubes as well as vessels. Tubes were solely found at Hasanlu and therefore represent a special feature of this site (Table 5.9, Figure 5.1). Core-formed vessels occur in Mesopotamia, the southern Levant and the Aegean (Carthage, Kameiros), and are thus distributed over the entire area of interest for this study (Table 5.9, Figure 5.1). Piriform bottles are spread throughout Babylonia, Assyria and the Levant and seem to be a common form among core-formed vessels. The group of large cylindrical bottles is also spread over a wide territory, since not only finds in Babylonia and Assyria have been made, but also in the Aegean (Carthage, Kameiros). Since the group of large cylindrical bottles is typologically very uniform, and examples of this type were found both in the Aegean (Kameiros, Carthage) and in Mesopotamia (Nippur, Assur, Nimrud), this could indicate a connection between these two regions. The group of large cylindrical bottles is obviously also a popular type of core-formed vessels, which were geographically widespread. In contrast to the two previous types, small jars and bottles with disc-base solely occur in Babylonia (Ur, Babylon, Nippur) (Table 5.9, Figure 5.1)

Even if at first glance the formed vessels are distributed over the entire region, a closer look reveals differences: in Babylonia all types of vessels occur, while in Assyria only ovoid bottles (Aššur), large cylindrical bottles (Nimrud, Aššur) and piriform bottles (Sultantepe) can be found. This is similar for the southern Levant, where only one piriform bottle (Busayra) and two globular bottles (Tell Jemmeh) are recorded (Figure 5.1). Furthermore, it can be noted that in Babylonian sites several different types of vessels appear and thus a large variety of types is observed at the individual sites. This is not the case in Assyria, the southern Levant and the Mediterranean, as only individual vessel types usually occur here at each site.

5.3.3 Primary products

Ingots, raw glass objects, as well as waste material, are attested in Assyria (Nimrud, Aššur, Nineveh) and Babylonia (Babylon, Duleym, Eridu); three pieces are also attested from sites outside Mesopotamia (Hasanlu, Beth-Shean) (Table 5.10). Only a very small number of ingots, raw glass and waste material can be dated at all.

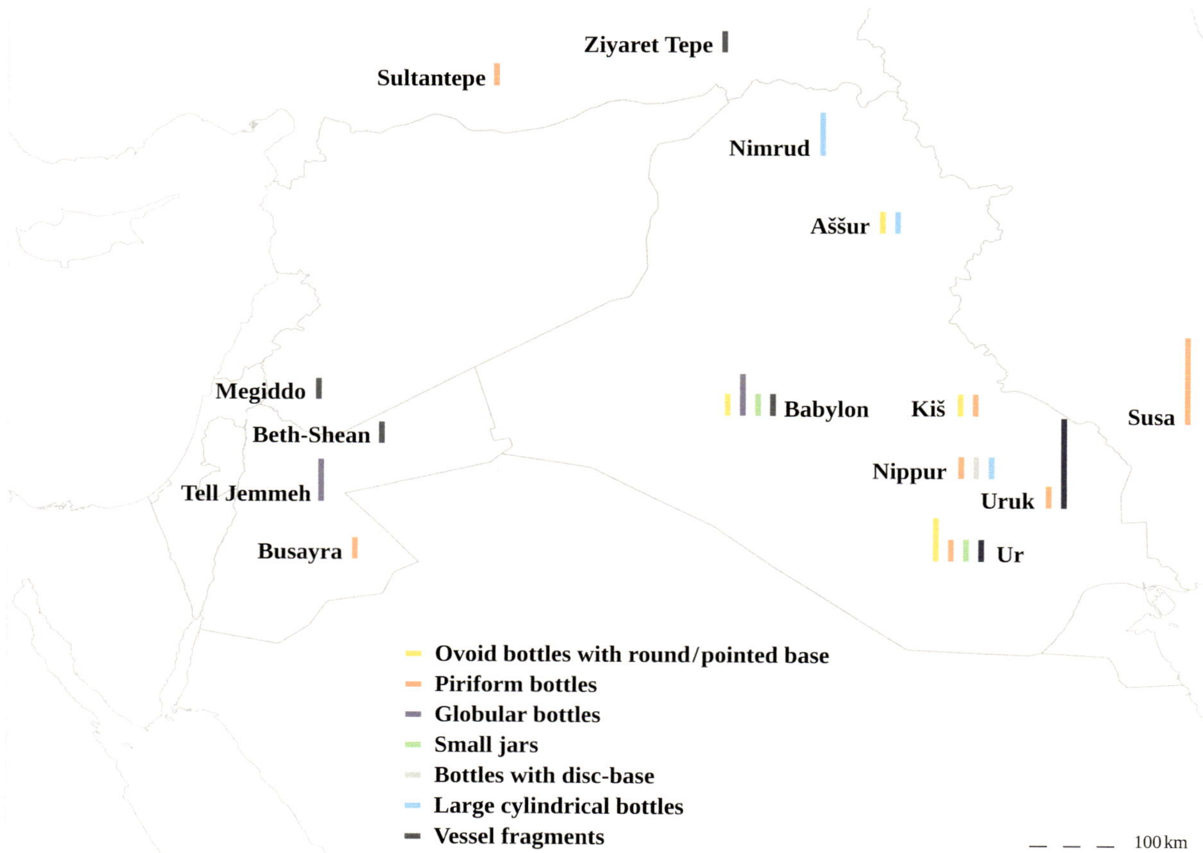

Figure 5.1: Geographical distribution of all core-formed vessel types and fragments contained in this study according to their number, with the exception of the pieces from Carthage and Kameiros (n= 49) (template: Katharina Schmidt, design: Mareike Walter).

	Vessel Fragments	Small jars	Large cyl. bottles	Bottles with disc-base	Globular bottles	Piriform bottles	Ovoid bottles/ round base	Ovoid bottles/ pointed base	Tubes
Aššur			1					1	
Babylon	1	1			2		1		
Beth-Shean	1								
Busayra						1			
Carthage			1						
Hasanlu									8
Tell Jemmeh					2				
Kameiros			2						
Kiš						1		1	
Megiddo	1								
Nimrud			2						
Nippur			1	1		1			
Sultantepe						1			
Susa						4			
Ur	1	1				1	1	1	
Uruk	3								
Ziyaret Tepe	1								
AM		2	3	3		1	1	5	

Table 5.9: Total number of core-formed vessels and tubes that occur at different sites.

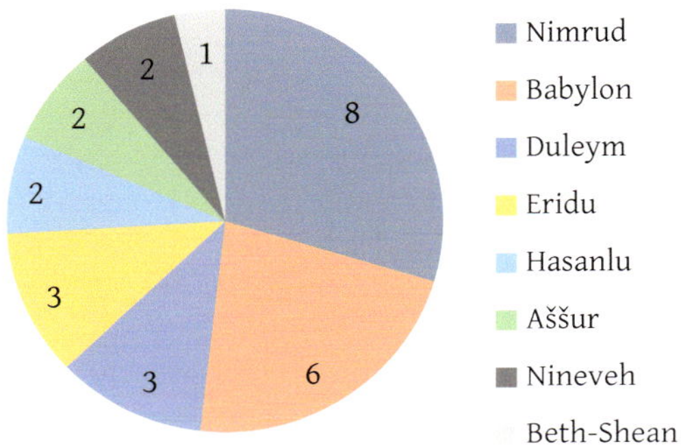

Table 5.10: Number of primary products and their occurrence at different sites (n = 27).

5.3.4 Summary

As well as cast-and-cut and core- and rod-formed objects, Figure 5.2 also shows the mosaic objects discussed in Chapter 4.1 to show the distribution of all objects included in this monograph. As indicated earlier, the mosaic technique no longer occurs in the first half of the 1st millennium (Chapter 4.1.4). Glass production in this period focuses solely on the production of core-formed and cast-and-cut objects.

Against the background of the objects collected in this work, it can be said that core-formed glass clearly dominates the assemblage of glass objects in Babylonia (Figure 5.2). Many different types of core-formed vessels were found there at various sites, which suggests that the production of core-formed vessels flourished in this region. Core-formed vessels are also attested in the southern Levant, even though they are less varied than in Babylonia (Figure 5.2). However, in this region a generally balanced picture between core-formed and cast-and-cut objects appears (Figure 5.2).

The distribution in Babylonia stands in contrast to Assyria, where only a small number of different core-formed vessels occur and only a small variety of types can be observed. This stands in sharp contrast to the distribution of cast-and-cut objects, as cast-and-cut vessels and inlays occur almost exclusively at sites in Assyria or at sites with an element of Assyrian influence, e.g. Samaria, Arslan Taş and Til Barsip (Figure 5.2). An

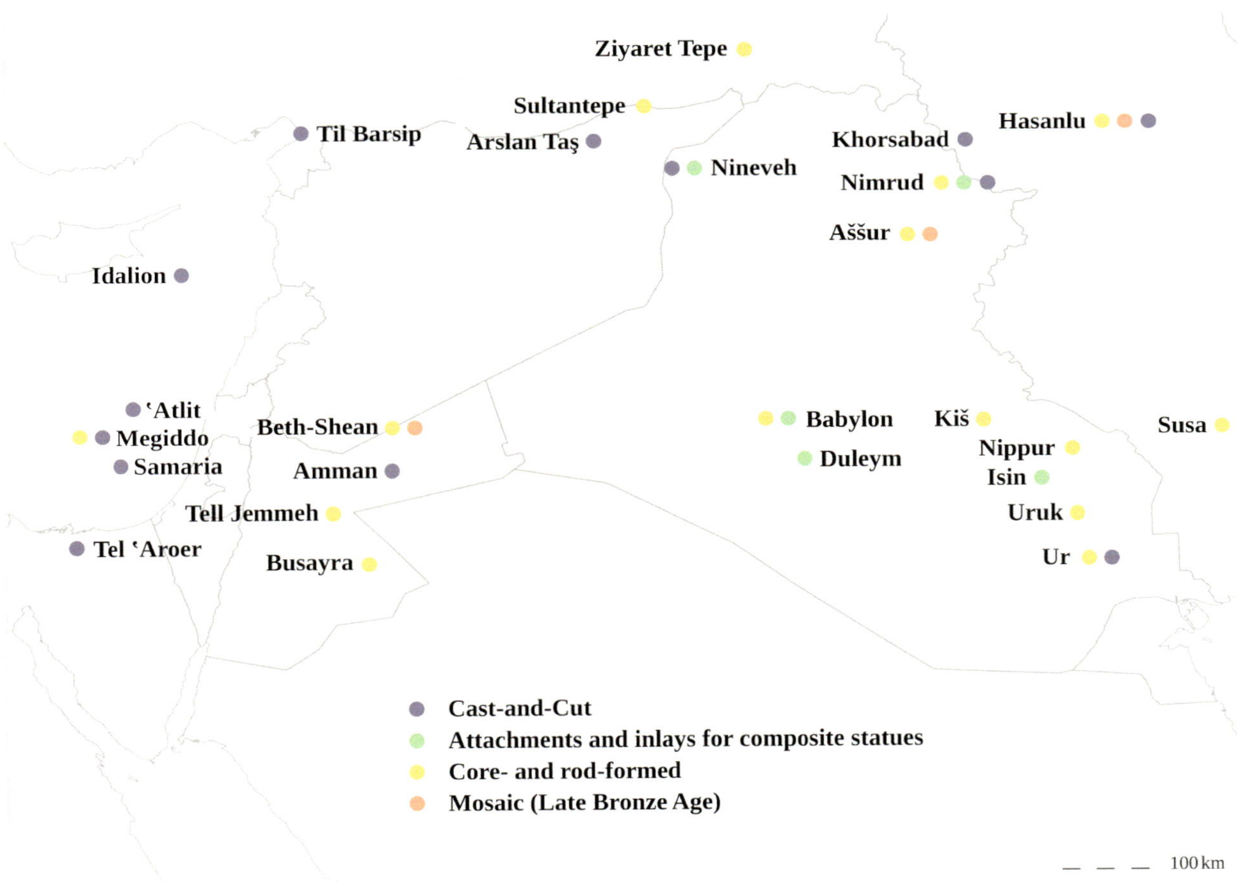

Figure 5.2: Geographical distribution of the different manufacturing techniques of glass in the Iron Age. Also included are the mosaic objects discussed in this monograph dated to the Late Bronze Age. The glass objects assigned to the group of attachments and inlays for composite statues are listed separately. The mosaic objects incorporated in this monograph identified as Late Bronze Age are also included. Gordion, Kameiros, Fortetsa, Praeneste and Carthage are not included, for these see Figure 3.1 (template: Katharina Schmidt, design: Mareike Walter).

exception is the group of attachments and inlays for composite statues, which occur almost exclusively in Babylonia (Figure 5.2).

Nimrud and Hasanlu show the widest range of different types of glass objects, which accounts, in general, for a thriving and diverse secondary glass industry. Raw glass finds occur both in Nimrud and Hasanlu, but this does not necessarily mean that there was also primary production there. Chemical analysis from Hasanlu coincides partly with glass from Nimrud, which shows that the same raw materials were used. It also shows, however, that each of the sites also used independent sources (Chapters 7.4.2, 7.4.3).

5.4. Distribution according to find contexts

The following chapter looks at the distribution of glass objects within different archaeological contexts. In terms of contextual affiliation, the general classifications 'palace', 'temple', 'funeral and 'domestic' are applied. An exception is the find context of the caravansary in 'Tel Aroer. The Fort Salmanasser in Nimrud, the 'Bâtiment aux Ivories' in Arslan Taş, the 'Ivory House' in Samaria and the 'Burnt Building II' in Hasanlu are also categorised here as 'palace' (Chapter 3.4.4.1).

5.4.1 Cast-and-cut objects

By far the largest number of inlays and vessels produced using the cast-and-cut technique were found in palatial and palace like structures (Table. 5.11). All inlays originate from palatial contexts, the vessels can also be almost exclusively assigned to this context. Only a smaller proportion were found in funerary and temple contexts (Table. 5.11). The mace-head, as well as all composite attachments and inlays, and one cast-and-cut vessel, can be assigned to temple contexts (Table 5.11). The palette was found in a domestic building and the cast-and-cut vessel fragment from 'Tel Aroer in a caravansary (Table. 5.11).

The cut-and-inlaid vessels were found – with the exception of the fragment from Tel 'Aroer – in palatial contexts. The majority of these vessels comes from Fort Salmanasser in Nimrud, but one was found in the Ammonite palace in Amman (Table. 5.12).

Most of the shallow undecorated, ribbed and petalled bowls come from palace contexts (Table. 5.12). In this respect all pieces from Nimrud originate from Fort Shalmaneser; the fragments from Hasanlu were found in Burnt Building II, which is also attributed here to a palace context (Chapter 3.4.4.1). One of the two bowls from Gordion can be assigned to either a palatial or a domestic structure. Apart from this find, the examples from Babylon and Gordion were found in burials. Regarding these two funerals, it is interesting that both can be identified as infant burials. Tumulus P in Gordion, on the basis of its rich grave goods, can be assigned to a child of rank. Whether or not the specific shape and material of the glass bowls had a certain relationship to infant burials cannot be determined.

The majority of the hemispherical bowls was found at Nimrud, where they can be attributed to the Fort Shalmaneser and Northwest Palace (Table. 5.12). The specimen from Khorsabad was also found in Residence K. In contrast, the hemispherical bowls from Praeneste (Italy) and Fortetsa (Crete) were both found in graves. It seems therefore likely that the function of the hemispherical bowls underwent a change, dependent on the geographical region in which they occur. While hemispherical bowls from Nimrud and Khorsabad were most likely used in connection with royal banquets, they instead served as grave goods in richly furnished inhumations in Italy and Crete (Chapter 4.2.2.4). Whether the hemispherical bowls were imported to serve as burial gifts, or had also been used there for

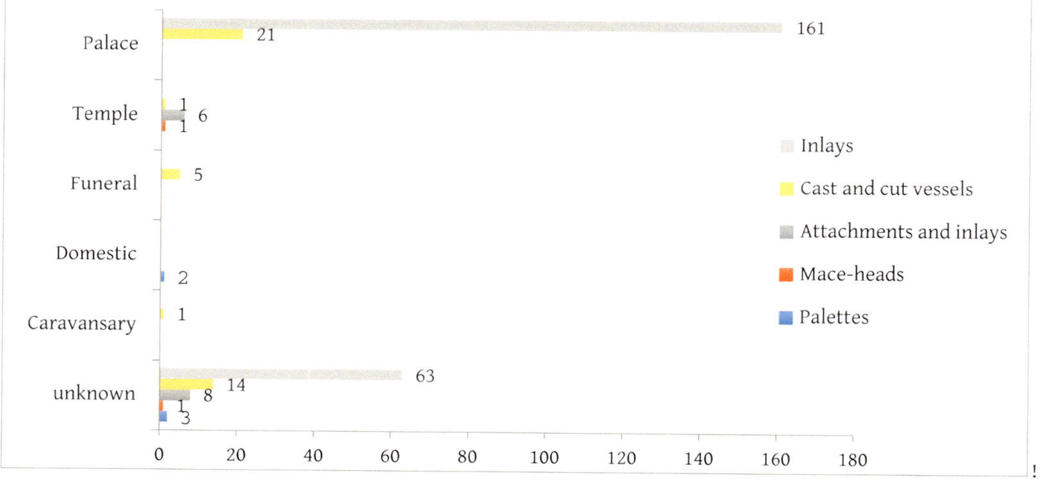

Table 5.11: Distribution of cast-and-cut objects among different find contexts (n= 285).

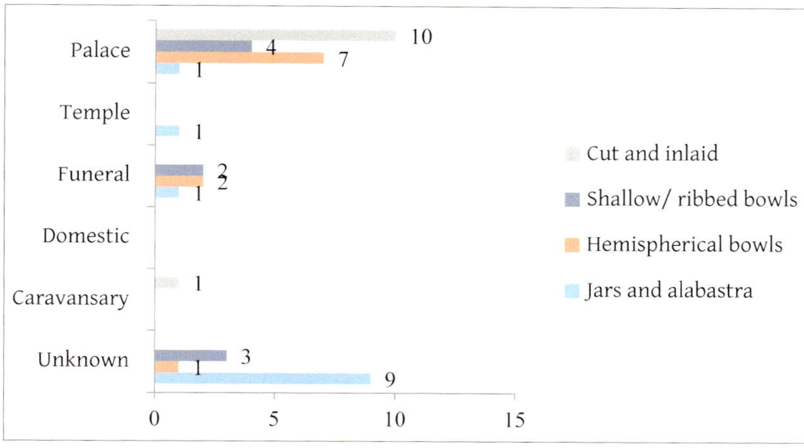

Table 5.12: Distribution of cast-and-cut vessels among different find contexts (n= 42).

banqueting before being deposited in graves, is an open question (Chapter 4.2.2.4).

Most of the vessels belonging to the group of jars and alabastra come from the art market (Table. 5.12). The 'Sargon Vase' from Nimrud was found in the Northwest Palace the piece from 'Atlit comes from a burial, and the alabstra from Idalion (Cyprus) can most likely be assigned to a temple complex.

To sum up, cast-and-cut vessels were predominantly found in palatial contexts. It should be emphasised that all pieces of this kind from Assyria – this especially includes Nimrud (Fort Shalmaneser, the Northwest Palace, but also Khorsabad (Residence K) – came to light exclusively in the palace environment, and can therefore in this region be linked to the Neo-Assyrian court. Cast-and-cut vessels were also part of palatial equipment in the palace of Amman and the Burnt Building II in Hasanlu. In Babylon, Gordion, Fortetsa, Praeneste and 'Atlit, cast-and-cut vessels were used as burial items. If they had served other purposes before cannot be determined. Only at Idalion did an alabastron most likely belong to a temple context.

5.4.2 Core- and rod-formed glass

By far the largest number of core-formed vessels discussed here were found in funerary contexts (19) (Table 5.13). The second largest group of vessels comes from domestic contexts (six) and only a small number from palatial (two) or temple contexts (one) (Table 5.13). 45% of the core-formed vessels cannot be attributed to any context.

Both of the sherds from palatial contexts come from the Burnt Palace in Nimrud; the sherd from the temple was found at Beth-Shean.

The burials in which core-formed vessels were found are located primarily in Babylonia (Babylon 3, Kiš 2,

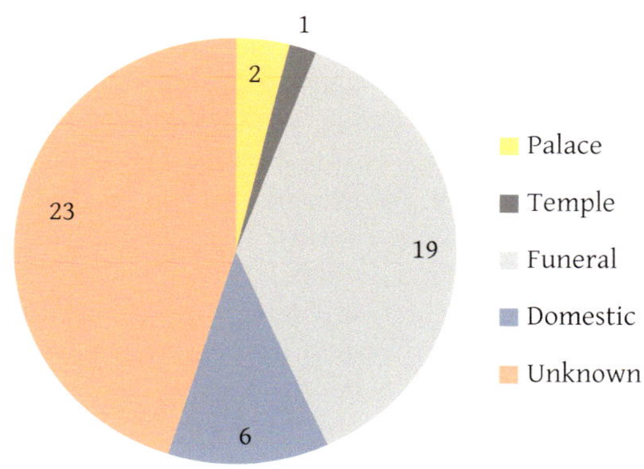

Table 5.13: Distribution of core-formed vessels among different find contexts (n= 51).

Nippur 2, Ur 1, Uruk 2), with only a small number found in graves of northern Mesopotamian sites, e.g. Aššur (two) and Ziyaret Tepe (one). The pieces from Carthage, Kameiros and Susa (four) also come from funerary contexts (Table 5.13).

Core-formed objects were found in dwelling contexts in Babylon (one), Ur (one), Busayra (one), Megiddo (one) and Sultantepe (Table 5.13).

Rod-formed tubes occur only at Hasanlu, where they were found in and around Burnt Building II and V.[3]

The distribution clearly shows that core-formed vessels at the beginning of the 1st millennium were mainly used as burial items. This was generally the case in Babylonia, but this specific function can also be observed at Susa. In palace contexts, however, hardly any core-formed vessels are documented, which shows that they obviously did not belong to the palace inventory.

[3] These tubes are not considered in the Table, as no comparisons to other sites can be drawn.

5.4.3 Primary products

The find contexts of ingots, raw glass, and waste material are largely unknown. Just two red ingots can be attributed each to the Burnt Palace and the South Eastern Palace at Nimrud. Two large blue ingots from Babylon were part of a hoard in a domestic context. The raw glass fragments from Hasanlu were all found in Burnt Building IV-V.

5.4.4 Summary

The evaluation of different find contexts depends strongly on the state of research and the relevant excavation areas at a certain site. However, the distribution of objects using different manufacturing techniques to the various find contexts demonstrates a clear distribution pattern, so that the previously formulated restriction is negligible: in palatial contexts, almost 100% of the cast-and-cut objects dominate the entire area of study (Table 5.14). In temples, burials and dwelling contexts this picture is reversed, even if not quite so clearly: core-formed vessels predominate here (Table 5.14). The contrast is particularly strong in domestic buildings, where only one cast-and-cut object was found, e.g. the palette from Megiddo. This distribution pattern shows impressively that cast-and-cut objects were part of the palace and its royal household, while in contrast, core-formed vessels were available to a wider group of people, which also includes their use as burial objects.

With regard to the objects of the cast-and-cut technique, it can therefore be assumed that this new technical development was closely associated with the institution of the palace. In this regard, not only the shapes, but the major characteristics of these vessels – transparency and translucency – have to be considered as important features, which will be further discussed in Chapter 8.2. Cast-and-cut inlays, except for inlays for composite statues, must be understood in conjunction with ivories and furniture closely connected to the royal palaces. Most of them were brought as booty to Assyrian cities, or were re-distributed to Neo-Assyrian governed sites, such as Arslan Taš (Chapters 8.1.2, 8.2).

Attachment and inlays for composite statues were only found in temples. It is therefore be tempting to think of their use as decorations for statues of gods (Chapter 4.2.2.11).

In addition to the cast-and-cut objects, ingots and raw glass material can also be attributed to the palace. Even if evidence is scarce, this can at least be considered regarding some of the pieces from Nimrud and Hasanlu. It can therefore be carefully suggested, at least regarding these two sites, that the glass industry there was connected to the institution of the palace and the Burnt Building II respectively.

5.5. Chronological developments during the Iron Age

This chapter summarises the chronological distribution concerning the different types and groups of glass objects by incorporating all available information discussed throughout the study. With regard to the dating of the objects, primarily their attribution to a certain archaeological find context is taken into account. This is shown in Table 5.3 with a black line. In some cases, and this is particularly true for the pieces found in Fort Shalmaneser and in the different palaces at Nimrud, only a *terminus ante quem* or a *terminus post quem* can be given (Chapter 3.1.4). These and similar cases are indicated with a grey line (Figure 5.3). This is due to the fact that in the respective buildings in Nimrud, as in other places, the find context cannot be attributed to a certain phase, but extends over several phases. In cases where there is no archaeological find context, a typological comparison is made to ensure a chronological classification. This is shown in figure 5.3

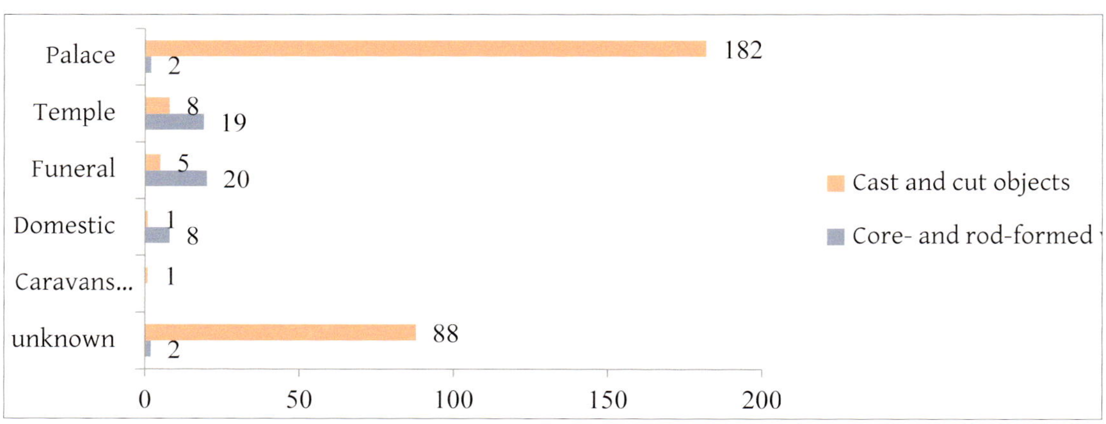

Table 5.14: Distribution of cast-and-cut objects and core- and rod-formed glass vessel among different find contexts. 'Tubes' and objects of primary production are not included (n=336).

with a grey dashed line. The typological comparisons of datable pieces are discussed in detail in Chapter 4. The data shown in figure 5.3 is not to be understood as fixed 'runtimes' of the types or groups of objects with a defined start and end point, but only represent the date of the objects investigated.

5.5.1 *Cast-and-cut glass*

5.5.1.1 Palettes, mace-heads and vessels

Cast-and-cut glass is the dominant type of glass being produced in the first half of the 1st millennium. In this chapter all groups of cast-and-cut glass are taken into account and are shown in Figure 5.3.

The glass palette from Megiddo was found in a context that dates to the second half of the 7th century (Chapter 3.3.7, Figure 5.3). The mace-head from Nineveh is attributed to a context that dates between the reign of Sargon II (721–705) and 612 (Chapter 3.1.5, Figure 5.3).

Among the group of jars and alabastra, the jars (Nim27, AM4) fall into the reign of Sargon II (721–705) (Chapter 4.2.2.3, Figure 5.3). Ur2 can most likely be attributed to the time of Nebuchadnezzar II (605–562) (Chapter 3.2.7, Figure 5.3). Alabastra are later in date and largely rely on typological comparisons drawn from other objects. Comparable finds point towards a date after the late 7th century (AM10, Id1, AM9) and even as late as the late 6th century (AM7, AM8, AM6, AM11, AM12). Regarding the large elongated alabastra (AM7, AM8), typological comparisons point to a post-Assyrian or Achaemenid date (Figure 5.3).

The hemispherical bowls from Nimrud were found in contexts with a *terminus ante quem* of 612 (Chapter 3.1.4, Figure 5.3). Khor1, however, was found in Residence K, which makes an attribution to the reign of Sargon II (722–705) likely (Chapter 3.1.2, Figure 5.3). The bowl from Praeneste can be attributed to a date prior to 650 (Chapter 3.4.7, Figure 5.3). The assumption that the hemispherical bowls could already have been produced at the beginning of the Neo-Assyrian period is supported by the chemical examination of some of the transparent bowls from Nimrud and Gordion. Chemical analysis on one of the transparent petalled bowls from Gordion shows that Phrygian glass does not exhibit a distinctive chemical signature, but rather clusters with glass from Nimrud (Privat *et al.* 2014; 2016) (Chapters 7.4.3, 7.4.5). Since the bowls from Gordion date back to the early 8th century, the production of the colourless primary glass, which was also used for the hemispherical bowls found in Nimrud, can be considered contemporaneous. It is therefore possible that hemispherical bowls occur already at the beginning of the Neo-Assyrian period, probably already during the reign of Ashurnasirpal II (883–859). This shows that the primary production of glass can also play an important role in the dating of glass objects. This is therefore discussed in detail in chapters 8.1 and 8.2.

The two bowls from Gordion (Gor1, Gor2), which belong to the group of shallow, undecorated and ribbed and petalled bowls, can be attributed to the early 9th and the first half of the 8th century (Chapter 4.2.2.5, Figure 5.3). From a stratigraphic point of view, Nim15 and Nim16 from Nimrud only provide a *terminus ante quem* for 612. The stratigraphic attribution of Bab3 points to a Neo-Babylonian or Achaemenid date (Chapter 3.2.1, Figure 5.3). Regarding the vessel fragments from Hasanlu (Has12), a date to stratum IVB (1000–800) can be supposed (Chapter 3.4.4, Figure 5.3).

Ar1, belonging to the group of 'cut-and-inlaid vessels', can stratigraphically be assigned to the 8th century (Chapter 3.3.2, Figure 5.3). Due to the typological similarity to Ar1, Nim22 and Nim23 could fall into the same period (Figure 5.3). The date for the incised bowl fragments Nim24, Nim25 and Nim26 suggest the 9th or 8th century, based on comparisons with a Phoenician bronze bowl from room AB of the Northwest Palace at Nimrud (Chapter 4.2.2.6, Figure 5.3). Whether cut-and-inlaid vessels were produced until the end of the Neo-Assyrian period cannot be decided on the basis of the available material.

5.5.1.2 Inlays

Most glass inlays, with the exception of inlays for composite statues and the large monochrome inlays, can be associated with ivory objects and furniture, and to a lesser extent with inlays for glass vessels. All the inlays from Nimrud were stored either in the palaces or in Fort Shalmaneser in Nimrud. For these pieces, solely a *terminus ante quem* of 612 can be assumed (Figure 5.3). Herrmann and Laidlaw (2013: 114) suggest that the great majority of ivories most likely arrived at Nimrud between 860 and 710, particular during the reigns of Ashurnasirpal II (883–859) and Sargon II (722–705). Ciafaloni (2009: 312) proposes that between the second half of the 8th and the first half of the 7th century a 'particularly propitious tendency to collect ivories and minor objects of Egyptian taste, Egyptianizing or even Egyptian (...)' must have existed. This would account for a later date for the deposition of the ivories at Nimrud, probably during the reign of Tiglath-pileser III (744–727), Sennacherib (704–681), and Esarhaddon (681–669), which would, however, not exclude an earlier date for their production (Ciafaloni 2009: 307). A date for the actual production of the ivories and for the glass inlays is more difficult to determine. With regard to the ivories, Herrmann and Laidlaw (2013: 115) suggest the period of 1150–710, since elephant ivory was used for Iron Age material, while hippopotamus ivory was common in the Late Bronze Age. If the ivory were produced during this period, it can be assumed that the glass inlays must also have been produced during this period.[4]

None of the painted inlays was found in a context that can be dated precisely. The decoration of the painted inlays matches the decoration on the ivory panels of the Phoenician style, but can also be combined with furniture panels decorated with ivory from the North Syrian style. Due to their close stylistic concordance, they can be associated with the incised bowl fragments Nim24, Nim25 and Nim26, which show typological parallels to bronze bowls of the 9th and 8th centuries. It is therefore conceivable that the painted inlays were already produced in the 9th or 8th century. However, the exact time period for the production of this type of inlay is unknown.

Rosette inlays belonging to groups 1 and group were found in rooms S10, S20, SW37, and T20 of Fort Shalmaneser, and in room V of the Northwest Palace. Whereas the finds from the Northwest Palace, as well as those from S10, S20, and SW37 cannot be dated precisely, the glass finds found in T20 come from a deposit sealed between two floors and dating, respectively, to the reign of Shalmaneser III (858–824) and Esarhaddon (681–669) (Chapter 3.1.4, Figure 5.3). Most of the rosette inlays that belong to group 3 can be attributed to the 'Bâtiment aux ivoires' at Arslan Taş, which dates to the first half of the 8th century (Chapter 3.1.3, Figure 5.3). The glass bowl A1 is decorated with glass frames of this type and was found in a context that dates prior to 700, thus supporting the date for the ivories from Arslan Taş. Finally, it can be assumed that group 3 rosettes were in use in the 8th century (Figure 5.3).

Almost all of the monochrome inlays from Nimrud were either found in room S10 in Fort Shalmaneser or in in room V in the Northwest Palace, therefore having a *terminus ante quem* of 612. None of the pieces can be dated on the basis of stratigraphic considerations. Only the ivories with the 'Lady at the Window' motif can be attributed to Arslan Taş and indicate the 8th century (Chapter 1.2.2.9, Figure 5.3).

Large monochrome inlays were only found at Khorsabad and can therefore be dated to the reign of Sargon II (721–705) (Chapter 3.1.2). Since the inlays from Khorsabad are unique of their kind, no further consideration can be given to their date (Figure 5.3).

The attachments and inlays for composite statues found in the Ninurta Temple at Nimrud (Nim1, Nim5, Nim6) were deposited sometime between the reign of Ashurnasirpal II (883–859) and the sack of the city in 612 (Chapter 3.1.4). Based on typological parallels elaborated in Chapter 4.2.2.11, it is likely that Nim1 and Nim2 fall into the time period, between the reign of Sargon II (721–705) and 612.[5] On the basis of typological similarities, Nim1 can be attributed between the 10th and late 8th century, as discussed in detail in Chapter 4.2.2.1. Is1 and Is2 show a *terminus ante quem* prior to the reign of Nebuchadnezzar II (605–562). The group of composite attachments and inlays range widely in date therefore (Figure 5.3).

5.5.2 Core- and rod-formed glass

Only a small number of core- and rod-formed vessels can be dated based on their archaeological context. Therefore the objects are mainly dated on the basis of typological comparisons with other vessels, which are discussed in detail in Chapter 4.3.3.2.

The group of ovoid bottles with pointed base have close parallels to glazed and unglazed ceramics from Assyria and Babylon from the 8th to 7th century (Chapter 4.3.3.1). Very close parallels occur among glazed pottery vessels from northern Syria and the Aegean of the same period. Bab4 is attributed to the time between the 7th and early 6th century. The find contexts of the glass vessels themselves, as well as their comparative pieces, allow a dating of these two types into the 6th century (Figure 5.3, Table 5.15).

The ovoid bottles with rounded base are not very typical, neither for the glazed pottery vessels, nor for the faience* vessels of that time (Chapter 4.3.3.1). Comparisons for this vessel type are more likely to be found among the smaller cast-and-cut alabastra, e.g. AM10, and their counterparts made of stone. One rare comparison made of bronze comes from Jordan, which dates to around 600. For Ur5 a *terminus post quem* for the 8th century can be established. The find context of Kiš2 is dated to the 5th century. Therefore, even if Ur5 has a *terminus ante quem* for the 8th century, the find spot of Kiš2 and the comparative finds rather indicate a date around 600 for this group (Figure 5.3, Table 5.15).

Among the piriform bottles there is a number of objects that can be attributed to specific find contexts (Chapter 4.3.3.1). However, the time period is quite broad: Bus1 can be linked to an Iron Age II or Achaemenid context, Nip1 to a Neo-Babylonian context, and Kiš1 to the 5th

[4] According to Herrmann and Laidlaw (2013: 114), Sennacherib (704–681) would have chosen his new capital, Nineveh, to store his booty, but almost none of the ivory objects was found there. And Esarhaddon (681–669), who restored Nimrud, probably also stored some of the stolen ivory there. The ivory found in Til Barsip and Arslan Taş was most likely brought there as donated booty via the Assyrian capitals.

[5] The attachments Nim1 and Nim2 show twisted elements between the curls of the beards, similar to those depicted on the reliefs of Sargon II (722–705), Sennacherib (704–681) and Ashurbanipal II (668–669). The design of Nim1 is dated by Strommenger (1970: 32) to the time of Sargon II.

Ovoid bottles, pointed	**Ovoid bottles, rounded**	**Piriform bottles**	**Globuar bottles**	**Small jars**	**Bottles with disc-base**	**Large cylindrical bottles**
8th – early 6th century	8th – 5th century	7th – 5 th century	late 7th – early 6th century	early 7th century	mid 6th – early 5th century	8th – late 7th century
	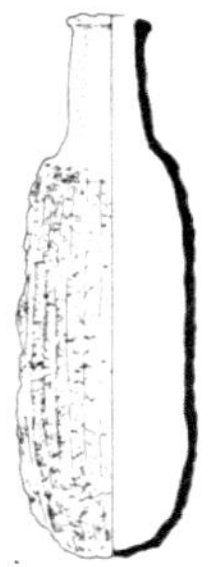		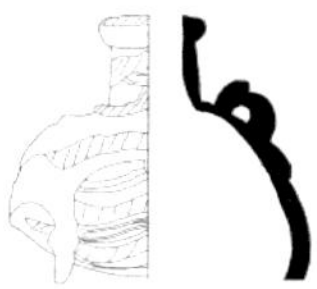		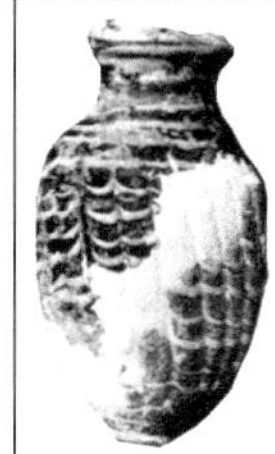	

Table 5.15: Chronological overview of the different types of core-formed glass vessels and their dates (not to scale).

century. In the case of the bottles from Susa (Sus1, Sus2, Sus3, Sus4), a close connection to faience vessels found in burial contexts at the same site (late 7th to 6th century) can be established (Figure 5.3, Table 5.15). Therefore the piriform bottles can be roughly dated from the 7th to the 5th century (Chapter 4.3.3.1).

The globular bottle Bab7 is attributed to a burial dating between the 7th and the early 6th century (Chapter 4.3.3.1). Bab6 is typologically so similar to Bab7 that they are probably contemporaneous. Both date to the late 7th or early 6th century (Figure 5.3, Table 5.15).

Bab5 belongs to the group of small jars and is cautiously attributed to an early 7th century burial (Chapter 4.3.3.1). Close comparisons can be found among glazed pottery vessels of the 7th century from Aššur (Andrae 1925: fig. 14; Hausleiter 2010: pl. 40d, 45i), Babylon (Reuther 1968: pl. 74, 135, 139) and Uruk (Boehmer *et al.* 1995: pl. 101e, 132b, 145a), which confirm this date (Figure 5.3, Table 5.15).

None of the bottles with disc-base was found in a datable context (Chapter 4.3.3.1). Because of their good comparison with core-formed vessels of the 'Mediterranean Group I' (mid 6th – early 5th century) it is likely that this group dates later, as the other vessel types presented in this study (Figure 5.3, Table 5.15).

The large cylindrical bottles form a particularly tight group, out of which three pieces were found in the Aegean (Kameiros (two), Carthage (one), and one in Mesopotamia (Aššur) (Chapter 4.3.3.1). Fragments from Nimrud and Nippur are to be added. As10 was found in a burial dating between the late 8th and late 7th century. The fragments from Nimrud come from a context with a *terminus ante quem* of 612. All the vessels – apart from the one from Aššur – have vertical, irregular tool marks on their necks, which could indicate that the pieces were made in the same workshop. The best comparisons for this type are among large storage jars (45–100 cm) made of ceramic, which were found at sites such as Beth-Shean (Mazar 2006: 457 pl. 37, no. 4), Hazor (Yadin 1961: pl. CCXXIX; Ben-Tor *et al.* 2012: Figure 4.13), Lachish (Tufnell 1953: pl. 95, no. 498) or Sarepta (Pritchard 1975: fig. 23, 20), and date to the 8th century (Figure 5.3, Table 5.15).

The tubes from Hasanlu are to be assigned to layer IVB and therefore date between 1000 and 800 (Chapter 3.4.4, Figure 5.3).

Kohl tubes can be attributed to the post-Assyrian and Achaemenid periods, and the head pendants cannot be dated before the middle of the 6th century (Figure 5.3).

5.5.3 Summary

The chronological summary of all glass objects discussed throughout this study shows first of all that glass production and processing was carried out in the 10th century, i.e. at the beginning of the period under study (Figure 5.3). This is confirmed by the finds from Hasanlu IVB, where both the group of tubes and a specimen of the shallow, undecorated, ribbed and petalled bowls occur (Figure 5.3, Chapter 3.4.4). Furthermore, one find of the group of attachments and inlays for composite statues from Nineveh can be attributed to the 10th century (Figure 5.3, Chapter 3.1.5). To these early finds also a number of glass beads from Pella can be added, which date to Iron Age IB/IIA (1050–850) (Bourke 1997: 112–113) (Chapter 7.4.4). Although not from Mesopotamia and its immediate surroundings, but because of their corresponding date, of importance are the glass objects from Nesikhons in Egypt, which show a *terminus ante quem* of 975/974, and which are therefore also mentioned here (Schlick-Nolte and Werthmann 2003). All these finds show that only a relatively small number of glass finds compared to later periods are present in the 10th century, but no interruption can be observed at the transition from the end of the second to the early 1st millennium (Figure 5.3).

In addition, it can be stated that the majority of objects produced using the cast-and-cut technique date earlier than the objects of the core-forming technique. All cast-and-cut vessels, with the exception of the jars and alabastra, were probably already used during the reign of Ashurnasirpal II (883–859), but at the latest from the 8th century onwards (Figure 5.3, Chapter 5.5.1.1). This is identical to the date of the inlays (Chapter 5.5.1.2). At this point, reference is made to the primary production of glass, which plays a role in determining the start of the group of cast-and-cut objects. It can be noted that the transparent glass from which the Gordion bowls are made is chemically identical to the transparent glass used among the glass objects from Nimrud (Chapters 7.4.3, 7.4.5). Since the bowls from Gordion can be attributed to the late 9th century, this shows that a generally early production of transparent glass can also be expected in Mesopotamia. In addition, a previous experimental phase must be expected, which must have ultimately led to the fully pronounced development of transparent glass objects as they are present in this study.

Production of vessels by the core-forming technique seemed to start later at around the 7th century (Table 5.15, Figure 5.3). Only the group of ovoid bottles are already present in the 8th century (Table 5.15, Figure 5.3). This picture is surprising, because even though the technique of core-forming was the common method used to produce glass vessels in the Late Bronze Age, this type of glass was – according to the material studied here – not produced in the 9th century. A production of core-formed vessels on a larger scale apparently only resumed in the middle of the 6th century with the beginning of the production of the 'Mediterranean I Group' vessels. During this period, core-formed glass is distributed throughout the Mediterranean (Chapter 4.3.1).

Due to their late date, the 'kohl tubes' and 'head pendants' have not been further considered in this work (Figure 5.3)

The picture presented here demonstrates that cast-and-cut glass occurs already in the 9th – and at Hasanlu in the 10th – century. This goes hand-in-hand with the invention and production of transparent, colourless glass. The tradition of core- and rod-forming never entirely disappeared, as is witnessed by the rod-formed tubes from Hasanlu. But it can clearly be stated that the majority of core-formed objects occur later than cast-and-cut objects in the 8th century.

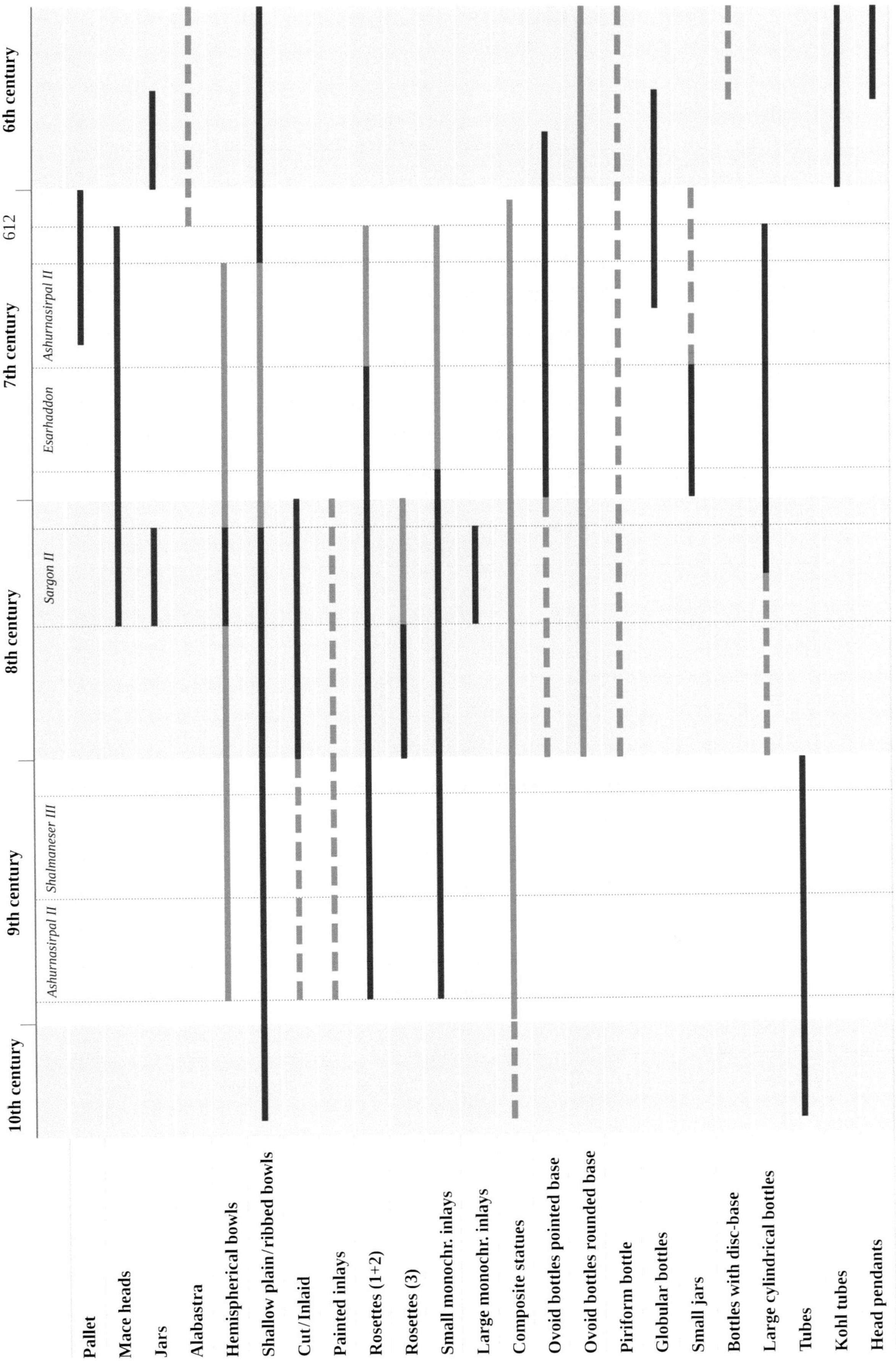

Figure 5.3: Running times of the different types of glass objects. The black line shows the date of the archaeological context, the grey line indicates a *terminus ante quem* or a *terminus post quem*, and the grey dashed line refers to typological comparisons of datable objects (template: Katharina Schmidt, design: Mareike Walter.

6. The Nineveh Glass Recipes

6.1. The understanding of the glass texts

The 'Nineveh glass recipes' comprise a group of cuneiform tablets found in the library of Ashurbanipal (668–631) in Nineveh, and which contain recipes for the production of raw glass in different colours. There are 38 fragments that can be joined to five clay tablets which are all kept in the British Museum. Apart from this group, three other isolated cuneiform texts from the mid to late 2nd millennium are known that deal with the production of raw glass, one Middle Babylonian text with unknown proveniance (BM 120960) (dated to the last third of the 2nd millennium) and one from Hattusha (BM 108561), both in the British Museum, and one from Babylon (VAT 16453) now in the Vorderasiatisches Museum, Berlin.

6.1.1 Previous studies on glass texts

The debate about the character and use of the Nineveh glass recipes and the other three glass texts focussed on the question how to classify them in the cuneiform textual record, and how to place these texts in the cultural and literary traditions of the period. Are the texts – in particular the Nineveh glass recipes – originals or copies? Did the texts serve as recipes, lexical lists, or ritual texts? Or do they combine features? These fundamental questions are closely linked to the way a text is interpreted, which in turn is strongly related to current trends in research. Using the example of the glass texts, it is impressively reflected that text interpretation is always a 'child of its time' and should be understood in this way. Against this background, the history of research is to be given adequate space in the following paragraphs.

In 1909, Virolleaud provided the first translation of a text that deals with the production of raw glass, i.e. K.203 (fragment of Tablet B after Oppenheim 1970) of the corpus of the Nineveh glass recipes. Landsberger was the first who identified almost all the texts that deal with the production of raw glass (Zimmern 1925: 178). Both Zimmern (1925) and Thompson (1925) published the texts almost simultaneously, approaching the subject from different angles: Zimmern in his article 'Assyrische chemisch-technische Rezepte, insbesondere für Herstellung farbiger glasierter Ziegel, in Umschrift und Übersetzung' emphasised the transliterations and literal translation of several fragments of the Nineveh glass recipes.[1] Thomson, as part of his monograph *On the chemistry of the ancient Assyrians* (1925) provided hand copies of the tablets, and focused on the determination of the stone names, as well as on the identification of the manufacturing process in these texts, which he set against the background of 'modern' technical processes, similar to Oppenheim in his 1970 study.[2] In the same year, 1925, Meissner published a partial translation of a Nineveh glass text, which he placed in the chapter 'Natural Sciences and Exact Sciences' ('Die Natur- und exakten Wissenschaften') of his book *Babylonien und Assyrien* (Meissner 1925: 383–385). All three authors positioned the texts into the broad field of chemistry, technology and science, which becomes evident from the titles of their publications. Zimmern (1925: 178) refers to their 'character of recipes' ('Charakter als Rezepte'), and Thompson also interprets them as realistic instructions for the production of raw glass:

> 'The first essential in solving any problems about ancient glass or glaze is, I take it, to compare the principles involved in the technique of modern methods. The fundamentals of all glassmaking are so simple, at any rate in theory, that we should be able to arrive at absolute certainty in the identification of most of the Assyrian components.' (Thompson 1925: 5)

An edition of the Middle Babylonian text BM 120960 was provided by Gadd and Thompson (1936), also referring to a chemical text in their title (1735080668 *A Middle-Babylonian chemical text*). The text (BM 108561) from Hattusha was published by Rosenkranz (1965) as an 'economic text' ('Wirtschaftstext').

Oppenheim (1970) wrote the seminal work on glass from Mesopotamia and included editions of all glassmaking texts then known (see also Chapter 1.3). Oppenheim defined the genre of the 'procedural instructions', to which he also assigned the texts dealing with the production of raw glass.[3] He went on further to define the text types that belong to the corpus of procedural instructions. Some of these would be considered today as examples of 'exact science', such as mathematical or astronomical texts, some others would fall into the domain of 'ritual' and 'medicine' (Oppenheim 1970: 5–6). The glass texts form – together with the texts that deal with the preparation of perfumes – a third group that comprises 'chemical texts', or texts dealing

[1] For an inventory of all fragments, see Zimmern 1925: 177. For a concordance list of all published fragments until 1970, see Oppenheim 1970: 29.

[2] It is interesting that the texts received attention at the same time; one reason might be their distinctive 'unique character'; see Zimmern 1925: 178.

[3] This term was first mentioned by Oppenheim (1970: 4); for further reference, see Jursa 2001: 299, footnote 4.

with 'alchemy' (Oppenheim 1970: 5).[4] Even though Oppenheim – like Zimmern, Thompson and Meissner – recognises the feasibility of the recipes, he brings a new dimension into the discussion, referring to the role of literary creation in the literary tradition:

> '[The 'chemical texts'] are meant to instruct scholars specializing in these fields (...). All of these are scholars – one should say rather than 'scribes'– who maintain the intellectual tradition by copying the texts which contain such instructions. Neither the perfume maker nor the craftsman producing colored glasses is likely to be a scribe. Under unknown circumstances, the technical lore of certain artisans which catered to the need of the court was fixed in writing, presumably upon a royal order. Once admitted to and incorporated into the corpus of traditional writings, these texts continued to be copied by tradition-conscious scribes and kept in private or royal libraries.' (Oppenheim 1970: 6)

The quote shows that Oppenheim clearly differentiates between the circumstances and the time of creation of the original text – which he labels as unknown – and the process of making copies of these originals. This differentiation is reflected by the use of the terms 'text' and 'manuscript' in this study (Chapters 6.3, 6.4). Even if Oppenheim accepts the feasibility of the instructions, he does not assume that these recipes were actually applied in practice and therefore reflect technical practice. Rather, he places the role of the scribes, the act of writing, and the maintenance of literary tradition in the centre of his argumentation, and indicates that the texts are 'primarily the work of scribes, or better, that of scholars bent on maintaining the literary tradition, rather than that of 'scientists' interested in producing a corpus of technical lore in writing.' (Oppenheim 1970: 40). This view is influenced by the mid-century Assyriological consensus about the word 'tradition', which was strongly shaped by what Oppenheim (1960: 410) himself called 'stream of tradition', and which understood the bulk of cuneiform literature as being conservative in nature, but less a product of individual creativity.

Moorey (1994) takes a sharp position by denying that the texts were of any practical use. He classifies the texts as 'lexical lists' and states that the texts 'served literary or administrative purposes, they did not represent a body of technical or scientific literature' (Moorey 1994: 210). He goes even further and denies that the literary tradition could have anything to do with technical history (Moorey 1994: 211). In this estimation, Moorey followed Muhly (1972: 181), who understood the text to be 'extended lexical lists', precursors of the classical handbook tradition. This approach to the texts must be seen against the background of research history, as research of Mesopotamian realia in that time relied heavily on the evidence of 'lexical lists', enumeration of names for objects (animals, plants, metals, stones, etc.), of which the most prominent scholar is Salonen (1966; 1970a; 1970b).

6.1.2 The distinction between 'manuscript' and 'text'

All previous approaches to the topic are taken into account here, however, emphasising a precise distinction between 'manuscript' and actual 'text' and their respective functions. A clear distinction between these two terms helps to avoid difficulties in classifying the texts, since formal aspects are mixed with content-related ones.

First of all, to make a precise separation between 'manuscript' and 'text', the terms must be defined. A relevant definition was proposed by Cooper (2005: 49), who illustrates the need for a definition by means of a conversation between two professors of Sumerology:

> 'Prof. X: Well, what text are you working on now?
>
> Prof. Y: 'Gilgamesh and the Stallion of Inana'.
>
> Prof. X: Interesting. How many texts do you have?
>
> Prof. Y: About 270, mainly from OB Uruk, but a few from Eridu.
>
> Prof. X. Hmm. Are there many textual variants?
>
> Prof. Y: Very few, though the Uruk version has more phonetic spellings than the Eridu recension.' (Cooper 2005: 49)

With this example, Cooper points out how differently the word 'text' is used, which refers to different things in detail, and proposes the following definitions: ''Gilgamesh and Stallion of Inana' is a 'composition' reconstructed from 270 'manuscripts', each of which contains a portion of the composition's text.' (Cooper 2005: 49). He understands 'text' as the 'words on a particular manuscript' and 'manuscript' as an 'exemplar of a composition (usually on a clay tablet)' (Cooper 2005: 50).

With reference to the Nineveh glass recipes, the term 'text' is used according to Cooper's definition. 'Manuscript' refers here to the copy of the original 'text' which is written on a clay tablet (e.g. the tablet on which manuscript A is written consists of eight fragments).[5] The term 'composition' is not expedient with regard to the Nineveh glass recipes and therefore

[4] For the texts on the production of perfume, see Chapter 6.3; for the texts, see Ebeling 1948. On alchemy, see Oppenheim 1960.

[5] Oppenheim (1970: 29) uses the words 'tablet' and 'manuscript' interchangeably.

not used in this study, as the contents of the individual (original) texts vary.

Furthermore, a clear distinction is made between the group of texts labelled here as the 'Nineveh glass recipes' found in the library of Ashurbanipal, and the remaining three isolated texts that deal with the production of raw glass (BM 120960, BM 108561, VAT 16453).[6]

6.2. The library of Ashurbanipal and its 'manuscripts'

Nineveh glass recipes were found in the palace of Ashurbanipal[7] which contained the large accumulation of 20,000 cuneiform tablets, not including the several thousand waxed writing boards which are not preserved.[8] Among the tablets two major groups can be distinguished. The first group contains texts of the 'stream of tradition' in the terminology of Oppenheim (Chapter 6.1), which are literary, lexical and historical texts, rituals, medical compendia, Sumerian prayers, and omen texts; the second group comprises archival documents, e.g. letters, contracts and administrative notes (Frahm 2004: 45). The clay tablets were acquired by Ashurbanipal (668–631), who was immensely interested in reading, writing and textual hermeneutics. The majority of the tablets were copies he ordered to be made.[9] Some of the tablets were inheritances, and about 17% of them were acquired on raids in Babylonia, during which even the Babylonian scholars themselves were often captured (Fincke 2004: 55). These tablet copies made for the library are called manuscripts.

Among this large number of clay tablets there are also the Nineveh glass recipes: a total of 38 fragments that could be joined to five manuscripts – designated by Oppenheim (1970) with the letters A, B, C, D and E – with texts describing the production of different coloured glasses (*zagindurû*, *tersītu*, 'fast bronze' and *dušû*).[10] Some of the manuscripts largely overlap with regard to their content, in some parts, however, there are variations, which are shown in the score edition to be found in the Appendix of this study.

6.3. The 'texts': genre and function

In contrast to 'manuscript', which represents the copy of an original text, 'text' refers to the original text and means the contents. It is unknown when the original texts were written. Due to the existence of the three singular glass texts from the mid 2nd millennium (BM 120960, BM 108561, VAT 16453), it can, however, be concluded that the tradition of writing down the glass texts dates back to this period.

The Nineveh glass recipes were copied during the reign of Ashurbanipal (668–631). Variations and parallel sequences to be found in the manuscripts make it likely that the group of Nineveh glass recipes represent the n-th generation of copies of the original texts: 'the essential step forward in the study of the glass texts lies in the realisation that they present a text composed of several sources.' (Oppenheim 1970: 24). Consequently, the information on the production of raw glass to be found in the Nineveh glass recipes would, at least partly, refer to earlier originals probably reaching back to the mid 2nd millennium.[11]

With regard to formal aspects, the Nineveh glass recipes are all similar in structure and use similar technical terms. Therefore they form a consistent group of texts. There are a number of rare or unique words, which can be interpreted as a technical jargon with reference to glassmaking. Some of the words are Assyrian dialect words, e.g. the Neo-Assyrian word *aḫussu* for Babylonian *uḫultu*. These rare and technical terms carry the risk for the modern researcher of resulting in circular reasoning: '(...) the translator looking to the glass technologist for the meaning of the word and the technologist looking to the translator for details of the raw materials for ancient glassmaking.' (Shortland 2008: 65).[12]

With regard to a genre definition, it was Oppenheim (1970: 40) who assigned the Nineveh glass recipes for the first time to the group of so-called 'procedural texts'.[13] It was he who coined the term of this genre (Oppenheim 1978: 649). As well as texts concerned with the production of raw glass, this category further includes instructions for the production of perfumes

[6] Oppenheim (1970: 4) made a similar subdivision and called the Nineveh glass recipes 'glass texts'.

[7] For the understanding of the term 'library', see Pedersén 1998.

[8] For the history of excavations and findings, see Perdersén 1998: 161–163 and Robson 2013: 41–42.

[9] How emphatic he was regarding these can be seen in some of the texts themselves; see Frame and George 2005.

[10] Six further fragments could not be joined and were therefore designated as such. Regarding the text discussed here, two new fragments that belong to manuscript B were identified (K. 20141, K. 18490). As indicated above, Oppenheim uses the words 'tablet' and 'manuscript' interchangeably; see Oppenheim 1970: 29.

[11] An aspect that could indicate an early date for the Nineveh glass recipes is that the chemical changes that took place in the primary production of glass in the early 1st millennium – choice of flux, even though plant was still widely used, the de-colourisers, and high lead, high copper red glass – are not reflected in the Nineveh glass recipes, as far as can be judged. This could either reflect the late 2nd-millennium glassmaking tradition reflected in the later manuscripts, or it is pure coincidence. Since only one of the Nineveh glass recipes was studied in detail in this monograph this cannot be fully assessed, but it would be interesting to investigate in the future.

[12] Within this group of Nineveh glass recipes, Oppenheim (1970: 27–28) distinguished between two different textual traditions, which he designated as 'Alpha Group' and 'Beta Group'. The groups differ solely in their ordering of the sections. The 'Alpha Group' comprises a detailed description of the procedures and standardised technical terms, the 'Beta Group', in contrast, contains an enumeration of ingredients and their measurements.

[13] Jursa 2001: 299; the expression 'procedural text' is based on Oppenheim 1970: 40.

(Ebeling 1948; 1949; 1950), or for dyes (Leichty 1979), and culinary texts (Bottéro 1995), in addition to others.

The main formal characteristic of the genre of 'procedural texts' is the use of the second-person singular, and the present tense, which implies that the acting person is addressed as 'you': 'you do that and that'. The texts are short, stereotyped with repeated use of the same verbs and nouns, and show similar phrase structure. They are also technical in nature, using a technical jargon (Jursa 2001: 299–300). With regard to the Nineveh glass recipes, all these characteristics apply. Particularly obvious is the use of a specialised technical jargon for glassmaking.

6.4. Function of 'manuscript' and 'text'

To explain the function of the texts in more detail, it is necessary, as mentioned in Chapter 6.3 and 6.4, to make a clear distinction between 'manuscript' and 'text', because both have different functions. The function of the 'text' asks for the contents of the words that were written down at some point in history before Ashurbanipal. The function of the 'manuscript' asks why the original texts were copied and placed in the library of Ashurbanipal.

Function of the 'text': The detailed analysis provided in Chapter 6.6 of one of the Nineveh glass recipes – that is concerned with the production of *zagindurû* glass – shows that this text follows a reasonable rational order by describing the production process in detail. Even though the identification of some words remains problematic, it is possible to follow the recipe. Most of the ingredients described can be identified as necessary for the production process of raw glass, which can be determined as a multi-stage process. Details on measurements and tools also seem to be realistic. The practicability of the production of *zukû*-glass was even tested by Brill (1970: 111–114), performing the described procedure step-by-step. By doing so, Brill demonstrated that the text explains a feasible method of the production of a pale coloured glass. With regard to the other recipes, this cannot be said for certain as they have not been subjected to a detailed analysis. When they are read, however, they seem to follow a similar principle, also because the group of the Nineveh glass recipes are very regular. It can therefore be stated that the processes described in the texts are feasible according to today's assessments (lastly Shortland 2008: 69). In conclusion, the Nineveh glass recipes should therefore be taken as a potential resource for technological information. How this precise knowledge came from the (glass-) workshop to the clay tablet in the scribe's office cannot be decided, not even if the glassworker himself had something to do with the process of capturing the text. Was the scribe himself in the workshop to record the knowledge of raw glass production? Unfortunately, these questions must remain unanswered.

Function of the 'manuscripts': The question of the function of the manuscripts, which are copies of the texts, is strongly related to the function of the library of Ashurbanipal itself, and the question why it was established. As mentioned above, the manuscripts in the library of Ashurbanipal are copies of earlier texts dealing with many different topics. The main aim of the library of Ashurbanipal, therefore, was to provide a large dataset of texts of very different character, even though a focus was placed on omens to aid royal decision-making and rituals. Even if one wants to associate the Nineveh glass recipes with the priestly sphere – as suggested, for example, by Robson (2013: 55), since the different colours can be assigned different magical functions, including beneficial and healing elements – they would also suit the corpus of the library texts.[14] The compilation of the library was 'neither static nor universally similar' (Robson 2013: 55), but also a little coincidental. Therefore, obviously the question of why the Nineveh glass recipes in particular were added to the library of Ashurbanipal cannot be answered.

6.5. Glass in cuneiform texts

The identification of the material glass in cuneiform texts is problematic, as neither a specific term for 'glass' nor for 'glassmaker' exists in the Akkadian language. Rather a multitude of expressions with different literal meanings refer to glass. The word's identification is then often based on the textual context.

Very often glass is designated with the logogram NA$_4$ (*abnu*), which means stone.[15] In many cases, only contextualisation enables a decision regarding whether reference is being made to the material stone or glass.[16] Regarding texts from the second half of the 2nd millennium, a distinction between 'genuine' and 'artificial' stone is often made.In this regard, the name and colour of the stone are used in association with the expression *kūru* (kiln) or *šadû* (mountain), indicating the 'artificial' (kiln) or 'genuine' (mountain) nature of the stone. The artificial lapis lazuli *uqnû ša kūri* (lapis lazuli from the kiln), to be identified as blue glass, occurs frequently. In the texts studied here,

[14] The cultural association of glass was very different in the ancient world from our modern understanding. Therefore, different colours of glass had special magical properties, including beneficial and healing elements. This, however, does not mean that the production of glass was necessarily either magical or technical. Rather, both concepts must be adopted simultaneously for ancient Mesopotamia. This means, in relation to the text, that the production process could contain ritual elements, but at the same time could function as a realistic recipe.

[15] NA$_4$ has a range of different meanings. Besides the natural state of stone, it can also signify metal and metrological weights made from any material; see Robson 2001: 40.

[16] For a summary on the text references, see Oppenheim 1970: 13.

also different kinds of 'stones' are mentioned that can be identified as different coloured glasses, according to the colours of the stone.

Oppenheim (1970) has compiled all relevant text types that contain evidence for coloured glasses, which are on the one hand economic records and letters, and on the other word lists, such as the ḪAR. ra= *ḫubullu*. Because of the intensive analysis provided by Oppenheim (1970: 9–21), a detailed discussion is not necessary here.[17]

6.6. The recipe for blue *zagindurû*-glass

This study provides a content-related commentary as well as an edition (see Appendix 1) about the production of blue *zagindurû*-glass. Since most manuscripts are available for this recipe, this particular example was chosen. By far the most common type of glass in the archaeological record is translucent blue, consequently this recipe is representative for this period.

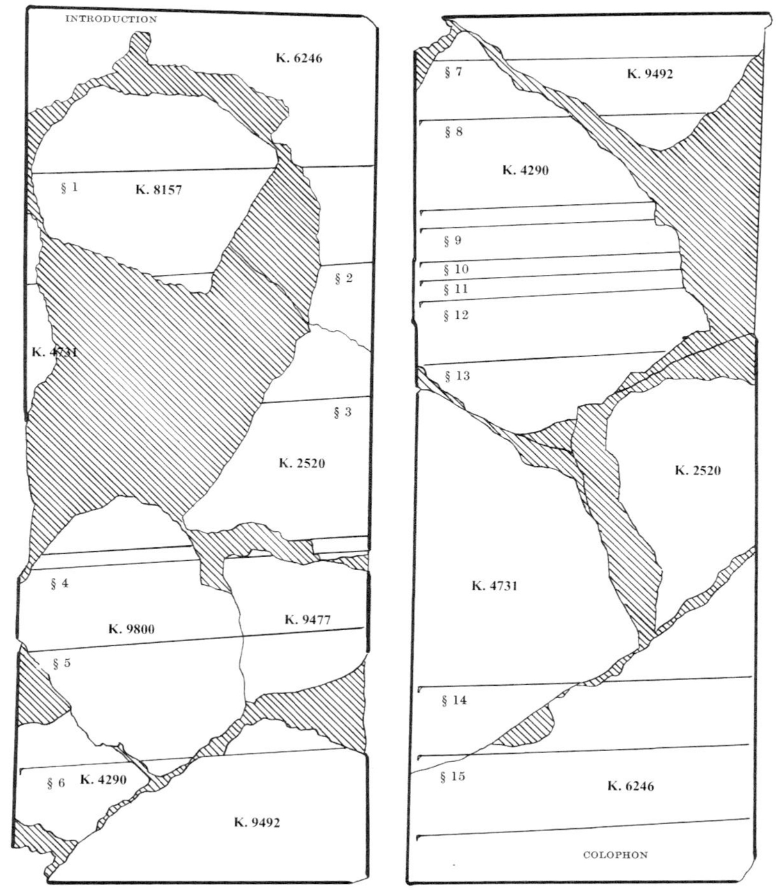

Figure 6.1: The different fragments joined to a clay tablet (here manuscript A in obverse and reverse) (after Oppenheim 1970: fig. 2, 3).

Six different versions of the *zagindurû*-recipe are preserved. Manuscripts A and B consist each of two recipes that follow one after the other. Further versions are manuscripts C and d. The recipes differ slightly, but all result in the same end product: *zagindurû*-glass. Regarding the recipes for *zagindurû*-glass, the following manuscripts and corresponding fragments are of interest.

The tablet with manuscript A[18] (one-columned) consists of eight fragments: K. 6246 + K. 8157 + K. 2520 + K. 4731 + K. 9800 + K. 9477 + K. 4290 + K. 9492 (Figure 6.1).

The tablet with manuscript B (Oppenheim Tablet B) (two-columned) can be reconstructed from 11 fragments in addition to two new fragments: K. 203 + K. 2311 + K. 4747 + K. 5839 + K. 5862 + K. 6891 + K. 9940 + K. 10493 + K. 13367 + K. 20141 (new join) + K. 18490 (new join) (Figure 6.2).

The tablet with manuscript C (Oppenheim Tablet C) (one-columned) can be reconstructed from five fragments: K. 6964 + K. 7619 + K. 13265 + K. 8452 + K. 81-2-4, 291. Finally, fragment 3 d is represented by K. 9551.

The singular recipes are organised by consistent division lines that demarcate the different intermediate products necessary to produce the end product (Figure 6.3). The single paragraphs within the texts describe the manufacturing processes. With reference to the text discussed here, the first intermediate product can be identified with *zukû*, the second with *tersītu* (Figure 6.3). The last paragraph finally mentions the end product: *zagindurû* (Figure 6.3).

The recipe explains the production of *zagindurû*-glass. This is indicated by the first sentence: 'If you want to produce *zagindurû*-glass' (**#2.1**) and '(out of the kiln) rises *zagindurû*-glass' in the last sentence of this text (**#4.9**). *Zagindurû*-glass represents the final product of the *chaîne opératoire* and is produced by various steps characterised by different intermediates, which are called *zukû* in **#2.7** and *tersītu* in **#3.9**. Each intermediate step is characterised by different operations, consisting

[17] Köcher 1957: 302: 18, 32, 37, 304: 3, 6, 8, 23, 26, 30, 31, 33, 306: 7; here, as well as genuine lapis lazuli, other artificial stones are listed. For a summary of further evidence, see Oppenheim 1970: 11–14. There are also other words that indicate a difference between artificial and genuine stone, e.g. *bašālu* (to cook) for an artificially produced stone, or *damqu* (fine) for a genuine stone; Bottéro 1949: 138: 7, 140: 30, 142: 43.

[18] Oppenheim (1970) called it Tablet A.

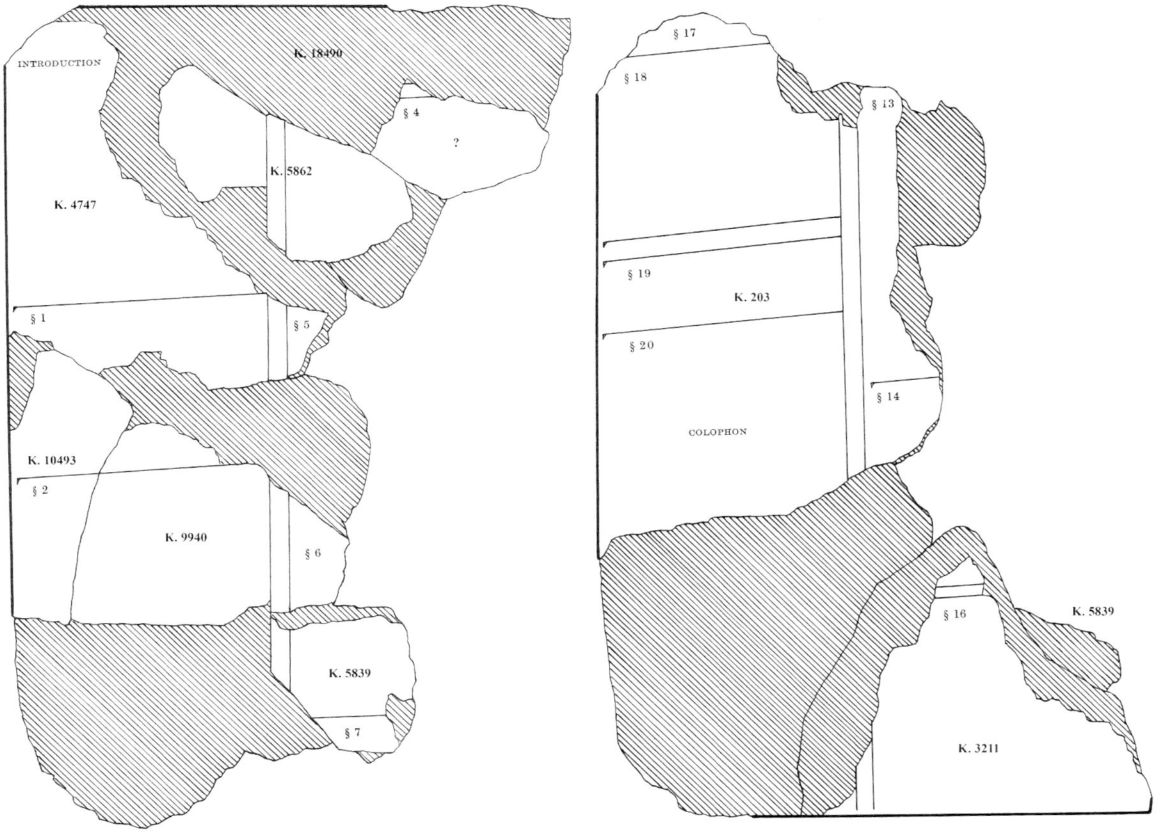

Figure 6.2a, b: The different fragments joined to a clay tablet (here manuscript B in obverse and reverse) (after Oppenheim 1970: fig. 4, 5)

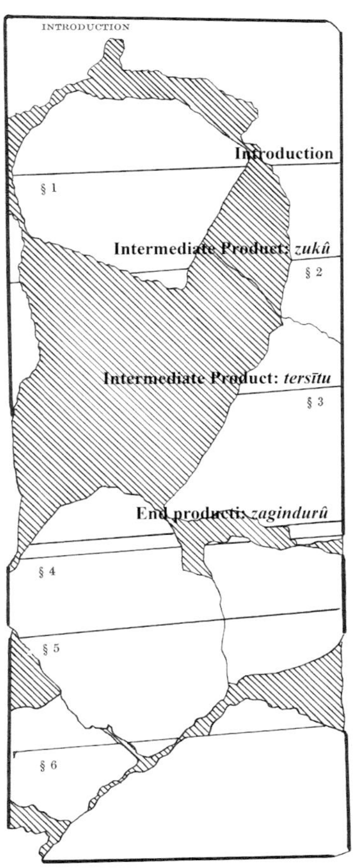

Figure 6.3: Division lines with intermediate products and end product of manuscript A (after Oppenheim 1970: fig. 2).

of repeated heating, crushing of sintered products, melting and annealing. These intermediate steps were necessary to release impurities from the batch* to make it more reactive and predictable for further processing. The complex and sophisticated composition of the manufacturing steps shows how carefully glass processing was carried out. This was necessary to avoid unintentional consequences, such as undesired discolouration or even waste materials, which may be caused by devitrification or overheating. Since any undesirable outcome would result in costly waste of batch material and fuel, it was imperative to minimise costs through great caution in the manufacturing process.

6.6.1 Coherent transcription and translation

The complete score edition of the text is attached in the Appendix 1. Here only the coherent transcription and translation of the recipe is presented.

enūma uššī kūri ša abni tanaddû [(1)] [(2)] *ina arḫi šalmi ūma magra teštene''e-ma uššī kūri tanamdi* [(3)] *adi kūra tuqtettû-ma tētepšu* (ms. B adds: *ina bīt kūri* [x]) [(4)] ^{d}*kūbē tuš*[*eššeb-m*] *a šanû aḫû lā irrub lā ellu ana panīšunu lā igger*

(5) *ginâ s*[*irqa*] *an*[*a p*]*ānīšunu tasarraq* (6) *ūm abna a*[*na libbi kūri tušerredu nīqa a*]*na maḫar* ^{d}*kūbē teppuš*

(7) *nignakki burāši tašakkan* (ms. B adds: *dišpa ḫimēta tanaqqī-ma*) (7) *išāta* [*ina*] *šapil kūri tanappaḫ-ma* (8) *abna* [*ana*] *libbi kūri tušerreda amēlū* [*ša ana muḫḫi kū*]*ri tuqarrabu* (9) [*ūtab*]*babū-ma ana muḫḫi kūri tuše*[*rreda*] (ms. C adds: *ina muḫḫīšun*[*u*]) (9) [*iṣu*] *ša ina šapal kūri* (10) [*t*]*ašarrapu ṣarbatu kabbartu qaliptu* (11) *qurû ša kiṣra lā nadû ina zumur api ṣabtū ina* ^{iti}NE *inakkašu* ms C adds: *ultu qištušu*) (12) *iṣu annû ina šapal kūri lillik*

(44) [*šu*]*mma zagindurû ana epēšīka* (44) 10 *mana immannak*[*u*] [*12/ 15 min*]*a aḫussa* (ms 1A and 1B adds: 1 2/3 *mana šamma peṣâ*) (44) *aḫ*[*ê tamarraq*] (45) *ištēniš taballal* (45) *ana kūri ša 4 īnīša kaṣīt*[*e tuš*]*erred-ma ina birīt īnāti ta*[*reddi*] (46) *išāta ṭābta lā qātirta tašarrap adi billuk*[*a iraš*]*šušu* (par. ms.1A;1B;4C: *ipeṣṣû*) (46) *ana ūme tušellâm-ma tu*[*kaṣṣ*]*i* (47) *tatâr-ma tamarraq ana dabti zakuti tessip* (47) *ana kūri ša takkanni kaṣiti tušerred* (48) *išāta ṭābta lā qātirta tašarrap adi iḫarraṣu tukattam* (49) *ultu iḫtarṣu ana muḫḫi agurri tanazzalam-ma zukû šumšu zukû šumš*[*u*]

(50) [10 *m*]*ana erâ nīḫa ana dabti zakūti tessip* (par. ms.2A, 2B: *tašakkan*) (50) *ana* [*k*]*ūri ša takanni emmeti tu*[*šerr*]*ed* (51) *bāb kūri tukattam* (51) *išāta danna*[*ta*] *lā* [*qā*]*tirta tašarrap* (52) [*adi erû*] *iraššušu* (52) *10 mana zakê taḫaššal-ma tamarraq* (53) [*bāb kū*]*ri tapettī-ma* (ms.5C adds: [*zuk*]*ê tušerred-ma*) *ana muḫḫi erî tanad*[*dī*]*-ma tatâr-ma bā*[*b*] *k*[*ūri tukattam*] (54) [*adi zukû eli*] *erû* (ms.5C adds: *u abnû*) *1-niš* (54) *immaḫḫaḫū-ma erû in*[*a šapal abni iššakkunu*] (ms.2A adds: *imaḫḫaḫu-ma*) (55) (ms. 2A adds: *erâ*) [*ina muterr*]*i 1-šu 2-šu 3-šu tabeḫḫeš ana ḫa*[*rāgi eššeti*] (ms. 5C adds: [*adi r*]*ēḫti erî ikkalu*) (56) *ina išāti tareddi ina appi muterri in*[*a īni m*]*amm*[*a tammar*] (ms. 5C adds: *tanazalašum-ma*) (57) *šumma abnu* (ms. 5C; 5d adds: *pan karāni bašlu/i itta*[*škin*]) *t*[*abaššil*] *ana muḫḫi agurri tan*[*a*]*zzala*[*m-m*]*a* (58) *tersītu šumšu*

(59) 10 *mana tersītu* 10 *mana b*[*ūṣ*]*u aḫussu harṣu lā tajāru* (60) 2/3 MA.NA *namrūtu ša tâm*[*ti sāndi*] *qalītu aḫê tamarraq* (ms. 3A adds: 1 2/3 *anzaḫḫu mesû*) (61) *ištēniš taballal* (61) *ana dabt*[*i*] *za*[*kū*]*ti* (ms. 3A instead: *eššēti*) *tessip ana kūri ša 4 īnīša kaṣiti* (62) [*tu*]*šerred-ma* (62) *i*[*na birī*]*ti īnē ina muḫḫi nēmedi tašakkan išid dabti kūra lā ikaššad* (63) [*išāta ṭ*]*ābta lā* [*qā*]*tirta tašarrap išātu ultu libbi īnī uṣṣâ* (ms B instead: *la qatirta ta šárrap*) (63) [*adi billūk*]*a iṣuddū išāta tašaddad ina kūrīšu tukaṣṣi tušellâm-ma tamarraq* (65) [*ana dapt*]*i zakūti tessip-ma ana kūri ša takkanni kaṣiti tušerre*[*d*] (66) [*išāta ṭā*]*bta lā qātirta tašarrap adi abnu iraššušū* (67) [*bāb kū*]*ri la tukattam ultu abnu irtaššu bāb kūri tukattam* (68) [*adi iḫarr*] *aṣu malani ina panīka tabeḫḫeš ultu iḫtarṣu* (69) [*ina īni ma*] *m*[*ma*] *tam*[*m*]*ar-ma šumma abnu uppūq ana dabti eššeti* (70) [*ta*]*nazza*[*lam-m*]*a* (70) *ina kūri tukaṣṣi illâm-ma zagindurû*

When you lay down the foundations of a kiln for 'stone', in a good month you search a propitious day and (then) you lay down the foundations of the kiln. Not before you have finished building the kiln, you [x] into the house of the kiln. You place the *kūbu*-images, another person (or) a stranger may not enter, an unclean (person) may not step in front of them. You pour regular strew offerings in front of them. On the day when you put the 'stone' in the middle of the kiln, you make a sheep sacrifice in front of the *kūbu*-images. You put an incense-burner of juniper, you sacrifice honey (and) ghee and you burn a fire in the lower part of the kiln. You bring down the 'stone' in the middle of the kiln, the persons, who you bring before the kiln, shall be purified before you bring them down to the kiln. The wood which you burn in the lower part of the kiln is thick, peeled poplar wood, [x], on which no 'knots' are attached, removed from the inner of the reed, which you cut from the forest in the month of Abu (August). (Only) this wood may go into the lower part of the kiln.

If you make *zagindurû*, you grind separately 10 minas of *immanakku*-stone, 12 minas of plant ash, 1 2/3 minas of 'white plant' and you mix them together. You bring it down into a cold kiln with its four 'eyes' and arrange it between the 'eyes'. You light a good, smokeless fire until your mixture glows (becomes white). You lift it (out of the kiln) into daylight, you let it cool and you grind it again. You gather (the powder) together for a clean *dabtu*-crucible. You put it down into a cold kiln with 'shelf'. You light a good, smokeless fire until it becomes ready then you cover it (crucible/kiln-opening). After it had become ready you pour it on baked bricks. This is zukû.

You gather (put) 10 minas (*c.* 5 kg) of 'slow' copper in a clean *dabtu*-crucible. You put it down into a hot kiln with 'shelf' and you close the door of the kiln. You light a good, smokeless fire until the copper-compound glows. You crush and grind 10 minas (*c.* 5 kg) *zukû* you open the door of the kiln and (you put *zukû* into it,) you put it on the copper compound and you close the door of the kiln again until *zukû* dissolves over the copper (and the stone) and the copper is deposited underneath the 'stone'. You stir the copper once, two times, three times with a rake (until the rest of the copper is 'eaten'), you arrange it in a new *ḫarāgu*-crucible until you see with the eye a 'crown' on the 'nose' of the rake. If the 'stone' turns into the colour of ripe grapes you cook the 'stone' in the copper (and) you pour it on baked bricks and this is called *tersītu.*

You grind separately 10 minas of *tersītu*, 10 minas of *būṣu*; plant ash not precisely measured and 2/3 minas of the roasted 'red' from the sea (3 A 32: 1 2/3 washed *anzaḫḫu*) is mixed together. You gather (the powder) together for a clean (new) *dabtu*-crucible, you put it down into a cold kiln with its four 'eyes'. You put it on a stand between the 'eyes', the bottom of the *dabtu*-crucible must not reach the kiln. You light a good, smokeless fire. The fire should come out of the 'eyes'. As soon as the mixture melts, you pull the fire away (and) you let it (mixture) cool in its kiln. You take it up and you grind (it). You gather (the powder) together for a clean *dabtu*-crucible and you put it down into a cold kiln with 'shelf'. You light a good, smokeless fire, you do not cover the door of the kiln until the 'stone' becomes red. After the 'stone' has become red, you cover the door of the kiln. You stir (it) [x] until it becomes clear. After it had cleared you will see a 'crown' with the eye. If the 'stone' is grown over (the rake) you pour it into a new *dabtu*-crucible (inside the kiln). You let it cool off in the kiln and (out of the kiln) rises *zagindurû*. Copy [x]

6.6.2 Introduction: the construction of the kiln and accompanying rituals

> **#1.1** When you lay down the foundations of a kiln for glass (lit. 'stone'),
> **#1.2** in a favourable month, you search a propitious day and (then) you lay down the foundations of the kiln.

In #1.1 and #1.2 the introduction begins with the construction of the foundations (*uššu*) for glassmaking kilns (*kūri ša abni*) (#1.1).

Literally *kūri ša abni* is 'kiln of stone'. Very often in cuneiform literature, glass is designated as *abnu* (often written NA$_4$), 'stone'. Therefore, glass together with faience* and glazed bricks counts among artificially produced 'stones' *abnu* (Schuster-Brandis 2008: 8–9).

No superstructure of the kiln is mentioned in the introduction. As there is insufficient archaeological evidence for glass kilns in Mesopotamia no assumptions can be made, either about the size, the material and shape of the foundation, or about the 'life time' of its use.

#1.2 is concerned with creating the most favourable circumstances to set up the glass kiln. It was important to choose the right month and day.

> **#1.3** Until you have finished building the kiln, you [...] into the house of the kiln. You place
> **#1.4** *kūbu*-images. Another person (or) a stranger may not enter, an unclean (person) may not step in front of them.
> **#1.5** You pour regular strew offerings (of flour) in front of them.

#1.3 is concerned with the magic protection of the kiln. Before the completion of the kiln it is not allowed to step close to it. Images of the '*kūbu*' demons are buried for the magic protection of the kiln.

The 'house of the kiln' (*bīt kūri*) (#1.3) can most likely be identified as the room or building where the kiln is placed.

Kūbu (#1.4) means 'Fötus' (AHw 498), 'Totgeburt' (Schwemer 1998: 164), and their 'demon' (CAD K 487; Lambert 1983: 265). Their function is described by Oppenheim (1970: 33) as 'vaguely apotropaic'. *Kūbū* regarding the production of glass could be understood as the 'birth' of the material out of the kiln and could in a figurative sense represent the meaning 'stillborn child'. Oppenheim (1970: 33) also considered the technological use of fire as point of reference. Unfortunately, there is no evidence, philological or archaeological, of how *kūbu*-images looked.

The *kūbu*-images were placed within the glassworkshop next to the (foundations of the) kiln itself after the construction of the furnace was completed. Each person had to be ritually clean before stepping in front of the kiln and the *kūbu*-images. This also applies therefore to the glassworkers. The ritual purification is considered the ideal state, which represents the perfect order and the absence of all evils (Sallaberger 2007). The ideal condition was therefore also decisive for the success of glass production, which is not surprising in view of the technological complexity of glass production (see Chapter 7).

#1.5: In addition, some flour was strewn as an offering before the *kūbu*-images.

> **#1.6** On the day when you put the glass (lit. 'stone') in the middle of the kiln, you make a sheep sacrifice in front of the *kūbu*-images.
> **#1.7** You put an incense-burner of juniper, and you burn a fire in the lower part of the kiln.
> **#1.8** You bring down the glass (lit. 'stone') in the middle of the kiln, the persons, whom you bring before the kiln, shall be purified and then you can bring them down to the kiln.

On the day production of glass should start, a sheep sacrifice was performed (#1.6), juniper was burned (#1.7) and sacrifices of honey and ghee (#1.7) were made in front of the *kūbu*-images. The ritual could only be carried out by ritually clean participants, and no 'unclean' person was allowed to approach the kiln (#1.8) (see #1.4).

The first day on which the production process should start is defined by the phrase 'to put the stone into the middle of the kiln' (#1.8), and also in #1.7 with the instruction to burn a fire.

This description of the offerings, as well as the detailed instructions on who was allowed to enter the glass workshop, show that the sacrifices were an inherent part of the installation of the glassmaking kiln and the establishment of the primary production process.[19]

[19] In §13 6–11 (Opp.) the performance of a ritual is mentioned that

The ritual act was meant to protect the glassmaking process from evil, for all the participants, including the glassmakers. It is important to stress here that the rituals and sacrifices that accompanied the construction of a glassmaking kiln were 'in no way atypical or extraordinary' (Oppenheim 1970: 33), but an important element of any kind of building work in ancient Mesopotamia. Therefore, the realisation of a ritual was considered as important as the construction work itself and necessary for its success (Ambos 2004: 3).

> **#1.9** The wood which you burn in the lower part of the kiln is peeled thick poplar wood, [...], on which no 'knots' are attached, taken from the reed-marsh (and) cut in the month of *Abu* (August).

#1.10 (Only) this wood may go into the lower part of the kiln.

The last two passages (#1.9; #1.10) of the introduction are dedicated to the treatment of wood that was used as fuel.

The tree species that is referred to in the text is *ṣarbatu,* which can be identified as *populus euphratica* (Euphrates poplar) (Postgate 2005: 329; CAD Ṣ 108).[20] *P. euphratica* grows even in salt-rich grounds and brackish waters and is the dominant species of poplar along the vegetation of the Babylonian Euphrates still today (Zohary 1983: 130). The Euphrates poplar also occurred in the form of deliberate cultivation in Mesopotamia, which is attested in Assyrian documents (Postgate 1992: 183). Poplar wood was known as 'good quality fuel' (Postgate 2005: 329), as it burned quickly, generating intensive heat over a short period of time (Willcox 1992: 5).

The text gives very specific information about the condition of the poplar wood (#1.9): The logs should be thick (*kabbarta*) and free of bark (literally 'peeled') (*qaliptu*) with no branching (literally 'knots') (*kiṣrum*). It had to be taken (*ṣabtu*) from the reed-marsh where it grew (*zumur api*), thus no old wood should be used. Furthermore, the poplars needed to be cut in the 'month of *Abu*' (August), when the wood contains low amounts of sap (Willcox 1992: 4). The importance of the right choice of wood is emphasised by the precative form '*lillik*' in the last paragraph '(Only) this wood may go (...)' (**#1.8**).

The wood was burned in the 'lower part of the kiln' (*šapil kūri*) (#1.7), which is the 'hearth of the kiln' or 'firebox.'[21]

The careful selection of the species of wood and its preparation is given great attention in the text. The right choice of fuel and preparation was essential, as very high temperatures (around 1000°C) had to be reached for the production of raw glass. Therefore, the wood needed to be dry and massive.

6.6.3 Production of the colourless primary glass zukû

#2 is concerned with the production of the first intermediate product (*zukû*), which is mentioned at the end of the #2 (#2.7).

> **#2.1** If it is blue *zagindurû*-glass for you to make

The first line draws reference to the end-product *zagindurû* of this recipe (#2.1).

Zagindurû is a loan of the Sumerian *zagin* (akk. *uqnu*), 'blue stone' ('Blaustein': Schuster-Brandis 2008: 455), and duru$_5$ 'wet'. Literally *zagindurû* means therefore 'wet blue stone'. More commonly, *zagindurû* is interpreted as 'a variety of lapis lazuli' (CAD Z 11; AHw 1502; Schuster-Brandis 2008: 455), however, *zagindurû* incorporates also other stones with a blue colour, not only lapis lazuli. Another important reference can be found in lexical lists. Here *zagindurû* is always described as 'pure' (*ellu, ebbu*) and 'shiny' (*namru*) (Schuster-Brandis 2008: 455). With regard to the texts discussed here, *zagindurû* can be identified as translucent, dark blue glass.[22] Thavaplan *et al.* (2016: 202) identify *zagindurû* as a 'vitreous material (...) that resembled the luster and hue of blue gemstones', and they further provide evidence that *zagindurû* in pulverised form served as the pigment Egyptian blue (Thavaplan *et al.* 2016: 201–202).

Contrary to expectations, the first line does not mention the product to be manufactured in this paragraph but mentions the final product. This clearly shows the function of *zukû* as an intermediate in glass production.

> **#2.2** you grind separately 10 minas of *immanakku*-stone, 12 minas of salicornia, you mix them together.

#2.2 lists all the ingredients needed to produce the batch*.

takes place before the production of tersītu-glass. Furthermore, in §L 34–38 (Opp.) offerings in front of the kūbu-images are described in connection with the production of golden zukû glass. Here a sheep sacrifice and 'offerings for the dead masters' are performed.

[20] Other species of poplar that are cultivated in Iraq today, i.e. p. alba and p. nigra, are not indigenous, and therefore do not need to be considered as referred to in the text.

[21] In text §13 20 the firebox is called našrapu, lit. 'place for burning' (maPRaS form from šarāpu 'to burn').

[22] Dark blue glass always appears more or less translucently in the archaeological record while turquoise glass is opaque (Chapters 7.2.3.1 and 7.2.2).

Immanakku: The first ingredient listed is *immanakku*, a substance almost exclusively attested in the glassmaking texts. Further evidence for *immanakku* can be found in standard Babylonian in the lexical list Ḫḫ XII 288 (MSL 10,12) [na4]im-ma-na = ŠU, in the Old Babylonian forerunner BM 92611 line 132, Urra = *ḫubullu* (MSL 10, 52), as well as in the stone list *Abnu šikinšu*. Here *immanakku* draws reference to pebbles from riverbeds:

A 36 NA$_4$ GAR-*šú* GIM *ṭe-ru-ut* ÍD-*ma* NA$_4$ *tuk-kup* [na4]IM. M[A.(AN). NA] MU. NI

'A stone, its form is like the deposit of the river, and the stone is dotted: immanakku is its name.'

The determinative NA$_4$ in Ḫḫ XVI 288 and BM 92611 and its description in *Abnu šikinšu* identify *immanakku* as a type of stone that can be found in riverbeds.[na234] In the glassmaking texts, the identification of *zukû* as glass in #2 makes the addition of silica imperative (Chapter 7.1.1). The interpretation of *immanakku* as quartzite pebble is therefore based on the Nineveh glass recipes, since it is necessary for the production of glass. It is a silica-bearing mineral.[24] The natural occurence of quartz in Mesopotamia in river pebbles is indicated in *Abnu šikinšu*.[25] Silica-rich minerals are present in different Mesopotamian soils and rocks (Jassim 2006: 167–168, 190, 219).[26]

Aḫussu: The second ingredient listed in #2.2 is *aḫussu*, a variant of *uḫullu*, *uḫūlu*, which is attested since the Old Babylonian period (*uḫūlum*) as 'a plant and its product (soda ash) used as a source of alkali'.[27] The text is therefore concerned with the production of plant ash glasses, using soda-rich plant as flux to lower the melting temperature from 1700°C (the melting point of silica) to about 1000°C (Chapter 7.1.3). As shown in Chapter 7.1.3 plant ash was still the common source of natron during the first half of the 1st millennium BCE. In none of the glassmaking texts has evidence for mineral natron as flux been identified so far. The amount of ash needed for the mixture is also given. In this regard, the quantity varies in different manuscripts, from 15 minas (Opp. 1 A, 1 B) (*c.* 7.5 kg) to 12 minas (Opp. 4 B, 4 C) (*c.* 6 kg). Regarding experimental studies carried out by Brill (1970), the amount of 6 kg of plant ash has turned out to be the most realistic amount.

Šammu peṣû, literally 'white plant', is listed as the third substance. This specific description could indicate another type of salt-tolerant plant, different from *aḫussu*. *Šammu peṣû* is absent in the manuscripts Opp. 4 B and 4 C – the manuscripts with the lower quantities of salicaronia – which makes it very likely that *šammu peṣû* was dispensable for the success of the glassmaking process.

In #2.2 not only the ingredients but also their weights in minas are given. 5 kg of crushed *immannaku*-stone are equivalent to two litres.[28] 7.5 kg of plant ash result in 25 litres, 5 kg of plant ash in 16.6 litres of ash.[29] All together, the recipe is concerned with a total volume of approximately 28 litres of batch* material. Regarding the shrinking of the batch during glass melting, a loss of *c.* 40% was observed in experiments (Taj-Eddin, pers. comm.).

The listed ingredients silica, plant ash, and white plant were ground and mixed together to a powdery batch*, which was reduced to 60%, i.e. *c.* 16.8 litres, because of the loss during glass melting (#2.2).

> **#2.3** You bring it down into a cold kiln with its four 'eyes' (i.e. openings) and arrange it between the 'eyes'.

#2.3 describes the placement of the batch* into a specific type of kiln.

kūru ša 4 īnīša: The batch was put into a specific type of kiln called *kūru ša 4 īnīša*, literally 'kiln with its four eyes' (#2.3). The name '*kūru ša 4 īnīša*' indicates that the kiln featured four openings, facilitating the control of the temperature and kiln atmosphere (Brill 1970: 113). Furthermore, the four openings enabled access to the vessels within the kiln chamber from different directions (openings), which facilitated the control and manipulation of the sintering* and melting process. Twice in the text reference is drawn to the 'kiln with its four eyes' (#2.3; #4.2). In both cases the kiln was used for the first heating of the batch. #2.3 describes the sintering and #4.2 the first melting of the material. Also in other glassmaking recipes the 'kiln with its four eyes' is optionally used for sintering (Opp. 16 B), as well as for melting processes during first heating (Opp. 7 A). It is interesting to note that a stand (*nēmedu*) is needed for the melting process (#4.2). Regarding the sintering process, no specific type of container is mentioned. Brill (1970: 113) suggests that the containers used for

[23] na4Brill (1970: 109) remarks that 'quartzite pebbles do not look like river silt dotted with pebbles but they are, in fact, one of the types of pebbles with which river silt itself is dotted'.

[24] For a detailed discussion and further literature, see Schuster-Brandis 2008: 419.

[25] For an ethnographical study of the collection of quartz-rich pebbles from the fields, see the video 'Glassmakers of Herat' https://www.youtube.com/watch?v=BMYE83DJU4Q (accessed: 31.08.2015).

[26] Oppenheim (1970: 74, fn. 85) draws attention to NA$_4$IM.AN.NA, which can also refer to the fact that AN.NA means 'tin' and therefore 'tin ore'. Also Brill (1970: 109) remarks in this regard that tin ore as well as silica could have been collected from a common source, as tin ore also occurs in stream gravels.

[27] CAD U 48, see also in AHw 1404 ,Salzkräuter(n) und deren Alkali (Natriumkarbonat)-haltige Asche'.

[28] The concentration of sand is approximately 2.5 kg/l; the formula for calculating the volume is: v=m/c.

[29] The concentration of ash is approximately 0.3.

sintering could have had a wider mouth as crucibles used for melting, so as to facilitate a maximum exposure of the powder to oxygen. Unfortunately the texts, as well as the archaeological record, lack direct information regarding the shape and size of kilns and crucibles. Experimental studies – prior to Roman tank furnaces – carried out on sintering and glassmaking showed that ancient glass kilns would have rather exhibited smaller diameters and small openings to help control heat and atmosphere in the kiln. In consequence, the crucibles used in the process are expected to have a small diameter as well as low crucible walls. This is because of the reduction of the batch* volume that occurred during the sintering process, amounting to *c.* 30% of the original batch material. In order to manipulate the compound after its reduction, low crucible walls, as well as a rake-shaped tool (see below *muterru*), seem to have been necessary (Taj-Eddin, pers. comm.). In Tell el-Amarna, flat, tray-like vessels with a low rim are attested as original parts of kiln furniture (Nicholson 1995: 15; for a discussion of crucibles see Chapter 8.1.2.1).

> **#2.4** You light a good, smokeless fire until your mixture glows.

Now, in #2.4 the fire is lit, and the batch* material is sintered.

The fire in the kiln is described in this passage, as well as in the rest of this recipe as 'good' (*ṭabtu*) and 'not smoking' (*lā qātirtu*) (#2.4, #2.6, #3.3, #4.4, #4.7). This can be expected after meticulous preparation of the wood in the introduction. The establishment of high temperatures in the kilns was necessary to sinter* and, in particular, to melt the batch. In the 'kiln with its four eyes' a temperature between 800° and 900°C for sintering was necessary, whereas during melting, a temperature of at least 1000°C was required.

The process of sintering* is described in the text as *peṣû,* 'to become white', in a technical sense also to 'sinter' (CAD P 334). In manuscript Opp. 4 A and 4 C, *rašāšu* is used for the same process. The word is almost exclusively known from the glassmaking recipes and can be translated as 'to glow' or 'to become red' (CAD R 191). Red or glowing probably means the colour that the kiln, and thus also the colour, the crucible has taken on when the kiln was heated. This colour indicates the high temperature reached in the kiln chamber. Both verbs *peṣû* and *rašāšu*, therefore, refer to the process of fritting: whereas *peṣû* refers to this process literally, *rašāšu* rather describes the temperature of the kiln and its colour (AHw II 960 'heiß glühend werden').

> **#2.5** You lift it (out of the kiln) into daylight, you let it cool and you grind it again. You collect it up for a clean *dabtu*-crucible.

In #2.5 the sintered mixture is taken out of the kiln in its hot state to cool off. The verb *kaṣûm* (#2.5) means 'to become cold, to cool off'. The product after the cooling process can be considered as a fused chunk of white, sintered* batch* material that was very hard and therefore needed to be ground for the further process (#2.5).

Dabtu: this word is only attested in the texts that deal with the production of raw glass. As indicated in #2.5, the processing of the sintered* material was carried out in a clean (*zakūtu*) *dabtu*-crucible. In this passage *dabtu* can be identified as crucible in which the sintered and crushed batch* was melted to be further processed into glass. The sole use of *dabtu* as a crucible for melting is demonstrated in several text passages in which *dabtu* is exclusively used during the melting process (#2.5, #3.1, #4.2, #4.6, #4.8) (see also §16 48´). It was imperative that the *dabtu*-crucible was either clean (*zakūtu*) or new (*eššetu*). Both adjectives are used interchangeably. This indicates that it was important not to run the risk of mixing residues that could be present in used crucibles. This becomes even more apparent regarding the existence of another crucible, the *ḫarāgu*-crucible in Opp 13 4´, 14 28´and 15. In contrast to *dabtu*-crucibles, *ḫarāgu*-crucibles are explicitly described as *ḫarāgu lā eššetu* ('not new') and are predominantly used for sintering. Later in our recipe, a new *ḫarāgu*-crucible (*ḫarāgi eššetu*) (#3.6) was used for melting the batch (for an overview Table 6.1).

> **#2.6** You put it down into a cold kiln with shelf. You light a good, smokeless fire until it becomes ready. Then you cover it (crucible/kiln-opening). After it had become ready you pour it on baked bricks.

In #2.6 the crucible is first placed into a specific kiln that is different from the 'kiln with its four eyes' (#2.3). The heat of the fire is intensified and the batch* melts.

Kūru ša takkanni: The crucible with the batch is positioned in a specific kiln (*kūru*) called *ša takkanni*, which is used repeatedly in the text (#2.6, #3.2, #4.6). The word *d/takkannu* was identified as 'room, bedroom' (Beaulieu 1992: 101–103), 'Türöffnung' as well as 'bankartiger Sockel' (AHw 151–152), which is likely to be the meaning here. Dietrich (2001: 76–77) referred to a kind of stand or podium to step on.[30] The translation suggested here is 'kiln with shelf', which refers to a so-called 'pot kiln' in which glass was melted in crucibles that stood on a shelf in the firing chamber (Figure 6.4).[31]

[30] *ana muḫḫi alāku* (Dietrich 2001: 76: 7).

[31] For an example of how a pot-kiln in Roman times could have looked, see Figure 6.4 http://www.romanglassmakers.co.uk/furnace2.htm (accessed: 20.08.2015). Oppenheim (1970: 34) suggests the translation 'chamber kiln'. The 'chamber' in this context could refer to an annealing chamber, in which the hot glass could have

Figure 6.4: Reconstruction of a Roman pot-kiln with shelf and crucibles. Its appearance could be similar to the 'kiln with shelf' (*kūru ša takkanni*) mentioned in the Nineveh glass recipe in #2.6. (http://www.theglassmakers.co.uk/archiveromanglassmakers/furnace4.htm, accessed 27.07.2018)

The existence and appearance of this type of pot-kiln is not attested archaeologically. In the 'kiln with shelf' only melting processes are carried out (#2.6, #4.6); the 'kiln with shelf' appears only in connection with crucibles, which indicates the presence of a bank inside the kiln chamber.

Glass only melts at temperature above 1000°C, which needed to be held for a certain time (see below *zukû*). A method to achieve this more easily was to cover (*kuttumu*) the 'door of the kiln' (*bāb kūri*) (#3.2, #3.5, #3.6). The 'door of the kiln' can rather be understood as a simple opening than an actual door; it was probably even only a fired brick or tile placed in front of the hole to cover it.[32]

At the moment the crucible was placed in the kiln chamber, the kiln could either be 'cold' (*kaṣītu*) (#2.6, #4.6) or hot (*emmetu*) (#3.2). Since one must assume that the kilns were heated for days in advance to achieve suitable temperatures, the reference to a cold kiln remains unsolved regarding this text passage.[33] The substance was then melted in the kiln until it 'becomes ready' (*iḫarraṣu*).[34]

#2.6, as well as #3.7, close with the process of pouring (*nazālu*) hot glass over fired bricks (*agurru*). *Agurru* means 'baked brick' with reference to the material (AHw 17). In this context, *agurru* could also be understood as a type of a fired brick platform onto which the melted glass was poured in its hot and fluid state. The text does not mention annealing* within this process and as the manufacturing steps are described in great detail throughout the text it is very likely that the slow cooling was not part of this particular step. If melted glass is not annealed slowly it cracks immediately. Regarding the succeeding step, carried out in #3, the grinding of glass cullet was much easier applicable as the crushing of a compact glass ingot.

> **#2.7** This is called *zukû*, 'the pure one' (a colourless primary glass).

Zukû: Etymologically, *zukû* (of *zakû*, 'pure') means 'the pure one' ('das Reine', AHw 1536). The fact that the fritted powder was melted leads to the assumption that the first intermediate product *zukû* (#2.7) is to be identified as glass. Only through the process of melting can a physical transformation into glass be achieved (see Chapter 2.1.1). *Zukû* never occurs as final product but only as an intermediate glass product.[35] As no colourants are added to the batch*, *zukû* can be identified as 'natural' base glass, a '(...) chemical

been cooled off. As it is imperative to anneal hot glass slowly to prevent it from cracking, a tempering chamber could have existed. For an example of an annealing chamber dating to the Roman period, see http://www.romanglassmakers.co.uk/furnace2.htm (accessed: 02.09.2015).

[32] For a Roman example, see http://www.romanglassmakers.co.uk/furnace10.htm (accessed: 15.09.2015).

[33] Once heated, the kilns ran for a considerable period to achieve a maximum utilisation of heat (Bachmann, pers. comm.).

[34] The translation suggested here follows the translation in CAD Ḫ 94 5 'to become ready', and specifically refers to a technical term in chemical and medical texts. Oppenheim (1970: 35) translates 'glows golden', derived from *ḫuraṣum* 'gold'; see also §48.

[35] With regard to §2, §5, §13 tersītu, and regarding §Q, i, iv Middle Babylonian text, *zukû* both serves as intermediate products in the manufacture of red-coloured glass. Furthermore, *zukû* occurs in §L 22-23 as *zukû ša ḫurāṣi* (su kù.gi), which can be identified as '*zukû* with a golden colour'; see Oppenheim 1970: 75–76.

intermediate in which the silica has been transformed into a workable, more reactive form by the reaction with plant ash.' (Brill 1970: 109) (Chapter 7.1.3). *Zukû* is almost exclusively attested in the glassmaking texts as 'a material for glassmaking' (CAD Z 153).

On the basis of the detailed description in the Nineveh glass recipes, Brill (1970: 110–114) conducted an experiment to investigate the likeliness of the quantities given in the text, and the feasibility of the instruction.[36] He indicated that the calculated composition of *zukû* lies 'in astonishingly close agreement with that of early Mesopotamian glasses' (Brill 1970: 111), and states that the ratio of 10:12 (see #2.2) of *immanakku* 'quartzite pebbles' to *aḫussu* 'salicaronia' is more realistic than the ratio of 10:15. Also the amount of silica generally appears in a relatively low proportion to the rest of the components.[37] As no time or temperature specifications of the different firings are mentioned in the text, the experiments provide useful evidence in this regard. Thus, during the first firing a temperature of 900°C was established for 24 hours producing a partially reacted frit, free of gaseous products. During the second heating a temperature of 1100°C was held for 16 hours. The annealing* process of the crucible was then carried out at a temperature of 450°C. The final glass product that resulted from this experiment is described as transparent glass with a slight amber tinge, a '(...) glass of unexpectedly high quality' (Brill 1970: 113).

Paragraph 2 therefore ends with the first intermediate product – a colourless primary glass named *zukû*, 'the pure one'.

6.6.4 Production of the blue primary glass tersītu

Paragraph 3 deals with the production of the second intermediate product – *tersītu.*

> **#3.1** You collect up 10 minas (5 kg) of 'slow' copper in a clean *dabtu*-crucible.

In #3.1, as a first step, a substance called 'slow copper' (*erû* [URUDU.ḪI.A] *nēḫu*) is melted in a clean *dabtu*-crucible.

erû nēḫu: The adjective *nēḫu, nīḫu* means literally 'calm', 'quiet', 'slow' (CAD N 151–152; AHw N 775). *Erû* is copper (Oppenheim 1970: 36; Reiter 1997: 150–152), so *erû nēḫu* literally means 'slow copper'.

The term 'slow copper' (*erû nēḫu*) can be matched with an almost opposite word pair 'fast bronze' (*arḫu* [UD.KA.BAR] *siparru*), which is mentioned in two of the Nineveh glass recipes (Opp. #13 l. 1, #15 l. 121). 'Fast bronze' (*arḫu siparru*) in Opp. #13 l. 1 occurs in connection with 'the production of red glass (literally red lapis lazuli)' (*epišti uqnî sāmi*). The passage in Opp. #15 l. 121 is less clear and needs further revision. Here 'fast bronze for work' (*arḫu siparru ša dulli*) is mentioned in connection with 'fine (x) lapis lazuli coloured glass'.[38] It is important that the two opposing word pairs result in different glass colours: whereas 'fast bronze' leads to red colouration, 'slow copper' was used to produce blue glass. Oppenheim (1970: 96) refers to both terms as 'a colouring agent'. This basic function of the two ingredients is considered to be correct but a more detailed explanation for the interpretation of the two substances can be made. Throughout the Early Dynastic, Old Akkadian and Ur-III periods, special cuneiform texts refer to the production of bronze. Bronze contained different proportions of tin and copper, from which different 'quality levels' were derived.[39] The amount of tin present in the bronze has an effect on the working properties of the material. The higher the amount of tin, the lower the melting point of the metal, the faster the melting.[40] Regarding the interpretation of 'fast bronze' and 'slow copper' this is fundamental, because for the production of red-coloured glass, the maintenance of a reduced kiln atmosphere* is imperative. Since it was difficult to maintain a reduced atmosphere in the furnace, it was of great interest to carry out the melting process as quickly as possible. 'Fast bronze' could therefore refer to a bronze with high levels of tin and a low melting point. The adjective 'fast' would in this case refer to the circumstance that the metal melts 'faster' as 'slow copper'. In contrast, the recipe under discussion is concerned with the production of *zagindurû* 'blue glass'. Blue glass can be produced in oxidising* kiln conditions. As an oxidising atmosphere was much easier to maintain there was no need to work the copper particularly quickly, therefore the use of 'slow' copper, that means a copper that melts later, was sufficient (for details on cuprite red glass, see also Chapter 7.2.3.2).[41]

[36] For immanakku he used as an ingredient quartzite pebble, and for aḫussu a kind of plant ash, which he bought at a market in Afghanistan (Brill 1970: 110–111, 122, table 1).

[37] In his experiment Brill used a ratio of 5 (192 gr): 6 (230.5 gr) of *immanakku* to *aḫussu*, with the quantity reduced to about 5% because of the scarcity of the available raw material; for a detailed description, see Brill 1970: 112.

[38] Brill (1970: 121) remarks that bronze would have not only introduced copper into the batch but also lead and tin. These elements promote the formation of cuprite, which is responsible for the red colouration of the glass (see Chapter 7.2.3.2). The text explicitly mentions 'the production of red glass' (*epišti uqnî sāmi*) (Opp. first line §13).

[39] For details on different ratios of copper and tin in different texts, see Reiter 1997: 298–307, 300 in particular.

[40] See Helwing (unpublished draft: 242, 246): 'Zinnbronze hat gegenüber reinem Kupfer den Vorteil, dass bei entsprechend hohen Zinnanteilen der Schmelzpunkt auf unter 1000°C herabgesetzt werden kann.' (Compared to pure copper, tin bronze has the advantage that the melting point can be reduced to below 1000°C if the tin content is correspondingly high.)

[41] The idea for this interpretation was developed in a discussion with Paola Paoletti in terms of the working properties of different metals and their corresponding designations, for which I am very grateful.

Similar to #2.5, a clean *dabtu*-crucible is also used here for the melting process. As indicated above, this means that the paragraph deals with the melting of glass.

> **#3.2** You put it down into a hot kiln with shelf and you close the door of the kiln.

In #3.2, the 'slow copper' in the *dabtu*-crucible (#3.1) is put into a 'kiln with shelf', which is pre-heated, and the door of the kiln is closed (see above #2.6).

Unlike in #2.6, the 'kiln with shelf' is pre-heated before the crucible is inserted. This could indicate that the same kiln is used as in #2, which is at this stage already hot. Here also the 'door of the kiln' (*bāb kūri*) is covered to reach higher temperatures, most likely with a fired brick or tile placed in front of the hole to cover it (see discussion #2.6).

> **#3.3** You light a good, smokeless fire until the copper glows.

Under the heat of a hot fire the glass-copper compound melts.

Rašāšu: See above #2.4 on *rašāšu* 'to glow', 'to become red', a word known mostly from the Nineveh glass recipes. Red or glowing here refers to the colour of the hot kiln and thus also to the colour the crucible has taken on. This colour therefore indicates the high temperature reached in the kiln chamber. In #2.4 this verb was used to describe the colour of the hot kiln during the sintering* process.

> **#3.4** You crush and grind 10 minas (*c.* 5 kg) of the primary colourless glass (*zukû*).
> **#3.5** You open the door of the kiln and you put it on the copper and you close the door of the kiln again until the primary colourless glass (*zukû*) dissolves over the copper and the copper is deposited underneath the glass (lit. 'stone').

In #3.4, 10 minas (*c.* 5 kg) of the primary colourless glass (*zukû*) – which was produced in #2 – are ground to a fine powder and in #3.5 mixed together with the melted copper compound in the kiln. Great care is taken when opening and closing the 'door of the kiln'. The door of the kiln is held closed until the *zukû* dissolves into the melted copper compound. Both components are melted together.

The addition of the primary colourless glass (*zukû*), most likely in form of crushed glass or powder, accelerated the process of melting, as the chemical reaction within the glass cullet is already completed. Another side-effect of using crushed glass for glass melting is the reduction of melting energy, which is reduced by 25% (Schaeffer 2014: 133).

> **#3.6** You stir the copper once, two times, three times, with a rake you arrange it in a new *ḫarāgu*-crucible until you see with the eye a 'crown' on the 'nose' of the rake.

In #3.6, the melted substance made of copper and glass is stirred with a rake and poured into another new crucible. In there the viscosity of the melt is tested with a rake.

Muterru : the word *muterru*, *mutirru*, literally 'turner', is also attested outside the glass texts and can also mean 'fire rake' (CAD M 300) ('Schürstange' in AHw M 688). This word is only attested in the Neo-Assyrian dialect.[42] As described above (#2.5), the mass of the viscous substance was reduced by melting. As a result, it became more difficult to reach the melt in the crucible. In order to continue to stir the viscous substance without any problems it was necessary to use the turner.

Ḫarāgu:The term *ḫarāgu* is solely attested in the Nineveh glass recipes,[43] and was interpreted by Oppenheim (1970: 89) as 'sagger' (CAD Ḫ 89) or 'crucible'. In these contexts only does the interpretation as 'crucible' makes sense. In #3.6, the crucible used is specified as new (*ḫarāgu eššetu*), which explicitly shows that the crucible needed to be unused. Similar to the clean (*zakūtu*) or new (*eššetu*) *dabtu*-crucible (#2.5), only these containers were used for glass melting, to avoid the risk of mixing residues that could be present in used crucibles (*ḫarāgu la eššetu*). The latter were used for the process of sintering (for an overview see Table 6.1). Crucibles for glass melting are very rarely attested in the archaeological record and were made of coarsely tempered clay (see Chapter 8.1.2.1).

The substance was only poured into the *ḫarāgu*-crucible when sufficiently stirred and melted, and was then further melted. The end of the melting process was tested with the rake (*muterru*). If the viscous glass formed a 'crown' (*mammu*), i.e. the moment the glass stuck to the tip of the rake, the melting process was completed (#3.6).

> **#3.7** If the glass (lit. 'stone') turns into the colour of ripe grapes you cook the glass in the copper (and) you pour it on baked bricks.

#3.7: The moment when the viscous glass was sufficiently hot was indicated by 'the colour of ripe grapes' (*karāni bašli*). When this happened, it could be poured onto a platform of fired bricks.

[42] For an illustration of a rake in the archaeological context, see the one found at Tell Abu Kharaz; see Fischer 2013: 245, fig. 226.

[43] Apart from this text here, see also §13 4 B, §14 28 B, §15 118 A, §15 119 A.

In #3.7, similar to #2.6, the hot glass is poured (*nazālu*) over fired bricks (*agurru*), probably a platform. Since in this passage, as in #2.6 above, the process of annealing* is not mentioned, it is likely that the bursting of the glass was intended, since in the following step the product was crushed again.

> **#3.8** And this is called *tersītu* 'preparation' (blue primary glass).

Tersītu: Etymologically, *tersītu* (*nomen actionis*) derives from *šutēršû*, which means 'bereitstellen' (AHw 1291), 'to make ready, prepare' (CAD T 390). Consequently, *tersītu* could be translated with 'preparation' ('Bereitstellung'). The term *tersītu* is also used as a covering term to describe various building materials readily prepared for construction work (CAD T 356. 1a). The term *tersītu* thus illustrates well its function as an intermediate product.

Based on the process described in #3, *tersītu* can be identified as a translucent blue primary glass that was coloured using copper (#3.1).[44] The function of this second intermediate product is similar to the primary colourless glass (*zukû*), as *tersītu* also forms a primary base glass needed for the manufacturing of the end product, *zagindurû*, now with the colouring agent added.

6.6.5 *Production of the end product: blue* zagindurû*-glass*

The last paragraph (#4) deals with the production of the end product: blue *zagindurû*-glass.

> **#4.1** You grind separately 10 minas of the blue primary glass *tersītu*, 10 minas of the primary glass *būṣu*; prepared salicorina, which does not return (i.e. is lost), and 2/3 minas of red mother-of-pearl (lit. 'shine from the sea') (ms. 3A adds: and washed *anzaḫḫu*). You mix it together.

In the first line (#4.1) the intermediate glasses from the previous manufacturing step (#3) are listed with other substances as ingredients for the end product. In addition, their precise measurements are given. The ingredients include 5 kg (10 minas) of crushed blue primary glass (*tērsītu*) (see #3.8), 5 kg (10 minas) of *būṣu*, salicornia (*aḫussu*) (not precisely measured: *lā tajāru*) and 333 gr (2/3 mina) of *namrūtu*.

Būṣu: Apart from its use in the Nineveh glass recipes, *būṣu* is already attested in texts from Nuzi and Assyria of the Late Bronze Age (CAD B 349 b), and refers, for instance, to a material of containers from Nuzi 'holding two silas of fine perfumed oil' (CAD B 349 b1), or from which the legs of a composite statue of two stags were made:

> Two stags, 'their legs are of gold and the coloured parts of their [...] are made of lapis lazuli from the kiln (i.e. blue glass) and white stone (*būṣu*)'.[45]

In passage #4.1, as well as in the other Nineveh glass recipes (Opp. §12, §14, §15, §20, §P), *būṣu* is a component in the process of glassmaking that is crushed and mixed with different primary glasses and plant ash. These references make the identification as a primary glass or frit likely. Regarding the function of *būṣu* in #4.1, it was used in a similar way as the colourless primary glass *zukû* and the blue primary glass *tērsītu*.

Aḫussu is plant ash and serves as the source of alkali (see #2.2).

Ḫarṣu: to become ready, therefore here prepared, i.e. ashed (see also #4.8).

Lā tayyāru, literally 'not returning', could indicate that the salicornia, which serves as a flux, would evaporate in the further process of glassmaking.

Namrūtu derives from the verb *nawārum, namēru* 'to shine' (AHw 768–770) and means therefore literally 'shine' or 'brilliance'. *Namrūtu* is thus often identified as 'Perlmutter' (AHw 729), mother-of-pearl. *Namrūtu ša tâmti* 'shine of the sea', as well as the reference to *sāndu, sāmtu* (CAD S 121–124)[46] indicates that a red product, gathered from the sea, was added here as ingredient. Oppenheim (1970: 39) proposed the translation 'coral'. The use of coral is rather unlikely. First, because no other ritual or procedural text uses coral as an ingredient, and also because it can be ruled out that a material as difficult to obtain as coral should be added to a recipe that does not need this substance. In contrast, shell, or mother-of-pearl could have been used. In this regard, the red colour could refer to its reddish glaze. The use of shell is attested in Roman natron glasses as a source for lime, which served as a stabiliser (see Chapter 7.1.5).[47] Brill (1970: 113–114) proposes *namrūtu ša tâmti* to be an oxidising agent, preventing any unwanted colouration of the base glass which could be caused by impurities. *Namrūtu ša tâmti* is listed in all versions of the recipes for the production of *zagindurû*-glass.

[44] Also identified by Oppenheim (1970: 77) as blue glass.

[45] 31 (...) ša guŠkin pu-ri-[da]-tu-šu 32 ša NA_4za.gìn ku-ri ù bu-ṣi te-qi-a-tu xxxx- šu –nu; see Köcher 1957: 31–32, VAT 16462.

[46] Red stone, mostly designated as 'cornelian', see also the entry in *Abnu šikinšu*: 'the stone whose appearance is like that of (the berry of) the boxthorn'; for further reference see Schuster-Brandis 2008: 413–414.

[47] Coral also contains high percentages of lime.

> **#4.2** You collect up (the powder) for a clean *dabtu*-crucible, you put it into a cold kiln with its four 'eyes' (i.e. openings).

The ingredients listed in #4.2 are ground to a powder, put in an unused *dabtu*-crucible, and placed in a 'kiln with four eyes'.

Dabtu: this word is only attested in the glass texts and means crucible (see #2.5). It is interesting to mention here that the *dabtu*-crucible is placed on a stand (*nēmedu*), in the following paragraph (#4.3), so that the substance in the crucible melts (for an overview of the different types of crucibles, see Table 6.1).

Kūru ša 4 īnīša: similar to #2.3, the batch* was put into the cold 'kiln with its four eyes', i.e. featuring four openings to control the temperature and kiln atmosphere (#2.3) (for a detailed discussion of this type of kiln, see #2.5).

> **#4.3** You put it on a support between the 'eyes' (i.e. openings). The bottom of the *dabtu*-crucible must not touch the kiln.
> **#4.4** You light a good, smokeless fire. The fire should come out of the 'eyes' (i.e. openings).

This passage (#4.3) describes that the crucible was placed on a stand in the middle of the kiln chamber and then the crushed substances melted under the heat of a strong fire. The crucible should not touch the bottom of the furnace.

Nēmedu means 'support, base' (CAD N), 'Stütze' (AHw 776). The use of a support in connection with 'kiln with its four eyes' shows that – unlike in the 'kiln with shelf' – there must be sufficient space for the installation of such a support in the middle of the furnace chamber. The use of a support allows the substance to melt faster, as the heat also reaches the bottom of the crucible. Apart from this passage, *nēmedu* appears only in two other Nineveh glass recipes in connection with an *atūnu*-kiln (#18, #U).

The fire needed to be particularly strong as the flames should come out of the openings (#4.4). This, combined with the use of a support to speed up the melting process, indicates that melting should occur, which is then described in the subsequent paragraph #4.5:

> **#4.5** As soon as the mixture melts, you draw the fire away (and) you let it (the mixture) cool in its kiln. You take it up and you grind (it).

In this paragraph the melting process is described. When the substance melted, the heat in the kiln chamber was lowered by pulling out the burning logs from the firebox. The cooling in the furnace chamber entailed the annealing* process of the glass within the chamber of the kiln (#4.5).

The cold and hard glass lump in its crucible was then taken out from the kiln and removed from the crucible to be ground to a fine powder.

> **#4.6** You collect up (the powder) for a clean *dabtu*-crucible and you put it down into a cold kiln with shelf.

The powder produced in #4.5 was again collected in a clean *dabtu*-crucible and placed into a cold 'kiln with shelf'; here the final melting process is carried out.

For the discussion of *dabtu*, see #2.5 and #4.2; and for '*kūru ša takkanni*' 'kiln with shelf', see #2.6.

> **#4.7** You light a good, smokeless fire until the glass (lit. 'stone') becomes red. You do not cover the door of the kiln. After the glass has become red, you cover the door of the kiln.

#4.7 describes that strong heat was generated in the kiln by lighting a good fire. At the moment the colour of the glass in the kiln started to 'turn red' or to 'glow', the door of the kiln was closed, the air supply was cut off, and thus the temperature in the kiln was maintained.

> **#4.8** Until it becomes ready, you stir it once in front of you. After it had become ready, you will see a 'crown' with the eye. If the glass (lit. 'stone') is grown over (the rake) you pour it into a new *dabtu*-crucible (inside the kiln).

The melted substance was stirred until all the ingredients were dissolved. A rake was used to test whether the substance was completely melted. If the melted glass stuck to the top of the rake, the melting process was finished. The hot glass was then poured again into another new *dabtu*-crucible. This procedure was carried out inside the kiln.

Ḫarāṣu, *ḫarṣu* (see also #2.6, #4.1), 'to become ready', refers to the dissolving of all individual components and is used to indicate the moment of the complete mixing of all ingredients.

Finally, the term 'crown' (*mammu*) (#4.8) – already used in #3.6 – shows also here that the melting of the substance was complete.

Epēqu, 'to embrace', in stative D, 'is grown over' (AHw E), is used here to describe the substance of the viscous glass (lit. 'stone') that adheres to the tip of the rake when it has melted.

Nazālu: See #2.6. In this paragraph *nazālu* ('to pour out') is used also in connection with melted glass, which was in this case poured into a new *dabtu*-crucible inside the still hot kiln (for the discussion of *dabtu*, see #2.5 and #4.2).

> **#4.9** You let it cool off in the kiln and (out of the kiln) rises blue *zagindurû*-glass.

The glass is annealed within the kiln and finally the end product *zagindurû* appears in the kiln (#4.9).

> **#4.10** Copy [x]

In the colophon the word *gabarû*, which means 'copy' occurs. Colophons using *gabarû* inform about the origin and nature of the original text and are accompanied by the name of the country or city from which it originated (Hunger 1968: 6). Unfortunately the tablet is damaged and provides no information on the origin of the text.

6.6.6 Summary

The text provides a detailed overview of the various methods and tools used in glass production. This is of particular importance, as neither glass workshops nor special tools for glass production have yet been identified in archaeological records.

Specific verbs are used in connection with the preparation of the batch* as well as with the manufacturing process. *Esēpu* ('to gather') and *šakānu* ('to put') occur in connection with the process of gathering* or placing the mixture into a crucible. *Šūrudu* ('to put down') is used in conjunction with all kinds of kilns (*kūru*),[48] always together with a crucible. Therefore the term only occurs in connection with glass melting but not in connection with the sintering* process. For sintering, the term *redû* ('to guide, direct') occurs, which could indicate that the mixture needed to be placed exactly between the four fire openings, the hottest point in the kiln. In **#4.6** even a support related to *redû* is mentioned.

Regarding the tools, the *muterru* ('rake') was used for stirring hot glass in a crucible and for testing its viscosity (**#3.6**). Two types of crucibles have been identified in the text, *dabtu* and *ḫarāgu*.[49] Both occur in connection with the process of glass melting. In this regard it is important to note that the crucibles needed to be *zukūtu* ('clean') or even *eššetu* ('new') for the melting process of glass. This shows how carefully this process was monitored. In one case, a support (*nēmedu*) is reported on which to place a crucible (**#4.3**). The support is only mentioned in connection with the 'kiln with its four eyes'. In contrast, *ḫarāgu lā eššetu*, a crucible that is 'not new', occurs only in connection with the sintering process (Opp. §13, §15). It is interesting to note that *ḫarāgu*-crucibles could be closed or covered (Opp. §15, 18). Evidently the glassmakers were aware of the principles of oxidation and reduction.

The introduction to the text shows that a certain type of furnace was built for the production of glass (*kūru ša abni*).[50] It is interesting to note here that *kūru*-kilns in the 1st millennium only occur in connection with metal and glass melting (Salonen 1964: 120). The *kūru*-kilns can be divided into two types: one is called *kūri ša 4 īnīša* ('kiln with its four eyes'), while the other is *kūri ša takkanni* ('kiln with shelf'). Whereas *kūru ša takkanni* is only used for melting glass, *kūri ša 4 īnīša* occurs in connection with sintering and melting processes (see Table 6.1). When the melting takes place in a *kūri ša 4 īnīša*, a support (*nēmedu*) is reported (**#4.3**) (see Table 6.1). Also, the fire during melting needed to be considerably strong, as it should come out of the eyes (*išātu ultu libbi īnī uṣṣâ*) (**#4.4**). *Kūru ša takkannu* had an opening that could be covered (*kattumu*, **#3.5**) and opened (*peḫû*, **#3.5**) by the use of the 'door of the kiln' (*bāb kūri*) (**#3.5**).[51] It can furthermore be observed that *kūru ša 4 īnīša* was explicitly not pre-heated (*kaṣû* 'cold') when the batch was placed in the kiln chamber, whereas *kūri ša takanni* occurs in pre-heated and not pre-heated state (see Table 6.1). This could have to do with the temperature level and the heating time required for the various processes.

	Cold	Hot	Sintering process	Melting process
kūru ša 4 īnīša 'kiln with its four eyes'	√		√	√ (with support)
kūru ša takkannu 'kiln with shelf'	√	√		√
dabtu zakūtu 'clean crucible'				√
dabtu eššetu 'new crucible'				√
ḫarāgu eššetu 'new crucible'				√
ḫarāgu la eššetu 'crucible, not new'			√	

Table 6.1: The use of different equipment in connection with sintering and melting processes.

[48] Oppenheim (1970: 72) remarks that this term is also used in connection with *atūnu*-kilns.

[49] Dabtu is most frequently used in the glassmaking text from Nineveh; *ḫarāgu* appears less frequently, *tamšiltu* is used rarely; see Oppenheim 1970: 70–71.

[50] The glassmaking texts mention three different types of kilns: *kūru*, *atūnu* (§8, §12, §18, §E) and *tenūru* (§U). It is, however, interesting to note that the introduction is always concerned with the setting up of a *kūru ša abni*.

[51] The *atūnu*-kiln mentioned in other glassmaking texts was used for long firings; it occurs with *tamšiltu*, which can be translated as 'mould'; see Oppenheim 1970: 70.

To summarise, the detailed analysis and interpretation of the *zagindurû*-glass recipes shows that the glassmaking texts can generally be considered as actual recipes for primary glass. The different procedural steps within the whole process are described in great detail. Remarkably, most of these operational steps are, in terms of chemical knowledge, as well as experimental studies, comprehensible in the broadest sense. Repeated heating and sintering processes, the slow stirring of the copper compound into the batch, as well as the different cooling process probably seem arbitrary when looking at them from a purely philological point of view, but by understanding the working properties of the material and the chemical composition of glass, the processes described become comprehensible. Expressions like 'slow copper' or 'fast bronze' can be understood on the basis of modern chemical knowledge. The recipe that describes the production of blue *zagindurû*-glass can even be confirmed on the basis of experimental studies. The use of a disproportionately high number of words exclusively attested in the glassmaking texts indicates the existence of a particular technical terminology for glassmaking.

7. Archaeometrical Evidence

The following section focuses on the different chemical components present in ancient glass. The chemical composition of glass can be determined on the basis of analyses of major, minor and trace elements that vary according to different materials present in the batch*. Analytical data can also help to understand and reconstruct the steps of the *chaîne opératoire* in ancient glassmaking. There is a particularly close connection between chemical analysis and the texts that deal with glassmaking, since without chemical knowledge it is not possible to understand and interpret these texts (Chapter 6).

This chapter aims to give an overview of the different ingredients of ancient Mesopotamian glass, including major constituents as well as colourants. In a second step, published chemical raw data collected from various sites is brought together for evaluation.

7.1. Major constituents of ancient glass

7.1.1 Silica

The main chemical constituent of glass is silicon dioxide or silica (SiO_2), which widely occurs as the mineral quartz. Silicate creates the three-dimensional structure of glass and is therefore also called a 'network former' (Chapter 2.1.1).[1] Silica makes up to 45–70% by weight of ancient glass and is the most common component of the earth's crust. Quartz is insoluble in water and therefore remains as a weathering product of rocks forming sand or quartz pebbles. It is then transported by water, wind and deposits to riverbeds, deserts and beaches (Freestone 1991: 39; Wedepohl 2003: 5).

Sand naturally contains impurities, mostly in the form of iron (Chaper 7.2.1). Because impurities greatly influence the glass melting process, there was a great interest in reducing these components. This was achieved by choosing sand that was naturally low in impurities or by using quartz pebbles.[2]

The raw materials chosen for a specific glass formula can be traced in the chemical results. In regard to the silica source, high amounts of aluminium oxide (Al_2O3) (above *c.* 1.5%) and iron oxide (FeO) suggest that sand was used. If the amounts of aluminium oxide (below 1.5%) and iron oxide are low, then purer quartz sources – e.g. quartz pebbles – have most probably been used.[3]

7.1.2 Flux

The melting point of pure silica lies at around 1700°C. This temperature was far too high to have been achieved in ancient pottery kilns or metallurgical furnaces. Thus in order to melt silica to a glass, a flux was added to the composition to lower the melting temperature to about 1000°C. Whereas silica is a network former, flux is known as a 'network modifier' which allows melting by lowering the melting point (Henderson 2013: 56; Pollard and Heron 1996: 156).

The most common flux to be used in the 2nd and 1st millennia was potash gained from plant ash (potassium oxide) and mineral natron (sodium carbonate), which form two characteristic glass groups commonly referred to as 'plant ash glass'* and 'natron glass'* (Chapters 7.1.3, 7.1.4).[4]

The use of flux in ancient glass changed in different periods throughout history (Table 7.1). Plant ash glass was common in the Late Bronze Age in Mesopotamia as well as in Egypt. The first glasses fluxed with mineral natron occur in the 10th century in Egypt among core-formed vessels found in the burial of Nesikhons, providing a *terminus ante quem* of 975/974 for glasses (Table 7.1) (Schlick-Nolte and Werthmann 2003).[5]

7.1.3 Plant ash glass

Plant ash glass contains two basic materials, ashes of salt-tolerant plants and silica. Glasses fluxed with plant ash are characterised by high levels of magnesia and potash. Therefore this type of glass is often also called 'high magnesia/high potash' glass, abbreviated HMHK (Sayre and Smith 1961). If the values of both components are higher than around 1.5%, the glass can be identified as plant ash glass.[6]

The source of potash and magnesia are salt-tolerant plants that are also known as halophytic plants*.

[1] For further details, see Newton and Davison 1989: 5–6 and Pollard and Heron 1996: 156, 162.

[2] A video filmed by Robert Brill in 1979 shows glass production in Herat, Afghanistan. Here glassworkers use quartz pebbles collected directly from the fields and dry riverbeds: https://www.youtube.com/watch?v=BMYE83DJU4Q (accessed: 25.3.2016).

[3] For a detailed study on silica and its components, see Henderson 2013: 56–64.

[4] The two groups were recognised for the first time by Sayre and Smith 1963.

[5] The transition from plant ash to natron glass can already be observed among beads from Italy being made from both mixed alkali and plant ash glasses, therefore Gratuze et al. 1998 and Henderson 1988. For the spread of natron glasses in Egypt, see Rehren and Pusch 2007: 216.

[6] For the first systematic study on plant ash, see Turner 1956; for further later publications, see Barkoudah and Henderson 2006; Henderson et al. 2009; Henderson 2013: 85–90.

	Egyptian 15th cent. BC	Roman 1st cent. AD	European 13th cent. AD	Syrian 14th cent. AD	Modern
Silica, SiO_2	65	68	53	70	73
Soda, Na_2O	20	16	3	12	16
Potash, K_2O	2	0.5	17	2	0.5
Lime, CaO	4	8	12	10	5
Magnesia, MgO	4	0.5	7	3	3
Batch materials	plant ash quartz	natron sand	wood ash sand/quartz	plant ash sand/quartz	synthetic components
Glass category	high magnesia	low magnesia	forest glass	high magnesia	

Table 7.1: Compositions of different types of glasses (% in weight) (after Freestone 1991: 40 tab. 3.1)

Halophytes grow naturally at edges of deserts, in saline maritime environments and in inland salinas from Western Europe throughout the Mediterranean to Central Asia. They tolerate soils with a high content of salt, which then accumulated in the leaves. In order to use halophytes for glass production the plants need to be ashed by burning. This process releases sodium carbonate that then reacts with silica in the glass batch*.[7]

The relationship between the chemical composition of halophytic plants* and the composition of plant ash glasses* is complex: soils in which salt plants grow clearly contribute to the composition of their ashes. Up to now, it is not entirely clear which species of salt-tolerant plants were used for glassmaking in different periods. There have been few studies on the basis of chemical analyses investigating this relationship (Brill 1970; Turner 1956; lately Tite *et al.* 2006). More recently Henderson (2013: 22–51) focuses increasingly on halophytes, especially on those from Syria. Therefore he takes geologically different areas into account which cover the Balih valley, the surrounding areas of Damascus and Aleppo as well the Euphrates region.[8]

7.1.4 Natron glass

Natron glass contains mineral natron as flux and exhibits low levels of magnesia (0.5–1.5%), and low levels of potash (0.1%–1.0%) (Shortland and Tite 2000: 145). Natron glasses are therefore referred to as 'low-magnesia/low-potash' glass (abbreviated LMLK) (Purowski *et al.* 2012). Compared to plant ash glass*, natron glass is more predictable in the glassmaking process as it contains less impurities and it also possesses higher levels of alkali, therefore making it a more efficient flux (Freestone 1991: 39–40).

Furthermore, the chemical composition of natron glass has a crucial effect on the stability of the material: natron glass is naturally low in lime, which is needed to make glass stable (Chapter 7.1.5). The increasing amount of chemically unstable glass at the beginning of the 1st millennium is, according to Rehren and Rosenow (forthcoming), the reason why glass almost entirely disappears from the archaeological record of that time, particularly in Egypt.

The most extensive deposit of mineral natron is Wadi el-Natrun, a group of evaporated lakes in the Western Desert of Egypt. This is the main source exploited in the Late Bronze Age and Iron Age (Henderson 2013: 51; Sayre and Smith 1961; Shortland and Tite 2000: 149–150). Shortland *et al.* (2006a), who presented an extensive study on mineral natron, point to several additional sources that include sites such as al-Jabbul in northern Syria, and Lake Van in Armenia, which could also be considered as potential sources for mineral natron. Further sources have been recently taken into consideration that are located in the peripheries of Ankara, and in eastern Anatolia around Lake Van, Lake Erçek and Lake Arin (Dardeniz 2015: 199).[9]

The introduction of natron as flux* in glassmaking is commonly attributed to Egypt. The earliest natron-based glass identified so far comes from the tomb of Nesikohns in Thebes and dates to 974 BCE (Schlick-Nolte and Werthmann 2003). But also at Pella, Hasanlu and Nimrud early natron-based glass occurs (Chapters 7.4.2, 7.4.3, 7.4.4). From the early 1st millennium onwards, natron-based glass slowly spreads throughout the Mediterranean and Levant as far as Europe, and becomes soon the predominant type of flux. By around the 5th century BCE, natron was used as flux for most of

[7] For further chemical details, see Barkoudah and Henderson 2006: 297–298; Henderson 2013: 23.

[8] For maps with sampling locations, see Henderson 2013: 37, fig. 2.8. For some periods written sources give some hints. For example in 14th-century AD Europe a plant known as 'Rochetta' was well known for its high quality in regard to glassmaking. It was therefore even imported from Syria (Amouric and Foy 1991); for further examples, see Barkoudah and Henderson 2006; Henderson 2013: 24.

[9] Dardeniz (2015: 199) suggests that natron from central Anatolian sources could have been used by Phrygian, Lydian and Lycian glassmakers, while the sources in eastern Anatolia could have been exploited by the Urartians.

the glass west of the Euphrates.[10] Shortland *et al.* (2006: 522–523), however, point out that plant ash as flux has never been entirely replaced by natron in Mesopotamia, Iran and Central Asia, which is, according to them, due to the circumstance that mineral natron was not widely available in these regions.[11]

7.1.5 Stabiliser

Lime (CaO) and magnesia (MgO) served as stabilisers in ancient glass. Stabilisers were essential for the durability of glass, as it stopped the material from dissolving in water. Late Bronze Age, as well as Iron Age, plant ash glasses* usually contained levels of calcium in excess of 4%, and magnesia in excess of 3% naturally, and were therefore less effected by alteration (Fletcher *at al.* 2008: 47). In contrast, in glass fluxed with mineral natron*, the amount of calcium (CaO) and magnesia (MgO) is relatively low, because natron has a low content of impurities. Unless lime was added through another source - such as calcareous sand or shell - natron-based glass was unstable and much more effected by alteration (Reade *et al.* 2009). Plant ash glass is therefore much better preserved in the archaeological record than natron-based glass in the early 1st millennium. According to Rehren and Rosenow (in press), this could be the reason why glass almost entirely disappeared from the archaeological record of that time, particularly in Egypt.

7.1.6 Opacifiers

In general, three qualities of glass can be distinguished depending on the amount of light that passes through it: opacity, translucency and transparency (Figure 2.1). Coloured opaque glass is among the earliest of attested glass types in the archaeological record.

In opaque glass, the wavelengths of light are reflected by the glass, which is caused by the presence of inclusions (crystalline particles) which are mostly small (typically ~5 μm) and well dispersed throughout the glass. These create an immiscible phase that diffuses the light and gives it a milky or opaque appearance (Pollard and Heron 1996: 163; Shortland 2012: 26; Turner and Rooksby 1959). Substances which produce such phases are called opacifiers. Opacifiers therefore change the property of the base glass from transparent to opaque.

The most common opacifiers in the Late Bronze and Iron Age periods were lead antimonate ($Pb_2Sb_2O_7$) for opaque yellow glass, and calcium antimonate ($Ca_2Sb_2O_7$ and $Ca_2Sb_2O_6$) for opaque white glass. Opaque turquoise was achieved by the combination of copper and calcium antimonate. Opaque red glass was produced by the use of cuprite (Cu_2O), which is a red copper (I) oxide (Chapter 7.2.3.2). Opaque green is very rare for Iron Age glass and could be produced by mixing the compounds of opaque light blue and opaque yellow (Figure 7) (Shortland 2002; Turner and Rooksby 1959).

Colorant	*Opacifier*	*Colour produced*
None	None	Colourless, amber or brown
Cu^{2+}	None	Translucent blue
Cu^{+}	None	Opaque red
Co	None	Translucent dark blue
Mn	None	Translucent purple
Mn	None	Black
None	Calcium antimonate	Opaque white
None	Lead antimonate	Opaque yellow
Cu^{2+}	Calcium antimonate	Opaque turquoise
Co	Calcium antimonate	Opaque deep blue
Cu^{2+}	Lead antimonate	Opaque green

Figure 7.1: Opacifiers and colouring agents of Late Bronze Age glasses from Egypt (Shortland 2002: 518).

7.1.7 Decolourisers

In contrast to opaque* glass, translucent as well as transparent glass allows light to be transmitted. Transparent glass is therefore characterised by its (almost) absence of colour. In this study, transparent colourless glass refers to deliberately decolourised glass, distinguished from the natural colour of uncoloured glass called 'aqua', which 'would perhaps be better termed 'free of deliberately added colorant' glass' (Shortland 2002: 517). Natural non-coloured glass mostly has an 'aqua' or greenish tinge due to natural impurities of mainly iron.

Transparent colourless glass could have intentionally been made by adding the decolouriser antimony oxide to the batch. Antimony oxide reacts either as a decolouriser or as an opacifier, depending on the quantity and firing conditions (Brill 1970: 116). Therefore, the rise of temperature or an increased amount of antimony oxide turns glasses from opaque into transparent.

Elevated levels of antimony oxide therefore indicate the deliberate use of decolourisers to achieve colourless glass. Elevated levels of this oxide are attested among glass from Nimrud, which identifies these pieces as the earliest decolourised glass objects (Chapter 7.4.3.2). In contrast, colourless glass from Late Bronze Age Mesopotamia (e.g. unprovenanced seals from Chogha

[10] For further details on the spread of natron and plant ash glasses, see Henderson 2013: 26–54; Shortland et al. 2006: 522–523.

[11] Mineral natron was already exploited from the 4th millennium onwards, not only for the production of vitreous materials, but also for the fabrication of medicine, soap and dye, and in Egypt it was, moreover, used for mummification(Shortland et al. 2006: 52).

Zanbil) and objects from Egypt (Thebes, Lisht) are not decolourised deliberately, as the concentrations of antimonite in the batch* are too small. The effect of decolouration could instead be due to the furnace atmosphere or the use of pure sands (Bimson and Freestone 1988; Sayre 1963: 269; Shortland and Eremi 2006: 584, 591). Brill (1970: 116) is sceptical about the conscious use of a decolouriser in glass from Nimrud. He explains that antimonate is almost ubiquitous in every glass sample of the 1st millennium in the entire Mediterranean and Mesopotamia, with a proportion of about 1–2%. A conscious use of antimony oxide as a decolouriser can therefore generally not be proven in his eyes.

7.2. Colourants and their sources

By adding a colourant to the basic glass composition, glass takes on a specific colour. Mesopotamian glass of the first half of the 1st millennium was, in most cases, deliberately coloured. The different colouring agents were added at some point in the manufacturing process to the basic glass components (Newton and Davison 1989: 58). This is also described in the Nineveh glass recipes that produce blue *zagindurû*-glass by the addition of a copper compound (Chapter 6.6).[12]

The production of a particular colour was a complex process that depended on various factors. Not only the question of how different colourants interact, but also the composition of the base glass, the amount of impurities, the temperature and conditions of the furnace, as well as the duration of the melting and annealing* times, were all important. The exact point of time at which a colourant transforms into a certain colour is therefore decisive. The glassmakers had to know these factors in detail and adapt the respective production process to the glass recipe for certain colours to achieve the desired results.

7.2.1 Iron

Iron (Fe) was present either as an impurity or as an intentional colourant* in glasses. Iron as an impurity had the most pronounced influence on the colouration and workability of glass (Weyl 1951: 91–113). Iron was present in every ancient plant ash glass* at a level of around 0.5% because of impurities that naturally occur in sands and flux*. The presence of iron in the batch* leads to a very characteristic pale greenish tinge.

Iron also served as a colourant*, with the outcome dependent on the amount, as well as on the state of oxidation. Under reduced conditions*, iron leads to an amber or yellowish translucent coloured glass. Elevated levels of 0.5–1.0% of iron, under reducing conditions, would colour glass in a strong blue (Dussubieux *et al.* 2018; Shortland 2012: 25). In addition, if iron was poorly ground it appeared as dark brown or as black spots on the surface of the glass object (Newton and Davison 1989: 58; Weyl 1951: 95).

7.2.2 Cobalt

Cobalt (Co) is amongst the strongest colourants* of ancient glass, which is why it often cancels out other colourants present in the compound. Even 0.1% of cobalt could result in a dark blue glass; in higher concentrations it even turns glass entirely dark (Shortland 2012: 25).[13]

Since there has been much discussion about a possible connection between Mesopotamian and Egyptian cobalt coloured glass, it is essential to discuss the Egyptian cobalt glass composition at this point. Evidence for the use of cobalt as a colourant can be found in Egypt from the 2nd millennium onwards. Cobalt-based glass associated with the New Kingdom (16th–11th centuries) contains particularly high proportions of alumina, magnesia, manganese, iron, nickel and zinc, and is therefore associated with the cobaltiferous alum deposits of the Western Desert in Egypt, namely the oases Dakhla and Kharga (Kaczmarczyk 1986; Rehren 2001; Shortland *et al.* 2006; for textual evidence, see Shortland 2012: 109). With reference to glasses from the Late Period in Egypt (mid 7th century), Kaczmarczyk and Hedges (1983: 41–54) state that the cobalt composition is not identical in that of the previous period. They therefore point to a change in the cobalt source in the 1st millennium. Analyses of cobalt glass in the tombs of Nesikhons, Thebes, from the 10th century, on the other hand, point to an Egyptian origin of the cobalt ore used in this glass (Schlick-Nolte and Werthmann: 2003). These results point to a continuous use of the cobalt sources of the Egyptian Western Desert, as well as to the exploitation of new cobalt sources at the beginning of the 1st millennium.

The elemental signature of Egyptian Western Desert cobaltiferous alums was identified among glass from northern Mesopotamia (Tell Brak, Nuzi) and Mycenaean Greece, as well as among ingots found around the shipwreck in Uluburun, all dating to the Late Bronze Age. This corresponding signature was used to demonstrate the trade link between these Late Bronze Age sites and Egypt. In contrast, cobalt-coloured glass from Nippur (Late Bronze Age) shows different constituents than those from Late Bronze Age Egyptian cobalt glass that are unique to Mesopotamia

[12] Another decolouriser is manganese, resulting either in opaque brown/purple, or if the amount of manganese is increased results in colourless glass. The use of manganese as a decolouriser is not attested before the 1st century AD;Newton and Davison 1989: 59.

[13] Henderson (2013: 69) also reports that Co^{3+} ions can create a pink colour.

(Walton *et al.* 2012). Analysis of a cobalt-coloured glass lump from Eridu seemingly confirms this supposition, as analyses indicate a distinctive cobalt composition unique for cobalt-blue glass used in this lump (Garner 1956). Cobalt-coloured glass from Nippur and Eridu demonstrates that beyond the Western Desert in Egypt further cobalt ore sources were exploited in the Late Bronze Age, which are so far unknown.

For the Iron Age period, studies on cobalt glass, especially from Nimrud, have been carried out, suggesting that cobalt used in this period could have been gathered from Egyptian sources (Bimson and Freestone 1985; Reade *et al.* 2005; Turner 1955; 1956). With regard to cobalt-coloured glasses from Nimrud, Reade *et al.* (2005) showed that the cobalt used in some of the glasses does not correlated with magnesia and iron and is rarely accompanied by more than a trace of nickel or zinc present in cobalt sources from Egypt. Therefore the cobalt-blue glasses from Nimrud were coloured using a pigment derived from cobaltiferous alum similar to that used in New Kingdom Egypt, but which varies slightly.[14]

Cobalt sources other than those in Egypt are not yet known. Cobalt ores are relatively rare in the Near East, but exist in Iran for example. Iranian cobalt sources were exploited in Islamic times, but whether these sources were used for Mesopotamian cobalt-coloured glass has not yet been investigated (Henderson 2013: 73; Kaczmarczyk 1986: 373). The same applies to possible sources in Turkey, about which it is not known whether they were exploited in the Late Bronze and Iron Ages (Henderson 2013: 74).[15] Further research in this regard would be worthwhile.

In summary, cobalt ores are comparably rare in the eastern Mediterranean and Mesopotamia. The most common source, exploited in the Late Bronze and Iron Ages, was the Western Desert in Egypt. Cobalt from this source can be connected with glass from the Late Bronze Age sites of Nuzi, Tell Brak, Mycenae, Uluburun, and probably also from Iron Age Nimrud. This shows a complex and distant trade, originating in Egypt. Finds from Late Bronze Age Nippur and Eridu, as well as from the early Iron Age site of the tomb of Nesikhons, indicate, however, the exploitation of an additional cobalt ore source that is unknown to us.

[14] Analysis indicate that most likely mineral natron was used as flux; Reade et al. 2005: 26.

[15] Kaczmarczyk (1986: 373) suggests that the cobalt used as colourant in Mesopotamian glasses throughout all periods, and in Egypt from the 1st millennium BC onwards it came from Iran. In this regard he refers to analyses taken from objects from the Early and Late Bronze Age, as well as from Sassanian and Hellenistic times.

7.2.3 Copper

Copper (Cu) was the most common colourant* of Late Bronze and Iron Age glass, and resulted in many different colours, depending on the kiln atmosphere as well as on the oxidation state of copper (metal, cuprous Cu or cupric Cu^{+2+}). Copper could be gathered from ores, minerals, metal alloys, or slags from copper smelting and melting operations. The use of a specific copper source, both ore and processed copper, can be determined on the basis of chemical analysis. Therefore the amount of arsenic, tin, zinc, lead and iron either indicate the use of copper ore or processed copper (Stapleton 2011: 97). Copper is a strong colourant, with 0.5% already resulting in a strong blue or green colour.

7.2.3.1 Copper and blue glass

Turquoise is one of the most common colours in Mesopotamian glass and was also extensively utilised in the production of faience* and Egyptian blue. Since copper ores are very widely spread in the eastern Mediterranean, the source of the copper colourant in glass is therefore difficult to determine (Moorey 1994: 242–277).

In translucent turquoise glass, copper occurs in the oxidised copper state. Opaque* turquoise glass also contains calcium antimonate as an opacifier* (Chapter 7.1.6). Copper was also used to achieve a dark blue colour. In contrast to cobalt, however, a significantly higher copper content (2–5%) was required (Shortland 2012: 25).

7.2.3.2 Copper and red glass

The mechanisms that led to opaque red-coloured glass are diverse. Generally it can be stated that under reducing conditions* copper turns glass red. Two different types of copper result in different types of red colouration: copper metal and cuprous oxide, also known as cuprite. Both of these types also vary on the basis of their lead contents. Red glass coloured with copper metal is low in lead and low in copper. Glass of this kind is therefore called 'low-lead/low-copper glass' (Bimson 1987; Freestone 1987). Cuprite-coloured red glass on the other hand is due to large branching of crystals that develop during cooling of the glass. The evolved crystals block the transmission of light, which turns glass opaque*.[16] The two different red glass groups can also be distinguished with the naked eye. Copper metal-coloured glass forms a dull red colour, whereas cuprite red glass turns glass bright red, which is called 'sealing wax' or 'cuprite red' (Barber *et al.* 2009: 116; Bimson and Freestone 1985: 120). In addition, there

[16] The glass cannot be re-melted in air, as the red cuprite crystals oxidise and turn to greenish-black; Bimson and Freestone 1985: 121.

is an orange-yellow form of cuprite which can appear as a thin layer on the surface of the glass (Bimson and Freestone 1985: 121, 180).

Low-lead/low-copper glass was very common in the 2nd millennium. Then, sometime in the first half of the 1st millennium, high-lead/high-copper glass appeared. The earliest high-lead/high-copper glass was identified at Arslan Taš (AM28), Nimrud (Nim198), and Toprakkale (Bimson and Freestone 1985: 120). The lead content in glasses from these sites is between 23–27% and the copper content between 8–13%, which indicates an intentional addition of lead ore.

7.2.3.3 The emergence of high-lead/high-copper red glass

High-lead/high-copper red glass has many advantages over low-lead/low-copper red glass. High-lead/high-copper glass was much less susceptible to devitrification, and was more predictable during the manufacturing process. Furthermore, the high content of lead had a positive influence on the colour, resulting in a more brilliant, 'gem-like' shiny colour (Freestone 1987: 186–187).

The production of lead-containing, high-copper glass requires a high technical performance, which mainly consists in maintaining a reduced furnace atmosphere. It can be observed, however, that once the value of high lead contents was recognised, this type of glass was produced almost exclusively. Maintaining a reduced kiln atmosphere* was a difficult task and was probably facilitated by the use of 'reductants', which promoted a reduction from the cupric to the cuprous state (Freestone 1987: 183). Possible reductants could have been carbon (charcoal), antimony, iron, and also lead, which were added to the melt. Another way to maintain a reduced atmosphere was to use crucibles with bound lids (Freestone 1987: 183). Crucibles with lids are attested in a glassmaking text concerned with the production of 'fast bronze' (*ḫarāgu šaktumtu*) (Tablet A §15: 118 Opp) (Chapter 6.6.6). This text passage refers, most likely, to the production of red-coloured glass, which is achieved in reduced conditions by using a crucible with lid (for the identification of 'fast bronze' Chapter 6.6.4).

7.2.4 Lead antimonate and calcium antimonate

Antimonate was used as an opacifier*, but served also as colourant* for glass. Lead antimonate ($Pb_2Sb_2S_3$) caused a yellow colour of the glass, and was most likely added in the form of roasted stibnite. Calcium antimonate ($Ca_2Sb_2O_7$) resulted in a white opaque glass.

With regard to the workability of lead antimonate during the production of yellow glass, Shortland (2002: 525) carried out experimental replication work that indicated that yellow pigment roasted together from lead and antimonate was unstable. This means that, at high temperatures, the pigment dissolved in the glass and resulted in a creamy-white coloured glass instead of a yellow colour. Applied to the production process of yellow glass, this means that it was imperative to work yellow glass carefully and quickly. Therefore the pigment was either carefully folded to the melted glass at relatively low temperature, or was processed extremely quickly to avoid high temperatures for too long (Shortland 2002: 525). Furthermore, the inhomogeneity of the lead and particles within the glass found in the backscattered SEM micrograph, indicates that the lead antimonate was added to raw glass in fluid state and not to the glass batch* materials when cold (Shortland 2002: 523).

The difficulty in handling yellow colouration can, for instance, be observed in core-formed vessels, particularly those of the later 'Mediterranean Groups' (Chapter 4.3.1). Therefore the colour pattern of most core-formed vessels contains yellow only for the feather decoration, handles and rims, and never as the base colour. This is because the second colour was not exposed to the heat again, as with the base colour.

The most probable source for antimonate throughout the Late Bronze and Iron Ages, and up until Roman times, was the antimony ore stibnite (Sb_2S_3), out of which antimonate was extracted by firing. Stibnite is a lead-free ore which occurs in the region of the eastern Mediterranean and the Near East (Mass *et al.* 2002: 70; Shortland 2002: 524). It has been suggested that stibnite used in Egyptian glasses of the Late Bronze Age came from sources of the Eastern Desert in Egypt, probably from Gebel Zeit on the Red Sea coast. It was therefore suggested that stibnite from Egypt could have been imported to Mesopotamia in this period (Shortland 2002: 524). In addition to Egypt, stibnite also occurs in Iran, Anatolia and the Caucasus. In all these regions, modern antimony mines are exploited, but there is no evidence that they were utilised in the past. However, the latest research confirms that stibnite used in Mesopotamian glasses originated most likely from the Caucasus (Shortland forthcoming).

7.2.5 Manganese

Manganese has long been used as colourant* in vitreous materials. Throughout different periods, manganese dioxide was used as a colourant for violet, red, dark green, dark blue and yellow (Henderson 2013: 75; Shortland 2012: 116–118). It even served as a decolouriser* in later periods. Generally, manganese-coloured glass is considerably rare in the Iron Age period. When it occurs, it was mainly used to colour glass purple-brown and violet (0.55–2.6%). Naturally, manganese occurs in its pure form as pyrolusite (Shortland 2012: 116–117).

7.3. Summary and conclusion

This chapter depicts the complex chemistry of ancient glasses and illustrates the underlying factors involved in the glassmaking process. All ancient glass recipes are based on silica (network former), soda (network modifier) and lime (stabiliser), which is why they belong to the group of soda-lime-silica glass.

The major sources of silica in Mesopotamia were most likely sand or quartz pebbles, the use of which is indicated in the Nineveh glass recipes with the word *immanaku*-stone (Chapter 6.6.3). The degree of contamination of the silica source with impurities can be measured by the level of aluminium. If the amount exceeds 1.5%, sand was probably used as a silica source.

While the use of plant ash* as flux* was widespread in the 2nd millennium, the use of mineral natron slowly commenced at the beginning of the 1st millennium (Table 7.1). The change in flux should be understood as a subtle change over a longer period of time, rather than as a sudden and abrupt event occurring at a specific workshop. Even if natron glass* possesses better working properties, as it is more chemically predictable, plant ash glass remained widespread, and never disappeared entirely in Mesopotamia (Table 7.1).

7.4. Re-evaluation of chemical data of Mesopotamian glass

7.4.1 Approach and methodology

The following chapter is concerned with the archaeometrical research carried out on Iron Age and Late Bronze Age glass. To demonstrate chronological developments and show compositional similarities between different regions, it is necessary to include chemical data from both the Late Bronze Age and Iron Age periods.

Solely data that has previously been published is taken into consideration. The four sites that have produced this evidence are for Iron Age Hasanlu (Chapter 7.4.2), Nimrud (Chapter 7.4.3), Pella (Chapter 7.4.4) and Gordion (Chapter 7.4.5).

Available analyses of 183 objects from Iron Age Hasanlu and Nimrud, and from Late Bronze Age sites Nippur, Nuzi and Tell Brak, which offer comparative material, are shown in Appendix 2. Since the different analyses were carried out at different times and with different equipment, the components recorded often vary. SiO_2, TiO_2, Fe_2O_3, MgO, CaO, Na_2O, K_2O and P_2O_5, for example, which are not detected for Nuzi and Tell Brak are analysed for Hasanlu. The analysis technique and the bibliography are mentioned for each site. Throughout the chapter, the data are investigated using bivariate graphs.

Another difficulty is that the analytical data cannot always be correlated with the individual archaeological object. Only 11 analyses of the objects contained in the catalogue are available for this study: five are glass tubes from Hasanlu (Has6, Has8, Has9, Has10), the others are from Nimrud and consist of an ingot (Nim198) and inlays (Nim119, Nim28, Nim153, Nim289).

7.4.2 Hasanlu

Out of 81 analyses from Hasanlu, 66 are of glass beads, a category of objects not considered in this monograph. The remaining samples were taken from 13 tubes and a mosaic beaker dating to the Late Bronze Age. Raw

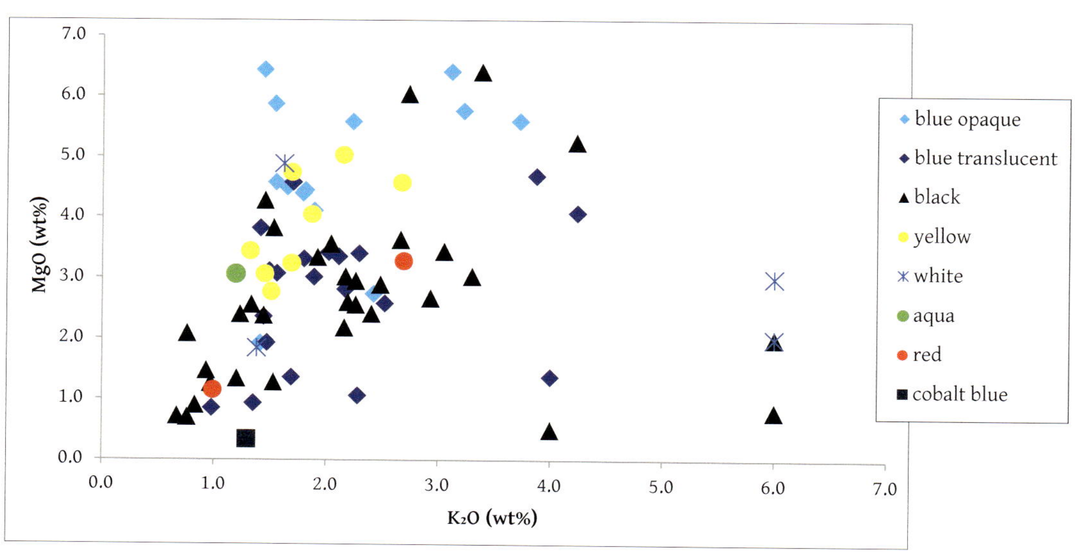

Table 7.2: Scatter plot of potash (K_2O) versus magnesia (MgO) for Hasanlu glass of different colours. Most of the values of both components are higher than around 1.5%, suggesting the use of plant ash as flux (raw data Appendix 2 with bibliography of each sample)

data on glass from Hasanlu was presented by Brill (1999: 43–44) and Brill and Stapleton (2012: 214–221). An extensive study with raw data from glass and glazed objects that can be associated with stratum IVB was published by Stapleton (2003; 2011).

7.4.2.1 Basic glass compositions

Hasanlu glass is of the soda-lime-silica type. By far the largest amount of glass attested at the site was fluxed with plant ash*, which is indicated by the high contents of potash (K_2O) and magnesia (MgO) (over 1.5%) (Chapter 7.1.3). A small number of black and translucent blue glass samples contain less than 1.5% each of potash and magnesia, which indicates the use of mineral natron (Table 7.2). Stapleton (2003: 119, 149) suggested that the low levels of potash and magnesia are the result of the lower proportion of plant ash present in the batch*, but are not an indication of the use of mineral natron. With regard to glass from Nimrud, comparable low levels of magnesia and potash are present. Reade *et al.* (2005: 26) and Brill and Stapleton (2012: 217) suggested a mix of mineral natron and low amounts of plant ash as flux.

Hasanlu black glass is compositionally very close to Pella black glass, in particular regarding the high levels of iron, the content being at an average of 9.7% (Reade *et al.* 2009: 49, 52). The high content of alumina and iron oxide, with the minerals alkali feldspar (about 10–30 wt%) and spinel grains in the glass composition indicate the use of sand as the source of silica (Stapleton 2003: 148–149).

7.4.2.2 Colouring agents

The objects from Hasanlu include dark blue, light blue, red, yellow and black-green coloured glass. The colouring agents are typical for 2nd and 1st millennium glass.

Regarding the translucent blue glass objects, the copper colourant* was most likely a copper carbonate or oxide ore, or mineral bearing iron, a metal alloy, or a slag from smelting and melting operations (Stapleton 2003: 77; 2011: 97). These propositions can be made on the basis of correlations between copper and arsenic, tin, zinc, lead and iron, as these are the elements associated with natural copper ores and processed copper.[17] However, the antimonate crystals in these transparent blue glasses are rare, indicating that the proportion of recycled material was very low (Stapleton 2003: 66–67).

Opaque* blue glass was almost exclusively used for the manufacture of tubes (Chapter 4.3.4). In contrast to translucent blue glass, opaque blue glass was made from another source of copper. Factors that show this are the positive correlation of arsenic (As) and copper, which indicates that these two components entered the glass in the same raw material. Arsenic and tin are common alloying elements in copper bronze and were found in the bronze artefacts of Hasanlu. However the bronze metals from the Hasanlu IVB levels do not seem to have been the exact copper source for the blue opaque glasses (Stapleton 2003: 138–139; 2011: 98, 113, table 8.3). The possibility of using alloyed metal in copper-coloured glass allows a connection between glass and metalworking workshops that existed at Hasanlu.

The yellow colouration is due to lead antimonate, calcium antimonate, and sodium antimonate crystals in the glass. As observed with translucent blue glass, alkali feldspar is present in yellow glass, with the content being at an average of 10–30%. This is due to the use of quartz-rich sands (Brill and Stapleton 2012: 217; Stapleton 2003: 148). An idea about the manufacturing process of yellow glass can be gained from its composition: lead (PbO) and potassium (K_2O) were most likely added to the batches as separate ingredients, since they have a negative correlation. Antimony was probably also added as a separate component, which is suggested because metallic antinomy was found at Hasanlu in the form of buttons and beads, which could have served as a source (Stapleton 2011: 99). Furthermore, the inhomogeneity of the batch* is due to a number of unreacted inclusions – like calcite – which shows that they did not undergo long processing during manufacture (Stapleton 2011: 99).

Green-black glass is rare in the archaeological record of the first half of the 1st millennium. The only known comparisons come from Pella (Reade *et al.* 2009). Chemically, the dark colouration is due to iron, lead and antimony (Chapter 7.4.4). The general compositions of the black beads vary in a wide range, with K_2O concentrations ranging from 0.83–2.66%, MgO concentrations from 0.90–3.82% and CaO concentrations from 1.56–5.02% (Brill 1999b: 43–44). The presence of iron, lead and antimony indicates that slag from processed copper ore was used.[18] A link between the glass- and metalworking, as indicated with regard to opaque blue glass is plausible.

Two pieces of red glass were analysed, one from the mosaic glass beaker from Hasanlu (Chapter 4.1.4), and the other from a chunk of red glass (Brill 1999b: 43). While the beaker fragment fall into the range of plant ash glass, the chunk reveals the use of mineral natron. According to Brill and Stapleton (2012: 217) the red colouration of Hasanlu glass was achieved by the use of an arsenical copper mineral added to the composition,

[17] For a detailed analysis, see Stapleton 2003: 113, table 8.3, 114–115, 137–142.

[18] For details on the ratios, see Brill and Stapleton 2012: 218; Stapleton 2003: 66–68.

and not from a copper alloy. The copper present in the mosaic beaker, in contrast, most likely derives from leaded bronze (Brill and Stapleton 2012: 218). Therefore, two different sources of red colourants are likely to have been used. The results of lead, oxygen and strontium isotope analysis confirm a link between the mosaic beaker and red glass from Babylonia and Iran (Babylon, Nippur, Susa, Persepolis) (Brill and Stapleton 2012: 219). The leads in the Hasanlu samples are isotopically very similar to the leads in a number of Mesopotamia objects characterised as 'Group M', and are particularly similar to some from Babylonia and Iran (Brill and Stapleton 2012: 219, Ref. A-77).

7.4.3 Trace elements

Brill (1999: 1: 43–44, 2: 44) analysed some glass beads from Hasanlu which show high levels of boron (up to 0.5% of B_2O_3). High boron glass from the Byzantine period has been discussed by Schibille (2011) and lately by Swan *et al.* (2017), who consider high concentrations of this element (> 300ppm) as a marker of glass manufactured in Turkey. However, at this stage there is no evidence for the existence of a high-boron glass industry in Turkey already in the Iron Age. The analysis of Brill (1999) shows, however, that also Iron Age glass from Hasanlu exhibits high levels of boron. As well as Turkey, boron deposits are found in northwestern Iran (Rahimpour and Kazemi 2003)[19] and in the western part of Lake Urmia (Assadpour *et al.* 2017). Hasanlu is located directly south of Lake Urmia, therefore it seems possible that soda with significant concentrations of boron was extracted locally (Dussubieux *et al.* 2018).

7.4.3.1 Conclusion

The analysis of Hasanlu reveals a broad spectrum of chemical compositions. The wide range of oxygen isotope compositions (12.3–17.3) shows that more than one silica source was used (Stapleton 2003: 97). The existence of different compositional groups is also supported by the diverse use of colourants that come from different sources and take into account different production processes (use of alloyed copper and copper ore in blue glass, use of arsenical copper ores and lead-containing bronze in red glass). The existence of local primary glass production is therefore likely, which is also underlined by the high boron content in some of the samples, a worthwhile field for future research.

Regarding the manufacturing processes of glass, Stapleton (2003: 154) notes that the compositions do not show simple correlations in oxides. This indicates the mixing of the batch material in several steps, with separate steps for the processing of colourants. The variety of copper colourants also shows that the craftsmen at Hasanlu had sophisticated technical knowledge of the material properties and its behaviour under firing. This shows that the pyrotechnic industry most likely interacted (Stapleton 2003: 154).

7.4.4 Nimrud

Altogether there are 31 accessible analyses from Nimrud, of which 21 derive from inlays, eight from bowls, one from an ingot and one from a fragment. Regarding most of the objects analysed, a correlation of the analytical data and the specific object cannot be reconstructed. Chemical raw data on Nimrud glass has been published by Barber and Freestone (2009), Brill (1978; 1999: 47–49), Reade *et al.* (2005) and Turner (1955).

7.4.4.1 Basic compositions

Glass from Nimrud is of the soda-lime-silica type. The majority of the samples shows high contents of potash (K_2O) and magnesia (MgO) (above 1.5% each), indicating the use of plant ash* (Table 7.3, Chapter 7.1.3). But this is not always distinct: red, light blue and colourless glasses form one group that was most likely fluxed with plant ash, but which lies right on the edge of levels typical for mineral natron (Chapter 7.1.4). In contrast, three of the four yellow glass samples show low levels of potash (0.5–1%) and magnesia (0.5–1%), indicating the use of mineral natron. Cobalt-based glass falls into a distinct group. Based on the values for potash and magnesia the use of a specific type of flux* is difficult to determine, as the amount of magnesia is far too high for natron glass. Regarding cobalt-coloured glass, Reade *et al.* (2005: 74) suggest the use of mineral natron as flux, which, according to them, is likely because of the low levels of lime.[20]

By plotting the values of alumina (Al_2O_3) and iron oxide (FeO) of the same samples, the three compositional groups implied above become evident (Table 7.4): Here, too, cobalt glass forms a distinct group with high amounts of alumina – which could be linked to the cobalt source – and yellow glass also falls into a distinct group high in iron. Colourless, red and copper-based glass falls tightly together. Reade *et al.* (2005: 24) suggested on the basis of their analysed data that four different compositional groups can be identified at the site.[21]

[19] The boron mine of Gharah-Gol is located c. 80 km southwest of Zanjan in northwestern Iran. It is the only active boron mine in Iran.

[20] Calcium is a major constituent of plant ash and would be present at a high level (5% or more), if one considers the use of plant ash as flux.

[21] The raw data is unpublished and therefore unavailable for direct comparison with the data in Appendix 2.

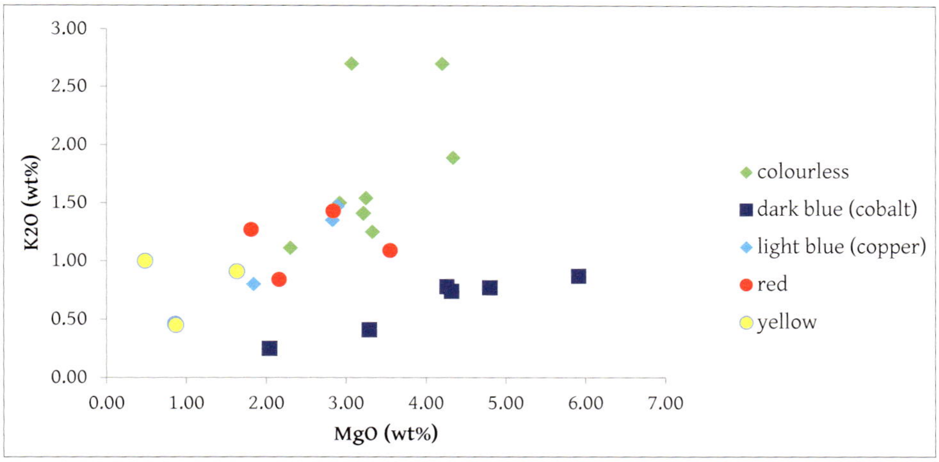

Table 7.3: Scatter plot of magnesia (MgO) versus potash (K_2O) for yellow glass, cobalt glass and a 'mixed group' consisting of colourless, red and light blue glass. The red box shows values usually connected with mineral natron as flux (raw data Appendix 2 with bibliography of each sample).

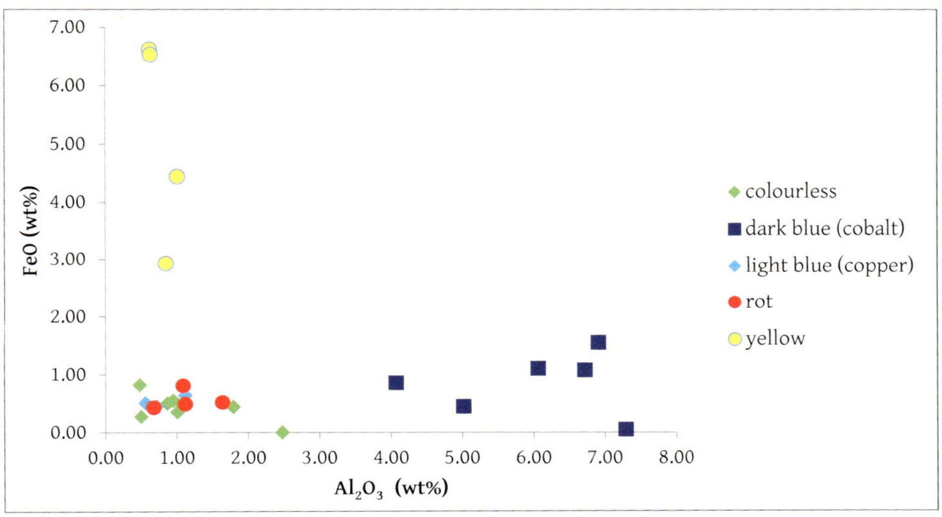

Table 7.4: Scatter plot of alumina (Al_2O_3) versus iron (FeO) for Nimrud glass groups of yellow glass, cobalt glass and red, colourless and light blue glass (raw data Appendix 2 with bibliography of each sample).

7.4.4.2 Colourless glass groups

The significance of colourless glass is discussed in Chapters 7.1.7, 8.2.1 . Analysis of colourless glass were taken from hemispherical bowls (objects unidentified), a painted plaque (Nim119) and the 'Sargon Vase' (Nim27). The average value of antimony oxide, for both the colourless bowls as well as the painted plaques, lies between 0.1% and 0.8%, which suggests the deliberate use of antimony as a decolouriser (Appendix 2).[22]

Some of the colourless glass was fluxed by the use of plant ash*. Reade *et al.* (2005: 24, fig. 1) could identify further colourless glass groups (Figure 7.2). 'Group 1' forms a tight compositional group of colourless glass with potash (K_2O) around 1.5% and magnesia (MgO) around 3%, which conforms to plant ash, but with potash being at the lower end (Figure 7.2). Also the 'Sargon Vase' (Nim27) belongs to this group, with values of 1.37% potash and 3.44% magnesia (Yoshinari 2013: 365). 'Group 2' depicts a standard plant ash group with typically high levels of potash and magnesia. 'Group 3' has low levels of magnesia (1.7%) and potash (0.95%), suggesting the use of mineral natron or low levels of plant ash used as flux*.

Three different composition groups of colourless glass can be distinguished from Nimrud, however the existence of other groups cannot be ruled out. The group containing samples of colourless hemispherical

[22] Further unpublished data on colourless glasses from Nimrud show the same levels of antimony (Sb) (Freestone, pers. comm.).

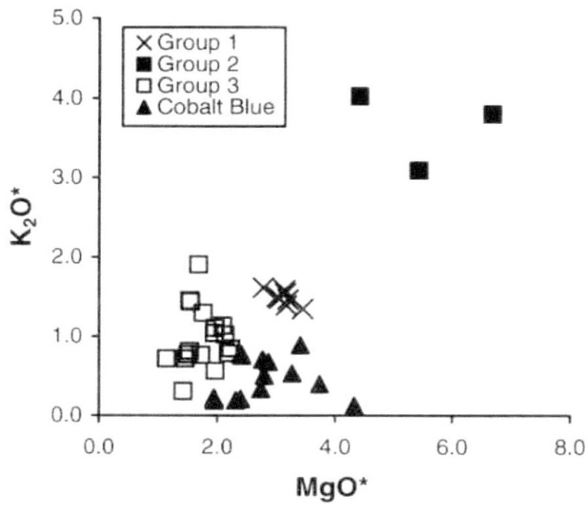

Figure 7.2: Different glass groups from Nimrud identified by Reade et al. (2005: 24). Group 1) Colourless glass; Group 2) Colourless and light blue glass; Group 3) Colourless and copper-based glass.

bowls was fluxed* with plant ash*. Another group containing hemispherical bowls and the 'Sargon vase' show low potash contents and therefore only partially confirm the use of plant ash. The third recipe probably employed mineral natron as flux. The existence of at least three different groups implies the existence of a dynamic production of colourless glass in Nimrud.

7.4.4.3 Colouring agents

At Nimrud both copper and cobalt were used as colourants for blue glass. As already noted, cobalt-coloured glass forms a distinct compositional group (Table 7.4). Based on the high levels of alumina, magnesia, manganese, iron, nickel and zinc, and the low levels of potash and magnesia, origins of the cobaltiferous alums from the Kharga and Dakhla Oases in the Egyptian Western Desert are likely, however slight variances in the compositions more likely account for another source of cobalt (Reade *et al.* 2005: 24).

A group of blue glasses coloured with copper, and another group coloured with a mix of copper and cobalt, can be identified. For the copper-based glass, copper oxide is present at a level of 1.8–5.5%. For the copper/cobalt-based glass samples, cobalt oxide ranges from 0.1–0.2% and copper oxide from 0.3–0.8%.

For both red and yellow-coloured glass, no analytical data for major, minor or trace elements have been published so far, except for one red glass ingot (Nim198) that shows high levels of copper oxide (9.6%) and lead (25%), which identifies it as high-lead/high-copper glass (Chapter 4.2.3.3).

Lead isotope analyses of yellow inlays suggest that two yellow wing-shaped inlays were made from lead coming from ore deposits geologically different from the other pieces. This shows that different sources of lead were used in Nimrud to produce yellow glass (Brill and Stapleton 2012: 267–269).

7.4.4.4 Conclusion

Available chemical raw data on glass from Nimrud is scarce (Appendix 2). Most of the glass in Nimrud is fluxed with plant ash, but mineral natron was also used. Reade *et al.* (2005: 24) identified four different composition groups, three of which have also been found in this study. One red ingot (Nim198) can be identified as high-lead/high-copper coloured glass. It is not clear whether the ingot was imported to Nimrud or whether it was produced at the site. Reade *et al.* (2005) determined three different composition groups for colourless glass, which shows that colourless glass was produced using different raw materials, probably even in different workshops. The existence of a primary workshop in Nimrud cannot yet be proven archaeologically, but the analysis of a colourless bowl from Gordion, which corresponds to compositions of Nimrud colourless glass, shows that colourless glass with the same composition was used at both sites (Chapter 7.4.5). The large number of colourless glass objects found in Nimrud makes the existence of a primary and/or secondary glass production at site in the first half of the 1st millennium likely (Chapters 4.2.2.4, 4.2.2.5, 4.2.2.6, 4.2.2.7). Further analytical data, in particular on the colourless glass objects found outside Nimrud, particularly from Praenestre and Fortetsa, would confirm or disprove Nimrud's unique position in the production of colourless glass. Although chemical evidence is scarce, Nimrud can be identified as one of the most important, probably even the most important, sites for glass production (primary production) and processing (secondary production) not only in Mesopotamia but throughout the eastern Mediterranean, including Egypt.

7.4.5 Pella

No chemical raw data on Pella glass has been published so far. This chapter solely refers to a preliminary report presented by Reade *et al.* (2009). Pella is one of the few sites where scientific analysis has been carried out on material from both the Late Bronze Age (38 samples), and the Iron Age (27 samples), providing good comparisons.

Glass from Pella is of a soda-lime-silica type. All 38 samples from the Late Bronze Age period show high levels of potash (K_2O) (3%) and magnesia (MgO) (4%), indicating the use of plant ash* as a flux* (Read *et al.* 2009: 49) (Figure 7.3). In contrast, only four pieces of the

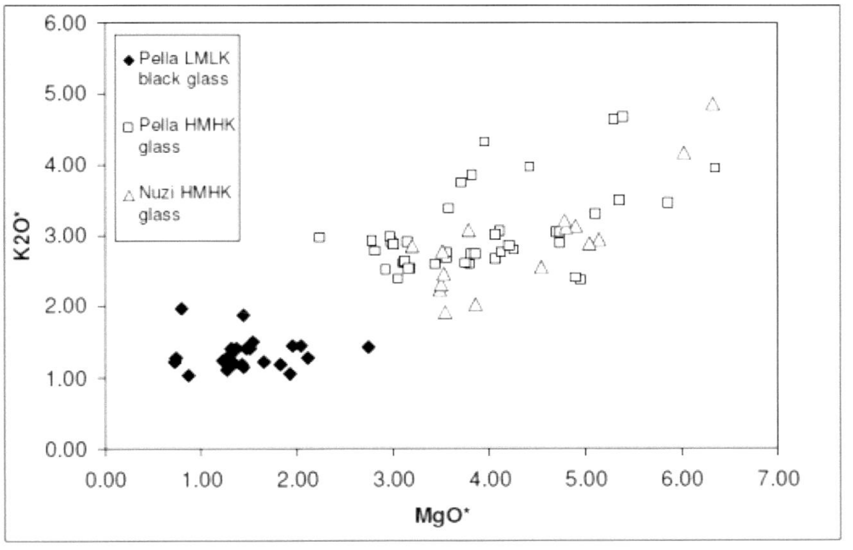

Figure 7.3: Scatter plot of magnesia (MgO) versus potash (K_2O) of Late Bronze Age and Iron Age glass from Pella (Reade et al. 2009: 49, fig. 1).

Iron Age glass objects are fluxed with plant ash (Read *et al.* 2009: 49) (Figure 7.3). The remaining samples show low levels of potash (1.3%), magnesia (1.5%), and lime (1%), and therefore indicate the use of mineral natron as a flux (Reade *et al.* 2009: 49). All examples that can be assigned to the natron type are (green-) black-coloured glasses (Figure 7.3).[23]

(Green-) black-coloured glass is rare in the archaeological record and has in the Near East so far been only recorded at Pella and Hasanlu (Chapter 7.4.2.2). Reade *et al.* (2009) analysed 23 (green) black glass beads from Pella, dating to the Iron Age period. The MgO and K_2O concentrations of the beads are below 1.5% suggesting the use of a soda flux* from mineral origin instead of plant ashes. Lime is very low with an average concentration of 2.26%. The average iron concentration is 9.72% as FeO. The black colouration of the Pella beads is due to an iron-bearing, chromium-rich mineral. Reade *et al.* (2009: 49, 51) indicate that it is unclear whether a separate iron oxide-rich colourant was added or if the glass batch* was made from impure sand containing heavy minerals (chrome) and high levels of iron oxide. The high levels of chromite and chromium indicate that the sand or colourant used for the manufacture of the (green) black beads originates from northern Syria or Anatolia, but does not derive from the Levantine coast or Mesopotamia (Reade *et al.* 2009: 51). The major composition matches with black glass from Hasanlu, and also with black glass found in France (Eyne) (Gratuze 2001).[24] On this basis, Reade *et al.* (2009: 53) proposed a long-distance glass trade between Pella, Hasanlu and France, which could have either incorporated beads or ingots during the early 1st millennium BCE.

7.4.6 Gordion

For Gordion, only preliminary reports on chemical analysis of glass objects have been published so far (Reade *et al.* 2012), without raw data. 48 samples of beads and colourless glass – 13 from Phrygian levels (9th–7th century), and 30 from Hellenistic levels – have been examined (Reade *et al.* 2012).

Glass from Gordion belongs to the soda-lime-silica group. Similar to the situation in Nimrud and Pella, the Iron Age glass samples also fall into two different groups: one showing high levels of potash (K_2O) and magnesia (MgO), indicating the use of plant ash*; the second (three samples) showing low levels of potash and magnesia, indicating the use of mineral natron (Reade *et al.* 2012: 82). Among plant ash glasses, low levels of alumina (0.7%) occur, which confirm to pure sources of silica. Plash glasses from Gordion, Nimrud and also Nuzi correlate regarding their basic compositions with those from Pella (Reade *et al.* 2012: 83–84). Natron-based glass from Pella corresponds with compositions from Nimrud but also with samples from glasses from the tomb of Nesikhons (Privat *et al.* 2014: 2016; Reade 1735094922 *et al.* 2012: 84).

One Iron Age cobalt-coloured bead shows the typical chemical fingerprint of cobalt ores coming from the Western Desert in Egypt (Reade *et al.* 2012: 85).

7.4.7 Late Bronze Age glass from Nippur, Nuzi, and Tell Brak

The following section gives an overview of the available chemical raw data of Late Bronze Age glass objects found at Nippur, Nuzi and Tell Brak. By comparing the Bronze

[23] It cannot be ruled out that plant ash with low levels of potassium was used as flux; see generally for this topic Tite et al. 2006.

[24] Reade et al. (2009: 53) report that iron levels of French natron glass fall between 9–20%, and are therefore close to the Pella results that exhibit an average composition of 9.72%; see Reade et al. 2009: 48, table 2.

Age to the Iron Age compositions, developments in glass technology can be recognised. Chemical analysis on Nippur (Walton *et al.* 2012) and Nuzi glass (Kirk 2009; Shortland *et al.* 2017) were discussed intensively, whereas samples from Tell Brak were only briefly mentioned by Brill (1999: 40).[25]

The group of glass objects excavated at Nippur comprise a unique group of Late Bronze Age ceremonial glass axes and mace-heads, which are the first significant group of cobalt coloured glasses in Mesopotamia.[26] Analytical data identify the pieces as soda-lime-silica glass, fluxed* by the use of plant ash (Walton *et al.* 2012). The colouring strategies of the blue glasses are split into two major groups: one coloured with copper oxide (1.7–2.3%); and the other a combination of copper and cobalt (Table 7.5).[27] The different types of blue colouration fall into different compositional groups. The differences in the rubidium contents indicate a certain variation in the raw materials, either the plant ash or the silica source (Walton *et al.* 2012: 850). The source of cobalt is of particular interest, because the levels of manganese, nickel and zinc clearly vary from Egyptian cobalt-coloured glass, which identifies the cobalt in the Nippur glasses as of non-Egyptian origin (Walton *et al.* 2009: 841).[28] This shows that Mesopotamian glassmakers used distinctive cobalt ores to colour dark blue glass. In contrast, the copper-coloured glasses are fairly similar to the glass compositions from Nuzi and Tell Brak (Walton *et al.* 2012: 851).

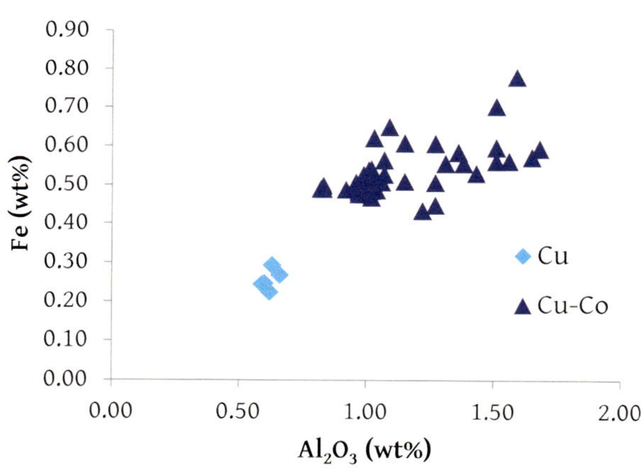

Table 7.5: Scatter plot of alumina (Al_2O_3) versus iron (Fe) for Nippur copper and copper-cobalt (Cu- and Cu-Co) based blue glass (raw data Appendix 2 with bibliography of each sample).

Also glass from Tell Brak and Nuzi belong to the soda-lime-silica type. The high levels of potash (K_2O) and magnesia (MgO) indicate the use of plant ash* as flux* at both sites. The greatest number of glass material from Nuzi was studied very recently by Shortland *et al.* (2017).[29] With regard to the beads and vessels, mainly blue glass (over 90%) was used. White, yellow and 'black' glass occurs almost exclusively as inlay material. Cobalt was not used in a significant amount in the Nuzi blue beads, which stands in sharp contrast to Egyptian blue glasses in which cobalt is common.[30] Translucent and opaque glass falls into two different compositional groups. Most of the translucent blue glasses have low alumina, less than 1.0% and contain less than 0.3% antimony oxide (Sb_2O_5). The opaque blue glasses, in contrast, contain between 1.5% and 3.1% antimony oxide (Sb_2O_5) (Shortland *et al.* 2017: 11). There is, furthermore, a striking difference within the opaque blue glasses containing, on average, lower amounts of soda and higher amounts of lime (13.8% Na_2O in the opaque glass, and 16.6% Na_2O in the translucent glass) (Kirk 2009: 85). These two distinctive compositional groups can be confirmed by isotope analysis of Sr/^{8786}Sr. Shortland *et al.* (2017: 15–16) suggest in consequence that the translucent and opaque blue glasses appear to have been made in different workshops, and probably were also potentially worked in different places. Glass from Nuzi is compositionally very similar to glass from Tell Brak, but differs distinctively from Late Bronze Age Egyptian glass.[31]

The scatter plot Table 7.6 shows two distinctive groups. One group contains the cobalt glass from Nippur – the close correlation could be due to FeO and Al_2O_3 present as cobalt additive. Apart from this group, however, a close correlation of glass from all three sites (Nuzi, Tell Brak, and copper-based glasses from Nippur) can be established that draws on similar raw material most likely originating from the same place.[32]

7.4.8 Conclusion

7.4.8.1 Mesopotamian Late Bronze Age and Iron Age glass compositions

Only by comparing Bronze Age with Iron Age glass compositions can similarities and differences become

[25] For isotope analysis of the three sites, see Degryse et al. 2010; 2015; Henderson et al. 2010.

[26] The objects were inscribed with the names of Kurigalzu II (1332–1308), Nazi-Maruttash (1307–1282) and Kashtiliashu IV (1232–1225), thus attributing the material to the 14th to 13th centuries; for the archaeological approach, see Clayden 2011.

[27] Containing also the following: cobalt (<10 ppm), lead (<20 ppm), nickel (<220 ppm) and arsenic (<100 ppm); Walton et al. 2012: 841.

[28] Walton et al. (2007: 845, 847) could identify four different compositional groups among the copper-cobalt based glasses from Nippur based on their trace elements.

[29] Beads, vessels and vessel sherds, amulets and raw glass fragments were examined in the Havard Semitic Museum collection (Shortland et al. 2017: 9). Chemical raw data from this publication were not considered in this monograph.

[30] The opaque turquoise glass contains an average of 2.0% antimony (Sb2O5).

[31] For detailed information on the varying elements, see Kirk 2009: 155.

[32] For details on the comparison of different trace elements from Nuzi and Tell Brak glass, see Shortland et al. 2007: 786–788. Raw data published in Shortland et al. 2017 were not considered here.

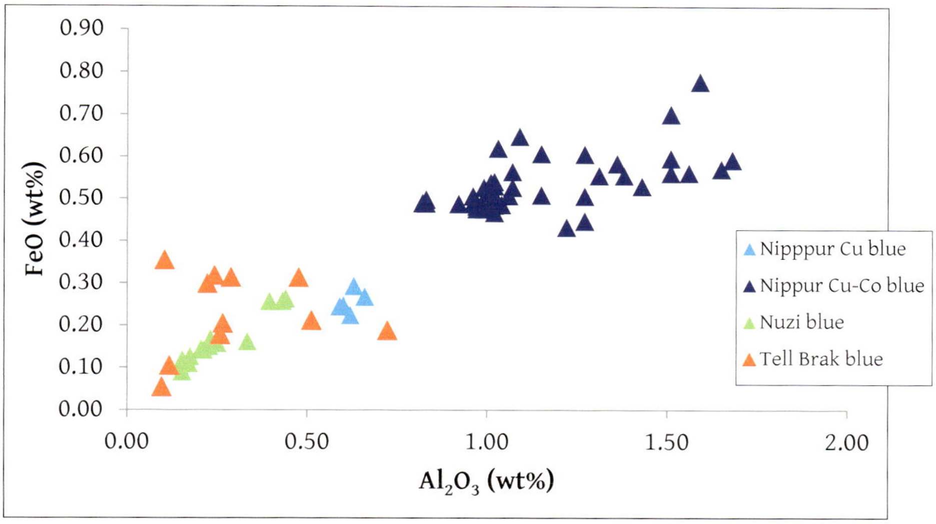

Table 7.6: Scatter plot of alumina (Al_2O_3) versus iron (FeO) for glass from the Late Bronze Age sites of Nippur, Nuzi, and Tell Brak, showing a characteristic Mesopotamian glass group and the particular group of glasses coloured with cobalt from Nippur (see raw data Appendix 2 with bibliography of each sample).

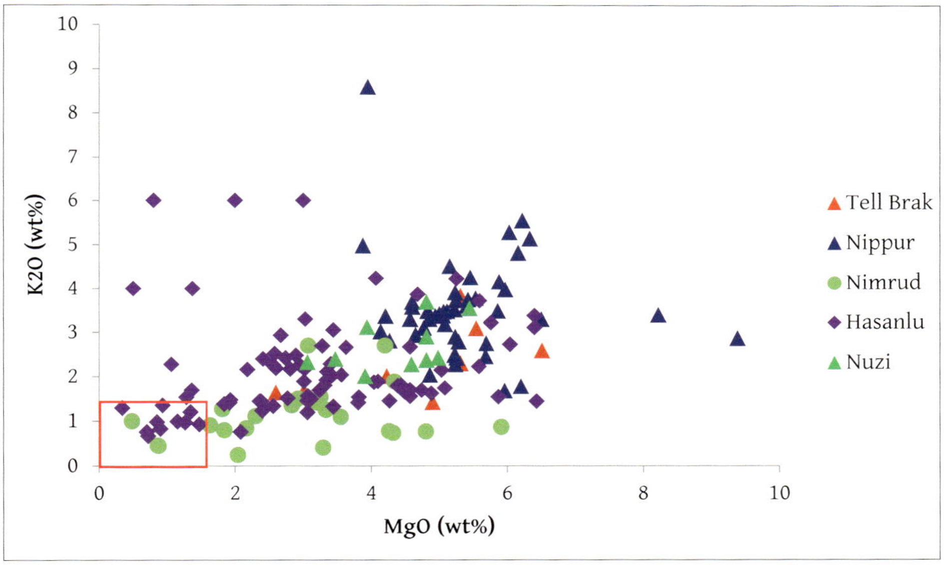

Table 7.7: Scatter plot of magnesia (MgO) versus potash (K_2O) for Late Bronze Age and Iron Age glass. The red box shows the group of natron glass (raw data Appendix 2 with bibliography of each sample).

evident and, consequently, characteristics for each period be determined.

Late Bronze Age as well as Iron Age glass belongs to the soda-lime-silica type. Late Bronze Age glasses can all be clearly identified as plant ash* glass with magnesia (MgO) $>4\%$ and potash (K_2O) $>1.5\%$ (Table 7.7). Only two samples of Nuzi and two samples of Tell Brak glasses show lower levels of MgO and K_2O.

With regard to Iron Age glass, the levels of potash (K_2O) and magnesia (MgO) are lower, as in the Late Bronze Age glasses. The majority of samples confirm to plant ash; the low levels of MgO and K_2O (both below 1.5%) indicate either the use of mineral natron as flux* or low levels of plant ash, or are due to iron oxide colourants in large amounts. It is striking that the Hasanlu samples fall into a wide range of MgO and K_2O levels, suggesting the composition of the Hasanlu glass being very variable. It can, however, be noted that the levels of MgO and K_2O in Late Bronze Age and Iron Age glasses

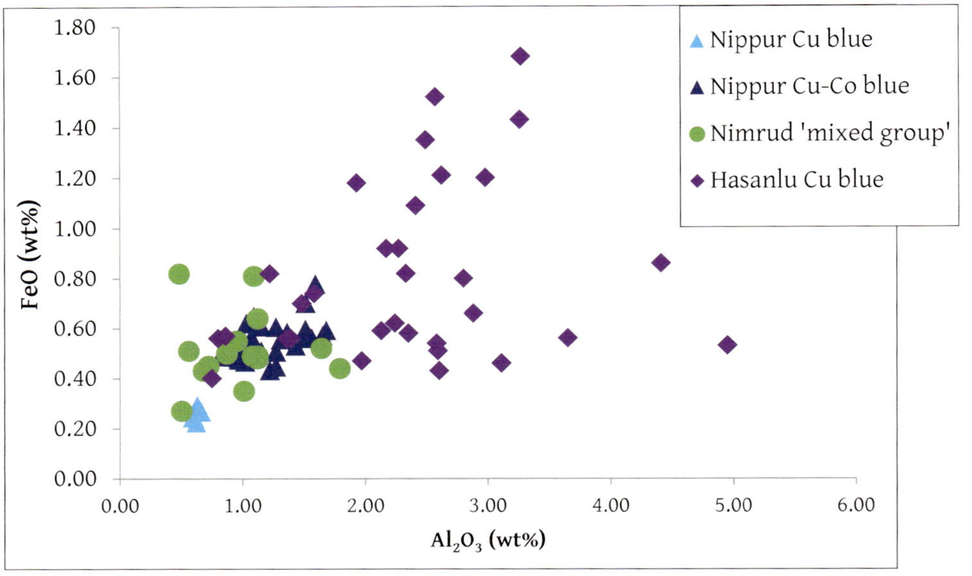

Table 7.8: Scatter plot of alumina (Al_2O_3) versus iron (FeO) for Nippur, Nimrud, and Hasanlu glass. Blue glasses from Nippur and the Nimrud mixed group show lower levels of FeO and Al2O3, as does the Hasanlu (blue) group, which only partly overlaps (raw data Appendix 2 with bibliography of each sample).

vary, which is due to the type, and/or amount of flux used in the composition.

Regarding the basic compositions, it was shown that glass from Tell Brak, Nuzi and Nippur is indistinguishable because of similar geology, creating a so-called 'Mesopotamian' glass group.[33] Also plant ash* glass from Nimrud matches largely the Tell Brak, Nuzi and Nippur compositions and can therefore also be added to the 'Mesopotamian' group (Table 7.8). According to Reade *et al.* (2009: 49, fig. 1), there is a coincidence in the basic composition of plant ash glass from Pella and Nuzi, and plant ash glass from Gordion, Nimrud and Nuzi (Reade *at al.* 2012: 83–84).[34] A colourless glass bowl from Gordion that chemically coincides with glass from Nimrud is of particular interest, as it indicates that colourless raw glass was traded from Nimrud to Gordion and was further processed there (secondary production) (Chapters 7.4.3, 7.4.5, 8.1.2).[35]

Cobalt-coloured glasses from Nimrud and Nippur differ from the basic 'Mesopotamian' compositions, which can be explained by the presence of cobalt and its accompanying elements (Chapter 7.2.2, Table 7.5). Nippur cobalt-based glass differs from Egyptian cobalt-coloured glasses in their significantly different trace elemental compositions. This shows that the glassmakers involved in the primary production in Nippur had clearly identified and utilised a distinctive cobalt ore source that was different from the Egyptian source (Walton *et al.* 2012). With regard to cobalt-coloured glass from Nimrud (Chapter 7.4.3.2) and the single cobalt-coloured glass object from Gordion (Chapter 7.4.5), the colouring pigment used for these objects shows similarities to cobaltiferious alums from the Western Desert in Egypt, although the values do not match precisely (Reade *et al.* 2005; Reade *et al.* 2009: 85). Whether or not Nimrud and Gordion cobalt-coloured glass used cobaltiferous alums from Egypt, or whether they drew on other sources, remains unknown at this stage of research.

Hasanlu glass is compositionally very variable and only partially coincides with the 'Mesopotamian' compositional group (Chapter 7.4.2). This implies the incorporation of several different sources of raw materials into the primary glass production at Hasanlu. or that the glass was made at several different sites which used different raw materials or recipes. However, the high level of Al_2O_3 among the Hasanlu copper blue glasses is significant and indicates that a distinct source of sand was used for the glass production. This shows that very distinct sources of raw materials were exploited for the production process of glass. Interestingly, cobalt does not occur at all among blue glass from Hasanlu. This could demonstrate that there was no access to cobalt ore or that the technology of colouring with cobalt was unknown to glassmakers at Hasanlu.

[33] The geology of the region is dominated by sediments deposited on the flood plains of the Tigris and Euphrates (Shortland et al. 2007: 787).
[34] Further conclusions cannot be drawn at present without raw data.
[35] Further conclusions cannot be drawn at present without raw data.

A comparison between Late Bronze Age and Iron Age glass shows that the composition of the former is very dense, while the group of Iron Age glass is much more diverse. These results indicate some significant changes between Late Bronze Age and Iron Age glass compositions. These changes could have taken place in the course of a major technological transformation in the glass industry at the beginning of the 1st millennium. These changes include:

1. The invention of colourless glass by the intentional addition of antimony as decolouriser.
2. A new type of red glass (the high-lead/high-copper glass), distinguished by its brilliance and colour.
3. The use of mineral natron as flux, which was, however, only common in Mesopotamia.

Analytical comparative data for the Iron Age is disproportionately rare in the region compared to Late Bronze Age data. The current state of research does therefore not allow an exact conclusion. Further archaeometric investigations in relation to the technological change that took place at the beginning of the Iron Age would be of great potential for glass research.

7.4.8.2 Exchange networks in the Late Bronze Age and Iron Age

In addition to Mesopotamia, Egypt was a major centre for glass production and manufacture. Evidence in this regard can mainly be drawn from the Late Bronze Age period. Shortland *et al.* (2007) indicated that glass from Malkata and Amarna, both in Egypt, is indistinguishable and varies greatly from Mesopotamian glass (Tell Brak and Nuzi). The differences in raw materials depend on the different geographical regions.[36] This confirms that glass workshops in Egypt and Mesopotamia produced primary glass by exploiting their own geological sources. Thus, it was suggested that exchange activities regarding raw glass did not take place during this period (Shortland *et al.* 2007: 787).

Evidence of the Late Bronze Age exchange mechanism is the cargo of the trading vessel known as the Uluburun shipwreck. 175 glass ingots were on board the ship, of which a number of different coloured samples (light blue, dark blue, purple, amber, colourless) have been analysed by Brill (1999, 17 samples), Brill and Stapleton (2012, 39 samples), and Jackson and Nicholson (2010). Trace-element compositions from the Uluburun ingots show correlations with Egyptian and Mycenaean glass but differ distinctly with those from Nuzi and therefore from the 'Mesopotamian' glass group. This indicates a movement of glass raw material from Egypt into the Mycenaean world (Jackson and Nicholson 2010: 300).[37] Whether glass from Mesopotamia was involved in this exchange cannot be fully ruled out.

There is strong evidence suggesting an import of raw glass ingots from Mesopotamia and Egypt into Mycenaean Greece in the Late Bronze Age period. This is because several pieces from Tell Brak and Nuzi, which form part of the 'Mesopotamian' glass group, to which also glass from Nimrud belongs, show major similarities with glass from Tiryns (Peloponnese) (late 7th century), and with Late Bronze Age beads of the Mycenaean type of unknown provenience (J. Paul Getty Museum).[38] The fact that characteristic Mycenaean-style beads appear in the typical Mesopotamian glass composition indicates that Mesopotamian ingots were imported to the West and were shaped by local craftsman into finished objects (Walton *et al.* 2009: 1497).

In summary, primary and secondary glass production in the Late Bronze Age was embedded in an exchange network linking Mesopotamia and Egypt with the Mycenaean world, but most likely not with each other (Rehren 2014). Both regions therefore used their own resources.

[36] Shortland et al. (2007) examined 54 samples of glass from Malakata, Amarna, Nuzi and Tell Brak.

[37] Provenance studies of metals and ceramics from the wreck show different origins around the Mediterranean. This indicates an anticlockwise voyage with several stops (e.g. Tyrins); see Shortland 2012: 146 with further literature.

[38] Analyses are based on trace elements; Walton et al. 2009.

8. Conclusion

8.1. Techniques and production

8.1.1 The different manufacturing techniques

Glass cannot be understood as a single category. Rather, glass objects must be grouped according to the various manufacturing techniques, namely mosaic (Chapter 4.1), cast-and-cut (Chapter 4.2) and core-forming (Chapter 4.3). The distinction between these different techniques is fundamental, with each technique resulting in a certain type of object and imparting its characteristic appearance. Also, regarding each manufacturing technique, different production processes are involved which required different technical prerequisites (Chapter 1.2). Furthermore, the contexts in which the different groups of objects occur vary, as well as the value attributed to them (Chapter 5). This is the reason why the manufacturing techniques occupy such an important place in this study, and why the pieces are described in such detail.

An important conclusion with regard to the manufacturing techniques of glass is, first of all, that mosaic glass was no longer produced at the beginning of the 1st millennium, and only starts later again in the Hellenistic period. Even though mosaic glass production flourished in the second half of the 2nd millennium, creating a wide variety of objects, it came to a halt at the beginning of the 1st millennium. Nevertheless, appreciation for objects of this kind did not disappear, as mosaic glass was reused or kept as heirlooms in contexts of the early 1st millennium (Chapter 4.1.1). Patterns using the mosaic technique are characterised by a detailed geometrical and figurative representation in different colours. Both the range of stylistic variation and degree of accuracy in the details of the mosaic glass patterns exceed the possibilities of other techniques in glassworking. The use of different colours to create polychrome objects should also be emphasised here (Chapter 4.1.2).

The 'cast-and-cut' technique, on the other hand, becomes absolutely predominant at the beginning of the early 1st millennium, already from the 10th century, and develops to the major branch of glassworking in Mesopotamia. Cast-and-cut glass includes the widest range of different objects, and comprises almost two thirds of all objects included in this study. As the term 'cast-and-cut' implies, objects of this kind were made by casting glass in, or over moulds*. In this regard, techniques already used in the Late Bronze Age period were applied, e.g. simple open moulds (Chapter 4.2.1). But also new techniques were introduced, such as slumping and sagging*, which became the decisive technique for cast-and-cut glass in the early 1st millennium (Chapter 4.2.1.4). The various casting techniques were either complex and time-consuming (casting in complex moulds), with the aim of producing glass objects with complex forms, or guaranteed the rapid production of certain glass objects (sagging and slumping). Often new forms were produced, such as hemispherical bowls. The introduction of this innovative and multi-faceted production technology shows that at the beginning of the 1st millennium new possibilities for glass processing were explored and further developed.

Cast-and-cut objects are mostly combined with cold-working techniques that comprise stone-working techniques (Chapter 4.2.1.6). This shows the close connection of these two production branches. In the production of cast-and-cut glass, hot processing must be strictly separated from cold processing, since the equipment of the workshops was very different and also the craftsmen involved had to possess different skills (pyrotechnical knowledge versus stone processing and inlay techniques). A separation of the two sectors becomes particularly clear regarding the group of cut vessels and painted inlays, which also sheds more light on the organisation of glass workshops (Chapter 4.2.2.6, 4.2.2.7).

The process of casting glass in moulds* is also closely related to the development and use of transparent glass that appears for the first time in the history of glass at the beginning of the 1st millennium. Transparent glass is used exclusively in connection with objects of the cast-and-cut technique; therefore the two innovations must be seen in close connection. Apart from transparent glass, translucent glass is also used for cast-and-cut glass objects, which are characterised by particularly deep and rich colour tones. Cast-and-cut objects are usually monochrome and therefore differ greatly from the appearance of mosaic and core-formed glass objects.

Core-forming, similar to the mosaic technique, has been known from the outset of glassmaking in the mid 2nd millennium. What can be observed, however, is that in the first half of the 1st millennium core-formed glass only occurs in the 8th, but then even more in the 7th century (Chapter 4.3.1). The principle of applying hot glass around a core is therefore paramount to glass forming and remains a common method throughout the history of glassworking, but was obviously uncommon at the beginning of the 1st millennium. As shown in Chapter 4.3.2 core-formed objects can be manufactured by applying different techniques i.e. coiling*, dipping*, and the addition of crushed glass onto a core. It cannot be ruled out that all of these techniques were used in the past, probably also simultaneously. Core-formed vessels are characterised by the use of different colours, even though the variety of forms and colours are less pronounced in the 1st millennium, as in the Late Bronze Age (Chapter 4.3.3.2).

Detailed examination of the manufacturing processes revealed that at the beginning of the 1st millennium both Late Bronze Age manufacturing techniques were used and also new techniques were developed. Therefore the assemblage of glass objects present at the beginning of the 1st millennium differs greatly from that of the Late Bronze Age. This is primarily due to the increase of cast-and-cut glass objects, which inevitably leads to an increase in monochrome glass. This stands in contrast to the Late Bronze Age, where polychrome glass dominated.

8.1.2 Glass workshops: identification of primary and secondary production

The production of glass is divided into a primary and secondary production, which comprise glassmaking and glassworking (Chapter 1.2). Evidence for both industries are almost absent in the archaeological record of the first half of the 1st millennium in Mesopotamia. Whether this is due to the general low number of glass workshops that existed, or the difficulties in identifying workshop areas in the archaeological context, cannot be decided here. Information on how pronounced the different glass workshops were, and how they were organised, can therefore only be gained on the basis of a combined study of glass artefacts and the texts that deal with glassmaking – in particular the Nineveh glass recipes, chemical analyses, experimental studies and analogies drawn from Late Bronze Age glassmaking sites in Egypt. Information on secondary production and workshops can mainly be gained from the glass objects themselves and their particular method of manufacturing. In this respect, the concept of *chaîne opératoire* provides an important framework, connecting all the various disciplines (Chapter 1.2).

8.1.2.1 Primary production

The primary production process

Ancient glassmaking cannot, as far as we know, be considered as a single process, but has to be understood as a string of repeated actions that created different intermediate products. Therefore different fritting and melting processes precede the final step of glass production. There is no doubt that the production of intermediate products facilitates the production of the final glass, but the intermediate steps in production have so far only been described in the Nineveh glass recipes and have, in the archaeological record, so far only been documented at Qantir-Piramesses and Tell el-Amarna (below). Today the production of glass relies on a single melting process and thus differs from the Late Bronze and Iron Age periods. For future research, an investigation with regard to this aspect would certainly be worthwhile.

In the Nineveh glass recipes, the importance of intermediate products in the course of Mesopotamian glassmaking becomes apparent (Chapter 6). Here, already the structure of the texts recalls the procedural sequence, as the texts are divided by consistent division lines. These lines demarcate the different intermediate products – the colourless primary glass *zukû* and the blue primary glass *tersītu*, which were needed to produce the end product: blue *zagindurû*-glass (Chapter 6.6). The translation of the end product *zagindurû* as *ebbum* ('the pure') therefore directly names the benefits of these procedures, i.e. the purification of the batch*.

The process of sintering* stands at the beginning of the *chaîne opératoire* of glass production and is described in the Nineveh glass recipes. This process was carried out in specific crucibles that differed from those used during glass melting. In the texts, *ḫarāgu la eššetu* (a crucible that is 'not new') occurs in connection with the sintering processes (Table 6.1, Chapter 6.6.4). Experimental studies have shown that sintering processes are best carried out in wide flat trays to facilitate a maximum exposure of powder. Archaeological evidence for this type of vessel comes probably from Tell el-Amarna in Egypt (Late Bronze Age), where flat, tray-like vessels with low rims could have probably been used for this procedure.[1] Based on the Nineveh glass recipes, it is known that different types of kilns were used for the sintering and melting processes. In this regard, a 'kiln with its four eyes'('*kūru ša 4 īnīša*') clearly refers to the sintering process (Table 6.1, Chapter 6.6.5).

The intermediate primary glasses result from a melting process most likely carried out at about 900°C, and therefore lower than the final melting temperature needed to produce the end product. Rehren and Pusch (2005: 1757) suggest, with regard to their research on glass production in Qantir-Piramesses, that the first intermediate product was most likely still rich in quartz grains and bubbles that needed to be further refined (Rehren and Pusch 2005: 1757, fig. 5). Refinement could be achieved by the crushing and repeated heating of the batch*, or by washing either the sintered* or primary glass product in water to remove nonreactive, but water-soluble components of the plant ash*. In the Nineveh glass recipes, the washing of primary glass is attested, which most likely therefore served the purpose of purification. With regard to the archaeological

[1] For this find, see Nicholson 1995: 15.

evidence in general, objects used for primary glass production can only be identified by means of those crucibles that still contain residues of glass melt or sinter production.

The Nineveh glass recipes indicate that crucibles for glass melting had to be 'clean' or 'new' during this process (*dabti zakūti, dabti eššeti, ḫarāgi eššeti*) (Table 6.1). With regard to the different kilns for glass melting, a 'kiln with shelf' (*kūri ša takkanni*), or a 'kiln with its four eyes' – and therefore the same kiln type as used for sintering* – was employed (Table 6.1, Chapter 6.6.3). A 'kiln with its four eyes' was, however, only used in connection with a stand (*nēmedu*), on which it was possible to place the crucible (Chapter 6.6.5). By using a stand the optimum heat utilisation could be achieved through a combination of radiant heat from above and from the base of the crucible (Rehren and Pusch 2005: 1757–1758). The two different types of kilns, with their partly different features mentioned in the text, confirms two different kinds of equipment used for sintering and melting.

In the ongoing melting processes, the batch*, in the form of powder or crushed glass, was added step-by-step, in some cases possibly by means of a funnel shaped tool, to facilitate the fusion and melting needed for glass of higher quality (Rehren and Pusch 2005: 1757, fig. 3). The addition of crushed and ground primary glass is also repeatedly mentioned in the Nineveh glass recipes (Chapter 6.6.4). Great care had also to be taken in the course of mixing the batch with the colouring agent, as becomes apparent in the course of the texts: the copper had to be carefully and repeatedly stirred underneath the melted glass batch. This was achieved by the use of a rake (*muterru*) (Chapter 6.6.4).

The manufacturing process of different colours required different steps of manufacture, which resulted in a change in the operational sequence. How different colouring substances react in the glass melt can nowadays be derived from chemical analysis, and can also be applied to the production of glass in the past (Chapter 7.2).

Reducing kiln atmospheres were necessary to produce red glass. This could be achieved by covering the openings of the kiln to stop the air supply, or by applying 'reductants' (charcoal, antimony, iron or lead) to the melt, by which this process could also be promoted. Remains of charcoal were found on one of the red ingots from Nimrud indicating the use of reductants (Chapters 4.5.1, 7.2.3.2). Another possibility was the use of crucibles with bound lids, as also attested in the Nineveh glass recipes (*ḫaragu šaktumtu*) (Opp. Tablet A §15: 118).

The last melting, that resulted in the final raw glass, required temperatures of at least 1000°C to produce an homogenously coloured and fully melted end product, i.e. in the form of an ingot. The resulting ingots had to be annealed* very slowly to prevent cracking (Chapter 2.1.2).

Crucibles and ingots

A crucible in which glass was melted can be identified on basis of glass remnants attached to its inner walls. As mentioned previously, different types of crucibles were identified in the Nineveh glass recipes – one for sintering* and another for the production of glass (Table 6.1). The majority of ingots presented in this monograph come from Nimrud and Babylon. They have a cylindrical shape with convex base, with a varying diameter of 16.4–22 cm. However cuboid ingots – even though few in number – can also be identified (Chapter 4.5.1). The thickness of these ingots does not exceed 3.6 cm. If one considers the loss of 40% of the batch* material during melting, a maximum height of 6 cm could be considered for the batch material in the crucible. A crucible height of 6 cm would therefore allow for an easy manipulation of the melted substance with a rake (stirring, mixing, testing of viscosity), as described in the Nineveh glass recipes, and an ideal utilisation of heat (see 6.6.4). Crucibles found at Qantir-Piramesse are also cylindrical in shape and exhibit convex inner bases, similar to ingots from Nimrud and Babylon. Almost all of the ingots from Qantir-Piramess exhibit a thin parting layer of lime, which was used as an adhesive to facilitate the removal of the glass from the crucible walls (Rehren and Pusch 2005: 1756, Figure 1).

Tools

No tool explicitly used for the production of glassmaking has been identified in the archaeological record so far, and also only few tools are attested in the Nineveh glass recipes. In this regard a rake (*muterru*) was used to stir the glass melt, as well as to check whether the melt had sufficiently mixed to a homogenous mass. Such a rake must have been made of copper to withstand the temperature of the hot melt (Rehren and Pusch 1997: 137). Further tools mentioned in the texts are tongs (*maššu*) and elastic rods (*tašnû*). The latter were used in to move the hot crucible (Oppenheim 1970: 71). Tools for secondary glass production and finishing are very likely to be linked to stone processing and ivory cutting (Chapters 4.2.1.6, 4.2.2.11).

Silica and plant ash

The individual ingredients of glass and their origin are dealt with in detail in Chapters 7.1, and 7.2. At this point only silica and plant ash* are mentioned in detail, since they can be securely identified in the Nineveh glass recipes. Plant ash used as flux* in ancient glass production occurs throughout the semi-desert environments in the Near East (Chapter 7.1.3). However not all species are useful for glassmaking. In this regard, the video 'The Glassmakers from Herat', showing glassmakers in Afghanistan, indicates, for example, that the right type of plant ash could be identified by its taste.2

Regarding Mesopotamian glass, chemical analysis show that silica was acquired either from sand or pebbles (Chapter 7.1.1). The Nineveh glass recipes refer to pebbles used for glassmaking. Pebbles rich in silica could be collected from riverbeds or even from fields, as shown by modern comparisons.

Fuel and firing temperatures

Very large amounts of wood were needed to run a glassmaking kiln. For an idea of the amount of fuel required, experiments on Roman glass production are mentioned here. These show that the amount of wood required to run a glass furnace for three weeks was approximately nine tons.[3] Once a fire was established in the kiln, it was kept for a considerable number of days to maintain heat. This was particularly important regarding primary production as very high temperatures were needed. This is also indicated in the glassmaking texts, in which fire had to be established for several days to generate the heat required (Chapter 6.6). The large amounts of wood needed for the primary production process must have had a major effect on the outline of the primary glass workshop, as the wood was most likely stored in close proximity to the workshop to ensure a constant fuel supply. The amount of charcoal was probably low, as the wood almost burned away entirely in the pit.[4]

In the Nineveh glass recipes, poplar wood was used for firing and this had to be chosen carefully (Chapter 6.6.2). Only thick logs of poplar, that had to be cut in August when the sap of the tree was low, had to be used. The text, furthermore, indicates that the wood was bound together and placed in a firebox in the lower part of the kiln. It can therefore be assumed that the firing chamber and the kiln chamber, in which glass was melted, were separate from each other. Regarding crucibles found at Qantir-Piramesse, it was assumed that a direct flame must have somehow been directed towards the bottom of the crucible, probably by the use of blow-pipes, in order to raise the maximum amount of heat for melting (Rehren and Pusch 2005).

The Nineveh manuscripts (Chapter 6.6) refer many times to a 'good and smokeless' fire (*išāta ṭābta lā qātirta*), pointing to the importance of this part of the procedure. The last step in the melting process of glass required a temperature of at least 1000°C. As glass exhibits no exact melting point, it was necessary to create heat as high as possible to let bubbles escape and homogenise the substance. The temperature of the kiln chamber, apart from other modern methods, could be measured by observing the chamber colour. Ancient glassmakers were most likely very well aware of the temperature in the kiln by observing the colour. This becomes also obvious regarding the Nineveh glass recipes, in which repeated reference is drawn to the colour of the glass melt at different stages of heating (see Chapters 6.6.3, 6.6.4, 6.6.5).

8.1.2.2 Secondary production

Secondary production techniques comprise core-forming*, mosaic glass manufacturing, and cast-and-cut techniques. In this regard, a differentiation has to be made between hot- and cold-working techniques.

Core-forming exclusively relies on hot-working techniques and is based on the principle of applying hot glass around a core (Chapter 4.3.2).

Mosaic glass, which dates back to the second half of the 2nd millennium, was also made under exposure to heat. Two different steps must be distinguished here: the production of the mosaic rods, and the fusion of the individual mosaic segments into an object (Chapter 4.1.2).

In contrast, the mosaic bowls discussed in this study, also attributed to the Late Bronze Age period, are composed of pieces not made of glass but of glassy faience*, or a similar material, and were therefore made in a cold-working process, only fusing them together in the very final stage (both techniques are explained in detail in Chapter 4.1.2.1).

Cast-and-cut objects were shaped under heat exposure into moulds*, which is described in detail in Chapter 4.2.1. After annealing*, the objects were further worked in cold state. In order to smooth the surface of cast-and-cut vessels, the pieces were often ground and polished. This was frequently accompanied by the final shaping of rims and handles, and the application of cut decorations. In this regard, engraving tools and turning wheels were used, combined with an abrasive. However, only among very few objects can traces of cold-working techniques be observed, as most of the surfaces are corroded heavily, making any final assumption difficult. In contrast, radiating toolmarks occur on the inside of a number of vessels, implying the use of a wheel (Chapter 4.2.1.6).

[2] https://www.youtube.com/watch?v=BMYE83DJU4Q (accessed: 8.4.2018).

[3] Experiments were carried out within the framework of the 'Roman Furnace Project' by Mark Taylor and David Hill; see http://www.romanglassmakers.co.uk/furnace13.htm (accessed: 4.4.2016).

[4] See http://www.romanglassmakers.co.uk/furnace13.htm (accessed: 4.4.2016).

The finishing of glass objects by cold-working techniques included smoothing, grinding, applying cut decorations or paint, and was undoubtedly carried out in specialised workshops. This is because these activities fall into the sphere of craftsmen engaged in inlay and engraving, for example for ivory work, but also stone cutting, wood carving, or similar decorative work.[5]

8.2. The role of the palace and the Neo-Assyrian Empire in Iron Age Mesopotamian glass production

8.2.1 Transparent cast-and-cut glass commissioned by the palace?

The first half of the 1st millennium is characterised by an important invention in glass technology – transparent colourless glass. Intentionally decolourised glass was produced by adding the decolouriser antimonate to the batch* (Chapter 7.1.7). Elevated amounts of antimonate are attested among analyses of hemispherical bowls from Nimrud, which identifies pieces from this site as the earliest decolourised glass objects in the history of glassmaking (Chapter 7.4.3). Altogether three different compositional groups of colourless glass objects could be identified at Nimrud, showing that transparent glass production flourished (Figure 7.2). At this stage it is not known where to locate these different production centres, whether they existed around Nimrud or far from the site.

Cast-and-cut technique is closely connected with transparent glass, and particularly with hemispherical bowls, cut-and-inlaid vessels, shallow, undecorated and ribbed and petalled bowls, and painted inlays (Chapters 4.2.2.4, 4.2.2.5, 4.2.2.6, 4.2.2.7). Colourless, transparent and cast-and-cut glass became the most widely spread type of glass at the beginning of the 1st millennium. This development had most likely already started in the 9th century, during the reign of Ashurnasirpal II or Shalmaneser III (Chapter 5.5.3, Figure 5.3). In Assyria, transparent cast-and-cut glass was solely found in palatial contexts (Chapter 5.4.4, Table 5.14). It is therefore likely that the impulse to develop this new form of glass, both regarding its transparent nature and its production technique, was driven by the Neo-Assyrian royal court.

Against the background of the geographical distribution of the objects collected in this work, it was shown that cast-and-cut vessels and inlays occur almost exclusively in cities in Assyria or in Assyrian dominated cities, e.g. Samaria, Arslan Taş or Til Barsip (Chapter 5.3.4, Figure 5.2). Apart from this group, cast-and-cut objects are almost non-existent in Babylonia, with the exception of the attachments and inlays for composite statues. This contrasts with the spread of core-formed vessels, which can only be observed in a small variety of types in Assyria but dominate the assemblage in Babylonia (Chapter 5.3.2, Figure 5.2). Here, many different types of core-formed vessels were found at different sites, suggesting that the production of core-formed vessels was varied and popular in the region. Core-formed vessels were also found in the southern Levant, albeit less diverse than in Babylonia (Chapter 5.3.2, Figure 5.2). Regarding their find context, almost all of the cast-and-cut objects known to this day come from palatial contexts. In contrast, in temples, graves and dwelling contexts, core-formed vessels dominate (Chapter 5.4.2, Table 5.14).

The geographical distribution of the glass objects raises the question of where the different types of objects were produced, or by whom they were commissioned. This is a difficult question and cannot be answered reliably due to the lack of finds of workshop areas in archaeological contexts. Nevertheless, some suppositions should be made at this point, which will have to be examined in the future.

With regard to the cast-and-cut glass objects, it is likely that, similar to their use, their production was also closely tied to the palace. This is probable because cast-and-cut glass was only found at palaces, and is strongly connected to the production of transparent glass, which, again, solely occurs within palatial contexts. Because both techniques were new and required a certain level of technical expertise, it is likely that it was commissioned and controlled by the palace. It is most probable that the organisation of the Neo-Assyrian Empire strongly promoted these new technological ambitions. In sharp contrast stands the wide distribution, in terms of find contexts, of the core-formed vessels, which indicates that workshops producing this vessel type were more widespread.

8.2.2 The question of 'Phoenician' glassworkers in the context of cold-working techniques

Apart from the transparent cast-and-cut vessels, those glass objects incised (cut vessels) (Chapter 4.2.2.6) or painted (painted inlays) (Chapter 4.3.3.7) with figurative designs are also interesting groups of objects to consider. Whereas primary glassmaking only involves hot-working methods, the secondary production of glass objects incorporates both hot- and cold-working techniques, which have to be clearly differentiated. This is because the *chaîne opératoire* of the primary and secondary production varies considerably, which leads to the assumption that the workshops differed with regard to their equipment, as well as to the craftsmen involved. This distinct separation of areas of responsibility is crucial for the reconstruction of the workshop environment and

[5] In contrast, raw glass production in any event falls into the area of hot-working with fire, similar to metal processing.

the incorporation of the individual production areas as outlined below (Chapter 1.2).

The finishing of glass objects by cold-working techniques includes smoothing, grinding, applying cut decorations or paint, and was undoubtedly carried out in specialised workshops. This is because these activities fall into the sphere of craftsmen engaged in inlay and engraving work, for example for ivories, but also stone cutting, wood carving or similar decorative work (Chapters 4.2.2.6, 4.2.2.7, 4.2.2.11).

On the basis of style of the painted inlays and incised bowl fragments Nim24, Nim25, and Nim26, far-reaching conclusions can be drawn. Stylistic comparisons of these motifs show close parallels to those on ivories and bronze objects of the Phoenician style, and to a lesser extent also to the north Syrian style. Also regarding their function, the painted inlays (and also the small monochrome glass inlays) served most likely as inlays for ivories, mostly of the Phoenician style (Phoenician Group I), and to a lesser extent the North Syrian style, see for details (Chapters 4.2.2.6, 4.2.2.7, 4.2.2.9). Stylistic similarity, as well as the similarity of the craft activity itself could lead to the conclusion that the painted and incised decoration on the glass objects was carried out by the same craftsmen who also cut the ivories (in Phoenician and north-Syrian style). In any case, a close connection between the glass objects and the ivories can be determined.

In this context, the question arises as to where this specific cold-working production took place. On the basis of all the evidence gathered, it is not possible at this stage to make a clear decision where these workshops should be located. If one assumes that craftsmen who work in the Phoenician style come from the Levant and work there, one could conclude that the painting and carving of the glass also took place in this geographical area. However, if one considers the political and social circumstances that the Neo-Assyrian Empire brought with it, it is not so much the geographical location as the institution of the palace with its various locations that plays a much more decisive role.

8.2.3 The impact of the Neo-Assyrian Empire on glass production by the displacement of specialists

In addition to opening up geographical territory, the Neo-Assyrian Empire fostered the large-scale relocation of people in order to serve state interests. The reasons for this are manifold and were described by Lanfranchi (1997: 81), Oded (1979) and Parpola (2004) among others. Radner (2012) indicated that deportees were often deliberately chosen. Therefore, in addition to the urban elites, craftsmen and specialists were also of interest in this regard. This resulted in an accumulation of experts in central Assyria, in particular in the cities of Nimrud, Nineveh and Aššur. Against this background, the painted plaques, the incised colourless bowl fragments and the monochrome inlays that have primarily been found set into ivories of the Phoenician type, have to be considered. These items clearly show a connection to the Levantine coast and northern Syria, but were only found at Assyrian cities, or at least in Assyrian dominated cities (Chapters 4.2.2.6, 4.2.2.7). The enhancement of human resources in the Assyrian centres therefore most likely had a great impact on craftsmanship, including glassmaking.

The expansions involved a general increase of diversity in *māt Aššur*, as annexed territories incorporated foreign customs, languages, material culture and local flora and fauna (Hunt 2015: 24). This could generally have created a certain degree of openness towards innovations, and new tastes, and could probably also have contributed to an increased appreciation of cast-and-cut glass instead of mosaic and core-formed glass, of monochrome, and also transparent glass instead of polychrome glass (Chaper 8.1.1).

8.3. Functions and values of glass objects and the material glass

8.3.1 Different forms of values

Glass is not just glass. During the course of this study it became obvious that glass cannot be considered as a uniform material, as it assumes a variety of types (manufacturing techniques) (Chapter 4), forms (types of glass objects) (Chapter 4), appearances (opaque, translucent, transparent, colour) (Chapter 2) and compositions (chemical) (Chapter 7). This is in particular true when it comes to the question of how different glass objects were used and how they were appreciated. The following chapter is intended to give an overview of the results in this work related to considerations on the subject of value. Regarding this aspect, a far-reaching theoretical approach to the topic is not provided here, but impulses for future studies are offered. As a general assumption it can be stated that it is a dangerous oversimplification to equate 'appreciation' solely with 'economic value'. These concepts might overlap, but generally have to be clearly differentiated. Many different concepts of value regarding material things exist. The evaluation of a particular object can, for example, depend on its rarity or uniqueness (Karpik 2011: 13–24), or it can be created by the specific social status of its owner in society (Bourdieu 1982: 277–332).

8.3.2 Use and significance of Iron Age Mesopotamian glass objects

The three major types of glass, cast-and-cut, core-formed, and mosaic glass have to be distinguished from one another, not only because they rely on different manufacturing techniques, but also because they show different patterns of distribution (Chapter 5.3). Distribution patterns – both in terms of geographical and contextual distribution – form an important source in terms of drawing conclusions about the function of a specific object, and, consequently, about the value that might had been assigned to it. The distribution across different contexts indicates that the different types of glass objects were used by different groups of people with different social status for different purposes, and were therefore most likely subjected to different forms of value. With regard to this study, it can be generally shown that mosaic glass no longer plays a role at the beginning of the 1st millennium (Chapter 4.1.4), cast-and-cut glass occurs almost exclusively in palace areas (Chapter 5.4.1, Table 5.14), while core-formed glass is widely used and also occurs in socially weaker contexts, such as simple houses and graves (Chapter 5.4.2, Table 5.14). This will be explained in more detail below.

Regarding their specific find contexts, almost all of the cast-and-cut objects known to this day come from palatial contexts; in contrast, in temples, graves and dwelling contexts, core-formed vessels dominate. The contrast is particularly strong in residential buildings where cast-and-cut objects are absent, apart from one object (Table 5.14). It is interesting that hemispherical bowls and shallow, undecorated and ribbed and petalled bowls outside the Assyrian heartland (Gordion, Praeneste, Fortetsa, Babylon) only occur in richly endowed graves, with the exception of the burial in Babylon, which dates relatively late into the Neo-Babylonian period (Figure 5.2). Obviously, the vessels were used differently outside Assyria. Only a few hemispherical bowls were found in the tumuli at Gordion, as well as in the graves from Praeneste and Fortetsa, which indicates their rarity. This, together with the fact that these kinds of bowls were probably reserved for kings in Mesopotamia, could suggest that hemispherical bowls were also highly valued outside of Mesopotamia (Chapter 4.2.2.4).

The distribution of finds shown here allows the conclusion that most of the cast-and-cut vessels and inlays for ivories and furniture were used in close connection with the palace. On the basis of typological and comparisons of vessels illustrated in other media (e.g. vessels in other materials, reliefs), which are explained in more detail in Chapters 4.2.2.4, 4.2.2.5, 4.2.2.6, the conclusion can be drawn that hemispherical bowls, shallow, undecorated and ribbed and petalled bowls, as well as cut-and-inlaid vessels, were all used as drinking vessels most probably used in connection with royal banquets at the Neo-Assyrian court, as the geographical distribution clearly indicates. Cast-and-cut inlays, with the exception of inlays for composite statues, must be seen in connection to ivory panels and furniture, which were kept and stored in palatial structures. Most of these objects were most likely brought as booty or as part of a gift exchange to the Assyrian cities (Chapters 4.2.2.7, 4.2.2.8, 4.2.2.9).

Often, the distribution of finds is associated with the fact that much of our information about the Neo-Assyrian period comes from capital cities excavated in the 19th and early 20th centuries, with the focus on monumental building structures; this must indeed be taken into consideration. In the case of the distribution of the cast-and-cut glass objects, however, it can be shown that the distribution outlined above is by no means random. First, the dwelling houses in Aššur have to be considered here as a reference. These have been published in detail by Miglus in *Das Wohngebiet von Assur* (1996), who in this study lists all finds present in these building structures;[6] if there were cast-and-cut glass objects among the finds in these houses, they would have been listed with certainty. As a next step the excavation databases of the recent excavations at Tall Halaf, Sincirli and Ziyaret Tepe were also consulted.[7] The searches revealed that no cast-and-cut glass objects were found at any of the sites, even though all of them are closely related to the Neo-Assyrian Empire and its capitals. Also with regard to core-formed glass objects, the amount of finds has been low. This negative evidence therefore shows that cast-and-cut objects were, indeed, very strongly tied to the palace and the king in the Neo-Assyrian period. This picture might change with continuous excavations but has to remain as it is at this stage of research.

The distribution of core-formed vessels stands in sharp contrast to this. Here the distribution pattern indicates that core-formed vessels were, at the beginning of the 1st millennium, obviously available to a wider group of people and were also used as burial objects (Chapter 5.4.2). The graves in which core-formed vessels were found, were, however, not particularly richly furnished. This is underlined by the poorer quality of these finds, which show a limited range of shapes and patterns (Chapter 4.3.3). Core-formed vessels mostly incorporate small bottles and pots, on average ranging between 5–8 cm. The low height and closed vessel shape, as well as the material characteristics of glass, stands in close connection to their use: it is likely that the vessels contained liquids or balms that could have been carefully poured through the narrow opening and the wide rim of the bottles (Chapter 4.3.3). One of these bottles (Bab7) was found covered with a piece of cloth, which could indicate that this vessel contained a

[6] Also in earlier studies of dwelling houses no glass finds at Aššur are listed by Preusser (1955).

[7] At this point I would like to thank the respective excavation directors and members of the team who granted me access to the excavation databases: for Tall Halaf, Lutz Martin; for Sincirli, David Schloen and Vincent van Exel; for Ziyaret Tepe, Tim Matney and Dirk Wicke.

fragrant substance, released through the cloth (Chapter 3.2). With regard to the material properties of glass, the surface is impermeable to water and therefore particularly useful for oily substances, such as the ones mentioned above. Due to the small size, and therefore the rather low volume, it is likely that the core-formed vessels contained substances of a certain value for its owner. The assessment of the prestige of a context depends on the observer and the comparisons made. With regard to the graves in which the core-formed vessels were found, it can certainly be argued that the burials were simply equipped compared to the palaces or monumental tumuli at Gordion (Chapter 3.4.3). For the deceased, however, the core-formed vessels probably possessed some value and served as a highly appreciated burial gift, most likely because of their material and/or contents. That glass was still a rare material at the beginning of the 1st millennium – in particular in comparison to the late 2nd millennium, also in the form of core-formed vessels, can be assumed from the overall small number of glass finds in comparison to other materials in this period. With regard to core-formed vessels, it can be summarised that this type of glass was much more widely spread in terms of geography and contexts, as in the Late Bronze Age, and was obviously also part of burials of lower status (Gries and Schmidt 2019). However, the overall low number of cast-and-cut glass vessels allows the conclusion that appreciation of vessels of this type was high among the relevant groups of individuals.

8.3.3 The material properties of glass and its value

Colour is one major characteristic of glass and has to be considered as an important factor regarding its appreciation. By looking at the monochrome and painted inlays that were set into ivory panels, it becomes clear that the colours of the glass inlays coincide with the colours of other materials, such as Egyptian blue, faience*, gold and lapis lazuli, which were used in the same context as glass (Chapters 4.2.2.7, 4.2.2.9). This shows that glass was obviously used simultaneously with these other materials in the same context and not 'in place of other materials', as for example the terms 'imitation' or 'substitute' would indicate. This suggests that it was not the material itself that was decisive in this specific context, but the material properties, such as colour and shine.

This understanding and interpretation of glass in relation to other materials becomes particularly clear with regard to attachments and inlays for composite statues (Chapter 4.2.2.11). Throughout different periods, a wide range of materials was used for attachments and inlays for composite statues; this also applied in the first half of the 1st millennium. In this regard, not only blue glass but also blue faience*, Egyptian blue and lapis lazuli occur. These different materials, artificially produced or natural, possess specific material properties that can be of chemical, mechanical and optical nature. As shown, these different materials could simply have been chosen to add variety to these statues, which would emphasise the appreciation of each materials' characteristics as important elements of the statue as a whole. The diversity of materials would create a certain effect on the observer that could not be achieved by any other decoration (paint, etc.). In this regard, glass plays a particularly important role because of its colourful and translucent nature, creating a particularly intensive and deep effect (Chapter 4.2.2.11) (Schmidt in press).

It can, however, also be noted that all materials selected for composite statues were used simultaneously and therefore served the same purpose. Similar to ivories, it seems therefore likely that it was the choice of any specific material that mattered, but rather the characteristics of the material, e.g. colour and shine. It can thus be concluded that artificial and genuine materials were appreciated similarly in this particular context. This reasoning is supported by the *mīs pî* ritual, the so-called 'mouth-opening' ritual, which gives us an idea about the process of inducing a cult statue, and which is described in detail with regard to the case made here in Chapter 4.2.2.11. The most important stage for this is the final ritual, in which the statue is transformed from its material state, made of different materials, into a divine being. Therefore, the actual materials used for production can be considered as secondary, as the ritual unifies the different parts into a uniform statue, with the emphasis being on the appearance and effect of the divine being, enhanced by the richness of colour and shine.

The aspect of imitation has to be mentioned in this connection, as glass is often referred to as an imitation of stone. The numerous examples consulted with regard to every single glass object included in this study shows that there is no specific media upon which glass objects draw. As indicated above, glass rather existed as an independent material, exhibiting independent characteristics and working properties. The simultaneous use of glass, stone and other artificial materials shows clearly that glass was not used as a substitute for other materials, but rather in conjunction with them. It would instead appear that an appreciation of colour inherent to glass and stone is a more important factor than the material itself.

8.4. Concluding remarks

Glass manifests itself in many different forms, both with regard to its material characteristics (primary production), and its outer shape (secondary production). With Iron Age Mesopotamian glass, this becomes obvious in terms of the range of colours glass can take, to which transparent colourless glass must also be added. This is complemented by various different new shapes that emerge in this period. The first half of the 1st millennium, therefore, and despite the very low number of finds, was a period when important technological developments took place that affected both the characteristics of the material itself – such as its workability – as well as its outer appearance.

These technological innovations were most likely fostered by a rising Neo-Assyrian Empire, which provided the necessary prerequisites for these developments. This shows clearly that the choice and selection of a specific material, colour, or technique, and therefore the development and acceptance of new objects and techniques, had both technological and cultural reasons. Both of these factors are therefore inseparably linked with each other and have to be considered equally when dealing with ancient materials. Changes in political and social structures, as well as in economies, can, but not necessarily have to, result in changes in technology. With regard to the glass industry, it can therefore be determined that both traditional techniques, already established in the previous period (mosaic, core-formed), existed, at least partly, simultaneously with newly developed technologies (cast-and-cut).

Glassmaking and glassworking in ancient Mesopotamia was not an experimental industry, producing desirable glass objects by accident. The glass objects themselves, the chemical data, and, in particular, the Nineveh glass recipes, all clearly demonstrate that glassmakers were very well aware of all the different procedural steps in the manufacturing process. The evidence confirms their great knowledge of the different raw materials and their effects on the batch, the different steps of purification achieved by intermediate products, as well as the impacts of metals and kiln atmosphere on colouration. Temperatures could be measured precisely and the degree of viscosity tested. The Nineveh glass recipes can therefore be interpreted as instructions for the production of primary glass.

It is important to understand and interpret the material beyond an emic perspective. Modern scholarship tends to claim glass as an imitation, and therefore glass is evaluated as a substitute. This view can by no means be applied to ancient Mesopotamian glass, which existed as an independent material used for the production of specific objects, creating particular shapes. It was used simultaneously with other artificial materials such as faience* or Egyptian blue, but also together with genuine stone, such as lapis lazuli. The appreciation of different types of glass and glass objects, arose for a variety of reasons closely connected to its function in royal, temple and funerary contexts. This clearly shows, particularly with regard to glass artefacts, that the value of things can only be defined by an embeddedness in specific contexts and activities, as well as the social integration of people attached to it. Concepts of value change over time, but they can also remain consistent, as shown by Andrae's quotation at the beginning of this study: 'The colour tones of the lands between the Euphrates and the Tigris are very light, dusty, and dull... Men have there unconsciously a strong need for expressing themselves in arrangements of colours' (1925: 1).

Index of Technical Terms

Annealing: The process of slow cooling a glass object in a heated environment, e.g. a heated chamber or the close proximity of the kiln. The slow cooling ensures that no compressive and tensile stresses are trapped in the glass during the cooling phase. If glass is not annealed properly, it may break.

Batch: The mixture of all raw materials to make glass. The basic ingredients are silica, flux, and a stabiliser, usually a colouring agent is also added. The ingredients are ground, mixed and put into a crucible in which the batch is melted at high temperature above 1000° in the kiln.

Casting: This technique comprises the forming of hot glass in a mould. Different types of moulds are used to form glass, such as open moulds, complex moulds and convex moulds. Also the lost-wax technique is included within this technique (see lost-wax).

Coiling: One of the ways to make core-formed glass objects. This technique includes the trailing of hot glass around the core. This could be achieved through viscous glass canes coiled around the core, or hot glass directly applied from the gathering rod.

Colourant: A colourant changes the colour of the base glass. Glass can be coloured by impurities or by the deliberate addition of colouring agents to the batch. Most colourants are based on metal oxides or minerals, or even waste materials (slag, bronze objects). Some examples include:

- Light blue: less copper oxide (a few wt%)
- Dark blue: high amounts of copper oxide (several wt%); small amounts of cobalt oxide
- (0.2 wt%)
- Blue opaque ('turquoise'): copper oxide and calcium antimonate
- Red opaque: copper metal or copper oxide or iron oxide particles
- Yellow opaque: lead antimonate; tin antimonate
- Green opaque: blue and yellow opaque mixed
- Purple: small amount of manganese (1–2 wt%)
- White opaque: air bubbles ('seeds'); quartz inclusions ('stones'); tin oxide; calcium antimonate; bone ash
- Yellow transparent ('amber'): reducing atmosphere (Fe-S complex)
- Colourless (=decoloured!): a little antimony or manganese oxide
- 'Aqua': light blue-green, due to natural impurity of iron oxide

Cooling marks: Fine veins that occur when the mould is too cold during the process of moulding glass.

Core-forming: The principal of core- and rod-forming incorporates the manipulation of hot glass around a core, which is removed after the glass has annealed.

Cutting: This process describes the removal of glass to shape or decorate an object. Cutting is only carried out in cold state after the glass has annealed, and is particularly frequently used for cast-and-cut glass objects. The cutting tool consists of a rotating wheel made of stone or metal, attached to a lathe, and an abrasive suspended in liquid.

Decolourant: The most common impurity in glass is iron, which causes a greenish-blue tint. By adding this component the natural and unwanted colour of the base glass is absorbed.

Dipping: One of the ways to make core-formed glass objects. This process incorporates the coating of the core by dipping it into melted hot glass. By turning the rod, and therefore the vessel, the viscous glass is evenly spread on the core, creating a uniform glass thickness.

Faience: A vitreous material made of powdered quartz, flux and lime, combined with water to make a clay-like paste. The paste is formed in a mould in cold state. The faience object is then heated to a temperature of *c.* 800–900°C, creating an alkaline glaze on the surface.

Fire-polishing: The practice of putting the object back into the kiln for a very short period of time, allowing the surface to slightly melt. In the broader sense this could also be applied to core-formed vessels to marver the threads.

Flux: The melting point of pure silica lies at around 1700°C. This temperature was far too high to have been achieved in ancient pottery kilns or metallurgical furnaces. Thus to melt silica to a glass, a flux was added to the composition to lower the melting temperature to *c.* 1000°C. Plant ash (potash) or mineral natron served as fluxes in antiquity.

Gathering iron: An implement which holds a gob of melted glass at its tip.

Glassy faience: Glassy faience refers to a more compact and less porous form of faience, and represents a stage between faience and glass. The major difference of glassy faience to faience is that the colour is evenly spread throughout the body core, and is not restricted to the surface of the object.

Halophytic plants: Salt-tolerant plants that grow in desert and semi-desert environments throughout the Near East. These plants were ashed and added to the batch to serve as flux. How these plants were collected and burnt can be watched in the video 'The glassmakers of Herat', shot by Robert Brill in 1968 (https://www.youtube.com/watch?v=BMYE83DJU4Q, accessed: 4.08.2018).

Iridescence: A result of weathering that causes a rainbow-like visual effect which changes according to the angle from which an object is viewed.

Lost-wax technique: A form of casting that originates from metalworking. In this process, a positive is produced from wax coated with the mould material. Both mould and wax are then heated to melt out the wax, which was then replaced by glass.

Marver/marvering: The surface of a glass object was often flattened and merged by rolling it on a marver, which is a smooth and flat surface.

Mosaic (glass): A glassworking technique in which a surface is created of many small adjoining glass pieces of different colours and/or patterns. The mosaic pieces are cut from canes and placed into a mould. The heat then causes the mosaic pieces to fuse.

Mould: Moulds are used to shape melted glass. A mould has the same shape as the desired object. In order to shape it, glass can be poured into the mould in viscous form, or it can be inserted as crushed powder or chunks and heated afterwards. When removed from heat, the viscous glass solidifies in the mould. To remove the cold glass object from the mould it was essential to use separators: i.e. a light coating of carbon or soot, talc, plaster or lime. These substances formed a barrier between the glass and mould to allow the glass to be removed easily. Different types of moulds include open moulds, multi-part moulds and convex moulds. The lost-wax technique has also to be considered in this regard.

Natron glass: Glass that contains mineral natron as flux and exhibits low levels of magnesia and potash. It is therefore referred to as 'low-magnesia/low-potash' glass (abbreviated LMLK). Natron glass is characterised chemically by low contents of potash and magnesia, i.e. below 1.5%.

Opacifiers: These change the transparency of the base glass, from transparent to translucent to opaque. Inclusions are due to unsolved crystalline particles or bubbles in the material. Substances that create such a phase are called opacifying agents or opacifiers. Opacifiers are based on anything which scatters light, from air, such as bubbles, to droplets of different glasses or salts in the glass, or crystals added to the glass, or grown in the glass.

Opaque glass: In opaque glass the wavelengths of light are reflected from the glass caused by the presence of inclusions. These create an immiscible phase that scatters the light and results in a milky or opaque appearance.

Oxidising atmosphere: An atmosphere in the kiln that has an excess of oxygen. An oxidising atmosphere will turn a glass batch containing copper into blue glass.

Pitting: A form of weathering, and describes the formation of small holes on the surface of a glass object.

Plant ash glass: Glass containing plant ash as flux, and characterised by high levels of magnesia and potash. This type of glass is thus often also referred to as 'high-magnesia/high-potash' (abbreviated HMHK).

Reduced atmosphere: An atmosphere in the kiln that has no oxygen. A reducing atmosphere will turn a glass batch containing copper into red glass.

Sintering: A process carried out at temperatures of 700–800°. The batch is heated to this temperature to create a coherent mass; the surface is melted together but not completely fused. This process makes the batch material more reactive. Sintering produces a frit that is further worked into glass. The process is also known as fritting.

Thread decoration

- Feather pattern: the viscous threads are pulled upwards and downwards. This pattern occurs as a broad or narrow design. If the upward and downward movement of the tool creating the pattern is not pronounced, the pattern appears as a zigzag or wavy decoration.
- Festoon pattern: the viscous threads are pulled upwards.

Undercut: While with open moulds the object could be removed through the opening, with complex shapes the glass object was trapped between the different parts of the mould, referred to as 'undercut'. Therefore the different parts had to be removed one after the other.

Weathering: This process describes a chemical reaction caused by the environment in which the glass object was placed. Weathering occurs on the surface of glasses, leaving behind siliceous weathering products that easily flake off. Typical traces of glass weathering are iridescence, pitting, flaking off, or the creation of a whitish corrosion layer.

Bibliography

Ådahl, K. and Strommenger, E. (eds) 1978. *Der Garten in Eden: 7 Jahrtausende Kunst und Kultur an Euphrat und Tigris; [Ausstellung] Prähistorische Staatssammlung München*. Berlin: Museum für Vor- u. Frühgeschichte der Staatl. Museen Preu. Kulturbesitz.

Albenda, P. 1988. The Gateway and Portal Stone Reliefs from Arslan Tash, *American Schools of Oriental Research* 271: 5–30.

Albenda, P. 2003. Dur-Sharrukin, the Royal City of Sargon II, King of Assyria, *The Canadian Society for Mesopotamian Studies* 38: 5–13.

Albenda, P. 1986. *The palace of Sargon, King of Assyria: Monumental wall reliefs at Dur-Sharrukin, from original drawings made at the time of their discovery in 1843-1844 by Botta and Flandin*. Editions Recherche sur les civilisations / Synthèse, 22. Paris: A.D.P.F.

Albright, W. F. 1961. The role of the Canaanites in the history of civilisation, in G. E. Wright (ed.), *The bible and the ancient Near East: Essays in honor of William Foxwell Albright*: 328–362. London: Routledge & Paul.

Algaze, G. *et al.* 1990. The Tigris-Euphrates Archaeological Reconnaissance Project: A preliminary report of the 1989–1990, *Anatolica* 17: 175–240.

Ambos, C. (ed.) 2004. *Mesopotamische Baurituale aus dem 1. Jahrtausend v. Chr*. Dresden: ISLET.

Amiet, P. 1966. *Elam*. Auvers-sur-Oise: Archée.

Amiet, P. and Mittmann, S. 1987. *Der Königsweg: 9000 Jahre Kunst und Kultur in Jordanien und Palästina: [Ausstellung, Rautenstrauch-Joest-Museum für Völkerkunde, Köln: 3.10.1987–27.3.1988, Schallaburg: April-November 1988, Prähistorische Staatssammlung, München: Januar-März 1989]: [Katalog-Handbuch]*. Mainz: von Zabern.

Amouric, H. and Foy, D. 1991. De la salicorne aux soudes facites; mutations techniques et variation de la demande, in *L'évolution des Techniques est-elle autonome (Ex-en-Provence 1989), Cahier d'histoire des techniques*: 39–75.

Andrae, W. 1913. *Die Festungswerke von Assur*. Deutsche Orient-Gesellschaft: Ausgrabungen der Deutschen Orient-Gesellschaft in Assur/A, 2. Leipzig: Hinrichs.

Andrae, W. (ed.) 1925. *Coloured ceramics from Ashur and earlier ancient Assyrian wall-paintings: From photographs and water-colours by members of the Ashur expedition organised by the Deutsche Orient-Gesellschaft*. London: Kegan Paul.

Andrae, W. 1967. *Die jüngeren Ischtar-Tempel in Assur*. 1935 edn. Deutsche Orient-Gesellschaft: Ausgrabungen der Deutschen Orient-Gesellschaft in Assur/A, 5. Osnabrück: Zeller.

Aruz, J. 2014. Introduction, in J. Aruz, Y. Rakic and S. B. Graff (eds), *Assyria to Iberia: At the dawn of the classical age [Exhibition on view at the Metropolitan Museum of Art, New York, from September 22, 2014, through January 4, 2015]*: 2–11. New Haven, CT: Yale University Press.

Aruz, J., Rakic, Y. and Graff, S.B. (eds) 2014. *Assyria to Iberia: At the dawn of the classical age [Exhibition on view at the Metropolitan Museum of Art, New York, from September 22, 2014, through January 4, 2015]*. New Haven, CT: Yale University Press.

Aruz, J. and Wallenfels, R. (eds) 2003. *Art of the first cities: The third millennium B.C. from the Mediterranean to the Indus*. New York: Metropolitan Museum of Art/Yale University Press.

Arveiller-Dulong, V. and Nenna, M.-D. (eds) 2011. *Les verres antiques du Musée du Louvre*. Paris: Réunion des Musées Nationaux (Catalogue / Musée du Louvre, Département des Antiquités Grecques, Étrusques et Romaines).

Assadpoura, M., Heuss-Aßbichlerb, S. and Jafari Baric, M. 2017. Boron contamination in the west of Lake Urmia, NW Iran, caused by hydrothermal activities, *Procedia Earth and Planetary Science* 17: 554–557.

Aubet, M. E. 1993. *The Phoenicians and the West: Politics, colonies, and trade*. Cambridge: Cambridge University Press.

Bagg, A. M. 2011. *Die Assyrer und das Westland. Studien zur historischen Geographie und Herrschaftspraxis in der Levante im 1 Jt. v. u. Z.* Leuven: Peeters.

Bagg, A. M. 2013. Palestine under Assyrian Rule: A new look at the Assyiran imperial policy in the west, *Journal of the American Oriental Society* 113: 119–144.

Baker, H. 1995. Neo-Babylonian Burial Revisited, in S. Campbell (ed.), *The archaeology of death in the ancient Near East*: 209–220. Oxbow Monograph, 51. Oxford: Oxbow Books.

Bär, J. 1996. *Der assyrische Tribut und seine Darstellung: Eine Untersuchung zur imperialen Ideologie im neuassyrischen Reich*. Kevelaer: Verlag Butzon & Bercke.

Barag, D. 1970. Mesopotamian Core-Formed Glass Vessels (1500–500 B.C.), in A. L. Oppenheim (ed.), *Glass and glassmaking in ancient Mesopotamia: An edition of the cuneiform texts which contain instructions for glassmakers with a catalogue of surviving objects*: 131–199. The Corning Museum of Glass Monographs, 3. Corning, NY: Corning Museum of Glass.

Barag, D. 1975. Rod-formed-kohl tubes of the mid-first-Millennium B.C., *Journal of Glass Studies* 17: 23–36.

Barag, D. 1982. Cosmetic glass palettes from the eighth – seventh centuries B.C., *Journal of Glass Studies* 24: 11–19.

Barag, D. 1983. Glass inlays and the classification and dating of ivories from the ninth – eighth centuries B.C., *Anatolian Studies* 33: 163–167.

Barag, D. 1991. Glass inlays in Phoenician ivories, glass and stone vessels, *L'Association International d'histoire du Verre* 12: 1–9.

Barag, D. 2011. A Neo-Assyrian Glass Bowl Fragment, in Y. Thareani, I. Taxel, D. Barag, D. Ilan and N.

Applbaum; (eds), *Tel 'Aroer: The Iron age II caravan town and the Hellenistic-early Roman settlement*: 259–260. Annual of the Nelson Glueck School of Biblical Archaeology/Hebrew Union College – Jewish Institute of Religion, 8. Jerusalem: Nelson Glueck School of Biblical Archaeology, Hebrew Union College – Jewish Institute of Religion.

Barag, D. (ed.) 1985. *Catalogue of Western Asiatic glass in the British Museum*. London: British Museum Press.

Barber, D. J., Freestone, I. C. and Moulding, K. M. 2009. Ancient copper red glasses: investigation and analysis by microbeam techniques, in A. J. Shortland, I. Freestone and T. Rehren (eds), *From mine to microscope: Advances in the study of ancient technology*: 115–127. Oxford: Oxbow Books.

Barkoudah, Y. and Henderson, J. 2006. Plant Ashes from Syria and the Manufacture of Ancient Glass: Ethnographic and Scientific Aspects, *Journal of Glass Studies* 48: 297–321.

Barnett, R. D. 1953. Assyrian objects from room M. 2, *Anatolian Studies* (VIII): 48–51.

Barnett, R. D. 1957. *A catalogue of the Nimrud ivories: with other examples of ancient near eastern ivories in the British Museum*. London: British Museum Press.

Barnett, R. D. 1963. A review of acquisitions 1955–62 of Western Asiatic Antiquities, II, *British Museum Quarterly*, 1963–4 (XXVII): 79–88.

Barnett, R. D. 1975. *A catalogue of the Nimrud Ivories: With other examples of Ancient Near Eastern Ivories in the British Museum; with a supplement by Leri Glynne Davies* (2nd edn). London: British Museum Press.

Barnett, R. D. (ed.) 1976. *Sculptures from the north palace of Ashurbanipal at Nineveh (668-627 B.C.)*. London: British Museum Press.

Barnett, R. D. (ed.) 1998. *Sculptures from the southwest palace of Sennacherib at Nineveh*. London: British Museum Press.

Bartelmus, A. and Sternitzke, K. (eds) 2017. *Karduniaš: Babylonia under the Kassites; The Proceedings of the Symposium held in Munich 30 June to 2 July 2011 = Tagungsbericht des Münchner Symposiums vom 30. Juni bis 2. Juli 2011*. Untersuchungen zur Assyriologie und Vorderasiatischen Archäologie, 11/1. Berlin: de Gruyter.

Beaulieu, P.-A. 1992. New Light on Secret Knowledge in Late Babylonian Culture, *Zeitschrift für Assyriologie und Vorderasiatische Archäologie* 82: 98–111.

Becker, J. (ed.) 2010. *Kulturlandschaft Syrien: Zentrum und Peripherie; Festschrift für Jan-Waalke Meyer*. Alter Orient und Altes Testament, 371. Münster: Ugarit-Verl.

Ben-Tor, A. *et al.* 2012. *Hazor VI: The 1990–2009 excavations: the Iron Age*. Jerusalem: Israel Exploration Society.

Ben-Tor, A. *et al.* 1997. *Hazor V: An account of the fifth season of excavation, 1968: text and illustrations*. Hazor final excavation reports, 5th v. Jerusalem: Israel Exploration Society; Hebrew University of Jerusalem.

Berlejung, A. 1998. *Die Theologie der Bilder: Herstellung und Einweihung von Kultbildern in Mesopotamien und die alttestamentliche Bilderpolemik*. Orbis biblicus et orientalis, 162. Freiburg, Schweiz: Univ.-Verl.

Berlejung, A., Meʾir, A. and Schüle, A. (eds) 2017. *Wandering Arameans: Aramaeans outside Syria: textual and archaeological perspectives*. Leipziger altorientalistische Studien, Band 5. Wiesbaden: Harrassowitz Verlag.

Bevan, A. 2007. *Stone vessels and values in the Bronze Age Mediterranean*. Cambridge: Cambridge University Press.

Bianchi, R. S. *et al.* 2002. *Reflections on ancient glass from the Borowski Collection: Bible Lands Museum Jerusalem*. Mainz: von Zabern.

Bienkowski, P. 2002. *Busayra: Excavations by Crystal-M. Bennett; 1971–1980*. British Academy Monographs in Archaeology, 13. Oxford: Oxford University Press.

Bimson, M. 1987. Opaque red glass: a review, in M. Bimson (ed.), *Early vitreous materials*: 165–171. Occasional Paper, 56. London: British Museum Press.

Bimson, M. and Freestone, I. C. 1985. Scientific examination of opaque red glass of the second and first millennia BC. in D. Barag (ed.), *Catalogue of Western Asiatic glass in the British Museum*: 119–122. London: British Museum Press.

Bimson, M. and Freestone, I. C. 1988. Some Egyptian glasses dated by royal inscriptions, *Archaeometry* 30: 11–15.

Bimson, M. and Werner, A. E. 1967. A survey of the Materials used in Ancient Glass, *Annales* 4: 256.

Bimson, M. (ed.) 1987. *Early vitreous materials*. Occasional Paper, 56. London: British Museum Press.

Blaylock, S. R. 1999. Iron Age pottery from Tille Höyük, South Eastern Turkey, in A. Hausleiter (ed.), *Iron Age pottery in Northern Mesopotamia, Northern Syria, and South-Eastern Anatolia: Papers presented at the meetings of the international 'table ronde' at Heidelberg (1995) and Nieborów (1997) and other contributions*: 263–286. Altertumskunde des Vorderen Orients, Bd. 10. Münster: Ugarit-Verl.

Boehmer, R. M. (ed.) 1987. *Uruk: Kampagne 38, 1985, Grabungen in J-K/23 und H/24-25*. Endberichte, 1. Mainz a.Rh: von Zabern.

Boehmer, R. M. and Pedde, F. (eds) 1995. *Uruk: Die Gräber*. Ausgrabungen in Uruk-Warka, 10. Mainz a.Rh: von Zabern.

Borger, R. 1956. *Die Inschriften Asarhaddons Königs von Assyrien*. Archiv für Orientforschung, Beihft. 9. Pl. V. Graz: Archiv für Orientforschung.

Bottéro, J. 1949. Les inventaires de Qatna, *Revue d' Assyriologie et d'archéologie orientale* 43: 105–223.

Bottéro, J. 1995. *Textes culinaires Mésopotamiens*. Mesopotamian Civilizations, 6. Winona Lake, Ind: Eisenbrauns.

Bouquillon, A. 2008. Glazed steatite, in M. S. Tite and A. J. Shortland (eds), *Production technology of faience and related early vitreous materials*: 23–36. Oxford University School of Archaeology Monograph, 72. Oxford: Oxford University School of Archaeology.

Bouquillon, A., Kaczmarczyk, A. and Vandiver, P. B. 2008. Faience production in the Near East and the Indus Valley, in M. S. Tite and A. J. Shortland (eds), *Production technology of faience and related early vitreous materials*: 93–109. Oxford University School of Archaeology Monograph, 72. Oxford: Oxford University School of Archaeology.

Bourdieu, P. 1982. *Die feinen Unterschiede: Kritik der gesellschaftlichen Urteilskraft*. Frankfurt am Main: Suhrkamp.

Bourke, S. J. 1997. Pre-classical Pella in Jordan: A conspectus of ten years' work (1985–1995), *Palestine Exploration Quarterly* 129: 94–115.

Braun, E. (ed.) 2013. *Early Megiddo on the East Slope (the 'Megiddo Stages'): A report on the early occupation of the East Slope of Megiddo; results of the Oriental Institute's excavations, 1925-1933*. Oriental Institute Publications, 139. Chicago, IL: University of Chicago Press.

Braun-Holzinger, E. A. 1991. *Mesopotamische Weihgaben der frühdynastischen bis altbabylonischen Zeit*. Heidelberger Studien zum alten Orient, Bd. 3. Heidelberg: Heidelberger Orientverlag.

Brill, R. H. 1962. A note on the scientist's definition of glass, *Journal of Glass Studies* 4: 127–138.

Brill, R. H. 1970. The chemical interpretation of the texts, in A. L. Oppenheim (ed.), *Glass and glassmaking in ancient Mesopotamia: An edition of the cuneiform texts which contain instructions for glassmakers with a catalogue of surviving objects*: 105–128. The Corning Museum of Glass Monographs, 3. Corning, NY: Corning Museum of Glass.

Brill, R. H. 1978. Some miniature glass plaques from Fort Shalmaneser, Nimrud, *Iraq* 40: 23–39.

Brill, R. H. (ed.) 1999a. *Chemical analyses of early glasses: Volume 1. Catalogue of Samples*. Corning, NY: Corning Museum of Glass.

Brill, R. M. (ed.) 1999b. *Chemical analysis of early glasses: Volume 2. Tables of Analyses*. Corning, NY: Corning Museum of Glass.

Brill, R. H. and Stapleton, C. P. 2012. *Chemical analyses of early glasses: Volume 3. The Years 2000-2011, Reports, and Essays*. Corning, NY: Corning Museum of Glass.

Brill, R. H. *et al.* 1999. Oxygen isotope analyses of early glasses, in R. H. Brill (ed.), *Chemical analyses of early glasses: Volume 1. Catalogue of Samples*: 303–322. Corning, NY: Corning Museum of Glass.

Brinkman, J. A. 1972. Foreign Relations of Babylonia from 1600 to 525 B.C.: The Documentary Evidence, *American Journal of Archaeology* (LXXVI): 271–281.

Brock, J. K. 1957. *Fortetsa; Early Greek tombs near Knossos*. British School at Athens. Supplementary Paper, 2. Cambridge: Cambridge Univ Press.

Bunnens, G. (ed.) 1990. *Tell Ahmar: 1988 season*. ABR-Nahrain/Supplements, 2. Leuven: Peeters.

Bunnens, G. 1997. Carved Ivories from Til Barsib, *American Journal of Archaeology* 101: 435–450.

Bunnens, G. (ed.) 2000. *Essays on Syria in the Iron Age*. Ancient Near Eastern Studies, Supplement 7. Louvain: Peeters.

Busz, R. and Gercke, P. (eds) 1999. *Türkis und Azur: Quarzkeramik in Orient und Okzident; [anläßlich der Sonderausstellung vom 18. Juli bis 3. Oktober 1999 im Ballhaus am Schloß Wilhelmshöhe und in Schloß Wilhelmsthal, Souterrain]*. Wolfratshausen: Ed. Minerva.

Cable, M. and Smedley, J. W. 1987. The replication of an opaque red glass from Nimrud, in M. Bimson (ed.), *Early vitreous materials*: 151–164. Occasional Paper, 56. London: British Museum Press.

Canciani, F. and Hase, F.-W.v. 1979. *La tomba Bernardini di Palestrina*. Latium vetus, 2. Roma: Consiglio Nazionale delle Ricerche.

Caubet, A. (ed.) 1995. *Khorsabad, le palais de Sargon II., roi d'Assyrie: Actes du colloque organisé au Musée du Louvre par le service culturel les 21 et 22 janvier 1994*. Musée National du Louvre, Conferences et colloques. Paris: La Documentation Française.

Caubet, A. 2014. Phoenician and East Mediterranean Glass, in J. Aruz, Y. Rakic and S. B. Graff (eds), *Assyria to Iberia: At the dawn of the classical age [Exhibition on view at the Metropolitan Museum of Art, New York, from September 22, 2014, through January 4, 2015]*: 167–170. New Haven, CT: Yale University Press.

Caubet, A. and Bouquillon, A. (eds) 2007. *Faïences et matières vitreuses de l'Orient ancien: Étude physico-chimique et catalogue des oeuvres du Département des Antiquités Orientales*. Gand: Éd. Snoeck.

Cecchini, S. M. 2009. Les ivories de Arslan Tash, in S. M. Cecchini, S. Mazzoni and E. Scigliuzzo (eds), *Syrian and Phoenician ivories of the early first millennium BCE: Chronology, regional styles and iconographic repertories, patterns of inter-regional distribution*: 87–132. Ricerche di archeologia del Vicino Oriente, 3. Pisa: Edizioni ETS.

Cecchini, S. M., Mazzoni, S. and Scigliuzzo, E. (eds) 2009. *Syrian and Phoenician ivories of the early first millennium BCE: Chronology, regional styles and iconographic repertories, patterns of inter-regional distribution*. Ricerche di archeologia del Vicino Oriente, 3. Pisa: Edizioni ETS.

Cecchini, S. and Venturi, F. 2012. A sounding at Arslan Tash. Re-visiting the 'Bâtiment aux Ivories, in R. Matthews, J. Curtis and M. Seymour (eds), *Proceedings of the 7th International Congress on the Archaeology of the Ancient Near East, 12 April - 16 April 2010, the British Museum and UCL, London: Mega-cities & mega-sites: The archaeology of consumption & disposal landscape, transport & communication*: 325–341. Wiesbaden: Harrassowitz Verlag.

Charpin, D. 2004. Histoire politique de Proche-Orient Amorrite (2002–1599), in P. Attinger and Merkus Wäfler (eds), *Die altbabylonische Zeit.* Orbis biblicus et orientalis, 160,4. Fribourg: Academic Press.

Cholidis, N. 2002. Kyros und die kriegerische Istar? Kritische Anmerkungen zu einer 'altorientalischen' Steintafel, in O. Loretz (ed.), *Ex Mesopotamia et Syria lux: Festschrift für Manfried Dietrich zu seinem 65. Geburtstag*: 105–113. Alter Orient und Altes Testament, 281. Münster: Ugarit-Verl.

Ciafaloni, D. 2009. A taste for Egypt: Egyptianizing ivories and other artifacts at the Neoassyrian court, in S. M. Cecchini, S. Mazzoni and E. Scigliuzzo (eds), *Syrian and Phoenician ivories of the early first millennium BCE: Chronology, regional styles and iconographic repertories, patterns of inter-regional distribution*: 307–317. Ricerche di archeologia del Vicino Oriente, 3. Pisa: Edizioni ETS.

Çilingiroğlu, A. (ed.) 1991. *Anatolian iron ages: The proceedings of the second Anatolian Iron Ages Colloquium held at İzmir, 4 - 8 May 1987.* Oxbow Monographs, 13. Oxford: Oxbow Books.

Clayden, T. 2011. The Nippur 'hoard'. *Al Rafidian*, 32: 1–57.

Clayden, T. and Schneider, B. 2015. Assurbanipal and the Ziggurat at Nippur, *KASKAL* 12: 349–382.

Cline, E. H. and Caploe, A. 2014. *1177 B.C: The year civilization collapsed.* Grand Haven, MI: Brilliance Audio.

Cohen, A. and Kangas, S. E. (eds) 2010. *Assyrian reliefs from the palace of Ashurnasirpal II: A cultural biography.* Hanover, NH: Hood Museum of Art, Dartmouth College.

Cooper, J. S. 2005. Right writing: talking about Sumerian orthography and texts, in J. A. Black (ed.), *Special Volume in honor of Professor Mamoru Yoshikawa*: 45–52. Acta Sumerologica, 22.2000. Hiroshima: Middle Eastern Culture Center in Japan.

Crowfoot, J. W., Crowfoot, G. M. and Kenyon, K. M. 1967. *The objects from Samaria.* Samaria-Sebaste, 3. London: Palestine Exploration Fund.

Crowfoot, J. W. and Crowfoot, G. M. 1938. *Early ivories from Samaria.* Samaria-Sebaste, 2. London: Palestine Exploration Fund.

Cummings, K. 1980. *The technique of glass forming.* London: Batsford.

Cummings, K. 1997. *Techniques of kiln-formed glass.* London: A & C Black/University of Pennsylvania.

Curtis, J. 2012. Assyrian and Urartian metalwork: Independence or interdependence, in S. Kroll, C. Gruber, U. Hellwag, M. Roaf, P. Zimansky (eds), *Biainili-Urartu: The proceedings of the symposium held in Munich 12-14 October 2007 = Tagungsbericht des Münchner Symposiums 12. - 14. Oktober 2007*: 427–443. Acta Iranica, 51. Leuven: Peeters.

Curtis, J. 1999 Glass inlays and Nimrud ivories, *Iraq* 61: 59–69.

Curtis, J. 2003. The Assyrian heartland in the period 612–539, in G. B. Lanfranchi, M. Roaf, R. Rollinger (eds), *Continuity of empire: Assyria, Media, Persia.* HANE. M, v. 5. Padova: S.a.r.g.o.n. editrice e libreria.

Curtis, J. 2013. *An examination of late Assyrian metalwork: With special reference to Nimrud.* Oxford: Oxbow Books.

Curtis, J., Collon, D. and Green, A. 1993. British Museum Excavations at Nimrud and Balawat in 1989, *Iraq* 55: 1–37.

Curtis, J. and Reade, J. E. (eds) 1995. *Art and Empire: Treasures from Assyria in the British Museum.* London: British Museum Press.

Da Riva, R. 2008. *The Neo-Babylonian royal inscriptions: An introduction.* Münster: Ugarit-Verlag.

Damerji, M. S. 1999. *Gräber assyrischer Königinnen aus Nimrud.* Mainz: Verl. des Römisch-Germanischen Zentralmuseums.

Dardeniz, G. 2015. Was ancient Egypt the only supplier of natron? New research reveals major Anatolian deposits, *Anatolica* (XLI): 191–202.

Degryse, P. *et al.* 2010. Isotopic discriminants between Late Bronze Age glasses from Egypt and the Near East, *Archaeometry* 52: 380–388.

Degryse, P., Henderson, J. and Hodgins, G. 2009. Isotopes in vitreous materials, a state-of-the-art and perspectives, in P. Degryse, J. Henderson and G. Hodgins (eds), *Isotopes in vitreous materials*: 15–30. Studies in Archaeological Sciences, 1. Leuven, Belgium: Leuven University Press.

Degryse, P. *et al.* 2006. Evidence for glass 'recycling' using Pb and Sr isotopic ratios and Sr-mixing lines: the case of early Byzantine Sagalassos, *Journal of Archaeological Science* 33: 494–501.

Dehm, R. 1997. Riche découverte funéraire dans la nécropole tumulaire d'Ihringen-Gündlingen, lieu-dit 'Nachtwaid-Ried', in P. Brun (ed.), *Vix et les éphémères principautés celtiques: Les VI et Ve siècles avant J.-C. en Europe centre-occidentale ; actes du colloque de Châtillon-sur-Seine (27 - 29 octobre 1993)*: 53–55: Archéologie aujourd'hui. Paris: Ed. Errance.

Dever, W. G. and Gitin, S. (eds) 2003. *Symbiosis, symbolism, and the power of the past: Canaan, ancient Israel, and their neighbors from the Late Bronze Age through Roman Palaestina: proceedings of the Centennial Symposium, W. F. Albright Institute of Archaeological Research and American Schools of Oriental Research, Jerusalem, May 29/31, 2000.* Winona Lake, IN: Eisenbrauns.

Dick, M. B. 2005. The Mesopotamian Cult Statue: A Sacramental Encounter with Divinity, in N. H. Walls (ed.), *Cult image and divine representation in the ancient Near East*: 43–67. American Schools of Oriental Research Books Series, 10. Boston, MA: American Schools of Oriental Research.

Dietrich, M. 2001. Babylonische Sklaven auf der Schreiberschule, in W. H. van Soldt *et al.* (eds), *Veenhof Anniversary Volume: Studies presented to Klaas R. Veenhof on the occasion of his sixty-fifth birthday:*

67–81. Leiden: Nederlands Instituut voor het Nabije Oosten.

Dornemann, R. H. 1983. *The archaeology of the Transjordan in the Bronze and Iron Ages*. Milwaukee, WI: Milwaukee Public Museum.

Dussubieux, L., Schmidt, K., Rowan, Y. M., Wasse, A. M. R. and Rollefson, G. O. 2018. Two Glass Beads from Wisad Pools in the Jordanian Black Desert, *Journal of Glass Studies* 4: 303–306.

Dyson, R. H. 1989. The Iron Age Architecture in Hasanlu: An Essay, *Expedition* 2-3: 107–127.

Dyson, R. H. and Muscarella, O. W. 1989. Constructing the Chronology and Historical Implications of Hasanlu IV, *Iran* 27: 1–27.

Dyson, R. H. and Voigt, M. M. 2003. A Temple at Hasanlu, in N. F. Miller, K. Abdi and W. M. Sumner (eds), *Yeki bud, yeki nabud: Essays on the archaeology of Iran in honor of William M. Sumner*: 219–236. Monograph, 48. Published in association with the American Institute of Iranian Studies and the University of Pennsylvania Museum of Archaeology and Anthropology. Los Angeles: Cotsen Institute of Archaeology at UCLA.

Ebeling, E. 1948. Mittelassyrische Rezepte zur Bereitung von wohlriechenden Salben, *Orientalia* 17: 129–145.

Ebeling, E. 1949. Mittelassyrische Rezepte zur Herstellung von wohlriechenden Salben (Fortsetzung), *Orientalia* 19: 265–278.

Ebeling, E. 1950. Mittelassyrische Rezepte zur Herstellung von wohlriechenden Salben (Fortsetzung), *Orientalia* 18: 404–418.

Elayi, J. 2000, Les sites phéniciens de Syrie au Fer III/ Perse: Bilan et perspective de recherche, in G. Bunnens (ed.), *Essays on Syria in the Iron Age*: 327–348. Ancient Near Eastern Studies, Supplement 7. Louvain: Peeters.

Elayi, J. 2008. On dating the reigns of Phoenician kings in the Persian period, in C. Sagona (ed.), *Beyond the Homeland: Markers in Phoenician chronology*: 97–112, Ancient Near Eastern Studies, Supplement 28. Leuven: Peeters.

Ertman, E. L. 2013. A Blue Glass Face Inlay of King Akhenaten, *Journal of Glass Studies* 55: 13–19.

Fales, F. M. and Postgate, J. N. (eds) 1992. *Imperial administrative records*. State Archives of Assyria, 7. Helsinki: Helsinki University Press.

Feldman, M. H. 2014. *Communities of style: Portable luxury arts, identity, and collective memory in the Iron Age Levant*. Chicago, IL: University of Chicago Press.

Fincke, J. 2004. The British Museum's Ashurbanipal Library Project, *Iraq* 64: 55–60.

Finkel, I. L. and Curtis, J. (eds) 2008. *Babylon: Myth and reality; [publ. to accompany the major exhibition 'Babylon: myth and reality' at the British Museum, 13.11.2008 - c 15.03.2009]*. London: British Museum Press.

Finkelstein, I. and Piasetzky, E. 2011. The Iron Age Chronology Debate: Is the Gap Narrowing?, *Near Eastern Archaeology*, 74: 50–54.

Fiorina, P. 2009. 'Nimroud, Fort Salmanasar: entrepôts et ateliers de la zone SW, in S. M. Cecchini, S. Mazzoni and E. Scigliuzzo (eds), *Syrian and Phoenician ivories of the early first millennium BCE: Chronology, regional styles and iconographic repertories, patterns of inter-regional distribution*: 27–46. Ricerche di archeologia del Vicino Oriente, 3. Pisa: Edizioni ETS.

Fischer, E. 2007. Ägyptische und ägyptisierende Elfenbeine aus Megiddo und Lachisch: Inschriftenfunde, Flaschen, Löffel. Alter Orient und Altes Testament, 47. Münster: Ugarit-Verl.

Fischer, P. M. 2013. *Tell Abu al-Kharaz in the Jordan Valley: The Iron age*. Vienna: Austrian Academy of Sciences.

Fischer, P. M. 2014. *Tell Abu Al-Kharaz in the Jordan Valley: Volume III: the Iron Age*. Contributions to the Chronology of the Eastern Mediterranean Ser, 34. Vienna: Austrian Academy of Sciences.

Fontan, E. (ed.) 2007. *La Méditerranée des Phéniciens de Tyr à Carthage: [cet ouvrage accompagne l'Exposition 'La Méditerranée des Phéniciens de Tyr à Carthage', Institut du Monde Arabe, 6 novembre 2007 - 20 avril 2008]*. Paris: Somogy Éd. d'Art.

Frahm, E. 2004. Royal hermeneutics: observations on the commentaries from Ashurbanipal's libraries at Nineveh, *Iraq* 66: 45–50.

Frahm, E. 2012. Headhunter, Bücherdiebe und wandernde Gelehrte: Anmerkungen zum altorientalischen Wissenstransfer im 1. Jahrtausend v. Chr., in H. Neumann and S. Paulus (eds), *Wissenskultur im Alten Orient: Weltanschauung, Wissenschaften, Techniken, Technologien*: 15–30. 4 Internationales Colloquium der Deutschen Orient-Gesellschaft 20. - 22. Februar, Münster. Wiesbaden: Harrassowitz Verlag.

Frame, G. and George, A. R. 2005. The royal libraries of Nineveh: New evidence for King Ashurbanipal's tablet collecting, *Iraq* 47: 265–184.

Frankfort, H. 1939. *Sculpture of the third millennium B.C. from Tell Asmar and Khafājah*. The University of Chicago Oriental Institute Publications, 44. Chicago, IL: University of Chicago Press.

Freestone, I. C. 1987. Composition and microstructure of early opaque red glass, in M. Bimson (ed.), *Early vitreous materials*: 173–191. Occasional Paper, 56. London: British Museum Press.

Freestone, I. C. 1991. Looking into Glass, in S. Bowman (ed.), *Science and the past*: 37–56. British Museum Press.

Freestone, I. C. 1997. Vitreous Materials: Typology and Technology, in E. M. Meyers (ed.), *The Oxford Encyclopedia of Archaeology in the Near East*: 306–309. Oxford: Oxford University Press.

Freestone, I. C., Stapleton, C. P. and Rigby, V. 2003. The production of red glass and enamel in the Late Iron Age, Roman and Byzantine periods, in C. Entwistle (ed.), *Through a glass brightly: Studies in Byzantine and medieval art and archaeology presented to David Buckton*. Oxford: Oxbow Books.

Fuchs, A. 2017. Die Kassiten, das mittelbabylonische Reich und der Zagros, in A Bartelmus and K. Sternitzke (eds), *Karduniaš: Babylonia under the Kassites; The Proceedings of the Symposium held in Munich 30 June to 2 July 2011 = Tagungsbericht des Münchner Symposiums vom 30. Juni bis 2. Juli 2011*: 123–165. Untersuchungen zur Assyriologie und vorderasiatischen Archäologie, 11/1. Berlin: de Gruyter.

Gadd, C. J. and Thompson, R. C. 1936. A middle-Babylonian chemical text. *Iraq* 3: 87–97.

Ganio, M. *et al.* 2012. The Sr–Nd isolation procedure for subsequent isotopic analysis using multi-collector ICP-mass spectrometry in the context of provenance studies on archaeological glass, *Journal of Analytical Atomic Spectrometry* 27: 1335–1341.

Garner, H. 1956. An early piece of glass from Eridu, *Iraq* 18: 147–149.

Gauckler, P. 1915. *Nécropoles puniques de Carthage*. Paris: Auguste Picard.

George, A. R. 1997. Assyria and the Western World, in S. Parpola (ed.), *Assyria 1995: Proceedings of the 10th Anniversary Symposium of the Neo-Assyrian Text Corpus Project, Helsinki, Sept. 7 - 11, 1995*: 69–75. Helsinki: Helsinki University Press.

Ghirshman, R. 1964. *Persia, from the origins to Alexander the Great*. London: Thames & Hudson.

Giberson, D. 2004. Sandcore vessel technology [videorecording] by Richard Remsen, Warner, NH : Joppa Press.

Gibson, M., Hansen, D. P. and Zettler, R. L. 2001. Nippur B. Archäologisch, *Zeitschrift für Assyriologie und Vorderasiatische Archäologie* 9: 546–565.

Gibson, M. 1972. *The city and area of Kish*. Coconut Grove, FL: Field Research Projects.

Gitin, S. 1997. The Neo-Assyrian Empire and its Western Periphery: The Levant, with a Focus on Philistine Ekron, in S. Parpola (ed.), *Assyria 1995: Proceedings of the 10th Anniversary Symposium of the Neo-Assyrian Text Corpus Project, Helsinki, Sept. 7 - 11, 1995*: 77–103. Helsinki: Helsinki University Press.

Goldstein, S. M. 1979. *Pre-Roman and Early Roman Glass in The Corning Museum of Glass*. Corning, NY: Corning Museum of Glass.

Gorelick, L. and Gwinnett, A. J. 1986. Further Investigation of the Method of Manufacture of an Ancient Near Eastern Cast Glass Vessel, *Iraq* 48: 15–18.

Graff, S. B. 2014. Kingdoms of Midas and Croesus: Western Anatolian States and Sanctuaries, in J. Aruz, Y. Rakic and S.B. Graff (ed.), *Assyria to Iberia: At the dawn of the classical age [Exhibition on view at the Metropolitan Museum of Art, New York, from September 22, 2014, through January 4, 2015]*: 104–109. New Haven, CT: Yale University Press.

Gratuze, B. 2001. Étude des perles protohistoriques en verre des dolmens d'Eyne - les Pascarets et la Borda - et d'Enveitg - Bragnoli - (Pyrénées Orientales). *IRAMAT, Centre E. Babelon, Internal Analytical Report*. Orléans: Institut de Recherches sur les Archéomatériaux.

Gratuze, B., Louboutin, C. and Billaud, Y. 1998. Les perles protohistoriques en verre du Musées des Antiquités nationales, *Antiquités Nationales* 30: 11–24.

Grayson, A. K. 1991. *Assyrian rulers of the early first millennium BC*. The Royal Inscriptions of Mesopotamia/Assyrian Periods, 2. Toronto: University of Toronto Press.

Gries, H. and Schmidt, K. 2019. Mittelassyrische kerngeformte Gefäße aus Assur (in preparation).

Grose, D. F. (ed.) 1989. *Early ancient glass: Core-formed, rod-formed, and cast vessels and objects from the late Bronze Age to the early Roman empire, 1600 B.C. to A.D. 50*. New York: Hudson Hills Press.

Gudenrath, W. 1991. Techniques of Glassmaking and Decoration, in H. Tait (ed.), *Five thousand years of glass*: 213–241. London: British Museum Press.

Gunter, A. C. 1991. *Gordion Excavations. Final reports III: The Bronze Age*. University Museum Monograph, 71. Philadelphia, PA: University of Pennsylvania, Museum of Archaeology and Anthropology/ University of Pennsylvania Press.

Gwinnett, A. J. and Gorelick, L. 1987. The change from stone drills to copper drills in Mesopotamia, *Expedition* 29: 15–24.

Hachmann, R. (ed.) 1983, *Frühe Phöniker im Libanon: 20 Jahre deutsche Ausgrabungen in Kāmid el-Lōz*. Mainz am Rhein: von Zabern.

Haevernick, T. E. 1968. Assyrisches Millefioriglas, *Forschungen und Berichte* 10: 63–70.

Haevernick, T. E. 1977. Gesichtsperlen, *Madrider Zeitschrift* 18: 152–231.

Hall, H. R. 1930. *A season's work at Ur, al-'Ubaid, Eridu and elsewhere*. London: Methuen.

Haller, A. 1954. *Die Gräber und Grüfte von Assur. Mit 1 Stadtpl. 31 Lichtdrucktaf. u. 197 Abb. i. Text*. Berlin: Mann.

Harden, D. B. (ed.) 1981. *Catalogue of Greek and Roman glass*. London: British Museum Press.

Harden, D. B. (ed.) 1968. *Masterpieces of glass: A selection*. London: British Museum Press.

Harden, P. 1940. *Glass vessels before glass-blowing*. Copenhagen: Munksgaard.

Harper, P. O. (ed.) 1995. *Assyrian origins: Discoveries at Ashur on the Tigris; antiquities in the Vorderasiatisches Museum, Berlin*. New York: Metropolitan Museum of Art.

Hauptmann, A. (ed.) 2008. *Archäometrie: Methoden und Anwendungsbeispiele naturwissenschaftlicher Verfahren in der Archäologie*. Veröffentlichungen aus dem Deutschen Bergbau-Museum Bochum, 156. Stuttgart: Schweizerbart.

Hauptmann, H. and Pernicka, E. (eds) 2004. *Die Metallindustrie in Mesopotamien von den Anfängen bis zum 2. Jahrtausend v. Chr. (Draft only)*. Orient-Archäologie, 3. Rahden/Westf: Verlag Marie Leidorf GmbH.

Hauser, S. R. 2012. *Status, Tod und Ritual: Stadt- und Sozialstruktur Assurs in neuassyrischer Zeit*. Abhandlungen der Deutschen Orient-Gesellschaft, 26. Wiesbaden: Harrassowitz Verlag.

Hausleiter, A. (ed.) 1999. *Iron age pottery in Northern Mesopotamia, Northern Syria, and South-Eastern Anatolia: Papers presented at the meetings of the international 'table ronde' at Heidelberg (1995) and Nieborów (1997) and other contributions*. Altertumskunde des Vorderen Orients, Bd. 10. Münster: Ugarit-Verl.

Hausleiter, A. 2010. *Neuassyrische Keramik im Kerngebiet Assyriens: Chronologie und Formen*. Abhandlungen der Deutschen Orient-Gesellschaft, Bd. 27. Wiesbaden: Harrassowitz Verlag.

Heeßel, N. P. 2002. *Pazuzu: Archäologische und philologische Studien zu einem altorientalischen Dämon*. Ancient Magic and Divination, 4. Leiden: Brill.

Heinz, M. (ed.) 2003 *Zwischen Erklären und Verstehen? Beiträge zu den erkenntnistheoretischen Grundlagen archäologischer Interpretation*. Tübinger archäologische Taschenbücher, 2. Münster: Waxmann.

Helbig, W. and Reisch, E. (eds) 1969. *Führer durch die öffentlichen Sammlungen klassischer Alterthümer in Rom: Dritter Band*. Leipzig: Teubner u. Baedeker.

Helwing, B. 2004. Klassifikation, Typologie und Datierung der untersuchten Metallobjekte, in H. Hauptmann and E. Pernicka (eds), *Die Metallindustrie in Mesopotamien (Draft only)*. Orient-Archäologie, 3. Rahden/Westf: Verlag Marie Leidorf GmbH.

Henderson, J. 2013. *Ancient glass: An interdisciplinary exploration*. Cambridge: Cambridge University Press.

Henderson, J. *et al.* 2005. The use of oxygen, strontium and lead isotopes to provenance ancient glasses in the Middle East, *Journal of Archaeological Science* 31: 665–673.

Henderson, J. 1988. Electron probe microanalysis of mixed-Alkali glasses, *Archaeometry*, 30(1): 77–91. doi: 10.1111/j.1475-4754.1988.tb00436.x

Henderson, J., Evans, J. A. and Barkoudah, Y. 2009. The provenance of Syrian plant ash glass: an isotopic approach, in P. Degryse, J. Henderson and G. Hodgins (eds), *Isotopes in vitreous materials*: 73 – 97. Studies in archaeological sciences, 1. Leuven: Leuven University Press.

Henderson, J., Evans, J., and Nikita, K. 2010. Isotopic evidence for the primary production, provenance and trade of Late Bronze Age glass in the Mediterranean, *Mediterranean Archaeology* 10: 1–24.

Herrmann, B. (ed.) 1994. *Archäometrie: Naturwissenschaftliche Analyse von Sachüberresten ; eine praktikumsbegleitende Veröffentlichung aus dem Arbeitskreis Umweltgeschichte der Georg-August-Universität Göttingen*. Berlin: Springer.

Herrmann, G. 2002. The Nimrud Ivories 5: The Ornate Group, in D. Oates and L. Al-Gailani Werr (eds), *Of pots and plans: Papers on the archaeology and history of Mesopotamia and Syria presented to David Oates in honour of his 75th birthday*: 128–142. London: NABU.

Herrmann, G. 1986. *Ivories from room SW37 Fort Shalmanesher*. Ivories from Nimrud 1949–1963, fasc 4. London: British School of Archaeology in Iraq.

Herrmann, G. 1989. The Nimrud Ivories, 1: The Flame and Frond School, *Iraq* 51: 85–109.

Herrmann, G. and Laidlaw, S. (eds) 2008. *Ivories from the North West Palace (1845-1992)*. London: British Institute for the Study of Iraq.

Herrmann, G. and Laidlaw, S. 2013. *Ivories from room SW11/12 and T10, Fort Shalmaneser*. Ivories from Nimrud 1949–1963, VII, 1–2. Gertrude Bell Memorial. London: British Institute for the Study of Iraq.

Herrmann, G. and Mallowan, M. 1974. *Furniture from SW7 Fort Shalmaneser*. Ivories from Nimrud (1949–1963, fasc 3. London: British School of Archaeology in Iraq.

Hess, C. and Wight, K. 2005. *Looking at glass: A guide to terms, styles, and techniques*. Looking at Series. Los Angeles: J. Paul Getty Museum.

Homes-Fredericq, D. and Hennessy, J. B. 1989. *Archaeology of Jordan*. Akkadica. Supplementum, 7–8. Leuven: Peeters.

Hrouda, B. (ed.) 1987. *Isin-Išān Baḫrīyāt III: Die Ergebnisse der Ausgrabungen 1983-1984*. Veröffentlichungen der Kommission zur Erschliessung von Keilschrifttexten. Ser. C, 4. München: Vlg der Bayerischen Akademie der Wissenschaften.

Hudson, M. (ed.) 1999. *Urbanization and land ownership in the ancient Near East: A colloquium held at New York University, November 1996, and the Oriental Institute, St. Petersburg, Russia, May 1997*. International Scholars' Conference on Ancient Near Eastern Economies, 2. Cambridge, MA: Peabody Museum of Archaeology and Ethnology.

Hudson, M. and Wunsch, C. 2004. *Creating economic order: Record-keeping, standardization, and the development of accounting in the ancient Near East*. Bethesda, MD: CDL Press.

Humbert, J.-B. and Zayadine, F. 1992. Trois Campagnes de Fouilles à Amman (1988–1991). Troisième Terrasse de la Citadelle (Mission Franco-Jordanienne), *Revue Biblique* 99: 214–260.

Hunger, H. 1968. *Babylonische und assyrische Kolophone*. Alter Orient und Altes Testament, 2. Kevelaer: Butzon & Bercker.

Hunt, Alice M. W. 2015. *Palace ware across the Neo-Assyrian imperial landscape: Social value and semiotic meaning*. Culture and history of the ancient Near East, 78. Leiden: Brill.

Huot, J.-L. 2014. *L'E.babbar de Larsa aux IIe et Ier millénaires: Fouilles de 1947 à 1985*. Beyrouth: Presses de l'Ifpo.

Hurowitz, A. 2003. The Mesopotamian God Image, from Womb to Tomb, *Journal of the American Oriental Society, Vol. 123, No. 1*, 123(1): 147–157.

Hussein, M. H. 2016. *Nimrud the Queens' Tombs*. Bagdad: Iraqi State Board of Antiquities and Heritage; Chicago: Oriental Institute of the University of Chicago.

İstanbul arkeoloji müzeleri yıllığı 1960. *Annual of the archaeological museums of Istanbul* (9). İstanbul.

Jones, J. D. 2005. Glass vessels from Gordion: Trade and influence along the royal road, in L. Kealhofer (ed.) *The archaeology of Midas and the Phrygians: Recent work at Gordion*: 101–116. Philadelphia, PA: University of Pennsylvania, Museum of Archaeology and Anthropology/University of Pennsylvania Press.

Jackson, C. M. and Nicholson, P. T. 2010. The provenance of some glass ingots from the Uluburun shipwreck, *Journal of Archaeological Research* 37: 295–301.

Jakob-Rost, L. (ed.) 1992. *Das Vorderasiatische Museum*. Mainz am Rhein: von Zabern.

James, F. W. 1966. *The iron age at Beth Shan: A study of levels VI - IV*. University Museum Monograph, 28. Philadelphia, PA: University of Pennsylvania, Museum of Archaeology and Anthropology/ University of Pennsylvania Press.

James, P. and van der Sluijs, M. A. 2008. Ziggurats, Colors, and Planets: Rawlinson Revisited, *The American School of Oriental Research* 60: 57–79.

Jassim, S. Z. (ed.) 2006. *Geology of Iraq*. Prague: Dolin.

Joannès, F. 1993–1997. Metalle und Metallurgie. A., *Zeitschrift für Assyriologie und Vorderasiatische Archäologie*, 8: 96–112.

Johns, C. N. 1933. Excavations in 'Atlit (1930–31): The South-Eastern Cemetery, *Quarterly of the Department of Antiquities in Palestine* 2: 41–104.

Jones, J. D. 2009. Did the Phrygians make glass? Sources of moulded glass in Iron Age and Hellenistic Gordion, in *AIHV Annales du 17e Congrès*: 21–26.

Jursa, M. 2001. Rezension zu: Bottéro, Jean: Textes culinaires Mésopotamiens. Mesopotamian Culinary Texts, *Zeitschrift für Assyriologie und Vorderasiatische Archäologie* 91: 298–302.

Jursa, M. 2005. Parfüm(rezepte). A: In Mesopotamien, *Reallexikon der Assyriologie und Vorderasiatischen Archäologie* 8: 335–336.

Jursa, M. 2014. The Neo-Babylonian Empire, in M. Gehler and R. Rollinger (eds), *Imperien und Reiche in der Weltgeschichte: Epochenübergreifende und globalhistorische Vergleiche*: 121–148. Wiesbaden: Harrassowitz Verlag.

Kaczmarczyk, A. 1986. The source of cobalt in ancient Egyptian pigments, *Proceedings of the 24th International Archaeometry Symposium*: 369–376.

Kaczmarczyk, A. and Vandiver, P. B. 2008. Faience production in Egypt, in M. S. Tite and A. J. Shortland (eds), *Production technology of faience and related early vitreous materials*: 5–92. Oxford University School of Archaeology Monograph, 72. Oxford: Oxford University School of Archaeology.

Kaczmarczyk, A. and Hedges, Robert E. M. (eds) 1983. *Ancient Egyptian faience: An analytical survey of Egyptian faience from predynastic to Roman times*. Warminster: Aris & Phillips.

Karpik, L. 2011. *Mehr Wert: Die Ökonomie de Einzigartigen*. Frankfurt am Main: Campus-Verl.

Kealhofer, L. (ed.) 2005. *The archaeology of Midas and the Phrygians: Recent work at Gordion*. Philadelphia, PA: University of Pennsylvania, Museum of Archaeology and Anthropology/University of Pennsylvania Press.

Kempinski, A. 1989. *Megiddo: A city state & royal centre in North Israel*. Tel-Aviv: Hakkibutz Hameuchad Pub. House.

Kertai, D. 2015. *The architecture of Late Assyrian royal palaces*. Oxford: Oxford University Press.

Kirk, S. 2009. *The vitreous materials from the 2nd millennium BC city of Nuzi: their preservation, technology and distribution*. Cranfield, UK: Cranfield University.

Kisa, A. 1908. *Das Glas im Altertume*. Hiersemanns Handbücher, 3. Leipzig: Hiersemann.

Kistler, E. 2010. Großkönigliches Symbolon im Osten - exotisches Luxusgut im Westen: Zur Objektbiographie der achämenidischen Glasschale aus Ihringen, in R. Rollinger (ed.), *Interkulturalität in der Alten Welt: Vorderasien, Hellas, Ägypten und die vielfältigen Ebenen des Kontakts*: 63–90. Philippika, 34. Wiesbaden: Harrassowitz Verlag.

Klengel-Brandt, E. 1977. *Reise in das alte Babylon*. Leipzig: Prisma-Verlag.

Klengel-Brandt, E. 1978. *Die Terrakotten aus Assur im Vorderasiatischen Museum Berlin*. Berlin: Dt. Verl. der Wiss.

Koch, L. C. 2011. *Früheisenzeitliches Glas und Glasfunde Mittelitaliens: Eine Übersicht von der Villanovazeit bis zum Orientalizzante und eine Analyse der Glasperlen als Grabbeigabe des Gräberfeldes Quattro Fontanili in Veji*. Bochumer Forschungen zur ur- und frühgeschichtlichen Archäologie, 4. Rahden/Westf: Verlag Marie Leidorf GmbH.

Köcher, F. 1957. Ein Inventartext aus Kar-Tukulti-Ninurta, *Archiv für Orientforschung* 18: 300–313.

Kohler, E. L. 1995. *The Gordion Excavations, 1950-1973: Final Reports II: The lesser Phrygian Tumuli*. University Museum Monograph, 88. Philadelphia, PA: University of Pennsylvania, Museum of Archaeology and Anthropology/University of Pennsylvania Press.

Koldewey, R. 1913. *Das wieder erstehende Babylon: Die bisherigen Ergebnisse d. dt. Ausgrabungen*. 2nd edn. Deutsche Orient-Gesellschaft: Sendschrift, 6. Leipzig: Hinrichs.

Koldewey, R. 1990. *Das wiedererstehende Babylon*. 5th edn. Becksche Sonderausgaben. München: Beck.

Koldewey, R. and Wetzel, F. 1932. *Die Königsburgen von Babylon*. Bagdad: Min. of Culture & Information, State Organization of Antiquities & Heritage.

Kopytoff, I. 1986. The cultural biography of things, in A. Appadurai (ed.), *The Social life of things: Commodities in cultural perspective*: 64 – 94. Cambridge: Cambridge University. Press.

Kreppner, F. J., Daszkiewicz, M.·l. and Kühne, H. 2006. *Die Keramik des 'Roten Hauses' von Tall Šēḫ Ḥamad, Dūr-Katlimmu: Eine Betrachtung der Keramik Nordmesopotamiens aus der zweiten Hälfte des 7. und aus dem 6. Jahrhundert v. Chr.* Berichte der Ausgrabung Tall Šēḫ Ḥamad/Dūr-Katlimmu (Batsh), 7,2. Wiesbaden: Harrassowitz Verlag.

Kühne, H. 1971. Glas, *Zeitschrift für Assyriologie und Vorderasiatische Archäologie*, 3: 410–426.

Kühne, K. 1999. Antike kieselkeramische Sinterwerkstoffe, in R. Busz (ed.), *Türkis und Azur: Quarzkeramik in Orient und Okzident; [anläßlich der Sonderausstellung vom 18. Juli bis 3. Oktober 1999 im Ballhaus am Schloß Wilhelmshöhe und in Schloß Wilhelmsthal, Souterrain]*: 104–110. Wolfratshausen: Ed. Minerva.

Labindo, D. 1966. The Egyptian Sand-core Technique: a new interpretation, *Journal of Glass Studies* 8: 124–127.

Lambert, W. G. 1983. Kūbu, *Zeitschrift für Assyriologie und Vorderasiatische Archäologie* (6). Berlin: de Gruyter.

Lamon, R. S. and Shipton, G. M. 1939. *Megiddo I: Seasons of 1925-34, strata 1-5*. Oriental Institute Publications, 42. Chicago, IL: University of Chicago Press.

Lanfranchi, G. 1997. Consensus to Empire: Some Aspects of Sargon II's Foreign Policy, in H. Waetzoldt and H. Hauptmann (eds), *Assyrien im Wandel der Zeiten: XXXIX Rencontre Assyriologique Internationale*: 81–87. Heidelberg: Heidelberger Orientverlag.

Lang, R. H. 1875. *Narrative of excavations in a Temple at Dali (Idalium) in Cyprus*. London: Harrison & Sons.

Launay, L. de 1895. Note sur la nécropole de Camiros dans l'ile de Rhodes, *Revue Archéologique* 27: 182–197.

Layard, A. H. 1856. *Nineveh und Babylon: Nebst Beschreibung seiner Reisen in Armenien, Kurdistan und der Wüste*. Leipzig: Dyk.

Lehmann, G. 1996. *Untersuchungen zur späten Eisenzeit in Syrien und Libanon: Stratigraphie und Keramikformen zwischen ca. 720 bis 300 v. Chr.* Altertumskunde des Vorderen Orients, Bd. 5. Münster: Ugarit.

Lehrer, G. 1974. A Phoenician Glass Bowl from Nimrud, *Journal of Glass Studies* 25(16): 9–13.

Leichty, E. 1979. A collection of recipes for dyeing, *Alter Orient und Altes Testament* 203: 15–20.

Lilyquist, C. 1994. The Dilbat Hoard, *Metropolitan Museum Journal* 29: 5–36.

Lilyquist, C. and Brill, R. H. (eds) 1993. *Studies in early Egyptian glass*. New York: Metropolitan Museum of Art.

Limper, K. 1989. *Uruk: Perlen, Ketten, Anhänger; Grabungen 1912-1985*. Endberichte, 2. Mainz am Rhein: von Zabern.

Lipschitz, O. 2005. *The Fall and Rise of Jerusalem: Judah under Babylonian Rule*. Winona Lake, IN, Eisenbrauns.

Liritzis, I. and Stevenson, C. M. (eds) 2012. *Obsidian and ancient manufactured glasses*. Albuquerque: University of New Mexico Press.

Liverani, M. 2011. From City-State to Empire: the Case of Assyria, in J. P. Árnason and K. A. Raaflaub (eds), *The Roman Empire in context: Historical and comparative perspectives*: 251–269. The ancient world: comparative histories. Chichester: Wiley-Blackwell.

Lloyd, S. and Gökçe, N. 1953. Anglo-Turkish Joint Excavations, 1952, *Anatolian Studies*, 3: 27–47.

Loud, G. 1948. *Megiddo 2. Seasons of 1935 - 1938: Text and plates*. Oriental Institute Publications, 62. Chicago, IL: University of Chicago Press.

Loud, G. and Altman, C. B. 1938. *Khorsabad II: The citadel and the town*. Oriental Institute Publications, 40. Chicago, IL: University of Chicago Press.

Lucas, A. and Harris, J. R. 1962. *Ancient Egyptian materials and industries*. 4th edn. London: E. Arnold.

Luckenbill, D. D. 1927. *Ancient records of Assyria and Babylonia*. Chicago, IL: University of Chicago Press.

Luschey, H. 1939. *Die Phiale*. Bleicherode am Harz: Nieft.

MacCown, D. E., Haines, R. C. and Hansen, D. P. 1967. *Nippur: Temple of Enlil, Scribal quarter, and soundings: excavations of the joint expedition to Nippur of the University Museum of Philadelphia and the Oriental Institute of the University of Chicago*. Oriental Institute Publications, 78. Chicago, IL: University of Chicago Press.

MacCown, D. E., Haines, R. C. and Biggs, R. D. 1978. *Nippur II: The North Temple and Sounding E: Excavations of the Joint Excavation to Nippur of the American Schools of Oriental Research and the Oriental Institute of the University of Chicago*. Oriental Institute Publications, 97. Chicago, IL: University of Chicago Press.

MacCown, D. E., Zettler, R. L. and Cole, S. W. 1978. *Nippur: Excavations of the Joint Expedition to Nippur*. Oriental Institute Publications, 97. Chicago, IL: University of Chicago Press.

MacDonald, B., Adams, R. and Bienkowski, P. (eds) 2001. *The archaeology of Jordan*. Levantine Archaeology, 1. Sheffield: Sheffield Academic Press.

Mackay, D. (ed.) 1951. *A guide to the archaeological collections in the University Museum (American University of Beirut)*. Beirut: Impr. Catholique.

Madhloom, T. A. 1970. *The chronology of Neo-Assyrian art*. London: Athlone Press.

Magen, U. 1986. *Assyrische Königsdarstellungen, Aspekte der Herrschaft: Eine Typologie*. Baghdader Forschungen, Bd. 9. Mainz am Rhein: von Zabern.

Makharadze, G. and Saginashvili, M. 1999. An Achaemenian glass bowl from Sairkhe, Georgia, *Journal of Glass Studies* 25(41): 11–17.

Mallowan, M. E. L. 1952. The Excavations at Nimrud (Kalḫu), 1949–1950. Ivories from the N. W. Palace, *Iraq* 14: 45–53.

Mallowan, M. E. L. 1954. The Excavations at Nimrud (Kalḫu), 1953, *Iraq* 16/1: 59–114.

Mallowan, M. E. L. (ed.) 1966. *Nimrud and its Remains, II*. London: Collins.

Mallowan, M. E. L and Herrmann, G. 1974. *Furniture from SW. 7 Fort Shalmaneser: Commentary, catalogue and plates*. Ivories from Nimrud (1949–1963), fasc. 3. London: British School of Archaeology in Iraq.

Marcus, M. I. 1991. The Mosaic Glass Vessels from Hasanlu, Iran: A Study in Large-Scale Stylistic Trait Distribution, *Art Bulletin* 73/4: 536–560.

Markoe, G. 1985. *Phoenician Bronze and silver bowls from Cyprus and the Mediterranean*. University of California Publications, 26. Berkley, CA: University of California Press.

Martinón-Torres, M. and Rehren, T. (eds) 2008. *Archaeology, history and science*. London: Publications of the Institute of Archaeology, University College London.

Marzahn, J. (ed.) 2003. *Wiedererstehendes Assur: 100 Jahre deutsche Ausgrabungen in Assyrien*. Mainz am Rhein: von Zabern.

Mass, J. L., Wypyski M. T and Stone, R. E. 2002. Malkata and Lisht glassmaking technologies: towards a specific link between second millennium BC metallurgists and glassmakers, *Archaeometry* 44: 67–82.

Matney, T. 2010. Material Culture and Identity: Assyrians, Aramaeans, and the Indigenous Peoples of Iron Age Southeastern Anatolia, in S. R. Steadman (ed.), *Agency and identity in the ancient Near East: New paths forward*: 128–147. Approaches to Anthropological Archaeology. London: Equinox.

Matney, T. *et al.* 2015. Excavations at Ziyaret Tepe, Diyarbakır Province, Turkey, 2011–2014 Seasons, *Anatolica* 41: 125–176.

Matoian, V. 2002–2003. Matières vitreuses au royaume d'Ougarit, *Annales Archéologiques Arabes Syriennes*: 153–162.

Matoian, V. and Bouquillon, A. 2003. Vitreous materials in Ugarit: new data, in P. R. S. Moorey, M. Roaf and D. Stein (eds), *Culture through objects: Ancient Near Eastern studies in honour of P. R. S. Moorey*: 333–346. Oxford: Griffith Institute.

Matson, F. R. 1957. Analyses of various substances from Persepolis: A Study of Glass, in E. F. Schmidt (ed.), *Persepolis*: 127–132. Oriental Institute Publications. Chicago, IL: University of Chicago Press.

Matthiae, P. 1998. *Ninive*. Centri e monumenti dell'antichità. Milano: Electa.

Mattila, R. 1995. *Niniveh, 612 BC: The glory and fall of the Assyrian empire; catalogue of the 10th Anniversary Exhibition of the Neo-Assyrian Text Corpus Project*. Helsinki: Helsinki University Press.

Maxwell-Hyslop, K. R. 1956. Urartian Bronzes in Etruscan Tombs, *Iraq* 18: 150–167.

Mayer-Opificius, R. 1995. Das Relief des Šamaš-rēš-uṣur aus Babylon.,in: M. Dietrich and O. Loretz (eds), *Vom Alten Orient zum Alten Testament. Festschrift für Wolfram Freiherrn von Soden zum 85. Geburtstag am 19. Juni 1993*: 331–348. Kevelaer: Verlag Butzon & Bercker; Neukirchen-Vluyn: Neukirchener Verlag.

Mazar, B. 2011. The Iron Age Chronology Debate: Is the Gap Narrowing? Another Viewpoint, *Near Eastern Archaeology* 74: 105–111.

Mazar, B. 2006. *Excavations at Tel Beth-Shean: 1989–1996*. The Beth-Shean Valley Archaeological Project publication, 1. Jerusalem: Israel Exploration Society.

McClellan, M. 1984. *Core-formed glass from dated contexts*. Dissertations available from ProQuest. Paper AAI8505107.

McGovern, P. E., Stuart, J. and Fleming, J. 1993. The Late Bronze Egyptian Garrison at Beth Shan: Glass and Faience Production and Importation in the Late New Kingdom, *Bulletin of the American Schools of Oriental Research* 290/291: 1–27.

McGovern, P. E. (ed.) 2000. *Ancient wine: The search for the origins of viniculture*. Princeton: Princeton University Press.

Mecquenem, M. R. 1922. Fouilles de Suse: Campagnes des années 1914–1929–1922. *Revue d' Assyriologie et* d›archéologie orientale 19: 110–140.

Meissner, B. 1920. *Babylonien und Assyrien*. Kulturgeschichtliche Bibliothek Reihe 1, 3. Heidelberg: Winter.

Meissner, B. 1925 (Ausg. 1924). *Babylonien und Assyrien*. Kulturgeschichtliche Bibliothek Reihe 1, 4. Heidelberg: Winter.

de Mequenem, R. 1931. Excavations at Susa (Persia), 1930–1931, *Antiquity* 5: 330–343.

Meyer, G. R. (ed.) 1965. *Altorientalische Denkmäler im Vorderasiatischen Museum zu Berlin*. Leipzig: Seemann.

Meyer, L.d. 1980. *Tell ed-Der III: Sounding at Abu Habbah (Sippar)*. Publicaties van het Belgisch Comité voor Historish, Epigrafisch en Archeologisch Onderzoek in Mesopotamië, 4. Leuven: Peeters.

Meyer-Opificius, R. 1995. Das Relief des Šamaš-rēš-uṣur aus Babylon, in M. Dietrich, O. Loretz and W. v. Soden (eds), *Vom alten Orient zum alten Testament: Festschrift für Wolfram Freiherrn von Soden zum 85. Geburtstag am 19. Juni 1993*: 331–348: Alter Orient und Altes Testament, 240. Kevelaer: Verlag Butzon & Bercker; Neukirchen-Vluyn: Neukirchener Verlag.

Miglus, P.A. (ed.) 1996. *Das Wohngebiet von Assur: Stratigraphie und Architektur*. Wissenschaftliche Veröffentlichung der Deutschen Orient-Gesellschaft. Berlin: Mann.

Miglus, P. A. 2003. Neue Forschungen in Assur, in J. Marzahn (ed.), *Wiedererstehendes Assur: 100 Jahre deutsche Ausgrabungen in Assyrien*: 183–190. Mainz am Rhein: von Zabern.

Moorey, P. R. S. 1979. *Kish excavations 1923-1933: With a microfiche catalogue of the objects in Oxford excavated by the Oxford-Field Museum, Chicago expedition to Kish in Iraq, 1923-1933*. Oxford: Clarendon Press.

Moorey, P. R. S. 1980. *Cemeteries of the first millennium B.C. at Deve Hüyük, near Carchemish, salvaged by T. E. Lawrence and C. L. Wooley in 1913: With a catalogue raisonné of the objects in Berlin, Cambridge, Liverpool, London and Oxford*. British Archaeological Reports International Series, 87. Oxford: British Archaeological Reports.

Moorey, P. R. S. 1994. *Ancient Mesopotamian materials and industries: The archaeological evidence*. Oxford: Clarendon Press.

Moorey, P. R. S *et al.* (eds) 2003. *Culture through objects: Ancient Near Eastern studies in honour of P. R. S. Moorey*. Oxford: Griffith Institute.

Muhle, B. 2008. *Vorderasiatische Keulen und ihr Umfeld vom 9. bis ins frühe 1. Jt. v. Chr. Typologie und Deutung*. Elektronische Dissertation, LMU München.

Muhly, J. D. 1972. Review: A. Leo Oppenheim, Robert H. Brill, Dan Barag, and Axel von Saldern, Glass and Glassmaking in Ancient Mesopotamia. An Edition of the Cuneiform Texts which contain instructions for Glassmakers with a catalogue of surviving objects, *Journal of Cuneiform Studies* 24: 178–182.

Müller, U. 2009. Netzwerkanalysen in der Historischen Archäologie. Begriffe und Beispiele, in S. Brather (ed.), *Historia archaeologica: Festschrift für Heiko Steuer zum 70. Geburtstag*: 735–754. Ergänzungsbände zum Reallexikon der germanischen Altertumskunde, 70. Berlin: de Gruyter.

Mullins, R. A. 2012. Late Bronze and Iron Age Temples at Beth-Sean, in J. Kamlah and H. Michelau (eds), *Temple building and temple cult: Architecture and cultic paraphernalia of temples in the Levant (2.-1. mill. B.C.E.): proceedings of a conference on the occasion of the 50th anniversary of the Institute of Biblical Archaeology at the University of Tübingen (28-30 May 2010)*: 127–157. Abhandlungen des Deutschen Palästina-Vereins, 41. Wiesbaden: Harrassowitz Verlag.

Muscarella, O. W. 1980. *The catalogue of ivories from Hasanlu, Iran*. University Museum Monographs, 40. Philadelphia, PA: University of Pennsylvania, Museum of Archaeology and Anthropology/ University of Pennsylvania Press.

Musche, B. 1992. *Vorderasiatischer Schmuck von den Anfängen bis zur Zeit der Achaemeniden: Ca. 10.000 - 330 v. Chr*. Handbuch der Orientalistik, 7, 1,2,2,7 = 1,2,B,7: Der alte vordere Orient, Die Denkmäler, Vorderasien. Leiden: Brill.

Nakhai, B. A. 2001. *Archaeology and the religions of Canaan and Israel*. ASOR Books, 7. Boston, MA: American Schools of Oriental Research.

Nasrabadi, B. M. 1999. *Untersuchungen zu den Bestattungssitten in Mesopotamien in der ersten Hälfte des ersten Jahrtausends v. Chr*. Baghdader Forschungen, 23. Mainz am Rhein: von Zabern.

Nenna, M.-D. (ed.) 2000. *La route du verre: Ateliers primaires et secondaires du second millénaire av. J.-C. au Moyen Âge*. Travaux de la Maison de l'Orient, 33. Lyon: Maison de l'Orient Méditerranéen.

Neumann, H. and Paulus, S. (eds) 2012. *Wissenskultur im Alten Orient: Weltanschauung, Wissenschaften, Techniken, Technologien*. 4 Internationales Colloquium der Deutschen Orient-Gesellschaft 20. –22. Februar, Münster. Wiesbaden: Harrassowitz Verlag.

Newton, R. G. and Davison, S. 1989. *Conservation of glass*. Butterworths Series in Conservation and Museology. London: Butterworths.

Nicholson, P. T. 1995. Glassmaking and Glassworking at Amarna: Some New Work, *Journal of Glass Studies* 37: 11–19.

Nicholson, P. T. 2012. 'Stone...that flows': Faience and Glass as Man-Made Stone in Egypt, *Journal of Glass Studies* 25(54): 11–23.

Nicholson, P. T., Jackson, C. M. and Trott, K. M. 2009. The Uluburun glass ingots, cylindrical vessels and Egyptian glass, *Journal of Archaeological Science* 36: 1496–1503.

Nissen, H. J., Bernbeck, R. and Bartl, K. (eds) 1999. *Fluchtpunkt Uruk: Archäologische Einheit aus methodischer Vielfalt : Schriften für Hans Jörg Nissen*. Internationale Archäologie. Studia honoraria, Bd. 6. Rahden/Westf: Verlag Marie Leidorf GmbH.

Nissen, H. J. and Renger, J. (eds) 1982. *Mesopotamien und seine Nachbarn: Politische und kulturelle Wechselbeziehungen im alten Vorderasien vom 4. bis 1. Jahrtausend v. Chr*. 2nd edn. Berliner Beitraege zum Vorderen Orient, Bd. 1. Berlin: D. Reimer.

Noble, J. V. 1969. The Technique of Egyptian Failence, *American Journal of Archaeology*, 473: 435–439.

Núnez Calvo, F. J. 2008. Phoenicia, in C. Sagona (ed.), *Beyond the Homeland: Markers in Phoenician chronology*: 19–95. Ancient Near Eastern studies: Supplement, 28. Leuven: Peeters.

Nunn, A. 1988. *Die Wandmalerei und der glasierte Wandschmuck im alten Orient*. Handbuch der Orientalistik. Siebente Abteilung, Kunst und Archäologie, 1., Der Alte Vordere Orient, 2. Abschnitt, Die Denkmäler, B, Vorderasien, Lfg. 6. Leiden: Brill.

O'Hea, M. 2011. Another Look at the Origins of Iron Age II Cast Glass Vessels in the Levant, *Levant*, 43: 153–172.

Oates, D. 1959. Fort Shalmaneser: An Interim Report, *Iraq* 21: 98–129.

Oates, D. 1961. The Excavations at Nimrud (Kalȟu), 1960, *Iraq* 23: 1–14.

Oates, D. 1963. The Excavations at Nimrud (Kalḫu), 1962, *Iraq* 25: 6–37.

Oates, D. 1970. The excavations at Tell al Rimah, 1968, *Iraq*, 32: 1–26.

Oates, D. 1965. The Excavations at Tell al Rimah, 1964, *Iraq* 27: 62–80.

Oates, D., Oates, J. and McDonald, H. 1997. *Excavations at Tell Brak: Vol. 1: The Mitanni and Old Babylonian Periods.* McDonald Institute monographs. London: British School of Archaeology in Iraq.

Oates, J. and Oates, D. 2001. *Nimrud: An Assyrian imperial city revealed.* The Ancient World online: other publications. London: British School of Archaeology in Iraq.

Oded, B. 1979. *Mass deportations and deportees in the Neo-Assyrian empire.* Wiesbaden: Reichert.

Oliver, A. 1970. Persian export glass, *Journal of Glass Studies 1970* 12: 9–16.

Onasch, H.-U. 2010. Ägyptische und assyrische Alabastergefäße aus Assur. Wissenschaftliche Veröffentlichungen der Deutschen Orient-Gesellschaft, 128. Wiesbaden: Harrassowitz Verlag.

Oppenheim, A. L. 1960. Mesopotamia in the early history of Alchemy, *Revue d' Assyriologie et d›archéologie* orientale 60: 29–45.

Oppenheim, A. L. (ed.) 1970. *Glass and glassmaking in ancient Mesopotamia: An edition of the cuneiform texts which contain instructions for glassmakers with a catalogue of surviving objects.* The Corning Museum of Glass monographs, 3. Corning, NY: Corning Museum of Glass.

Orchard, J. J. 1978. Some miniature painted glass plaques from Fort Shalmaneser, Nimrud, *Iraq* 40: 1–21.

Orthmann, W. 1975. *Der alte Orient.* Propyläen Kunstgeschichte, Bd14. Berlin: Propyläen Verlag.

Paley, S. M. 1976. *King of the world: Ashur-nasir-pal II of Assyria 883-859 B. C.* New York: Brooklyn Museum.

Panighello, S. E. *et al.* 2012. Analysis of polychrome Iron Age glass vessels from Mediterranean I, II and III groups by LA-ICP-MS, *Journal of Archaeological Science* 39: 2945–2955.

Di Paolo, S. (ed.) 2018. *Composite artefacts in the Ancient Near East: Exhibiting an imaginative materiality, showing a genealogical nature.* Archaeopress Ancient Near Eastern Archaeology, 3. Oxford: Archaeopress.

Parpola, S. 2003. Assyria's Expansion in the 8th and 7th centuries and its long-term repercussions in the West, in W. G. Dever and S. Gitin (eds), *Symbiosis, symbolism, and the power of the past: Canaan, ancient Israel, and their neighbors from the Late Bronze Age through Roman Palaestina: proceedings of the Centennial Symposium, W.F. Albright Institute of Archaeological Research and American Schools of Oriental Research, Jerusalem, May 29/31, 2000*: 99–111. Winona Lake, IN: Eisenbrauns.

Parpola, S. 2004. National and Ethnic Identity in the Neo-Assyrian Empire and Assyrian Identity in Post-Empire Times, *JAAS* 18/2: 15–22.

Parzinger, H. and Eichmann, R. (eds) 2001. *Migration und Kulturtransfer: Der Wandel vorder- und zentralasiatischer Kulturen im Umbruch vom 2. zum 1. vorchristlichen Jahrtausend : Akten des Internationalen Kolloquiums, Berlin, 23. bis 26. November 1999.* Kolloquien zur Vor- und Frühgeschichte, Bd. 6. Bonn: Habelt.

Paynter, S. and Tite, M. 2001. The evolution of glazing methods in Ancient Near East and Egypt, in A. Shortland (ed.), *The social context of technological change in Egypt and the Near East, 1650-1550 BC*: 239–254. Oxford: Oxbow Books.

Pedde, F. 2000. *Vorderasiatische Fibeln: Von der Levante bis Iran.* Deutsche Orient-Gesellschaft: Abhandlungen der Deutschen Orient-Gesellschaft, 24. Saarbrücken: Saarbrücker Dr. und Verl.

Pedde, F. 2015. *Gräber und Grüfte in Assur II: Die mittelassyrische Zeit.* Ausgrabungen der Deutschen Orient-Gesellschaft in Assur. Reihe D, Allgemeines, 3. Wiesbaden: Harrassowitz Verlag.

Pedde, F. and Lundström, S. (eds) 2008. *Der alte Palast in Assur: Architektur und Baugeschichte; mit interaktiven Architekturplänen und Fotos auf CD-ROM.* Wiesbaden: Harrassowitz Verlag.

Pedersén, O. 1998. *Archives and libraries in the ancient Near East.* Bethesda, MD: CDL Press.

Peltenburg E. J. 1969. Al Mina glazed pottery and its relations, *Levant* 1: 73–96.

Peltenburg E. J. 1987. Early faience: recent studies, origins and relations with glass, in M. Bimson (ed.), *Early vitreous materials*: 5–25. Occasional Paper, 56. London: British Museum Press.

Peters, J. P. 1898. *Nippur or explorations and adventures on the Euphrates: Narrative of the University of Pennsylvania expedition to Babylonia, 1888-'90.* 2nd edn. New York: G. P. Putnam's Sons.

Petrie, William M. Flinders 1928. *Gerar.* British School of Archaeology in Egypt, and Egyptian Research Account, 43. London: British School of Archaeology in Egypt.

Polanyi, K. 1979. Ökonomie und Gesellschaft. Suhrkamp-Taschenbuch Wissenschaft, 295. Frankfurt am Main: Suhrkamp.

Pollard, A. M. and Heron, C. 1996. *Archaeological chemistry.* RSC paperbacks. Cambridge: Royal Society of Chemistry.

Postgate, C., Oates, D. and Oates, J. 1997. *The excavations at Tell al Rimah: The pottery.* Iraq Archaeological Reports, 4. Warminster: British School of Archaeology in Iraq.

Postgate, J. N. 2005. Pappel, *Reallexikon der Assyriologie und Vorderasiatischen Archäologie* 10: 329.

Postgate, J. N. 1992. Trees and timber in the Assyrian texts, in *Trees and timber in Mesopotamia*: 177–192. Bulletin on Sumerian Agriculture, 6. Cambridge: Sumerian Agriculture Group.

Poulsen, F. 1912. *Der Orient und die fruhgriechische Kunst.* Roma: ‚L'Erma' di Bretschneider.

Preusser, C. 1955. *Die Paläste in Assur: Mit einem Stadtplan, 26 Tafeln und 3 Abb. im Text.* Ausgrabungen der Deutschen Orient-Gesellschaft in Assur: A, 8. Berlin: Mann.

Pritchard, J. B. 1975. *Sarepta*. University Museum Monograph. Philadelphia, PA: University of Pennsylvania, Museum of Archaeology and Anthropology/University of Pennsylvania Press.

Privat, K., Reade, W. and Jones, J. D. 2014. Analysis of Molded and Core-formed Glass from 1st Millennium BC Gordion, Anatolia, *Microscopy and Microanalysis* 20: 2016–2017.

Purowski, T. *et al.* 2012. A study of glass beads from the Hallstatt C-D from southwestern Poland: implications for glass technology and provenance, *Archaeometry* 54: 144–166.

Radner, K. 1999. Money in the Neo-Assyrian empire, in J. G. Dercksen (ed.), *Trade and finance in ancient Mesopotamia: Proceedings of the First MOS Symposium (Leiden 1997)*: 127–157. Uitgaven van het Nederlands Historisch-Archaeologisch Instituut te İstanbul Publications de l'Institut historique-archéologique néerlandais de Stamboul, 84. Istanbul: Nederlands Historisch-Archaeologisch Instituut te Istanbul; Nederlands Instituut voor het Nabije Oosten.

Radner, K. 2001. Eine Bronzeschale mit neuassyrischer Inschrift, *State Archives of Assyria Bulletin* 13: 17–25.

Radner, K. 2004. Assyrische Handelspolitik: die Symbiose mit unabhängigen Handelszentren und ihre Kontrolle durch Assyrien, in R. Rollinger (ed.), *Commerce and monetary systems in the ancient world: Means of transmission and cultural interaction; proceedings of the Fifth Annual Symposium of the Assyrian and Babylonian Intellectual Heritage Project, held in Innsbruck, Austria, October 3rd - 8th 2002*: 152–169. Oriens et Occidens, 6. Stuttgart: Steiner.

Radner, K. 2006–2008. Provinz. C. Assyrien, *Zeitschrift für Assyriologie und Vorderasiatische Archäologie* 11: 42–68.

Radner, K. 2009. The Assyrian King and his Scholars: The Syro-Anatolian and the Egyptian Schools, in M. Luukko *et al.* (eds), *Of gods, trees, kings and scholars: Neo-Assyrian and related studies in honour of Simo Parpola*: 221–238. Studia Orientalia, 106. Helsinki: Finnish Oriental Society.

Radner, K. 2013. Sultantepe, *Reallexikon der Assyriologie und Vorderasiatischen Archäologie* 13: 287–288.

Radner, K. 2011. The Assur-Nineveh-Arbela triangle: Central Assyria in the Neo-Assyrian Period, in P. A. Miglus and S. Mühl (eds), *Between the cultures: The central Tigris region from the 3rd to the 1st millennium BC; conference at Heidelberg, January 22nd - 24th, 2009*: 321–464 Heidelberger Studien zum alten Orient, Bd. 14. Heidelberg: Heidelberger Orientverlag.

Radner, K. 2012. *Mass deportation: the Assyrian resettlement policy*. University College London. http://www.ucl.ac.uk/sargon/essentials/governors/massdeportation/ (accessed: 21.11.2018).

Radner, K. 2014. The Neo-Assyrian Empire, in M. Gehler *et al.* (eds), *Imperien und Reiche in der Weltgeschichte: Epochenübergreifende und globalhistorische Vergleiche: Imperien des Altertums, mittelalterliche und frühneuzeitliche Imperien*: 101 – 119. Wiesbaden: Harrassowitz Verlag.

Rahimpour, B. H. and Kazemi, H. A. 2003. Geology, mineralogy and genesis of the Gharah-Gol Boron deposit, SW of Zanjan, Iran, *Journal of Science (University of Teheran)* 29: 1–29.

Rau, P. 1991. 2. Oberflächenfunde: 2.2. Archäologische Befunde, in U. Finkbeiner and A. Becker (eds), *Uruk: Kampagne 35 - 37, 1982 - 1984; die archäologische Oberflächenuntersuchung (Survey)*: 57–72. Mainz am Rhein: von Zabern.

Reade, J. E. 2002. The Ziggurrat and Temples of Nimrud, *Iraq* 64: 135–216.

Reade, W., Freestone, I. C. and Bourke, S. 2009. Innovation and continuity in Bronze Age and Iron Age glass from Pella in Jordan, in: Annales du 17e Congrès de l'Association Internationale pour l'Histoire du Verre, 2006. Antwerp: 47–54.

Reade, W., Jones, G. O. and Privat, K. 2012. Iron Age and Hellenistic Monochrome Glasses from Gordion, in: *Annales du 18e Congrès de l'Association Internationale pour l'Histoire du Verre, Thessaloniki 2009*: Thessaloniki. 81–86.

Reade, W., Freestone and I. C., Simpson, J. 2005. Innovation or continuity? Early first millennium BCE glass in the Near East: the cobalt blue glasses from Assyrian Nimrud, in: *Annales du 16e Congrès de l'Association Internationale pour l'Histoire du Verre, 2003.* Nottingham: 23–25.

Rehm, E. (ed.) 1997. *Kykladen und Alter Orient: Bestandskatalog des Badischen Landesmuseums Karlsruhe.* Karlsruhe: Badisches Landesmuseum.

Rehren, T. 1997. Ramesside glass-colouring crucibles, *Archaeometry*, 39: 355–368.

Rehren, T. 2001. Aspects of the Production of Cobalt-blue Glass in Egypt, *Archaeometry* 43: 483–489.

Rehren, T. 2014. Glass production and consumption between Egypt, Mesopotamia and the Aegean, in P. Pfälzner, H. Niehr, E. Pernicka, S. Lange and T. Köster (eds), *Contextualising grave inventories in the ancient Near East: Proceedings of a workshop at the London 7th ICAANE in April 2010 and an International Symposium in Tübingen in November 2010, both organised by the Tübingen Post-Graduate School 'Symbols of the Dead'*: 217–223. Qaṭna Studien Supplementa, 3. Wiesbaden: Harrassowitz Verlag.

Rehren, T. and Pusch, E. 1997. New Kingdom Glass-Melting Crucibles from Qantir-Piramesses, *The Journal of Egyptian Archaeology* 83: 127–141.

Rehren, T. and Pusch, E. 2005. Late Bronze Age Glass Production at Qantir-Piramesses, Egypt, *Science*, No. 5729(308): 1756–1758.

Rehren, T. and Pusch, E. 2007. Glas für den Pharao - Glasherstellung in der Spätbronzezeit des Nahen Ostens, in G. A. Wagner (ed.), *Einführung in die Archäometrie*: 215–235. Berlin: Springer.

Reiter, K. 1997. *Die Metalle im alten Orient: Unter besonderer Berücksichtigung altbabylonischer Quellen*. Alter Orient und Altes Testament, 249. Münster: Ugarit-Verl.

Reuther, O. 1968. *Die Innenstadt von Babylon (Merkes)*. Ausgrabungen der Deutschen Orient-Gesellschaft in Babylon. Leipzig: Hinrichs.

Rice, P. M. 1987. *Pottery analysis: A sourcebook*. Chicago, IL: University of Chicago Press.

Roaf, M. 1990. *Cultural Atlas of Mesopotamia and the Ancient Near East*. Oxford: Equinox Books.

Roaf, M. 1998. Multiple Rabbets on Doors in Iron Age Assyria and Iran, *Iranica Antiqua* 33: 57–80.

Roaf, M. 2012. The Iron Age architecture of Hasanlu: An interim review, in S. Elliyoun and A. Sadraie (eds), *The Architecture of Hasanlu: Collected Papers: Persian Translation*: 103–115. Tehran: Hasanlu Translation Project Publications.

Roaf, M. 2001. Continuity and Change from the Middle to the Late Assyrian Period, in H. Parzinger and R. Eichmann (eds), *Migration und Kulturtransfer: Der Wandel vorder- und zentralasiatischer Kulturen im Umbruch vom 2. zum 1. vorchristlichen Jahrtausend : Akten des Internationalen Kolloquiums, Berlin, 23. bis 26. November 1999*: 357–369. Kolloquien zur Vor- und Frühgeschichte, Bd. 6. Bonn: Habelt.

Robson, E. 2001. Technology in Society. Three textual case studies from Late Bronze Age Mesopotamia, in A. Shortland (ed.), *The social context of technological change in Egypt and the Near East, 1650-1550 BC*: 39–57. Oxford: Oxbow.

Robson, E. 2013. Reading the libraries of Assyria and Babylonia, in G. Woolf, J. König and K. Oikonomopoulou (eds), *Ancient libraries*: 38–56. Cambridge: Cambridge University Press.

Rollinger, R. (ed.) 2010. *Interkulturalität in der Alten Welt: Vorderasien, Hellas, Ägypten und die vielfältigen Ebenen des Kontakts*. Philippika, 34. Wiesbaden: Harrassowitz Verlag.

Rollinger, R. 2014. Das teispidisch-achaimenidische Großreich Ein 'Imperium' avant la lettre?, in M. Gehler and R. Rollinger (eds), *Imperien und Reiche in der Weltgeschichte: Epochenübergreifende und globalhistorische Vergleiche*: 149–192.

Roobaert, A. and Bunnens, G. 1999. Excavations at Tell Ahmar – Til Barsib, in G. del Olmo Lete and J.-L. Montero Fenollós (eds), *Archaeology of the Upper Syrian Euphrates, the Tishrin Dam area: Proceedings of the international symposium held at Barcelona, January 28th-30th 1998*: 163–178. Aula orientalis-supplementa, 15. Sabadell: AUSA.

Rosenkranz, B. 1965. Ein hethitischer Wirtschaftstext, *Zeitschrift für Assyriologie und Vorderasiatische Archäologie* 57: 236–248.

Roßberger, E. 2018. Shining, Contrasting, Enchanting: Composite Artefacts from the Royal Tomb of Qaṭna, in S. Paolo (ed.), *Composite artefacts in the Ancient Near East: Exhibiting an imaginative materiality, showing a genealogical nature*: 39–50. Archaeopress Ancient Near Eastern Archaeology, 3. Oxford: Archaeopress.

Routledge, B. 1997. Mesopotamian 'Influence' in Iron Age Jordan: Issues of Power, Identity and Value, *Bulletin of the Canadian Society for Mesopotamian Studies* 32: 33–41.

Rowe, A. 1940. *The Four Canaanite temples of Beth-Shan: Part 1: The temples and cult objects*. Publications of the Palestine section of the University Museum University of Pennsylvania, Vol. 2. Part. I. Philadelphia, PA: University of Pennsylvania, Museum of Archaeology and Anthropology/ University of Pennsylvania Press.

Safar, F., Muṣṭafā, M. and Lloyd, S. 1981. *Eridu*. Bagdad: Ministry of Culture and Information.

Saldern, A. von 1959. Glass finds at Gordion, *Journal of Glass Studies* 1: 23–50.

Saldern, A. von 1966a. Glass, in M. E. L. Mallowan (ed.), *Nimrud and its Remains, II*. London: 623–634. London: Collins.

Saldern, A. von 1966b. Mosaic Glass from Hasanlu, Marlik, and Tell al- Rimah, *Journal of Glass Studies* 8: 9–25.

Saldern, A. von 1970. Other Mesopotamian Glass Vessels (1500–600 B.C.), in A. L. Oppenheim (ed.), *Glass and glassmaking in ancient Mesopotamia: An edition of the cuneiform texts which contain instructions for glassmakers with a catalogue of surviving objects*: 203. The Corning Museum of Glass Monographs, 3. Corning, NY: Corning Museum of Glass.

Saldern, A. von 2004. *Antikes Glas*. Handbuch der Archäologie. München: Beck.

Sallaberger, W. 2007. Reinheit. A. Mesopotamien, *Reallexikon der Assyriologie und Vorderasiatischen Archäologie* 11: 295–299.

Salonen, A. 1964. Die Öfen der Alten Mesopotamier, *Baghdader Mitteilungen* 3: 100–124.

Salonen, A. 1966. *Die Hausgeräte der alten Mesopotamier*. Suomalaisen Tiedeakatemian toimituksia, B. Helsinki: Suomalainen Tiedeakat.

Salzmann, A. 1875. *Nécropole de Camiros: Journal des fouilles exécutées dans cette nécropole pendant les années 1858 à 1865*. Paris: Detaille.

Sams, G. 1994. *The Gordion excavations, 1950-1973: Final reports: The Early Phrygian Pottery*. University Museum Monograph, 79. Philadelphia, PA: University of Pennsylvania, Museum of Archaeology and Anthropology/University of Pennsylvania Press.

Sayre, E. V. 1963. The intentional use of antimony and manganese in ancient glasses, in F. R. Matson and G. E. Rindone (eds), *Advances in Glass Technology, II*: 263–282. New York: Plenum.

Sayre, E. V. and Smith, R. W. 1961. Compositional categories of ancient glass, *Science* 133: 1824–1826.

Schaeffer, H. A. 2014. *Werkstoff Glas: Alter Werkstoff mit groer Zukunft*. Berlin: Springer.

de Schauensee, M. 2001. A note on three glass plaques from Hasanlu, *Iraq* 63: 99–106.

de Schauensee, M. (ed.) 2011. *Peoples and crafts in period IVB at Hasanlu, Iran.* Hasanlu Special Studies, 4. Philadelphia, PA: University of Pennsylvania, Museum of Archaeology and Anthropology/ University of Pennsylvania Press.

Schibille, N. 2011. Late Byzantine Mineral Soda High Alumina Glasses from Asia Minor: A New Primary Glass Production Group, *PLoS ONE*, 6/4. Available at: e18970.doi://10.1371/journal.pone.0018970 (accessed: 01.05.2018).

Schlanger, N. 2005. The chaîne opératoire, in C. Renfrew and P. G. Bahn (eds), *Archaeology: The key concepts.* Routledge Key Guides. London: Routledge.

Schlick-Nolte, B. 1999. Ägyptische Fayence und Ägyptisch Blau im Alten Ägypten, in R. Busz (ed.), *Türkis und Azur: Quarzkeramik in Orient und Okzident ; [anläßlich der Sonderausstellung vom 18. Juli bis 3. Oktober 1999 im Ballhaus am Schloß Wilhelmshöhe und in Schloß Wilhelmsthal, Souterrain]* 12–51. Wolfratshausen: Ed. Minerva.

Schlick-Nolte, B. (ed.) 1968. *Die Glasgefäße im alten Ägypten.* Münchener Universitätsschriften : Münchner ägyptologische Studien, 14. Berlin: Hessling.

Schlick-Nolte, B. and Werthmann, R. 2003. Glass vessels from the burial of Nesikhons, *Journal of Glass Studies* 45: 11–34.

Schmidt, E. F. (ed.) 1957. *Persepolis.* The University of Chicago Oriental Institute Publications. Chicago, IL: University of Chicago Press.

Schmidt, K. (forthcoming) The Use and Significance of Glass in the Context of Mesopotamian Composite Statues, in *Innovations in the Technologies of Glass* (ed.). Berlin: Edition Topoi.

Schmitt, A. W. 2012. *Die Jüngeren Ischtar-Tempel und der Nabû-Tempel in Assur: Architektur, Stratigraphie und Funde ; mit interaktiven Architekturplänen und Fotos auf CD-ROM.* Wissenschaftliche Veröffentlichungen der Deutschen Orient-Gesellschaft, 137. Wiesbaden: Harrassowitz Verlag.

Schuler, F. 1962. Ancient Glassmaking Techniques. The Egyptian Core Vessel Process, *Archaeology* 15: 32–37.

Schuster-Brandis, A. 2008. *Steine als Schutz- und Heilmittel: Untersuchung zu ihrer Verwendung in der Beschwörungskunst Mesopotamiens im 1. Jt. v. Chr.* Alter Orient und Altes Testament, Bd. 46. Münster: Ugarit-Verlag.

Schütte-Maischatz, A. 2011. *Die Phiale: Zur zeichenhaften Funktion eines Gefäßtyps.* Wissenschaftliche Schriften der WWU Münster / 10, 9. Münster: Monsenstein und Vannerdat.

Schweizer, F. 2003. *Glas des 2. Jahrtausends v. Chr. im Ostmittelmeerraum.* Remshalden: Greiner.

Schwemer, D. 1998. *Akkadische Rituale aus Ḫattuša: Die Sammeltafel K Bo XXXVI 29 und verwandte Fragmente.* Texte der Hethiter, Heft 23. Heidelberg: C. Winter.

Searight, A., Reade, J. and Finkel, I. L. 2008. *Assyrian stone vessels and related material in the British Museum.* Oxford: Oxbow Books.

Seefried Brouillet, M. 1982. *Les pendentifs en verre sur noyau des pays de la Méditerranée antique.* Collection de l'Ecole Française de Rome, 57. Rome: École Française.

Seidl, U. 1983. Kultbild. B. Archäologisch, *Reallexikon der Assyriologie und Vorderasiatischen Archäologie*: 314–319.

Seidl, U. 1989. *Die babylonischen Kudurru-Reliefs: Symbole mesopotamischer Gottheiten.* Orbis biblicus et orientalis, 87. Freiburg: Univ.-Verl.

Shortland, A. J. (ed.) 2001. *The social context of technological change in Egypt and the Near East, 1650-1550 BC.* Oxford: Oxbow Books.

Shortland, A. J. 2002. The use and origin of antimonate colorants in early Egyptian glass, *Archaeometry* 44: 517–530.

Shortland, A. J. 2006. Application of lead isotope analysis to a wide range of Late Bronze Age Egyptian materials, *Archaeometry* 4: 657–669.

Shortland, A. J. 2008. Cuneiform Glass Texts. A Question of Meaning?, in M. Martinón-Torres and T. Rehren (eds), *Archaeology, history and science*: 61 – 75. London: Publications of the Institute of Archaeology, University College London.

Shortland, A. J. 2009. The implications of lead isotope analysis for the source of pigments in Late Bronze Age Egyptian vitreous materials, in P. Degryse, J. Henderson and G. Hodgins (eds), *Isotopes in vitreous materials*: 99–111. Studies in Archaeological Sciences, 1. Leuven: Leuven University Press.

Shortland, A. J. 2012. *Lapis Lazuli from the Kiln: Glass and glassmaking in the late Bronze Age.* Studies in Archaeological Sciences, 2. Leuven: Leuven University Press.

Shortland, A. J. *et al.* 2006. Natron as a flux in the early vitreous materials industry: sources, beginnings and reasons for decline, *Journal of Archaeological Science* 33: 521–530.

Shortland, A. J. *et al.* 2017. The analysis of late Bronze Age glass from Nuzi and the question of the origin of glass-making, *Archaeometry* 59: 1–20.

Shortland, A. J. and Eremin, K. 2006. The analysis of second millennium BC glass from Egypt and Mesopotamia, part 1: new WDS analyses, *Archaeometry* 48: 581–605.

Shortland, A. J., Freestone, I. and Rehren, T. (eds) 2009. *From mine to microscope: Advances in the study of ancient technology.* Oxford: Oxbow Books.

Shortland, A. J., Rogers, N. and Eremin, K. 2007. Trace element discriminants between Egyptian and

Mesopotamian Late Bronze Age glasses, *Journal of Archaeological Science* 34: 781–789.

Shortland, A. J. and Tite, M. S. 2000. Raw materials of glass from Amarna and implications of the origins of Egyptian glass, *Archaeometry* 42: 141–152.

Sievertsen, U. 2012. Die eisenzeitliche Keramik, in A. Baghdo, W. Orthmann, M. Novák and L. Martin (eds), *Vorbericht über die dritte bis fünfte syrisch-deutsche Grabungskampagne auf dem Tell Halaf*: 139–184. Ausgrabungen auf dem Tell Halaf in Nordost-Syrien, 2. Wiesbaden: Harrassowitz Verlag.

Sillar, B. and Tite, M. S. 2000. The Challenge of 'Technological Choices' for Materials Science Approaches in Archaeology, *Archaeometry* 42: 2–19.

Spaer, M. 2003. Reconsidering the Iron Age Glass Inlays Found in Association with Carved Ivories, in: *Annales du 16e Congrès de l'Association Internationale pour l'Histoire du Verre, 2003.* Nottingham: 28–31.

Spaer, M. 2001. *Ancient glass in the Israel Museum: Beads and other small objects.* Catalogue/Israel Museum, 447. Jerusalem: Israel Museum.

Spycket, A. 1968. *Les statues de culte dans les textes mesopotamiens des origines a la Ire dynastie de Babylone.* Revue biblique / Cahiers, 9. Paris: Gabalda.

Squitieri, A. 2017. *Stone vessels in the Near East during the Iron Age and the Persian period (c. 1200–330 BCE).* Archaeopress Ancient Near Eastern Archaeology, 2. Oxford, Archaeopress.

Stapleton, C. P. 2003. *Geochemical analysis of glass and glaze from Hasanlu, Northwestern Iran: Constraints on manufacturing technology.* PhD dissertation, University of Georgia, Athens, GA.

Stapleton, C. P. 2011. Glass and glaze analysis and technology from Hasanlu, Period IVB, in M.d. Schauensee (ed.), *Peoples and crafts in period IVB at Hasanlu, Iran*: 87–102. Hasanlu Special Studies, 4. Philadelphia, PA: University of Pennsylvania, Museum of Archaeology and Anthropology/ University of Pennsylvania Press.

Stapleton, C. P. and Swanson S. E. 2002a. Batch material processing and glassmaking technology of 9th century B.C. Artifacts excavated from the site of Hasanlu, Northwest Iran, in P. B. Vandiver, M. Goodway and J. L. Mass (eds), *Materials issues in art and archaeology VI: Symposium held November 26-30, 2001, Boston, Massachusetts, U.S.A.*: 315–321. Materials Research Society Symposium Proceedings, 712. Warrendale, PA: Materials Research Society.

Stapleton, C. P. and Swanson S. E. 2002b. Chemical analysis of glass artefacts from Iron Age levels at Hasanlu, Northwestern Iran, *Journal of glass technology* (43C): 151–157.

Starr, R. F. 1939. *Nuzi: Report on the excavations at Yorgan Tepa near Kirkuk, Iraq, conducted by Harvard Univ. in conjunction with the American Schools of Oriental Research and the Univ. Museum of Philadelphia 1927-1931. With app. by Ruth Sears Chute and others.* Harvard-Radcliffe Fine Arts Series. Cambridge, MS: Harvard University Press.

Stern, E. M. and Schlick-Nolte, B. (eds) 1994. *Frühes Glas der alten Welt: 1600 v. Chr. - 50 n. Chr. ; Sammlung Ernesto Wolf.* Stuttgart: Hatje.

Strommenger, E. 1970. *Die neuassyrische Rundskulptur.* Abhandlungen der Deutschen Orient-Gesellschaft, 15. Berlin: *Deutsche Orient-Gesellschaft.*

Strommenger, E. 1962. *Fünf Jahrtausende Mesopotamien: Die Kunst von den Anfängen um 5000 v. Chr. bis zu Alexander dem Großen.* München: Hirmer.

Stronach, D. 2000 The Imagery of the Wine Bowl, in P. E. McGovern (ed.), *Ancient wine: The search for the origins of viniculture*: 175–195. Princeton: Princeton University Press.

Suter, C. E. 1992. Die Frau am Fenster in der orientalischen Elfenbeinschnitzkunst des frühen 1. Jahrtausends v. Chr., *Jahrbuch der Staatlichen Kunstsammlungen in Baden-Württemberg.* Berlin: Deutscher Kunstverlag.

Suter, C. E. (ed.) 2005. *Crafts and images in contact: Studies in Eastern Mediterranean art of the first millennium BCE.* Orbis biblicus et orientalis, 210. Fribourg: Academic Press.

Swan, C. M., Rehren Th., Dussubieux, L. and Eger, A. A. 2017. High-Boron and High-Alumina Middle Byzantine (10th–12th century CE) Glass Bracelets: A Western Anatolian Glass Industry, *Archaeometry.* Available at: http://onlinelibrary.wiley.com/wol1/doi/10.1111/arcm.12314/full (accessed: 7.1.2019).

Tadmor, H. 1982 The Aramaization of Assyria: Aspects of Western Impact, in H. J. Nissen and J. Renger (eds), *Mesopotamien und seine Nachbarn: Politische und kulturelle Wechselbeziehungen im alten Vorderasien vom 4. bis 1. Jahrtausend v. Chr.*: 449–470. 2nd edn. Berliner Beitraege zum Vorderen Orient, Bd. 1. Berlin: D. Reimer.

Tait, H. 1979. *The golden age of Venetian glass.* London: British Museum Press.

Tait, H. (ed.) 1991. *Five thousand years of glass.* London: British Museum Press.

Taj-Eddin, Z. 2014. *Egyptian Faience; Ancient Making Methods and Consideration of Technical Challenges in Sculptural Practice.* PhD Dissertation, University of Westminster.

Tappy, R. E. 1992. *The archaeology of Israelite Samaria.* Harvard Semitic Studies, 44. Atlanta, GA: Scholars Press.

Tatton-Brown, V. 1981. Rod-formed Glass Pendants and Beads of the 1st Millennium BC, in D. B. Harden (ed.), *Catalogue of Greek and Roman glass.* London: British Museum Press.

Tatton-Brown, V. 1985. Appendix: Rod-Formed Pendants, in D. Barag (ed.), *Catalogue of Western Asiatic glass in the British Museum*: 115–117. London: British Museum Press.

Taylor, M. and Hill, D. 2005 *Core-formed Glass*, http://www.theglassmakers.co.uk/archiveromanglassmakers/nl7text.htm (accessed: 15.01.2016).

Thareani, Y. *et al.* (eds) 2011. *Tel 'Aroer: The Iron Age II caravan town and the Hellenistic early Roman settlement.* Annual of the Nelson Glueck School of Biblical Archaeology / Hebrew Union College - Jewish Institute of Religion, 8. Jerusalem: Nelson Glueck School of Biblical Archaeology, Hebrew Union College–Jewish Institute of Religion.

Thavaplan, S., Stenger, J. and Snow, C. 2016. Color and meaning in ancient Mesopotamia: The case of Egyptian blue, *Zeitschrift für Assyriologie und Vorderasiatische Archäologie* 106/2: 198–214.

Thimme, J. 1973. *Phönizische Elfenbeine: Möbelverzierungen des 9. Jahrhunderts v. Chr. Eine Auswahl aus den Beständen des Badischen Landesmuseums.* Karlsruhe: Badisches Landesmuseum Karlsruhe.

Thompson, R. C. 1925. *On the chemistry of the ancient Assyrians.* London: Luzac and Co.

Thompson, R. C. 1936. *A dictionary of Assyrian chemistry and geology.* Oxford: Clarendon Press.

Thompson, R. C. and Hutchinson, R. W. 1929. *A century of exploration at Nineveh.* London: Luzac.

Thureau-Dangin, F. (ed.) 1931. *Arslan-Tash.* Paris: Geuthner.

Tite, M. S. and Bimson, M. 1987. Identification of early vitreous materials, in J. Black (ed.), *Recent advances in the conservation and analysis of artefacts*: 81–85. London: Summer Schools Press, the University of London Institute of Archaeology.

Tite, M. S. and Shortland, A. J. (eds) 2008. *Production technology of faience and related early vitreous materials.* Oxford University School of Archaeology monograph, 72. Oxford: Oxford University School of Archaeology.

Tite, M. S., Shortland, A. J. and Maniatis, Y. 2006. The composition of the soda-rich and mixed alkali plant ashes used in the production of glass, *Journal of Archaeological Science* 33.9: 1284–1292.

Toker, A. and Öztürk, J. 1992. *Museum of Anatolian civilizations: Metal vessels.* Ankara: Turkish Republic, Ministry of Culture, General Directorate of Monuments and Museums.

Tufnell, O. 1953. *Lachish III - Tell ed-Duweir. The Iron Age.* Publications, v. 3. London, New York, Toronto: published for the Trustees of the late Sir Henry Wellcome by the Oxford University Press.

Turner, W. E. S. 1955. Glass fragments from Nimrud of the eighth to the sixth century, *Iraq* 17: 57–68.

Turner, W. E. S. 1956. Studies of ancient glasses and glassmaking processes. Part II. The composition, weathering characteristics and historical significance of some Assyrian glasses of the eighth to sixth centuries B. C. from Nimrud. *Journal of the Society of Glass Technology* 38: 445–456.

Turner, W. E. S. and Rooksby, H. P. 1959. A study of the opalising agents in ancient opal glasses throughout three thousand four hundred years. *Glastechnische Berichte* 8: 17–29.

Ussishkin, D. 1985. Level VII and VI at Tel Lachish and the end of the Late Bronze Age in Canaan, in J. N. Tubb (ed.), *Palestine in the bronze and iron ages: Papers in honour of Olga Tufnell*: 213–230. Institute of Archaeology Occasional Publications, 11. Abingdon: Routledge.

Ussishkin, D. 2006. Sennacherib's Campaign to Philistia and Judah, Ekron, Lachish, and Jerusalem, in Y. Amit *et al.* (eds), *Essays on ancient Israel in its Near Eastern context: A tribute to Nadav Na'aman*: 339–385. Winona Lake, IN: Eisenbrauns.

van Ess, M. and Pedde, F. 1992. *Uruk: Kleinfunde 2 : Metall und Asphalt, Farbreste, Fritte/Fayence, Glas...* Ausgrabungen in Uruk-Warka, 7. Mainz am Rhein: von Zabern.

Vandiver, P. B. 2008. Raw materials and fabrication methods used in the production of faience, in M. S. Tite and A. J. Shortland (eds), *Production technology of faience and related early vitreous materials*: 37–56. Oxford University School of Archaeology Monograph, 72. Oxford: Oxford University School of Archaeology.

Vickers, M. J. 1972. An Achaemenid glass bowl in a dated context. *Journal of Glass Studies* 14: 15–16.

Virolleaud, C. 1909. Une formule chimique: K 203, *Babyloniaca* 3: 221–222.

Wäfler, M. 1975. *Nicht-Assyrer neuassyrischer Darstellungen.* Alter Orient und Altes Testament. Bd. 26. Kevelaer: Butzon & Bercker.

Wagner, G. A. (ed.) 2007. *Einführung in die Archäometrie.* Berlin: Springer.

Walker, C. B. F. and Dick, M. 2001. The induction of the cult image in ancient Mesopotamia. The Mesopotamian Mīs Pî ritual. State Archives of Assyria Literary Texts, 1. Helsinki: Neo-Assyrian Text Corpus Project.

Walton, M. S. *et al.* 2012. Analysis of Late Bronze Age glass axes from Nippur - a new cobalt colourant. *Archaeometry* 54: 835–852.

Walton, M. S. *et al.* 2009. Evidence for the trade of Mesopotamian and Egyptian glass to Mycenaean Greece, *Journal of Archaeological Science* 36: 1496–1503.

Wartke, R.-B. (ed.) 1982. *Glas im Altertum: Zur Frühgeschichte und Technologie antiken Glases; Sonderausstellung des Vorderasiatischen Museums.* Berlin: Staatliche Museen.

Wartke, R.-B. 1999. Quarzkeramik in Vorderasien, in R. Busz (ed.), *Türkis und Azur: Quarzkeramik in Orient und Okzident ; [anläßlich der Sonderausstellung vom 18. Juli bis 3. Oktober 1999 im Ballhaus am Schloß Wilhelmshöhe und in Schloß Wilhelmsthal, Souterrain]*: 52–65. Wolfratshausen: Ed. Minerva.

Wartke, R.-B. 2012. High Tech-Keramik des Alten Orients. Experimentell gestützte Überlegungen zur

Herstellung von Mosaikschalen aus Quarzkeramik, in H. Neumann and S. Paulus (eds), *Wissenskultur im Alten Orient: Weltanschauung, Wissenschaften, Techniken, Technologien*: 401–416. 4 Internationales Colloquium der Deutschen Orient-Gesellschaft 20. – 22. Februar, Münster. Wiesbaden: Harrassowitz Verlag.

Wedepohl, K. H. 2003. *Glas in Antike und Mittelalter: Geschichte eines Werkstoffs : mit 29 Tabellen*. Stuttgart: Schweizerbart.

Weippert, H. 1988. *Palästina in vorhellenistischer Zeit*. Handbuch der Archäologie/Vorderasien, 2,1. München: Beck.

Werner, P. 2009. *Der Sîn-Šamaš-Tempel in Assur*. Wissenschaftliche Veröffentlichungen der Deutschen Orient-Gesellschaft, 122. Wiesbaden: Harrossowitz Verlag.

Wetzel, F., Schmidt, E. and Mallwitz, A. 1957 *Das Babylon der Spätzeit*. Deutsche Orient-Gesellschaft: Ausgrabungen der Deutschen Orient-Gesellschaft in Babylon, 8. Berlin: Mann.

Weyl, W. A. 1951. *Coloured glasses*. Monographs on Glass Technology. Sheffield: Society of Glass Technology.

Wicke, D. 2009 'Intermediate Tradition' – dreifach problematisch, in S. M. Cecchini, S. Mazzoni and E. Scigliuzzo (eds), *Syrian and Phoenician ivories of the early first millennium BCE: Chronology, regional styles and iconographic repertories, patterns of inter-regional distribution*: 239–284. Ricerche di archeologia del Vicino Oriente, 3. Pisa: Edizioni ETS.

Wicke, D. 2013. Itti nišē kurAššur amnūšunūti 'Zu den Leuten Assyriens zählte ich sie' Beobachtungen zum kulturellen Austausch am Oberen Tigris in neuassyrischer Zeit, in T. R. Kämmerer (ed.), *Patterns of urban societies*. Alter Orient und Altes Testament, 390/2. Münster: Ugarit-Verl.

Wicke, D. 2008. *Vorderasiatische Pyxiden der Spätbronzezeit und der Früheisenzeit*. Alter Orient und Altes Testament, Bd. 45. Münster: Ugarit-Verlag.

Wight, K. B. 2011. *Molten color: Glassmaking in antiquity*. Los Angeles, CA: J. Paul Getty Museum.

Willcox, G. 1992. Timber and trees: ancient exploitation in the Middle East: evidence from plant remains, in *Trees and timber in Mesopotamia*: 1 – 31. Bulletin on Sumerian Agriculture, 6. Cambridge: Sumerian Agriculture Group.

Winter, I. 1976. Phoenician and North Syrian Ivory Carving in Historical Context: Questions of Style and Distribution, *Iraq* 38: 1–22.

Winter, I. 1981. Is there a South Syrian Style of ivory carving in the Early First Millennium BC? *Iraq* 43: 101–130.

Winter, I. 2010. *On art in the ancient Near East, II: From the Third Millennium BCE* Culture and History of the Ancient Near East, 34/2. Leiden: Brill.

Winter, I. J. 1973. *North Syria in the early first millennium B.C., with special reference to ivory carving*. Ann Arbor, MI: University Microfilms International.

Wittmann, O. 1969. *Art in glass: A guide to the glass collections, The Toledo Museum of Art*. Toledo, Ohio: Toledo Museum of Art.

Woolley, C. L., Mallowan, M. E. L., Mitchell, T. C. 1976. *Ur excavations: The Old Babylonian Periods*. Joint Expedition of the British Museum and of the University Museum, University of Pennsylvania, Philadelphia, to Mesopotamia. London: Published for the Trustees of the two Museums.

Woolley, L. 1965. *Ur excavations, Vol. VIII. The Kassite period and the period of the Assyrian kings* (VIII). Joint Expedition of the British Museum and of the University Museum, University of Pennsylvania, Philadelphia, to Mesopotamia. London: Published for the Trustees of the two Museums.

Woolley, L. and Mallowan, M. E. L. 1962. *The Neo-Babylonian and Persian periods*. Ur Excavations, v. 9. Joint Expedition of the British Museum and of the University Museum, University of Pennsylvania, Philadelphia, to Mesopotamia. London: Published for the Trustees of the two Museums.

Wullen, M. and Marzahn, J. (eds) 2008. *Babylon, Mythos & Wahrheit: [Publikation erscheint anlässlich der Doppelausstellung 'Babylon. Mythos und Wahrheit'; eine Ausstellung der Staatlichen Museen zu Berlin, des Musée du Louvre und der Réunion des Musées Nationaux, Paris, und des British Museum, London. Pergamonmuseum, Museumsinsel Berlin, 26. Juni – 5. Oktober 2008]*. München: Hirmer.

Yadin, Y. 1960. *Hazor II: An account of the 2nd season of excavations*. Jerusalem: Magnes Press, Hebrew University.

Yadin, Y. 1961. *Hazor III-IV: An account of the third and fourth seasons of excavations, 1957-1958: plates*. Jerusalem: Magnes Press, Hebrew University.

Yamada, S. 2000. *The construction of the Assyrian empire: A historical study of the inscriptions of Shalmanesar 3 (859–824 B.C.) relating to his campaigns to the West*. Leiden: Brill.

Yon, M. (ed.) 1981. *Dictionnaire illustré multilingue de la céramique du Proche Orient ancien*. Collection de la Maison de l'Orient Méditerranéen, 10. Paris: Boccard.

Yoshinari, A. 2013 *Kodai garasu: Shikisai no kyōen = Ancient glass: feast of color*. Exhibition Miho Museum, Kyoto: Miho Museum.

Young, R. S. 1957. Gordion 1956: Preliminary Report, *American Journal of Archaeology* 61/4: 319–331.

Young, R. S. 1981. *The Gordion Excavations. Final Reports I.: Three Great Early Tumuli*. University Museum Monograph, 43. Philadelphia, PA: University of Pennsylvania, Museum of Archaeology and Anthropology/University of Pennsylvania Press.

Zachariasen, W. 1932. The atomic arrangement of glass, *Journal of the American Chemical Society*, 54: 3841–3851.

Zayadine, F. 1973. Recent on the Citadel of Amman, *Annual of the Department of Antiquities* 18: 17–36.

Zayadine, F., Humbert J.-B. and Najjar, M. 1976. The 1988 Excavations on the Citadel of Amman: Lower Terrace, Area A, *Annual of the Department of Antiquities* 33: 357–363.

Zimmern, H. 1925. Assyrische chemisch-technische Rezepte,: insbesondere für Herstellung farbiger glasierter Ziegel, in Umschrift und Übersetzung, *Zeitschrift für Assyriologie und Vorderasiatische Archäologie* 36: 177–208.

Zohary, M. 1983. *Pflanzen der Bibel: Vollständiges Handbuch*. Stuttgart: Calwer.

Catalogue

Objects are listed in the catalogue according to archaeological sites, which are arranged alphabetically. The object number consists of the abbreviated name of the site and object number (for instance Nim1 for Nimrud, object 1). Acquired objects are labelled with AM (Art Market) and listed at the end of the catalogue.

The catalogue contains information on the findspot, context, date, museum number, and object number. In addition to measurements in centimetres, a description of each item, in which the properties of the glass itself and the state of overall preservation of the object, is included.

Each object can be attributed to a category ('Type') comprised of the overall glass group – mosaic, cast-and-cut, core-formed, ingot – as well as a subcategory (for instance cast-and-cut/rosette inlays).

The site names relate to the 'commonly' used names, regardless of their modern or ancient designation, in order to find consensus between the different disciplines and facilitate reading. The use of 'Tell' ('Tel', 'Tepe', etc.) is omitted.

The figure reference in the catalogue refers to the published pictures elsewhere. The references for the picture illustrated in the plates section can be found in a separate picture reference. All images in the catalogue are shown at a scale of 1:1, unless otherwise specified. Information on objects studied by the author are listed in the catalogue. With regard to objects not studied by the author, information is taken from references, which are cited. The chronological attribution is based on the date of the find contexts discussed in Chapter 3, and on typological considerations, discussed in Chapter 4. In this regard, if possible, year dates are given, but if this is not possible then archaeological periodisations indicate the approximate range in date (Neo-Assyrian, IA II, etc.). The year dates, of course, only indicate approximate values, and aim to facilitate the comparison of objects.

Tel 'Aroer
Ar1 (Plate 21)
Type: cast-and-cut/cut-and-inlaid vessels
Dimensions: ht.: 4.6/w.: 0.3–0.4
Description: transparent; colourless; yellowish corroded; heavy pitting; strongly iridescent; convex fragment of a wall sherd; incised decorative band with diagonal pattern
Context: Caravansary; Area A; Locus 40; Phase A2; Stratum III; northern section of W 1006
Date: 8th century
Museum Number/Object Number/Field Number: F/47/1
Reference: Barag 2011: 259
Figure Reference: Barag 2011: 468 pl. VIII; 48, 2

'Atlit
At1 (Plate 13)
Type: cast-and-cut/jars and alabastra
Dimensions: ht.: 5.5/w.: 4.1
Description: transparent; greenish tinge; heavily corroded; round bottom
Context: south-east cemetery; Tomb L 21 B, e-iv
Date: *c.* 6th Century
Location: Rockefeller-Museum, Jerusalem
Museum Number/Object Number/Field Number: 32.653
Reference: Saldern 1970: 227, no. 54
Figure Reference: Saldern 1970: fig. 49

Amman
A1 (Plate 23)
Type: cast-and-cut/cut-and-inlaid vessels
Dimensions: ht.: 6.7/w.: 0.4–0.6/rim: 11
Description: half of a hemispherical glass bowl corroded to green, formerly probably blue; wall with surrounding recessed bands, these inlaid with rosette (group 3) and rectangular glass inlays
Context: destruction layer of the Ammonite palace, citadel Amman
Date: *terminus ante quem* 700
Reference: Humbert and Zayadine 1992: 257, 263; O´Hea 2011: 161, 162
Figure Reference: Humbert and Zayadine 1992: pl. 14b; O´Hea 2011: 161, 162

Arslan Taş
AT1
Type: cast-and-cut/rosette inlays (group 3) with frame
Dimensions: ht.: 1.5x1.7/w.: 0.3
Description: opaque; formerly red, frame corroded to green; irregular; rectangular; inserted in the middle with rosette (type 3)
Context: 'Bâtiment aux ivories'; room 14; floor
Date: 8th century
Location: Musée du Louvre, Paris
Museum Number/Object Number/Field Number: AO 11499a
Reference: Thureau-Dangin 1931: 138

Figure Reference: Curtis 1999: 63, fig. 9 a–b, d–e; Thureau-Dangin 1931: pl. XLVII, 116; Caubet and Bouquillon 2007: 49, no. 181, 189

AT2 (Plate 32)
Type: cast-and-cut/rosette inlays (group 3) with frame and bronze frame (QC 1)
Dimensions: ht.: 2.1 x 2.1/w.: 0.5
Description: opaque; formerly red, frame corroded to green; irregular; rectangular; inserted in the middle with rosette (type 3)
Context: 'Bâtiment aux ivories'; room 14; floor
Date: 8th century
Location: Musée du Louvre, Paris
Museum Number/Object Number/Field Number: AO 11499b
Reference: Caubet and Bouquillon 2007: 189
Figure Reference: Caubet and Bouquillon 2007: 49, no. 181, 189; Curtis 1999: 63, fig. 9c

AT3 (Plate 32)
Type: cast-and-cut/rosette inlays (group 3) with frame and bronze frame
Dimensions: ht.: 2.2/w.: 2.3
Description: opaque; formerly red, frame corroded to green; irregular; rectangular; inserted in the middle with rosette (type 3)
Context: 'Bâtiment aux ivories'; room 14; floor
Date: 8th century
Reference: Thureau-Dangin 1931: 138
Figure Reference: Thureau-Dangin 1931: pl. XLVII, 113

AT4 (Plate 32)
Type: cast-and-cut/rosette inlays (group 3) with frame and bronze frame
Dimensions: ht.: 2.2/w.: 2.3
Description: opaque; formerly red, frame corroded to green; irregular; rectangular; inserted in the middle with rosette (type 3)
Context: 'Bâtiment aux ivories'; room 14; floor
Date: 8th century
Reference: Thureau-Dangin 1931: 138
Figure Reference: Thureau-Dangin 1931: pl. XLVII, 114

AT5 (Plate 31)
Type: cast-and-cut/rosette inlays (group 3) with frame
Dimensions: ht.: 1.5/w.: 1.7
Description: opaque; formerly red, frame corroded to green; irregular; rectangular; inserted in the middle with rosette (type 3)
Context: 'Bâtiment aux ivories'; room 14; floor
Date: 8th century
Reference: Thureau-Dangin 1931: 138
Figure Reference: Thureau-Dangin 1931: pl. XLVII, 117

Aššur

As1 (Plate 1)
Type: mosaic/bowls
Dimensions: rim: 14/base: 7
Description: opaque; mosaic bowl
Context: burial 'Scherbengrab 311'; Ass 12481; hB7I
Date: Middle-Assyrian
Location: Istanbul Arkeoloji Müzeleri
Museum Number/Object Number/Field Number: Ass 12481b/C 8828
Reference: Haller 1954: 29; Kühne 1971: 424; Miglus 1996: 385; Saldern 1970: 226, no. 47
Figure Reference: Haller 1954: pl. 12 d–e; Saldern 1970: fig. 42, 43

As2 (Plate 2, 3)
Type: mosaic/bowls
Dimensions: ht.: 5.5/w.: 14.4/rim: 13.2/base: 5.5/th.: 0.8
Description: opaque; hexagonal mosaic pieces; dome shaped; rounded walls pulling inwards; circulated incised line directly below the edge; base-ring; wall thickness becomes thicker from the edge to the base; mosaic pieces are slightly drawn towards the rim; surface is regular
Context: dB5IV; close to the 'Kanal d. Tukultī-Ninurta I' (channel of Tukultī-Ninurta I)
Date: Middle-Assyrian
Location: Vorderasiatisches Museum Berlin
Museum Number/Object Number/Field Number: VA Ass 3655a/Ass 7858
Reference: Haevernick 1968: 66, 70; Saldern 1970: 215, no. 7; Wartke 1982: 24–27; Wartke 2012: 401–416
Figure Reference: Busz 1999: 335, no. 142; Haevernick 1968: colour pl. III, 12; Kühne 1971: 424; Saldern 1970: fig. 2H; Saldern 2004: pl. 2, 8; Wartke 1982: fig. 13a; Wartke 2012: 404, fig. 1, 2

As3 (Plate 4)
Type: mosaic/bowls
Dimensions: ht.: 2.7/w.: max. 7.2/th.: 0.6
Description: opaque; red, floral decoration; pattern thus interlocks with one petal each assigning two flowers; decoration is visible on both sides of the vessel; surface faint; round bowl; ring bottom
Context: eD7I; 'Suchgraben, östlich der Lehmziegelkante'
Date: Middle-Assyrian
Location: Vorderasiatisches Museum Berlin
Museum Number/Object Number/Field Number: VA Ass 3655b/Ass 12757b
Reference: Haevernick 1968: 66, 70; Schmitt 2012: 197, no. 1121
Figure Reference: Haevernick 1968: colour pl. III; Saldern 1970: fig. 2H

As4 (Plate 4)
Type: mosaic/bowls
Dimensions: ht.: 1.2/w.: 3.9/th.: 0.8
Description: opaque; beige-yellow; yellow-red-green rafter ribbon or feather decor; pitting at top and bottom; soft, almost floury surface; fragment of a bowl
Context: dC6II; 'auf Palastterrasse d. Tukulti-Ninurta'

Date: Middle-Assyrian
Location: Vorderasiatisches Museum Berlin
Museum Number/Object Number/Field Number: VA 5517/Ass 7995
Reference: Haevernick 1968: 67, 70
Figure Reference: Saldern 1970: fig. 2H

As5
Type: mosaic/bowls
Dimensions: ht.: 2/w.: 1.4/th.: 0.7–1.2
Description: opaque; red, yellow, blue; pitting; fine traces of grinding on top; irregular fragment
Date: Middle-Assyrian
Location: Vorderasiatisches Museum Berlin
Museum Number/Object Number/Field Number: VA 08034/Ass 17273

As6 (Plate 5)
Type: mosaic/inlay
Dimensions: ht.: 2/w.: 2.5/th.: 0.7–0.8
Description: opaque; translucent; dark green with red spots; on surface only two yellowish white petals and parts of the flower interior preserved; slightly iridescent in fracture; strongly eroded on the back; colour strongly faded
Context: i 15; 'Südwall, östliche Wallkrone'
Date: Middle-Assyrian/Neo-Assyrian
Location: Vorderasiatisches Museum Berlin
Museum Number/Object Number/Field Number: VA 5703/Ass 1420
Reference: Haevernick 1968: 64; Saldern 1970: 215, no. 7
Figure Reference: Haevernick 1968: colour pl. I, 4; Saldern 2017: fig. 2F–H

As7 (Plate 5)
Type: mosaic/inlay
Dimensions: ht.: 2.7/w.: 2.6/th.: 0.6
Description: opaque; white base glass with square rosette inlay; petals translucent dark green with red streaks; obverse smooth; reverse irregular
Context: iC4V; 'nördliche Prothyse, Schutt'
Date: Middle-Assyrian/Neo-Assyrian
Location: Vorderasiatisches Museum Berlin
Museum Number/Object Number/Field Number: VA 5702/Ass 1527
Reference: Haevernick 1968: 65; Saldern 1970: 215, no. 7
Figure Reference: Haevernick 1968: colour pl. I, 4

As8
Type: mosaic/inlay
Dimensions: th.: 0.7
Description: opaque; light green, partly red, light blue streaks; floral decoration made of white-light yellow inlays; decoration visible on both sides; deepened hollow on the outside, formerly filled with red and white glass
Date: Middle-Assyrian/Neo-Assyrian
Location: Vorderasiatisches Museum Berlin
Museum Number/Object Number/Field Number: VA Ass 3657/Ass 19950
Reference: Haevernick 1968: 63–64; Saldern 1970: 215, no. 7
Figure Reference: Haevernick 1968: colour pl. I, I

As9 (Plate 44)
Type: core-formed/ovoid bottles with pointed base
Dimensions: ht.: 16.5/w.: 7
Description: opaque; dark brown; yellow, green, white horizontal lines on body and neck; steep neck; rim bent outwards; wall tapering downwards; pointed base; 'duck-head' handle on shoulder; directly below neck base
Context: burial 961; 'Kompositgrab'; Ass 10708; bE5V
Date (Context): Neo-Assyrian
Date Typology: 8th – early 6th century
Location: Istanbul Arkeoloji Müzeleri
Museum Number/Object Number/Field Number: Inv. No. 12897/12879
Reference: Barag 1970: 155, no. 7; Haller 1954: 88; Kühne 1971: 424; Miglus 1996: 382
Figure Reference: Barag 1970: fig. 43; Haller 1954: fig. 116, pl. 19 d; İstanbul Arkeoloji Müzesi 1960: fig. 22, 23, 24

As10 (Plate 53)
Type: core-formed/large cylindrical bottles
Description: opaque; dark; light garland pattern; light wavy line on neck; light line around the bottom, light line on handles; short, broad neck; almost horizontal shoulders; cylindrical body, slightly widening downwards; thickened rim, rounded; two 'duck-head' handles on shoulder; convex bottom
Context: Burial 70; Erdgrab; Ass 12191; eA10I; Innenkante Binnenwall
Date: late 8th – late 7th century
Reference: Barag 1970: 155, no. 8; Haller 1954: 15; Hauser 2012: 145; Kühne 1971: 424
Figure Reference: Barag 1970: fig. 44

As11 (Plate 67)
Type: raw glass fragment
Dimensions: ht.: 3.1/w.: 5.2/th.: 1.2–1.4
Description: opaque; light blue; beige-brown corrosion layer; in some places strong pitting; in some places round bubbles; glass surface smooth, shiny; objects consist of three glued fragments; one side is straight, which probably represents the backside, other side irregular, probably surface
Context: fC6III; 2.50 m below the surface of the tell
Location: Vorderasiatisches Museum Berlin
Museum Number/Object Number/Field Number: VA Ass 4726/Ass 20951 a-e
Reference: Werner 2009: 32, no. 116

As12 (Plate 67)
Type: raw glass fragment
Dimensions: ht.: 4.1/w.: 5.6
Description: opaque; white, light green streaks in

places; thick beige-brown corrosion layer; slightly glossy, iridescent surface
Location: Vorderasiatisches Museum Berlin
Museum Number/Object Number/Field Number: VA 5825

As13 (Plate 6)
Type: mosaic/tiles
Dimensions: ht.: 5/w.: max. 4.2/th.: 2.5–2.9
Description: opaque; light blue inlaid with yellow stripes with black frame; many round bubbles on surface; pitting; underside irregular, brown corrosion layer
Context: around the Ziggurat and the Ištar Temple
Date: Middle-Assyrian
Location: Vorderasiatisches Museum Berlin
Museum Number/Object Number/Field Number: VA 5162/78

As14 (Plate 6)
Type: mosaic/tiles
Description: opaque; light blue, yellow, black bar decoration forming part of a hexagon; yellow rosette on red background; pitting corrosion; irregular fragment
Context: around the Ziggurat and the Ištar Temple
Date: Middle-Assyrian
Reference: Haevernick 1968: 66, no. 11; Saldern 1970: 215, no. 7
Figure Reference: Haevernick 1968: colour pl. II; Harper 1995: pl. 13; Saldern 1970: fig. 2G; Saldern 2004: pl. 2, 10

As15 (Plate 6)
Type: mosaic/tiles
Description: opaque; light blue, yellow, black staff decoration; rosette inlay, red, yellow; irregular fragment
Context: around the Ziggurat and the Ištar Temple
Date: Middle-Assyrian
Reference: Haevernick 1968: 66, no. 11; Saldern 1970: 215, no. 7
Figure Reference: Haevernick 1968: colour pl. II; Harper 1995: pl. 13; Saldern 1970: fig. 2G; Saldern 2004: pl. 2, 10

As16 (Plate 6)
Type: mosaic/tiles
Dimensions: ht.:7.6/w.: max. 7.1/th.: 2.9
Description: opaque; hexagonal fields, each bordered with thin black stripes, separated by wider yellow stripes, turquoise filled base, central red circle with eight-leaf yellow rosette with black (?) circle; irregular fracture edges; heavily corroded
Context: around the Ziggurat and the Ištar Temple
Date: Middle-Assyrian
Location: Vorderasiatisches Museum Berlin
Museum Number/Object Number/Field Number: VA 5820
Reference: Busz 1999: 339, 340; Haevernick 1968: 66, no. 11; Saldern 1970: 215, no. 7
Figure Reference: Busz 1999: 340, no. 150; Haevernick 1968: colour pl. II; Harper 1995: pl. 13; Saldern 1970: fig. 2G; Saldern 2004: pl. 2, 10

As17 (Plate 6)
Type: mosaic/tiles
Description: opaque; light blue, yellow, black bar decoration; pitting corrosion; irregular fragment
Context: around the Ziggurat and the Ištar Temple
Date: Middle-Assyrian
Reference: Haevernick 1968: 66, no. 11; Saldern 1970: 215, no. 7
Figure Reference: Haevernick 1968: colour pl. II; Harper 1995: pl. 13; Saldern 1970: fig. 2G; Saldern 2004: pl. 2, 10

As18 (Plate 6)
Type: mosaic/tiles
Dimensions: ht.: 7.7/w.: max. 8.2
Description: opaque; two hexagonal fields, each bordered with thin black stripes, separated by wider yellow stripes; turquoise-filled base, in the middle red, dark blue circle with eight-leaved white rosettes; broken edges and underside heavily corroded
Context: around the Ziggurat and the Ištar Temple
Date: Middle-Assyrian
Location: Vorderasiatisches Museum Berlin
Museum Number/Object Number/Field Number: VA 8257
Reference: Busz 1999: 339, 340; Haevernick 1968: 66, no. 11; Saldern 1970: 215, no. 7
Figure Reference: Busz 1999: 340, no. 151; Haevernick 1968: colour pl. II; Saldern 1970: fig. 2G

Babylon

Bab1 (Plate 44)
Type: cast-and-cut/composite attachments and inlays
Dimensions: ht.: 3/w.: 6
Description: opaque; dark blue, to grey corroded; two rows of curls separated by a horizontal line; back uneven
Location: British Museum, London
Museum Number/Object Number/Field Number: BM 122061; 1881,1103.32
Reference: Barag 1985: 76, no. 67
Figure Reference: Barag 1968: pl. 9, 67

Bab2 (Plate 42)
Type: cast-and-cut/composite attachments and inlays
Dimensions: ht.: 4.3/w.: max. 5.2
Description: opaque; dark blue; surface almost completely light grey weathered; convex fragments of a wig; flat, rounded corners on underside; drilled shaft hole; bubbles in various sizes
Location: British Museum, London
Museum Number/Object Number/Field Number: BM 122062; 1881, 1103.33
Reference: Barag 1985: 76, no. 68

Figure Reference: Barag 1985: pl. 9, 68

Bab3 (Plate 17)
Type: cast-and-cut/shallow undecorated and ribbed and petalled bowls
Dimensions: ht.: 4/rim: ca. 15.2/th.: 0.2–0.4
Description: transparent; colourless with greenish tinge; surface milky weathered; flat; rim draws strongly outwards; prominent transition from neck to vessel body; round base
Context: Merkes; burial 109 (Ovalsarg); child's burial; left-sided; bowl was in front of the face and the hands raised in front of it
Date: Neo-Babylonian/Achaemenid
Location: Vorderasiatisches Museum Berlin
Museum Number/Object Number/Field Number: VA Bab 04378/Bab35613
Reference: Fossing 1945: 45, 46; Kühne 1971: 423; Reuther 1968: 210, 211; Saldern 1970: 226, no. 46
Figure Reference: Reuther 1968: 211, fig. 107b; pl. 65, fig. 109a; Saldern 1970: fig. 41

Bab4 (Plate 44)
Type: core-formed/ovoid bottles with pointed base
Dimensions: ht.: 9.3/w.: 3.8
Description: opaque; black-green; white wavy lines; body tapering downwards; neck and loops broken off
Context: burial 119
Date (Context): 7th – early 6th century
Reference: Barag 1970: 160, no. 1; Kühne 1971: 423; Reuther 1968: 220, 221
Figure Reference: Barag 1970: fig. 59; Reuther 1968: pl. 74, 119a; Saldern 2004: pl. 9, 46

Bab5 (Plate 49)
Type: core-formed/small jars
Dimensions: ht.: 6.7/w.: 5.9/base: 3.5
Description: opaque; white; blue thread support on rim; decoration of blue circles with red circle centre 'eyes'; circles distributed over body and neck; long neck; slightly outward drawing rim with blue thread support; prominent transition from neck to body; spherical body; round slightly convex base; 'duck-head' handle with blue thread support
Context: Merkes; Burial 109; child burial
Date: early 7th century
Location: Vorderasiatisches Museum Berlin
Museum Number/Object Number/Field Number: VA 8452
Reference: Barag 1970: 160, no. 3; Kühne 1971: 423; Reuther 1968: 210
Figure Reference: Barag 1970: fig. 60; Jakob-Rost 1992: 135, no. 74; Kühne 1971: 423 fig. 9; Meyer 1965: fig. 146; Reuther 1968: pl. 65, fig. 109b, fig. 107a; Saldern 2004: pl. 9, 46; Wartke 1982: 18, 7

Bab6 (Plate 48)
Type: core-formed/globular bottles
Dimensions: ht.6.7/w.: 7/rim: 2.7/th.: 0.5–0.8
Description: opaque; grey; yellow-white thread decoration; yellow thread decoration on edge and handle; thread decoration on neck drawn to feather decoration, wavy on body; strongly corroded colours; surface smooth outside; slightly glossy; uneven surface inside, dull, brown; straight neck; round edge with thread decoration; prominent shoulders; irregular, oval body; wall thickness irregular; 'duck-head' handle
Context: Merkes; 23l2 +6,00
Date (Typology): late 7th – early 6th century
Location: Vorderasiatisches Museum Berlin
Museum Number/Object Number/Field Number: VA Bab 4205/Bab 45228

Bab7 (Plate 49)
Type: core-formed/globular bottles
Dimensions: ht.: 9.2/w.: 7
Description: opaque; dark blue; white thread decor; thick yellow-orange thread overlay; both threads drawn to feather decor on the neck; only horizontal decoration on the body; strongly wavy surface; 'duck-head' handles in basic colour with yellow thread; white thread on the edge; steep neck, spreading slightly downwards; spherical body; outwardly extending, elongated rounded rim; two 'duck-head' handles below the shoulders; round bottom; slight vertical grooves over the entire body
Context: Merkes; Burial 119
Date: late 7th – early 6th century
Location: Vorderasiatisches Museum Berlin
Museum Number/Object Number/Field Number: VA 8449
Reference: Barag 1970: 160, no. 2; Jakob-Rost 1992: 135, no. 75; Kühne 1971: 423; Reuther 1968: 220. 221; Wartke 1982: 18
Figure Reference: Barag 1970: fig. 60; Klengel-Brandt 1977: 91, no. 39; Kühne 1971: 423 fig. 10; Meyer 1965: fig. 146; Reuther 1968: pl. 74, 119b; Wartke 1982: 18, 7

Bab8
Type: core-formed/vessel fragments
Dimensions: ht.: 6.7/w.: 5.9/base: 3.5
Description: translucent; dark blue; neck fragment
Context: Merkes; Burial 109
Date: late 7th century
Location: Vorderasiatisches Museum Berlin
Reference: Fossing 1945: 45, 46; Reuther 1968: 220, 221
Figure Reference: Reuther 1968: pl. 65c

Bab9 (Plate 63, 64)
Type: ingot/rectangular
Dimensions: ht.: 5.5/w.: 25/th.: 7.5
Description: translucent; dark blue; much pitting in fracture; iridescent corrosion layer in places; sandy-grey top and bottom side
Context: hoard ('Haus des Perlenfabrikanten)
Date: hoard contained objects from the Kassite up to the Parthian period
Location: Vorderasiatisches Museum Berlin

Museum Number/Object Number/Field Number: VA Bab 7612/Bab 6566
Reference: Wullen and Marzahn 2008: 603
Figure Reference: Wullen and Marzahn 2008: 603, fig. 414

Bab10
Type: raw glass fragment
Dimensions: ht.: 2.7
Description: opaque; light blue, green spots in places; beige-brown corrosion layer; slightly iridescent; pitting; surface strongly furrowed, irregular chunks
Location: Vorderasiatisches Museum Berlin
Museum Number/Object Number/Field Number: VA Bab 7616

Bab11 (Plate 65)
Type: ingot/rectangular
Dimensions: ht.: 6.6/w.: 12/th.: 5.8
Description: translucent; dark blue; strongly corroded; white, strongly iridescent corrosion layer; pitting; many round bubbles of different sizes; underside sandy, rough; upper side straight; smooth
Location: Vorderasiatisches Museum Berlin
Museum Number/Object Number/Field Number: VA Bab 7633; VA Bab 7612

Bab12 (Plate 68)
Type: raw glass fragment
Dimensions: ht.: 2.8/w.: 2
Description: opaque; dark green, few yellow spots, red streaks; irregular glass fragment; convex upper side
Location: Vorderasiatisches Museum Berlin
Museum Number/Object Number/Field Number: VA Bab 7622
Reference: Wullen and Marzahn 2008: 603
Figure Reference: Wullen and Marzahn 2008: 603, fig. 414

Bab13
Type: raw glass fragment
Dimensions: ht.: 8.8/w.: 8.4
Description: transparent to translucent; greenish tinge; beige-brown corrosion layer, glass strongly attacked by pitting, corrosion in places; very few small round bubbles; irregular shape
Location: Vorderasiatisches Museum Berlin
Museum Number/Object Number/Field Number: VA Bab 7613
Reference: Wullen and Marzahn 2008: 603
Figure Reference: Wullen and Marzahn 2008: 603, fig. 414

Bab14
Type: ingot/round with crucible
Dimensions: ht.: 15/w.: 13/th.: 2.3–3.4
Description: translucent; dark blue, red, green streaks; blue, strongly blue-green iridescent in corroded areas; various large, round bubbles; glass bars adhered to crucible; crucible slightly thickening towards the centre; outside with circulating trough; glass surface smooth, glossy
Location: Vorderasiatisches Museum Berlin
Museum Number/Object Number/Field Number: VA Bab 7614

Til Barsip
TB1 (Plate 29)
Type: cast-and-cut/rosette inlays (group 2)
Dimensions: ht.: 1/w.: 1/th.: 0.2
Description: opaque; blue; six-leafed rosette with inner circle; square
Context: Area C; building Cl; room I; Stratum II, phase B
Date: *terminus ante quem* 600
Museum Number/Object Number/Field Number: TAH 91 C29/F30610.365
Reference: Bunnens 1997: 449, no. 13
Figure Reference: Bunnens 1997: 450, fig. 17

Beth-Shean
BS1 (Plate 62)
Type: ingot/round
Dimensions: ht.: 2.8/w.: 4.4/rim: *c.* 20
Description: opaque; red core, light green layer; white brittle corrosion layer; underside straight; top side sloping towards the middle
Context: South Temple; Level V; Room 1028; '(…) room was in fact, almost entirely removed in the constructions of Cisterns 10-a and b.'
Date: Iron Age – Hellenistic
Location: Oriental Institute of the University of Chicago
Museum Number/Object Number/Field Number: 29-105-791

BS2 (Plate 55)
Type: core-formed/vessel fragments
Dimensions: th.: 0.5
Description: opaque; grey, light blue, white, yellow feather finish; surface uneven; corrosion layer
Context: South Temple; Level V; Room 1028; '(…) room was in fact, almost entirely removed in the constructions of Cisterns 10-a and b.'
Date: 1150–925 (Iron Age IB/IIA)
Location: Institute of Archaeology (UCL), London
Museum Number/Object Number/Field Number: 29-105-792/25-10-41

BS3 (Plate 3)
Type: mosaic/bead or pendant
Dimensions: ht.: 1.2
Context: surroundings of the Temple of Early Seti I, room 1062
Date: 12th century – 1140 (Iron Age IA)
Location: Israel Museum, Jerusalem
Museum Number/Object Number/Field Number: IAA 36-1679
Figure Reference: Rowe 1940: pl. 33, 46; Spear 2001: 118, fig. 51

Busayra

Bus1 (Plate 46)
Type: core-formed/piriform bottles
Dimensions: ht.: 15
Description: opaque; dark blue; yellow, green, white feather decor; pear-shaped body; pointed bottom; 'duck-head' handle
Context: dwelling, ash layer
Date: Iron Age II/Achaemenid period
Museum Number/Object Number/Field Number: Reg.173
Reference: Bienkowski 2002: 363
Figure Reference: Bienkowski 2002: pl. 10, 23

Carthage

Car1 (Plate 53)
Type: core-formed/large cylindrical bottles
Dimensions: ht.: 15/w.: 5.5
Description: opaque; dark blue; white-glue thread decoration; straight, short neck; round rim; wide, slightly downward widening vessel; round base
Context: burial 27
Date 8th – late 7th century
Reference: Barag 1970: 167, no. 12; Fossing 1940: 37
Figure Reference: Grose 1989: 77, fig. 41

Tell ed-Duleym

Dul1 (Plate 42)
Type: cast-and-cut/composite attachments and inlays
Dimensions: ht.: 3.1/w.: 2.9/th.: 1.6
Description: opaque; turquoise; strongly eroded; three curls; flat back
Location: British Museum, London
Museum Number/Object Number/Field Number: 81-11-3, 1922
Reference: Barag 1985: 76, no. 65
Figure Reference: Barag 1985: pl. 8

Dul2 (Plate 44)
Type: cast-and-cut/composite attachments and inlays
Dimensions: ht.: 2.7/w.: 3/th.: 1.3
Description: opaque; dark blue; small spherical bubbles; convex fragment; cylindrical shaft hole with remains of bronze
Location: British Museum, London
Museum Number/Object Number/Field Number: 81-11-3, 1920
Reference: Barag 1985: 77, no. 69
Figure Reference: Barag 1985: colour pl. C, pl. 9

Dul3
Type: ingot/round
Dimensions: ht.: 3.5/w.: 21.5/rim: 38
Description: opaque; dark red-orange; many small and large bubbles; green weathering layer; flat bottom; straight, vertical sides; flat, uneven upper side
Location: British Museum, London
Museum Number/Object Number/Field Number: BM 136780; 1881, 0830.843
Reference: Barag 1985: 109, no. 168
Figure Reference: Barag 1985: pl. 19

Dul4
Type: ingot/round
Dimensions: ht.: 12/w.: 11/rim: c. 40/th.: 3.6
Description: opaque; dark red-orange; many small and large bubbles; green weathering layer; flat bottom; straight, vertical sides; flat, uneven upper side
Museum Number/Object Number/Field Number: BM 136781; 1881, 0830.844
Reference: Barag 1985: 109, no. 169
Figure Reference: Barag 1985: pl. 20

Dul5
Type: raw glass fragment
Dimensions: ht.: 4.1/w.: 4.5/th.: 3.8
Description: opaque; dark red-orange; irregular fragment
Location: British Museum, London
Museum Number/Object Number/Field Number: BM 136782; 1881, 0830.845
Reference: Barag 1985: 109, no. 170
Figure Reference: Barag 1985: no. 170

Eridu

Er1
Type: raw glass fragment
Dimensions: ht.: 3.1/w.: 1.7/th.: 1.3
Description: translucent; dark blue; irregular fragment
Location: British Museum, London
Museum Number/Object Number/Field Number: 1918-10-12, 496
Reference: Barag 1985: 112, no.
Figure Reference: http://www.britishmuseum.org/research/collection_online/collection_object_details.aspx?objectId=1599589&partId=1&searchText=1856,0908.319&page=1. (accessed: 08.08.2018)

Er2
Type: raw glass fragment
Dimensions: ht.: 1.8/w.: 2.8/th.: 1.5
Description: opaque; red, black streaks; irregular; smooth, glossy surface
Location: British Museum, London
Museum Number/Object Number/Field Number: 1856, 0908.317
Reference: Barag 1985: 112, no. 184
Figure Reference: http://www.britishmuseum.org/research/collection_online/collection_object_details.aspx?objectId=1599581&partId=1&searchText=1856,0908.317&page=1 (accessed: 08.08.2018)

Er3
Type: raw glass fragment
Dimensions: ht.: 4/w.: 3.5/th.: 2.6
Description: opaque; white; irregular fragment

Location: British Museum, London
Museum Number/Object Number/Field Number: 1856, 0908.319
Reference: Barag 1985: 112, no. 185
Figure Reference: http://www.britishmuseum.org/research/collection_online/collection_object_details.aspx?objectId=1599589&partId=1&searchText=ibrahim+-glass&page=1 (accessed: 08.08.2018)

Fortetsa
Fo1 (Plate 16)
Type: cast-and-cut/hemispherical bowls
Dimensions: ht.: 7.5/rim: 13.3/th.: 0.5
Description: transparent; light green; strongly weathered; slightly asymmetrical shape; slightly wider than hemispherical
Context: Burial P
Date: *terminus ante quem* 630
Location: Heraklion Museum, Crete
Museum Number/Object Number/Field Number: 1567
Reference: Barag 1985: 53; Brock 1957: 134, no. 1567; Fossing 1940: 36; Saldern 1970: 225, no. 43; Saldern 2004: 60
Figure Reference: Brock 1957: pl. 112, no. 1567; Fossing 1940: 36, fig. 23; Saldern 1970: fig. 38

Gordion
Gor1 (Plate 18)
Type: cast-and-cut/shallow, undecorated and ribbed and petalled bowls
Dimensions: ht.: 3.5–3.8/rim: 15.2–15.7/th.: 0.3–1
Description: transparent; colourless, light yellow-greenish tinge; pitting; brownish corrosion layer; few round, spherical bubbles unevenly distributed; low skin; slightly outwardly extending, rounded rim, slightly thickened on the inside; flat bottom with central concave depression; 32 radially arranged petals on the outside, these convex, turned outwards
Context: Tumulus P; found in a bronze bowl
Date: 827–803
Location: Museum of Anatolian Civilisations, Ankara
Museum Number/Object Number/Field Number: 4000 G 206, Tumulus P 48
Reference: Jones 2009: 106; Saldern 1959: 25, 26; Saldern 1970: 217, no. 16; Saldern 2004: 59; Young 1957: 325–331
Figure Reference: Jones 2009: 105, fig. 8–3; Young 1981: 81, fig. 18, pl. 15, A. B; Young 1957: pl. 94, fig. 32; Young 1957: pl. 94, fig. 32; Saldern 1959: 22, fig. 1, 25 fig. 2; Saldern 1970: fig. 14–16; Saldern 2004: fig. 11, pl. 53

Gor2
Type: cast-and-cut/shallow, undecorated and ribbed and petalled bowls
Dimensions: ht.: 7.8/w.: 24
Description: transparent; colourless; outside and inside of the fragment plastically decorated
Context: domestic or palatial structure, City Mound
Date: 8th century
Location: Istanbul Arkeoloji Müzeleri
Museum Number/Object Number/Field Number: 7177 G283
Reference: Jones 2009: 108
Figure Reference: Jones 2009: 107, fig. 8–4; http://www.ammrf.org.au/news-and-media/research-enabled/analysing-ancient-glass/(accessed: 08.08.2018)

Hasanlu
Has1 (Plate 5)
Type: mosaic/inlay
Dimensions: ht.:1.5 /w.: 1.4/th.: 0.5
Description: opaque; green marbled background; in it flower, alternating green-yellow; yellow strongly attacked by pitting; smooth surface
Context: Burnt Building II; room 5, 7; CC 31 (4) 46; secondary context, inlaid in an alabaster vessel
Date: 1450–1050 (stratum V-IVB)
Location: University of Pennsylvania, Museum of Archaeology and Anthropology
Museum Number/Object Number/Field Number: 65-31-23_A/HAS 64-127
Reference: de Schauensee 2001: 99–106
Figure Reference: de Schauensee 2001: 101, fig. 2, 102, fig.3

Has2 (Plate 5)
Type: mosaic/inlay
Dimensions: ht.: 1.1/w.: 0.9/th.: 0.5
Description: opaque; white-beige; dark blue-green square; dark red streak in some places; yellow circle in the middle, nine white petals
Context: Burnt Building II; room 5, 7; CC 31 (4) 46; secondary context, inlaid in an alabaster vessel
Date: 1450–1050 (stratum V-IVB)
Location: University of Pennsylvania, Museum of Archaeology and Anthropology
Museum Number/Object Number/Field Number: 65-31-23_B/HAS 64-127
Reference: de Schauensee 2001: 99–106
Figure Reference: de Schauensee 2001: 101, fig. 2, 102, fig.3

Has3 (Plate 5)
Type: mosaic/inlay
Dimensions: ht.: 1.1/w.: 1.4/th.: 0.5
Description: translucent; dark blue, green corroded in places; yellow opaque areas; rounded edges
Context: Burnt Building II; room 5, 7; CC 31 (4) 46; secondary context, inlaid in an alabaster vessel
Date: 1450–1050 (stratum V–IVB)
Location: University of Pennsylvania, Museum of Archaeology and Anthropology
Museum Number/Object Number/Field Number: 65-31-23_C/HAS 64-127
Reference: de Schauensee 2001: 99–106
Figure Reference: de Schauensee 2001: 101, fig. 2, 102, fig.3

Has4 (Plate 57)
Type: core-formed/tubes
Dimensions: ht.: 15/w.: 4.3
Description: opaque; white corroded; light blue in places; light blue-green streaks; surface smooth; surface smooth inside; wrinkled; eight-sided bar
Context: Burnt Building II; room 5, 7, 11 (?); CC 31(4)/277
Date: stratum IVB: 1050–800
Location: University of Pennsylvania, Museum of Archaeology and Anthropology
Museum Number/Object Number/Field Number: 65-31-281/HAS64-678

Has5 (Plate 60)
Type: core-formed/tubes
Dimensions: ht.: 6.9/w.: 1.6
Description: opaque; white corroded, blue, green corroded in places; surface smooth outside; surface wrinkled inside; four-sided tube, these oblique; rounded at ends
Context: Burnt Building II; room 2; CC31(4)/280 Room 2 fill
Date: stratum IVB: 1050–800
Location: University of Pennsylvania, Museum of Archaeology and Anthropology
Museum Number/Object Number/Field Number: 65-31-279/HAS 64-481

Has6 (Plate 59)
Type: core-formed/tubes
Dimensions: ht.: 15.2/w.: 3.1x2.8
Description: opaque; white corroded, light blue in places; heavily corroded; many small cracks on surface; round, oval bubbles in fracture; four lateral rods with hole; sides slightly concave; rod slightly curved
Context: Burnt Building II; room 7a; CC 31 (4)/278
Date: stratum IVB: 1050–800
Location: University of Pennsylvania, Museum of Archaeology and Anthropology
Museum Number/Object Number/Field Number: 65-31-282/HAS64-679
Reference: Stapleton 2003: 33, 108

Has7 (Plate 60)
Type: core-formed/tubes
Dimensions: ht.: 4.4/w.: 2.2
Description: opaque; turquoise, strongly white corroded; surface smooth; surface rough inside; fragment, three smooth sides preserved; bitumen residues
Context: Burnt Building V; room 3; Y33(4)[2]/63 LOT17
Date: stratum IVB: 1050–800
Location: University of Pennsylvania Museum of Archaeology and Anthropology
Museum Number/Object Number/Field Number: 75-29-442/HAS 74-N543

Has8 (Plate 59)
Type: core-formed/tubes
Dimensions: ht.: 3/w.: 3
Description: opaque; white corroded, small blue spot; round bubbles of different sizes; surface smooth outside, surface irregular inside; four-sided rod; rod tapering downwards; smooth, rounded tip; hole widens towards the end
Context: Burnt Building II; room 5; CC30(5)[1]/23
Date: stratum IVB: 1050–800
Location: University of Pennsylvania, Museum of Archaeology and Anthropology
Museum Number/Object Number/Field Number: 61-5-54/HAS 60-299
Reference: Stapleton 2003: 33, 108

Has9 (Plate 58)
Type: core-formed/tubes
Dimensions: ht.: 5.6
Description: opaque; white corroded; round bubbles; four-sided rod, irregular, central hole
Context: between the southern end of Burnt Building II and fortification wall; CC32(4)/20\ CC31(4)/20
Date: stratum IVB: 1050–800
Location: University of Pennsylvania, Museum of Archaeology and Anthropology
Museum Number/Object Number/Field Number: 65-31-280/HAS 64-84
Reference: Stapleton 2003: 33, 108

Has10 (Plate 57)
Type: core-formed/tubes
Dimensions: ht.: 7.1
Description: opaque; white, light blue in places; iridescent; many round, partly oval bubbles; surface smooth outside; surface wrinkled inside; four-sided; slightly tapering towards the bottom; hole strongly widened at the bottom
Context: Burnt Building II; room 7a; CC31 (4)[B2]/279
Date: stratum IVB: 1050–800
Location: University of Pennsylvania, Museum of Archaeology and Anthropology
Museum Number/Object Number/Field Number: 65-31-283/HAS 64-680
Reference: Stapleton 2003: 33, 82, 108

Has11 (Plate 58)
Type: core-formed/tubes
Dimensions: ht.: 4.1/w.: 3.3 x 2.9
Description: opaque; white corroded, light blue in places; black traces on the outside; pitting; surface glossy, smooth; four-sided stick; broken off in one place; slightly oval hole
Context: Burnt Building II; room 7a; CC31 (4)[B2]/279
Date: stratum IVB: 1050–800
Location: University of Pennsylvania, Museum of Archaeology and Anthropology
Museum Number/Object Number/Field Number: 45-31-993

Has12 (Plate 19)
Type: cast-and-cut/shallow, undecorated and ribbed

and petalled bowls
Description: 14 fragments of formerly one vessel; translucent; dark blue; dark corrosion layer; ribbed surface
Context: Burnt Building II; room 5; CC31(4)/54
Date: stratum IVB: 1050–800
Location: University of Pennsylvania, Museum of Archaeology and Anthropology
Museum Number/Object Number/Field Number: 65-31-284/HAS 64-135

Has13 (Plate 66)
Type: raw glass fragment
Description: opaque; light blue, white weathering layer; small isolated round bubbles; irregular chunks
Context: Burnt Building IV-V; room 4; W32 (4)/15
Date: stratum IVB: 1050–800
Location: University of Pennsylvania, Museum of Archaeology and Anthropology
Museum Number/Object Number/Field Number: 75-29-783c/HAS 74-S116A

Has14 (Plate 66)
Type: raw glass fragment
Dimensions: th.: 0.3–1
Description: opaque; red core; light green corroded; irregular pieces
Context: Burnt Building IV-V; room 4; W32 (4)/15
Date: stratum IVB: 1050–800
Location: University of Pennsylvania, Museum of Archaeology and Anthropology
Museum Number/Object Number/Field Number: 75-29-783a/HAS 74-S111A

Idalion
Id1 (Plate 13)
Type: cast-and-cut/jars and alabastra
Dimensions: ht.: 9.7/w.: 3.9/base: 3.5/th.: 0.4–1
Description: transparent; colourless, greenish-light blueish cast; slight pitting; many round bubbles; hole drilled regularly wide; prominent shoulder; slightly thickened bottom towards the bottom
Context: possibly near Apollo Amyklaios Temple
Date: *c.* 6th century
Location: British Museum, London
Museum Number/Object Number/Field Number: 73.3-20.182
Reference: Barag 1985: 68, no. 45; Saldern 1970: 226, no. 50
Figure Reference: Barag 1985: fig. 4, pl. 5; Saldern 1970: fig. 46r

Isin
Is1 (Plate 45)
Type: cast-and-cut/composite attachments and inlays
Dimensions: largest attachment: ht.: 2.8/w.: 6.2
Description: opaque/translucent; dark blue; beige brown weathering layer; six irregular fragments; beard beads on front side; straight back side
Context: Gula temple; 79-77N/97-95W
Date: *terminus ante quem* Nebuchadnezzar II (reign 605–562)
Museum Number/Object Number/Field Number: IB1366
Reference: Hrouda 1987: 43
Figure Reference: Hrouda 1987: pl. 19, 32

Is2 (Plate 45)
Type: cast-and-cut/composite attachments and inlays
Dimensions: largest attachment: ht.: 3.2/w.: 2.9
Description: translucent-opaque; dark blue; beige-brown corrosion layer; five fragments; carved, wavy, zigzag decoration; partially perforated
Context: Gula temple; 79-77N/97-95W; Court 2, in asphalt of plaster created by Nebuchadnezzar II
Date: *terminus ante quem* Nebuchadnezzar II (reign 605–562)
Museum Number/Object Number/Field Number: IB1366
Reference: Hrouda 1987: 43
Figure Reference: Hrouda 1987: pl. 19, 32

Tell Jemmeh
TJ1 (Plate 50)
Type: core-formed/globular bottles
Dimensions: ht.: 2.2/w.: 3.5/th.: 0.4
Description: translucent, formerly dark blue, green corroded on the outside; yellow-white feather decor; surface outside is glossy, smooth; surface inside sandy grey
Date (Typology): late 7th – early 6th century
Location: Institute of Archaeology (UCL), London
Museum Number/Object Number/Field Number: EXXXVI.18/11

TJ2 (Plate 50)
Type: core-formed/globular bottles
Description: translucent, formerly dark blue, today green eroded; yellow-white feather decor, underneath white thread support; very irregular wall thickness and vessel surface; surface sandy, reddish inside
Date: late 7th – early 6th century
Location: Institute of Archaeology (UCL), London
Museum Number/Object Number/Field Number: EXXXVI.24/14

Kameiros
Kam1 (Plate 52)
Type: core-formed/large cylindrical bottles
Dimensions: ht.: 15.6/w.: 4.7/rim: 2.6/base: 4.7
Description: opaque; dark blue; yellow thread decoration; this strongly protruding, not marvered; in places golden iridescence; straight, long neck, irregular round edge with thread support; straight, slightly outwardly pulling thick thread at the bottom; round bottom, 'duck-head' handle
Context: burial
Date: 8th–late 7th century

Location: British Museum, London
Museum Number/Object Number/Field Number: 1861, 1024.18
Reference: Barag 1970: 165, no. 1; Fossing 1940: 38, 39; Harden 1981: 56, no. 78
Figure Reference: Barag 1970: fig. 65; Fossing 1940: 37, fig. 24; Grose 1989: 77, fig. 42; Harden 1981: pl. 7, 78

Kam2 (Plate 52)
Type: core-formed/large cylindrical bottles
Dimensions: ht.: 15.6/w.: 4.4/th.: 0.3–0.4
Description: opaque; dark blue; dark yellow thread decoration; heavily weathered; light pitting, some bubbles on surface; surface very strongly grooved, rough, sandy; long, straight neck; prominent shoulder, sloping obliquely downwards; straight wall slightly widening downwards; pointed round bottom; 'duck-head' handle directly under shoulder; wall slightly widening downwards
Date: 8th – late 7th century
Location: British Museum, London
Museum Number/Object Number/Field Number: 1860, 0201.52
Reference: Barag 1970: 165, no. 2; Barag 1985: 70, no. 49; Fossing 1940: 38; Harden 1981: 56, no. 79
Figure Reference: Barag 1985: pl. 7; Harden 1981: pl. 7, 79

Khorsabad
Khor1 (Plate 14)
Type: cast-and-cut/hemispherical bowls
Dimensions: ht.: 3.3/w.: 1.5/th.: 0.1
Description: transparent; colourless; strongly golden corroded; slightly iridescent; thick white corrosion layer on the inside; pitting; surface smooth; round bubbles of various sizes
Context: Residence K, rooms 51/52
Date: 722–705
Location: Oriental Institute Museum, Chicago
Museum Number/Object Number/Field Number: A 17539/DS1180
Khor2 (Plate 39)
Type: cast-and-cut/large monochrome inlays
Dimensions: ht.: 3.7/w.: 6.6/th.: 1.2
Description: opaque; turquoise; pitting; upper side slightly convex, smooth; side edges inclined obliquely inwards; underside irregular, deep, irregular notches; hole 0.5 cm in diameter
Date: 8th century
Location: Musée du Louvre, Paris
Museum Number/Object Number/Field Number: AO 29550a

Khor3 (Plate 40)
Type: cast-and-cut/large monochrome inlays
Dimensions: ht.: 3/w.: 3.5/th.: 1.4
Description: opaque; turquoise; pitting; many small cracks; tops slightly convex, smooth; side edge, sloping inwards
Date: 8th century
Location: Musée du Louvre, Paris
Museum Number/Object Number/Field Number: AO 29550b

Khor4 (Plate 40)
Type: cast-and-cut/large monochrome inlays
Dimensions: ht.: 5.5/w.: 3.1/th.: 0.9
Description: opaque; turquoise; pitting; smooth surface; irregular underside rounded at an angle to the upper side
Date: 8th century
Location: Musée du Louvre, Paris
Museum Number/Object Number/Field Number: AO 29550c/

Khor5 (Plate 41)
Type: cast-and-cut/large monochrome inlays
Dimensions: ht.: 3.2/w.: 1.8/th.: 1.2
Description: opaque; turquoise; fragment; broken on all sides; rough underside
Date: 8th century
Location: Musée du Louvre, Paris
Museum Number/Object Number/Field Number: AO 29550d

Khor6 (Plate 41)
Type: cast-and-cut/large monochrome inlays
Description: opaque; turquoise; two fragments; one hole in the middle (0.3 cm); smooth upper side; rough underside
Date: 8th century
Location: Musée du Louvre, Paris
Museum Number/Object Number/Field Number: AO 29550e

Khor7 (Plate 40)
Type: cast-and-cut/large monochrome inlays
Dimensions: ht.: 4.3/w.: 3.7/th.: 0.8
Description: opaque; turquoise; surface smooth; pitting; underside irregular; surface very glossy; traces of processing on underside; bevelled, rounded edges
Date: 8th century
Location: Musée du Louvre, Paris
Museum Number/Object Number/Field Number: AO 29550g

Khor8 (Plate 41)
Type: cast-and-cut/large monochrome inlays
Dimensions: ht.: 4.8/w.: 3/th.: 1.2
Description: opaque; turquoise, green corroded; black spots on underside, possibly bitumen; small, round bubbles; fragment broken on all sides; upper side smooth, underside irregular, rough, recesses
Date: 8th century
Location: Musée du Louvre, Paris
Museum Number/Object Number/Field Number: AO 29550f

Kiš

Kiš1 (Plate 47)
Type: core-formed/piriform bottles
Dimensions: ht.: 9.5/w.: 4
Description: opaque; grey-green; white feather decor; neck pulling outwards; round body tapering downwards; pointed base
Context: Burial 23, Tell 'W'
Date: 5th century
Location: Iraq Museum, Baghdad
Museum Number/Object Number/Field Number: IM 2277
Reference: Barag 1970: 159; Moorey 1979: 52
Figure Reference: Barag 1970: fig. 57

Kiš2 (Plate 55)
Type: core-formed/ovoid bottles with rounded base
Dimensions: ht.: 9.2/w.: 3.4/rim: 1.8/th.: 0.3–0.5
Description: opaque; yellowish-light brown; white thread decoration on handle, rim; thread decoration forms pointed and round arches; round broken bubbles; smooth surface; long straight neck; straight wall; slightly tapering at the bottom; ovoid body; 'duck-head' handle; pointed round bottom
Context: Burial 54, Tell 'W'
Date: 5th century
Location: The Field Museum, Chicago
Museum Number/Object Number/Field Number: 230904
Reference: Barag 1970: 159; Moorey 1979: 52
Figure Reference: Barag 1970: fig. 58

Megiddo

Meg1 (Plate 7, 8)
Type: cast-and-cut/pallets
Dimensions: ht.: 2.2/w.: 10
Description: translucent; green; iridescent; heavily corroded; smooth top, central, semicircular recess; round, smooth underside
Context: Stratum II, SQ Q-8, Locus 1275 room of a dwelling
Date: 650–600
Location: Oriental Institute Museum, Chicago
Museum Number/Object Number/Field Number: A 18950/2132; M 4167
Reference: Barag 1982: 11
Figure Reference: Barag 1982: 12, fig. 1. 2

Meg2
Type: core-formed/vessel fragments
Dimensions: ht.: c. 2/w.: c. 2
Description: opaque; feather decoration; wall fragment
Context: dwelling; Stratum V; Locus 1636
Context: 1050–920
Museum Number/Object Number/Field Number: M 5190
Reference: Lamon and Shipton 1939: pl. 101, 18
Figure Reference: Lamon and Shipton 1939: pl. 101, 18

Nimrud

Nim1 (Plate43)
Type: cast-and-cut/composite attachments and inlays
Dimensions: ht.: max. 5.1/w.: 5.5/th.: 1.5
Description: dark blue; grey corroded; right side of a beard; decor carved; back slightly concave
Context: Ninurta Temple
Date: late 8th – late 7th century
Location: British Museum, London
Museum Number/Object Number/Field Number: BM 91572/N. 784
Reference: Barag 1985: 75, no. 64
Figure Reference: Barag 1985: pl. 8

Nim2 (Plate43)
Type: cast-and-cut/composite attachments and inlays
Dimensions: ht.: 5.1/w.: 3.6/th.: 1
Description: translucent; dark blue; grey weathering layer; upper part of a beard; convex form with seven diagonal lines, each of which has a curl in its diamonds
Context: Ninurta Temple
Date: *terminus ante quem* 612
Location: British Museum, London
Museum Number/Object Number/Field Number: BM 91574/N. 785
Reference: Barag 1985: 75, 76, no. 63
Figure Reference: Barag 1985: pl. 8

Nim3 (Plate44)
Type: cast-and-cut/composite attachments and inlays
Dimensions: ht.: 3.9/w.: 5/th.: 0.6–1.1
Description: opaque; blue; grey corrosion layer; convex, curved, raised band; carved herringbone pattern; round perforation (0.4–0.6 cm); area under band decorated with carved wavy pattern; upper corner is sharp-edged; reverse is flat
Location: British Museum, London
Museum Number/Object Number/Field Number: BM 90974
Reference: Barag 1985: 77, no. 70
Figure Reference: Barag 1985: fig. 6, pl. 9

Nim4 (Plate44)
Type: cast-and-cut/composite attachments and inlays
Dimensions: ht.: 1.9/w.: 3.4/th.: 0.8
Description: translucent; dark blue; grey corroded; wavy edge; engraved interior drawing
Location: British Museum, London
Museum Number/Object Number/Field Number: N. 792
Reference: Barag 1985: 77, no. 71
Figure Reference: Barag 1985: fig. 6, pl. 9

Nim5 (Plate42)
Type: cast-and-cut/composite attachments and inlays
Dimensions: ht.: 6.4/w.: 4.8/th.: 1
Description: translucent; dark blue; grey weathering; lower part of a beard; 12 plaits divided, six with diagonal lines

Context: Ninurta Temple
Date: *terminus ante quem* 612
Location: British Museum, London
Museum Number/Object Number/Field Number: BM 91573/N. 783
Reference: Barag 1985: 75, no. 62
Figure Reference: Barag 1985: pl. 8

Nim6
Type: cast-and-cut/composite attachments and inlays
Dimensions: ht.: 4.5/base: 1.2
Description: opaque; dark blue; rigid pitting; round inlay which probably sits in quartz-ceramic mass with the help of adhesives; mass sits in black eye-shaped stone which is open on the back
Context: Ninurta Temple
Date: *terminus ante quem* 612
Location: British Museum, London
Museum Number/Object Number/Field Number: BM 118042/N. 762

Nim7 (Plate14)
Type: cast-and-cut/hemispherical bowls
Dimensions: ht.: 5.4–7.2/w.: 10.4/rim: 9.6–10.3/th.: 0.3–0.5
Description: transparent; colourless, light brown weathering layer; pitting; iridescence; two stress cracks on base; many small round bubbles, evenly wedged; unthickened, rounded edge; wall slightly inward; slightly thicker wall towards base; trough on base
Context: NW palace; in the south of room AA
Date: *terminus ante quem* 612
Location: British Museum, London
Museum Number/Object Number/Field Number: BM 91521/N. 821
Reference: Barag 1985: 62, no. 28; Saldern 1966a: 627; Saldern 1970: 219, no. 21
Figure Reference: Barag 1985: fig. 3, 28; Grose 1989: 75, fig. 38

Nim8 (Plate15)
Type: cast-and-cut/hemispherical bowls
Dimensions: ht.: 6.2–7.4/rim: 12.3/th.: 0.3–0.8
Description: transparent; colourless, slightly greenish; many different bubbles, thick corrosion layer; edge strongly effected by corrosion; dull, iridescent surface; strong pitting in places; unthickened, rounded edge; round bottom; wall thicknesses slightly thicker towards the centre
Context: NW Palace; south in room AA
Date: *terminus ante quem* 612
Location: British Museum, London
Museum Number/Object Number/Field Number: BM 91534/N. 820
Reference: Barag 1985: 63, no. 29; Saldern 1966a: 627; Saldern 1970: 218, no. 20
Figure Reference: Barag 1985: fig. 3, no. 29, pl. C, 29

Nim9 (Plate15)
Type: cast-and-cut/hemispherical bowls
Dimensions: ht.: 8/rim: 14.4/th.: 0.2–0.5
Description: transparent; colourless, slightly greenish tinge; many small, round bubbles; golden iridescence on the outside; light pitting; beige-brown corrosion layer; smooth surface; rounded rim; very thin wall in places; slightly irregular base
Context: NW Palace; south in room AA
Date: *terminus ante quem* 612
Location: British Museum, London
Museum Number/Object Number/Field Number: BM 91523/N. 818
Reference: Barag 1985: 63, no. 30; Saldern 1959: 28; Saldern 1966a: 627; Saldern 1970: 218, no. 19
Figure Reference: Barag 1985: pl. 4; Saldern 1966a: 627, no. 586; Saldern 1959: 28, fig. 4; Saldern 1970: fig. 19

Nim10 (Plate 16)
Type: cast-and-cut/hemispherical bowls
Dimensions: ht.: 9.7/rim: 13.8–14.7/th.: 0.1–0.4
Description: transparent; colourless, greenish; round bubbles; brownish weathering layer; pitting, iridescent surface; slightly larger than hemispherical; wall thickness decreasing from edge to base
Context: NW '; south in room AA
Date: *terminus ante quem* 612
Location: British Museum, London
Museum Number/Object Number/Field Number: BM 91525/N. 818
Reference: Barag 1985: 63, no. 31
Figure Reference: Barag 1985: pl. 4

Nim11 (Plate 14)
Type: cast-and-cut/hemispherical bowls
Dimensions: ht.: 6/w.: 2.8 /th.: 0.2–0.4
Description: translucent; dark violet; many elongated bubbles; grey weathering layer; pitting; fragment of the lower area
Context: Fort Shalmaneser; room SW37
Date: *terminus ante quem* 612
Location: British Museum, London
Museum Number/Object Number/Field Number: BM 134897; 1966, 1217.5/ND 10249
Reference: Barag 1985: 63, no. 36
Figure Reference: Barag 1985: pl. 4

Nim12 (Plate 14)
Type: cast-and-cut/hemispherical bowls
Dimensions: ht.: 5.5/w.: 7.3/th.: 0.3
Description: transparent; colourless; strong pitting; iridescence; many small round bubbles distributed over the whole vessel; straight, slightly rounded edge; different wall thickness
Context: Fort Shalmaneser; room SW37
Date: *terminus ante quem* 612
Location: British Museum, London
Museum Number/Object Number/Field Number: BM 134893/ND 10249

Reference: Barag 1985: 63, no. 32
Figure Reference: Barag 1985: fig. 3

Nim13 (Plate 14)
Type: cast-and-cut/hemispherical bowls
Dimensions: th.: 0.3–0.6
Description: translucent; turquoise; very thin weathering layer
Date: *terminus ante quem* 612
Location: British Museum, London
Museum Number/Object Number/Field Number: N. 791
Reference: Barag 1985: 63, no. 35
Figure Reference: Barag 1985: fig. 3

Nim14 (Plate 17)
Type: cast-and-cut/shallow, undecorated and ribbed and petalled bowls
Dimensions: ht.: ca. 4.4/rim: ca. 15/th.: 0.2–0.4
Description: transparent; colourless, slightly greenish tinge; pinhole-like and small bubbles; pitting; iridescence; slightly outwardly inclined wall; straight, tapering edge; flat bottom, sharp upturn at transition from bottom to wall
Context: Fort Shalmaneser; room SW37
Date: *terminus ante quem* 612
Location: British Museum, London
Museum Number/Object Number/Field Number: BM 134894; 1966, 1217.2/ND 10249
Reference: Barag 1985: 64, no. 37; Saldern 1970: 210, no. 23
Figure Reference: Barag 1985: fig. 3, pl. 4; Saldern 1966a: 628, no. 587

Nim15 (Plate 18)
Type: cast-and-cut/shallow undecorated and ribbed and petalled bowls
Dimensions: ht.: 3.5/th.: 0.3–0.4
Description: translucent; colourless, greenish cast; corroded; iridescence; irregular, vertical ribs at a distance of 0.5 cm
Context: Fort Shalmaneser; room SW37
Date: *terminus ante quem* 612
Location: Iraq Museum, Baghdad
Museum Number/Object Number/Field Number: ND 10250
Reference: Saldern 1966a: 631; Saldern 1970: 222, no. 33
Figure Reference: Saldern 1970: fig. 29

Nim16 (Plate 18)
Type: cast-and-cut/shallow undecorated and ribbed and petalled bowls
Dimensions: th.: 0.2–0.3
Description: transparent/translucent; colourless, slight greenish tinge; corroded; iridescence; probably bottom fragment; irregular ribs
Context: Fort Shalmaneser; room SW37
Date: *terminus ante quem* 612
Location: Iraq Museum, Baghdad
Reference: Saldern 1966a: 631; Saldern 1970: 222, no. 33
Figure Reference: Saldern 1970: fig. 29

Nim17 (Plate 18)
Type: cast-and-cut/shallow, undecorated and ribbed and petalled bowls
Dimensions: ht.: 4.7/th.: 0.4–0.6
Description: transparent; colourless; pin-shaped and spherical bubbles; pitting; iridescence; steep, round wall; rounded; fluted vessel body; of radial ribs
Date: 9th century – 614
Location: British Museum, London
Museum Number/Object Number/Field Number: 1948, 114.293
Reference: Barag 1985: 65, no. 39
Figure Reference: Barag 1985: fig. 3, pl. 4

Nim18 (Plate 20)
Type: cast-and-cut/cut-and-inlaid vessels
Dimensions:/rim.: 7.2/th.: 0.3–0.5
Description: translucent; colourless; whitish weathering; iridescence; below the edge a recess into which rosettes are inserted
Context: Fort Shalmaneser; room SE13 (Gate chamber)
Date: *terminus ante quem* 612
Location: British Museum, London
Museum Number/Object Number/Field Number: BM 134899; 1966, 1217.7
Reference: Barag 1985: 65, 66, no. 40; Saldern 1966a: 630; Saldern 1970: 221, no. 30
Figure Reference: Barag 1985: fig. 3, pl. 4; Saldern 1966a: 630, no. 590; Saldern 1970: fig. 26

Nim19
Type: cast-and-cut/cut-and-inlaid vessels
Dimensions: ht.: 3.4
Description: almost transparent; colourless; corrosion layer; sequence of groove, strip, groove; very thin wall
Context: Fort Shalmaneser; room SE13 (Gate chamber)
Date: *terminus ante quem* 612
Location: British Museum, London
Reference: Saldern 1966a: 630; Saldern 1970: 220, no. 28a
Figure Reference: Saldern 1970: fig. 23

Nim20 (Plate 20)
Type: cast-and-cut/cut-and-inlaid vessels
Dimensions: ht.: 4.6/w.: 7/rim: 11/th.: 0.2–0.4
Description: transparent translucent; colourless; dull weathering layer; iridescence; decorated with recesses and raised lasts; rim drawn slightly inward
Context: Fort Shalmaneser; room SW37 (B)
Date: *terminus ante quem* 612
Location: British Museum, London
Museum Number/Object Number/Field Number: ND 10250
Reference: Saldern 1966a: 629; Saldern 1970: 220, no. 27
Figure Reference: Saldern 1966a: 629, no. 588; Saldern 1970: fig. 21, 22

Nim21 (Plate 21)
Type: cast-and-cut/cut-and-inlaid vessels
Dimensions: ht.: 5.6/rim: 12
Description: transparent-translucent; colourless; below the edge, wide low furrows; edges of the fields are bevelled; very thin wall; pitting; iridescence
Context: Fort Shalmaneser; room SW37
Date: *terminus ante quem* 612
Location: British Museum, London
Museum Number/Object Number/Field Number: 10,250/ND 10250
Reference: Saldern 1966a: 631; Saldern 1970: 221, no. 31
Figure Reference: Saldern 1966a: 631, no. 592; Saldern 1970: fig. 27

Nim22 (Plate 22)
Type: cast-and-cut/cut-and-inlaid vessels
Dimensions: ht.: 7.9/w.: 4.7/rim: 11/th.: 0.3–0.4
Description: transparent, colourless with greenish tinge; very few small round bubbles in places also oval; slight iridescence; pitting; surface smooth; wall sherd of a steep-walled bowl with almost straight wall; straight, tapering edge, below the edge 1.5 cm-wide band of diamond pattern
Context: Fort Shalmaneser; room SW37
Date: 8th century
Location: British Museum, London
Museum Number/Object Number/Field Number: BM 134895; 1966, 1217.3/ND 10249
Reference: Barag 1985: 65, no. 38; Saldern 1966a: 630–631; Saldern 1970: 222, no. 32
Figure Reference: Barag 1985: fig. 3, pl. 4; Saldern 1966a: 631, no. 591

Nim23 (Plate 22)
Type: cast-and-cut/cut-and-inlaid vessels
Dimensions: ht.: 4.7/w.: 4/th.: 0.3–0.5
Description: transparent; colourless, greenish tinge; elongated and small bubbles; pitting, iridescence; elongated weathering cracks; steep-walled bowl with almost straight wall; straight, tapering edge; below the edge is a 1.4 cm-wide band of diamond pattern
Context: Fort Shalmaneser; room SW37
Date: 8th century
Location: British Museum, London
Museum Number/Object Number/Field Number: BM 134896; 1966, 1217.4
Reference: Barag 1985: 65, no. 38A
Figure Reference: Barag 1985: pl. 4

Nim24 (Plate 24)
Type: cast-and-cut/cut-and-inlaid vessels
Dimensions: ht.: 3/w.: 4.7/rim: 10.4/th.: 0.1
Description: transparent; colourless, greenish stitch; incised motifs
Context: Fort Shalmaneser; room SW37
Date: 9th – 8th century
Location: Iraq Museum, Baghdad
Museum Number/Object Number/Field Number: ND 10250a
Reference: Mallowan 1966: 416; Saldern 1970: 222, no. 34a
Figure Reference: Mallowan 1966: 416, no. 345a; Saldern 1970: fig. 30

Nim25 (Plate 24)
Type: cast-and-cut/cut-and-inlaid vessels
Dimensions: ht.: 3/w.: 1.6/rim: 10.4/th.: 0.1
Description: transparent; colourless, greenish stitch; incised motifs
Context: Fort Shalmaneser; room SW37
Date: 9–8th century
Location: Iraq Museum, Baghdad
Museum Number/Object Number/Field Number: ND 10250b
Reference: Mallowan 1966: 416; Saldern 1970: 222, no. 34b
Figure Reference: Mallowan 1966: 416, no. 345b; Saldern 1970: fig. 31

Nim26 (Plate 24)
Type: cast-and-cut/cut-and-inlaid vessels
Dimensions: ht.: 2.3/w.: 2.1/rim: 10.4/th.: 0.1
Description: transparent; colourless, greenish stitch; incised motifs
Context: Fort Shalmaneser; room SW37
Date: 9th–8th century
Location: Iraq Museum, Baghdad
Museum Number/Object Number/Field Number: ND 10250c
Reference: Mallowan 1966: 416; Saldern 1970: 222, no. 34c
Figure Reference: Mallowan 1966: 416, no. 345c; Saldern 1970: fig. 32

Nim27 (Plate 11)
Type: cast-and-cut/jars and alabastra
Dimensions: ht.: 8.6/w.: 6.2/rim: 3.7/base: 5.7/th.: 0.7–1.7
Description: transparent translucent; colourless, light green stitch; very strong pitting; violet iridescence; round, small bubbles distributed throughout the whole vessel, highest concentration in handles; concentric, deepened circles; inscription with lion carved on one side; short neck inclined outwards; slightly thickened, angular rim; oval vessel body with strongly stepped shoulder; round bottom with base; rectangular handles below the shoulder
Context: NW Palace; north-eastern corner room 1
Date: 721–705
Location: British Museum, London
Museum Number/Object Number/Field Number: BM 90952/N. 2070; E. 12084
Reference: Barag 1985: 60, no. 26; Fossing 1940: 35–36; Saldern 1959: 27–28; Saldern 1966a: 626; Saldern 1970: 218, no. 17
Figure Reference: Barag 1985: fig. 2, 26, pl. B, 26; Fossing 1940: 35, fig. 22; Meissner 1920: 235, fig. 57;

Saldern 1959: 27, fig. 3; Saldern 1966a: 626, fig. 584; Saldern 1970: fig. 17

Nim28 (Plate 27)
Type: cast-and-cut/rosette inlays (group 1)
Dimensions: ht.: 1.6/w.: 1.4
Description: opaque; blue; six-leaf rosette without inner circle; square
Context: Fort Shalmaneser; room S10, S30, SW37
Date: *terminus ante quem* 612
Location: British Museum, London
Museum Number/Object Number/Field Number: BM 140351; 1987, 0131.24/ND 10229
Reference: Brill 1999a: 46, no. 3229, 3230; 48; Brill 2012: 625; Curtis 1999: 68
Figure Reference: Curtis 1999: fig. 6

Nim29 (Plate 27)
Type: cast-and-cut/rosette inlays (group 1)
Dimensions: ht.: 1.6 /w.: 1.4
Description: opaque; blue; six-leaf rosette without inner circle; square
Context: Fort Shalmaneser; room S10, S30, SW37
Date: *terminus ante quem* 612
Location: British Museum, London
Museum Number/Object Number/Field Number: BM 140351; 1987, 0131.24/ND 10229
Reference: Brill 1999a: 46, no. 3229. 3230; 48; Brill 2012: 625; Curtis 1999: 68
Figure Reference: Curtis 1999: fig. 6

Nim30 (Plate 27)
Type: cast-and-cut/rosette inlays (group 1)
Dimensions: ht.: 1.6/w.: 1.4
Description: opaque; blue; six-leaf rosette without inner circle; square
Context: Fort Shalmaneser; room S10, S30, SW37
Date: *terminus ante quem* 612
Location: British Museum, London
Museum Number/Object Number/Field Number: BM 140351; 1987, 0131.24/ND 10229
Reference: Brill 1999a: 46, no. 3229, 3230; 48; Brill 2012: 625; Curtis 1999: 68
Figure Reference: Curtis 1999: fig. 6

Nim31 (Plate 27)
Type: cast-and-cut/rosette inlays (group 1)
Dimensions: ht.: 1.6/w.: 1.4
Description: opaque; blue; six-leaf rosette without inner circle; square
Context: Fort Shalmaneser; room S10, S30, SW37
Date: *terminus ante quem* 612
Location: British Museum, London
Museum Number/Object Number/Field Number: BM 140351; 1987, 0131.24/ND 10229
Reference: Brill 1999a: 46, no. 3229, 3230; 48; Brill 2012: 62; Curtis 1999: 68
Figure Reference: Curtis 1999: fig. 6

Nim32 (Plate 27)
Type: cast-and-cut/rosette inlays (group 1)
Dimensions: ht.: 1.6/w.: 1.4
Description: opaque; blue; six-petalled rosette without inner circle; square
Context: Fort Shalmaneser; room S10, S30, SW37
Date: *terminus ante quem* 612
Location: British Museum, London
Museum Number/Object Number/Field Number: BM 140351; 1987, 0131.24/ND 10229
Reference: Brill 1999a: 46, no. 3229, 3230; 48; Brill 2012: 625; Curtis 1999: 68
Figure Reference: Curtis 1999: fig. 6

Nim33
Type: cast-and-cut/rosette inlays (group 1)
Dimensions: ht.: 1.6/w.: 1.4
Description: opaque; blue; six-petalled rosette without inner circle; square
Context: Fort Shalmaneser; room S10, S30, SW37
Date: *terminus ante quem* 612
Location: British Museum, London
Museum Number/Object Number/Field Number: BM 140351; 1987, 0131.24/ND 10229
Reference: Brill 1999a: 46, no. 3229, 3230; 48; Brill 2012: 625; Curtis 1999: 68
Figure Reference: Curtis 1999: fig. 6

Nim34 (Plate 27)
Type: cast-and-cut/rosette inlays (group 1)
Dimensions: ht.: 1.4/w.: 1.4
Description: opaque; blue; six-petalled rosette without inner circle; square
Context: Fort Shalmaneser; room S10, S30, SW37
Date: *terminus ante quem* 612
Location: British Museum, London
Museum Number/Object Number/Field Number: BM 140351; 1987, 0131.24/ND 10229
Reference: Curtis 1999: 68
Figure Reference: Curtis 1999: 61, fig. 6

Nim35 (Plate 27)
Type: cast-and-cut/rosette inlays (group 1)
Dimensions: ht.: 1.2/w.: 1.3
Description: opaque; blue; six-petalled rosette without inner circle; square
Context: Fort Shalmaneser; room S10, S30, SW37
Date: *terminus ante quem* 612
Location: British Museum, London
Museum Number/Object Number/Field Number: BM 140351; 1987, 0131.24/ND 10229
Reference: Curtis 1999: 68
Figure Reference: Curtis 1999: 61, fig. 6

Nim36 (Plate 27)
Type: cast-and-cut/rosette inlays (group 1)
Dimensions: ht.: 1.5/w.: 1.7
Description: opaque; blue; six-petalled rosette without inner circle; square

Context: Fort Shalmaneser; room S10, S30, SW37
Date: *terminus ante quem* 612
Location: British Museum, London
Museum Number/Object Number/Field Number: BM 140351; 1987, 0131.24/ND 10229
Reference: Curtis 1999: 68
Figure Reference: Curtis 1999: 61, fig. 6

Nim37 (Plate 28)
Type: cast-and-cut/rosette inlays (group 2)
Dimensions: ht.: 1.7/w.: 1.6
Description: opaque; blue; six-petalled rosette without inner circle; square
Context: Fort Shalmaneser; room S10, S30, SW37
Date: *terminus ante quem* 612
Location: British Museum, London
Museum Number/Object Number/Field Number: BM 140351; 1987, 0131.24/ND 10229
Reference: Curtis 1999: 68
Figure Reference: Curtis 1999: 61, fig. 6

Nim38 (Plate 28)
Type: cast-and-cut/rosette inlays (group 2)
Dimensions: ht.: 1.6/w.: 1.5
Description: opaque; blue; six-petalled rosette without inner circle; square
Context: Fort Shalmaneser; room S10, S30, SW37
Date: *terminus ante quem* 612
Location: British Museum, London
Museum Number/Object Number/Field Number: BM 140351; 1987, 0131.24/ND 10229
Reference: Curtis 1999: 68
Figure Reference: Curtis 1999: 61, fig. 6

Nim39 (Plate 28)
Type: cast-and-cut/rosette inlays (group 2)
Dimensions: ht.: 1.6/w.: 1.5
Description: opaque; blue; six-petalled rosette without inner circle; square
Context: Fort Shalmaneser; room S10, S30, SW37
Date: *terminus ante quem* 612
Location: British Museum, London
Museum Number/Object Number/Field Number: BM 140351; 1987, 0131.24/ND 10229
Reference: Curtis 1999: 68
Figure Reference: Curtis 1999: 61, fig. 6

Nim40 (Plate 28)
Type: cast-and-cut/rosette inlays (group 2)
Dimensions: ht.: 1.6/w.: 1.5
Description: opaque; blue; six-petalled rosette without inner circle; square
Context: Fort Shalmaneser; room S10, S30, SW37
Date: *terminus ante quem* 612
Location: British Museum, London
Museum Number/Object Number/Field Number: BM 140351; 1987, 0131.24/ND 10229
Reference: Curtis 1999: 68
Figure Reference: Curtis 1999: 61, fig. 6

Nim41 (Plate 28)
Type: cast-and-cut/rosette inlays (group 2)
Description: opaque; blue; six-petalled rosette without inner circle; square
Context: Fort Shalmaneser; room S10, S30, SW37
Date: *terminus ante quem* 612
Location: British Museum, London
Museum Number/Object Number/Field Number: BM 140351; 1987, 0131.24/ND 10229
Reference: Curtis 1999: 68
Figure Reference: Curtis 1999: 61, fig. 6

Nim42 (Plate 28)
Type: cast-and-cut/rosette inlays (group 2)
Description: opaque; blue; six-petalled rosette without inner circle; square
Context: Fort Shalmaneser; room S10, S30, SW37
Date: *terminus ante quem* 612
Location: British Museum, London
Museum Number/Object Number/Field Number: BM 140351; 1987, 0131.24/ND 10229
Reference: Curtis 1999: 68
Figure Reference: Curtis 1999: 61, fig. 6

Nim43 (Plate 28)
Type: cast-and-cut/rosette inlays (group 2)
Description: opaque; blue; six-petalled rosette without inner circle; square
Context: Fort Shalmaneser; room S10, S30, SW37
Date: *terminus ante quem* 612
Location: British Museum, London
Museum Number/Object Number/Field Number: BM 140351; 1987, 0131.24/ND 10229
Reference: Curtis 1999: 68
Figure Reference: Curtis 1999: 61, fig. 6

Nim44 (Plate 28)
Type: cast-and-cut/rosette inlays (group 2)
Description: opaque; blue; six-petalled rosette without inner circle; square
Context: Fort Shalmaneser; room S10, S30, SW37
Date: *terminus ante quem* 612
Location: British Museum, London
Museum Number/Object Number/Field Number: BM 140351; 1987, 0131.24/ND 10229
Reference: Curtis 1999: 68
Figure Reference: Curtis 1999: 61, fig. 6

Nim45 (Plate 28)
Type: cast-and-cut/rosette inlays (group 2)
Description: opaque; blue; six-petalled rosette without inner circle; square
Context: Fort Shalmaneser; room S10, S30, SW37
Date: *terminus ante quem* 612
Location: British Museum, London
Museum Number/Object Number/Field Number: BM 140351; 1987, 0131.24/ND 10229
Reference: Curtis 1999: 68
Figure Reference: Curtis 1999: 61, fig. 6

Nim46 (Plate 28)
Type: cast-and-cut/rosette inlays (group 2)
Description: opaque; blue; six-petalled rosette without inner circle; square
Context: Fort Shalmaneser; room S10, S30, SW37
Date: *terminus ante quem* 612
Location: British Museum, London
Museum Number/Object Number/Field Number: BM 140351; 1987, 0131.24/ND 10229
Reference: Curtis 1999: 68
Figure Reference: Curtis 1999: 61, fig. 6

Nim47 (Plate 28)
Type: cast-and-cut/rosette inlays (group 2)
Description: opaque; blue; six-petalled rosette without inner circle; square
Context: Fort Shalmaneser; room S10, S30, SW37
Date: *terminus ante quem* 612
Location: British Museum, London
Museum Number/Object Number/Field Number: BM 140351; 1987, 0131.24/ND 10229
Reference: Curtis 1999: 68
Figure Reference: Curtis 1999: 61, fig. 6

Nim48 (Plate 35)
Type: cast-and-cut/small monochrome inlays, double-triangle shaped
Dimensions: ht.: 1.1/w.: 2.1/th.: 0.2
Description: translucent; dark blue; irregular; double-triangle shaped; side edges sloping outwards
Context: Fort Shalmaneser; room S10
Date: Date: *terminus ante quem* 612
Location: British Museum, London
Museum Number/Object Number/Field Number: 1987-1-31; 140359/ND 8081

Nim49 (Plate 29)
Type: cast-and-cut/rosette inlays (group 2)
Dimensions: ht.: 3.2/th.: 0.5
Description: opaque; blue; six-petalled rosette without inner circle; square
Date: Neo-Assyrian
Location: British Museum, London
Museum Number/Object Number/Field Number: 140346; 1987, 0131.19/ND 10229
Reference: Brill 1999a: 46, no. 3249, 49; Curtis 1999: 58
Figure Reference: Curtis 1999: 61, fig. 6

Nim50 (Plate 29)
Type: cast-and-cut/rosette inlays (group 2)
Dimensions: ht.: 2.8/w.: 2.8/th.: 0.5
Description: opaque; blue; six-petalled rosette without inner circle; square
Date: Neo-Assyrian
Location: British Museum, London
Museum Number/Object Number/Field Number: 140346; 1987, 0131.19/ND 10229
Reference: Brill 1999a: 46, no. 3249, 49; Curtis 1999: 58
Figure Reference: Curtis 1999: 61, fig. 6

Nim51 (Plate 28)
Type: cast-and-cut/rosette inlays (group 2)
Dimensions: ht.: 0.5/w.: 0.5
Description: opaque; blue; six-petalled rosette without inner circle; square
Date: Neo-Assyrian
Location: British Museum, London
Museum Number/Object Number/Field Number: 140346; 1987, 0131.19/ND 10229
Reference: Brill 1999a: 46, no. 3249, 49; Curtis 1999: 58
Figure Reference: Curtis 1999: 61, fig. 6

Nim52 (Plate 27)
Type: cast-and-cut/rosette inlays (group 1)
Dimensions: ht.: 1.4/w.: 1.6/th.: 0.2
Description: opaque; blue; six-petalled rosette without inner circle; square
Context: Fort Shalmaneser; room T20
Date: mid 9th – mid 8th century
Location: British Museum, London
Reference: Curtis *et al.* 1993: 15, 16; Curtis 1999: 59
Figure Reference: Curtis et al. 1993: 14, fig. 14, 5; Curtis 1999: fig. 1, 2

Nim53 (Plate 27)
Type: cast-and-cut/rosette inlays (group 1)
Dimensions: ht.: 1.8/w.: 1.3/th.: 0.2
Description: opaque; blue; six-petalled rosette without inner circle; square
Date: Neo-Assyrian
Location: British Museum, London
Museum Number/Object Number/Field Number: BM 140356; 1987, 0131.29/ND 6406
Reference: Curtis 1999: 68

Nim54 (Plate 27)
Type: cast-and-cut/rosette inlays (group 1)
Dimensions: ht.: 1.8/w.: 1.3/th.: 0.2
Description: opaque; blue; six-petalled rosette without inner circle; square
Date: Neo-Assyrian
Location: British Museum, London
Museum Number/Object Number/Field Number: BM 140356; 1987, 0131.29/ND 6406
Reference: Curtis 1999: 68

Nim55 (Plate 27)
Type: cast-and-cut/rosette inlays (group 1)
Dimensions: ht.: 1.8/w.: 1.3/th.: 0.2
Description: opaque; blue; six-petalled rosette without inner circle; square
Date: Neo-Assyrian
Location: British Museum, London
Museum Number/Object Number/Field Number: BM 140356; 1987, 0131.29/ND 6406
Reference: Curtis 1999: 68

Nim56 (Plate 27)
Type: cast-and-cut/rosette inlays (group 1)

Dimensions: ht.: 1.8/w.: 1.3/th.: 0.2
Description: opaque; blue; six-petalled rosette without inner circle; square
Date: Neo-Assyrian
Location: British Museum, London
Museum Number/Object Number/Field Number: BM 140356; 1987, 0131.29/ND 6406
Reference: Curtis 1999: 68

Nim57 (Plate 32)
Type: cast-and-cut/rosette inlays
Dimensions: ht.: 1/w.: 1.2/th.: 0.2
Description: opaque; green; square leaves
Date: Neo-Assyrian
Location: British Museum, London
Museum Number/Object Number/Field Number: BM 140356; 1987, 0131.29/ND 6406

Nim58 (Plate 27)
Type: cast-and-cut/rosette inlays (group 1)
Dimensions: ht.: 1.3/w.: 1.3/th.: 0.1
Description: opaque; blue; six-petalled rosette without inner circle; square
Date: Neo-Assyrian
Location: British Museum, London
Museum Number/Object Number/Field Number: BM 140356; 1987, 0131.29/ND 6406

Nim59 (Plate 27)
Type: cast-and-cut/rosette inlays (group 1)
Dimensions: ht.: 1.5/w.: 1/th.: 0.1
Description: opaque; blue; six-petalled rosette without inner circle; square
Date: Neo-Assyrian
Location: British Museum, London
Museum Number/Object Number/Field Number: BM 140356; 1987, 0131.29/ND 6406

Nim60 (Plate 30)
Type: cast-and-cut/rosette inlays (group 3)
Dimensions: ht.: 0.5/w.: 0.5/th.: 0.2
Description: translucent; dark blue; white filling; square
Date: Neo-Assyrian
Location: British Museum, London
Museum Number/Object Number/Field Number: BM 140356; 1987, 0131.29/ND 6406

Nim61 (Plate 29)
Type: cast-and-cut/rosette inlays (group 2)
Dimensions: ht.: 1.7/w.: 1.8/th.: 0.3
Description: opaque; blue; six-petalled rosette without inner circle; square
Context: Fort Shalmaneser; room SW37; filling
Date: *terminus ante quem* 612
Location: Metropolitan Museum of Art, New York
Museum Number/Object Number/Field Number: 62.269.15b/ND 10229
Reference: Curtis 1999: 68
Figure Reference: Curtis 1999: fig. 5

Nim62 (Plate 27)
Type: cast-and-cut/rosette inlays (group 1)
Dimensions: ht.: 1.4/w.: 1.7/th.: 0.2
Description: opaque; blue; six-petalled rosette without inner circle; square
Context: Fort Shalmaneser; room SW37; filling
Date: *terminus ante quem* 612
Location: Metropolitan Museum of Art, New York
Museum Number/Object Number/Field Number: 62.269.15c/ND 10229
Reference: Curtis 1999: 68
Figure Reference: Curtis 1999: fig. 5

Nim63
Type: cast-and-cut/rosette inlays (group 1)
Dimensions: ht.: 1.4/w.: 1.7/th.: 0.2
Description: opaque; blue; six-petalled rosette without inner circle; square
Context: Fort Shalmaneser; room SW37; filling
Date: *terminus ante quem* 612
Location: Metropolitan Museum of Art, New York
Museum Number/Object Number/Field Number: 62.269.15d/ND 10229
Reference: Curtis 1999: 68
Figure Reference: Curtis 1999: fig. 5

Nim64 (Plate 29)
Type: cast-and-cut/rosette inlays (group 2)
Dimensions: ht.: 1.8/w.: 1.8/th.: 0.2
Description: opaque; blue; six-petalled rosette without inner circle; square
Context: Fort Shalmaneser; room SW37; filling
Date: *terminus ante quem* 612
Location: Metropolitan Museum of Art, New York
Museum Number/Object Number/Field Number: 62.269.15a/ND 10229
Reference: Curtis 1999: 68
Figure Reference: Curtis 1999: fig. 5; http://www.metmuseum.org/art/collection/search/325561?sortBy=Relevance&ft=62.269.15a&pg=1&rpp=20&pos=1 (accessed: 08.08.2018)

Nim65 (Plate 27)
Type: cast-and-cut/rosette inlays (group 1)
Dimensions: ht.: 1.4/w.: 1.6/th.: 0.2
Description: opaque; blue; six-petalled rosette without inner circle; square
Context: Fort Shalmaneser; room T20
Date: mid 9th – mid 8th century
Location: British Museum, London
Reference: Curtis *et al.* 1993: 15, 16; Curtis 1999: 59
Figure Reference: Curtis et al: 1993: 14, fig. 14, 6; Curtis 1999: fig. 1, 2

Nim66 (Plate 30)
Type: cast-and-cut/rosette inlays (group 3)
Dimensions: ht.: 1.3/w.: 0.75/th.: 0.3

Description: translucent; dark blue; white filling; rectangular
Context: Fort Shalmaneser; room S10
Date: *terminus ante quem* 612
Location: British Museum, London
Museum Number/Object Number/Field Number: BM 140359/ND 8081
Reference: Curtis 1999: 68
Figure Reference: Curtis 1999: 61, fig. 7

Nim67 (Plate 30)
Type: cast-and-cut/rosette inlays (group 3)
Dimensions: ht.: 1/w.: 0.7/th.: 0.2–0.3
Description: translucent; dark blue; white filling; rectangular
Context: Fort Shalmaneser; room S10
Date: *terminus ante quem* 612
Location: British Museum, London
Museum Number/Object Number/Field Number: BM 140359/ND 8081
Reference: Curtis 1999: 68
Figure Reference: Curtis 1999: 61, fig. 7

Nim68 (Plate 30)
Type: cast-and-cut/rosette inlays (group 3)
Dimensions: ht.: 1.1/w.: 0.9/th.: 0.2
Description: translucent; dark blue; white filling; rectangular
Context: Fort Shalmaneser; room S10
Date: *terminus ante quem* 612
Location: British Museum, London
Museum Number/Object Number/Field Number: BM 140359/ND 8081
Reference: Curtis 1999: 68
Figure Reference: Curtis 1999: 61, fig. 7

Nim69 (Plate 30)
Type: cast-and-cut/rosette inlays
Dimensions: ht.: 0.9/w.: 1.1/th.: 0.2
Description: opaque; greenish corroded; glossy surface; white rest of insert; top and bottom straight; side edges straight
Context: Fort Shalmaneser; room S10
Date: *terminus ante quem* 612
Location: British Museum, London
Museum Number/Object Number/Field Number: BM 140359/ND 8081
Reference: Curtis 1999: 68
Figure Reference: Curtis 1999: 61, fig. 7

Nim70 (Plate 27)
Type: cast-and-cut/rosette inlays (group 1)
Dimensions: ht.: 1.1/w.: 1.2
Description: opaque; blue; six-petalled rosette without inner circle; square
Date: Neo-Assyrian
Location: Metropolitan Museum of Art, New York
Museum Number/Object Number/Field Number: 58.31.43
Reference: http://www.metmuseum.org/art/collection#!?q=58.31.43&sortBy=Relevance&sortOrder=asc&page=1 (accessed: 08.08.2018)

Nim71 (Plate 27)
Type: cast-and-cut/rosette inlays (group 1)
Dimensions: ht.: 1.4/w.: 2.5
Description: opaque; blue; six-petalled rosette without inner circle; square
Date: Neo-Assyrian
Location: Metropolitan Museum of Art, New York
Museum Number/Object Number/Field Number: 58.31.44
Reference: Saldern 1966a: 632, 633; Saldern 1970: 224, no. 39
Figure Reference: Saldern 1966a: 633, no. 594; Saldern 1970: fig. 36a–e

Nim72 (Plate 29)
Type: cast-and-cut/rosette inlays (group 2)
Dimensions: ht.: 1.2/w.: 1.2
Description: opaque; blue; six-petalled rosette without inner circle; square
Context: NW Palace; room V ('Layard's dump')
Date: *terminus ante quem* 612
Location: Iraq Museum, Baghdad
Museum Number/Object Number/Field Number: N. 368
Reference: Curtis 1999: 60, 67; Mallowan 1952: 51
Figure Reference: Curtis 1999: fig. 3; Mallowan 1952: pl. XIV

Nim73 (Plate 29)
Type: cast-and-cut/rosette inlays (group 2)
Dimensions: ht.: 1.2/w.: 1.2
Description: opaque; blue; six-petalled rosette without inner circle; square
Context: NW Palace; room V ('Layard's dump')
Date: *terminus ante quem* 612
Location: Iraq Museum, Baghdad
Museum Number/Object Number /Field Number: N. 368
Reference: Curtis 1999: 60, 67; Mallowan 1952: 51
Figure Reference: Curtis 1999: fig. 3; Mallowan 1952: pl. XIV

Nim74 (Plate 29)
Type: cast-and-cut/rosette inlays (group 2)
Dimensions: ht.: 1.2/w.: 1.2
Description: opaque; blue; six-petalled rosette without inner circle; square
Context: NW Palace; room V ('Layard's dump')
Date: *terminus ante quem* 612
Location: Iraq Museum, Baghdad
Museum Number/Object Number/Field Number: N. 368
Reference: Curtis 1999: 60, 67; Mallowan 1952: 51
Figure Reference: Curtis 1999: fig. 3; Mallowan 1952: pl. XIV

Nim75 (Plate 29)
Type: cast-and-cut/rosette inlays (group 2)
Dimensions: ht.: 1.2/w.: 1.2
Description: opaque; blue; six-petalled rosette without inner circle; square
Context: NW Palace; room V ('Layard's dump')
Date: *terminus ante quem* 612
Location: Iraq Museum, Baghdad
Museum Number/Object Number/Field Number: N. 368
Reference: Curtis 1999: 60, 67; Mallowan 1952: 51
Figure Reference: Curtis 1999: fig. 3; Mallowan 1952: pl. XIV

Nim76 (Plate 29)
Type: cast-and-cut/rosette inlays (group 2)
Dimensions: ht.: 1.2/w.: 1.2
Description: opaque; blue; six-petalled rosette without inner circle; square
Context: NW Palace; room V ('Layard's dump')
Date: *terminus ante quem* 612
Location: Iraq Museum, Baghdad
Museum Number/Object Number/Field Number: N. 368
Reference: Curtis 1999: 60, 67; Mallowan 1952: 51
Figure Reference: Curtis 1999: fig. 3; Mallowan 1952: pl. XIV

Nim77 (Plate 29)
Type: cast-and-cut/rosette inlays (group 2)
Dimensions: ht.: 1.2/w.: 1.2
Description: opaque; blue; six-petalled rosette without inner circle; square
Context: NW Palace; room V ('Layard's dump')
Date: *terminus ante quem* 612
Location: Iraq Museum, Baghdad
Museum Number/Object Number/Field Number: N. 368
Reference: Curtis 1999: 60, 67; Mallowan 1952: 51
Figure Reference: Curtis 1999: fig. 3; Mallowan 1952: pl. XIV

Nim78 (Plate 29)
Type: cast-and-cut/rosette inlays (group 2)
Dimensions: ht.: 1.2/w.: 1.2
Description: opaque; blue; six-petalled rosette without inner circle; square
Context: NW Palace; room V ('Layard's dump')
Date: *terminus ante quem* 612
Location: Iraq Museum, Baghdad
Museum Number/Object Number/Field Number: N. 368
Reference: Curtis 1999: 60, 67; Mallowan 1952: 51
Figure Reference: Curtis 1999: fig. 3; Mallowan 1952: pl. XIV

Nim79 (Plate 30)
Type: cast-and-cut/rosette inlays (group 3)
Description: translucent; dark blue; white filling; square
Context: Fort Shalmaneser; room S10
Date: *terminus ante quem* 612
Location: British Museum, London
Museum Number/Object Number/Field Number: BM 140359; 1987, 0131.32/ND 8081
Reference: Curtis 1999: 68
Figure Reference: Curtis 1999: fig. 7

Nim80 (Plate 30)
Type: cast-and-cut/rosette inlays (group 3)
Description: translucent; dark blue; white filling; square
Context: Fort Shalmaneser; room S10
Date: *terminus ante quem* 612
Location: British Museum, London
Museum Number/Object Number/Field Number: BM 140359; 1987, 0131.32/ND 8081
Reference: Curtis 1999: 68
Figure Reference: Curtis 1999: 61, fig. 7

Nim81 (Plate 30)
Type: cast-and-cut/rosette inlays (group 3)
Description: translucent; dark blue; white filling; square
Context: Fort Shalmaneser; room S10
Date: *terminus ante quem* 612
Location: British Museum, London
Museum Number/Object Number/Field Number: BM 140359; 1987,0131.32/ND 8081
Reference: Curtis 1999: 68
Figure Reference: Curtis 1999: 61, fig. 7

Nim82 (Plate 30)
Type: cast-and-cut/rosette inlays (group 3)
Description: translucent; dark blue; white filling; square
Context: Fort Shalmaneser; room S10
Date: *terminus ante quem* 612
Location: British Museum, London
Museum Number/Object Number/Field Number: BM 140359; 1987, 0131.32/ND 8081
Reference: Curtis 1999: 68
Figure Reference: Curtis 1999: 61, fig. 7

Nim83 (Plate 30)
Type: cast-and-cut/rosette inlays (group 3)
Description: translucent; dark blue; white filling; square
Context: Fort Shalmaneser; room S10
Date: *terminus ante quem* 612
Location: British Museum, London
Museum Number/Object Number/Field Number: BM 140359; 1987, 0131.32/ND 8081
Reference: Curtis 1999: 68
Figure Reference: Curtis 1999: 61, fig. 7

Nim84 (Plate 30)
Type: cast-and-cut/rosette inlays (group 3)
Description: translucent; dark blue; white filling;

square
Context: Fort Shalmaneser; room S10
Date: *terminus ante quem* 612
Location: British Museum, London
Museum Number/Object Number/Field Number: BM 140359; 1987, 0131.32/ND 8081
Reference: Curtis 1999: 68
Figure Reference: Curtis 1999: 61, fig. 7

Nim85 (Plate 30)
Type: cast-and-cut/rosette inlays (group 3)
Description: translucent; dark blue; white filling; square
Context: Fort Shalmaneser; room S10
Date: *terminus ante quem* 612
Location: British Museum, London
Museum Number/Object Number/Field Number: BM 140359; 1987, 0131.32/ND 8081
Reference: Curtis 1999: 68
Figure Reference: Curtis 1999: 61, fig. 7

Nim86 (Plate 30)
Type: cast-and-cut/rosette inlays (group 3)
Description: translucent; dark blue; white filling; square
Context: Fort Shalmaneser; room S10
Date: *terminus ante quem* 612
Location: British Museum, London
Museum Number/Object Number/Field Number: BM 140359; 1987, 0131.32/ND 8081
Reference: Curtis 1999: 68
Figure Reference: Curtis 1999: fig. 7

Nim87 (Plate 30)
Type: cast-and-cut/rosette inlays (group 3)
Description: translucent; dark blue; white filling; square
Context: Fort Shalmaneser; room S10
Date: *terminus ante quem* 612
Location: British Museum, London
Museum Number/Object Number/Field Number: BM 140359; 1987, 0131.32/ND 8081
Reference: Curtis 1999: 68
Figure Reference: Curtis 1999: fig. 7

Nim88 (Plate 30)
Type: cast-and-cut/rosette inlays (group 3)
Description: translucent; dark blue; white filling; square
Context: Fort Shalmaneser; room S10
Date: *terminus ante quem* 612
Location: British Museum, London
Museum Number/Object Number/Field Number: BM 140359; 1987, 0131.32/ND 8081
Reference: Curtis 1999: 68
Figure Reference: Curtis 1999: fig. 7

Nim89 (Plate 31)
Type: cast-and-cut/rosette inlays (group 3) with frame
Dimensions: ht.: 0.9/w.: 1.7/th.: 0.2–0.3
Description: opaque; formerly red, frame corroded to green; irregular; rectangular; inserted in the middle with rosette (type 3)
Context: Fort Shalmaneser; room S10
Date: 8th century
Location: British Museum, London
Museum Number/Object Number/Field Number: BM 140359/ND 8081
Reference: Curtis 1999: 68
Figure Reference: Curtis 1999: 61, fig. 7

Nim90 (Plate 30)
Type: cast-and-cut/rosette inlays (group 3)
Dimensions: w: 0.8/th.: 0.2–0.3
Description: opaque; dark blue; white filling; round
Context: Fort Shalmaneser; room S10
Date: *terminus ante quem* 612
Location: British Museum, London
Museum Number/Object Number/Field Number: BM 140359; 1987, 0131.32/ND 8081
Reference: Curtis 1999: 68
Figure Reference: Curtis 1999: 61, fig. 7

Nim91 (Plate 30)
Type: cast-and-cut/rosette inlays (group 3)
Dimensions: w.: 0.8/th.: 0.2–0.3
Description: translucent; dark blue; white filling; round
Context: Fort Shalmaneser; room S10
Date: *terminus ante quem* 612
Location: British Museum, London
Museum Number/Object Number/Field Number: BM 140359; 1987, 0131.32/ND 8081
Reference: Curtis 1999: 68
Figure Reference: Curtis 1999: 61, fig. 7

Nim92 (Plate 30)
Type: cast-and-cut/rosette inlays (group 3)
Dimensions: w.: 0.8/th.: 0.2–0.3
Description: translucent; dark blue; white filling; round
Context: Fort Shalmaneser; room S10
Date: *terminus ante quem* 612
Location: British Museum, London
Museum Number/Object Number/Field Number: BM 140359; 1987, 0131.32/ND 8081
Reference: Curtis 1999: 68
Figure Reference: Curtis 1999: 61, fig. 7

Nim93 (Plate 30)
Type: cast-and-cut/rosette inlays (group 3)
Dimensions: w.: 0.8/th.: 0.2–0.3
Description: translucent; dark blue; white filling; round
Context: Fort Shalmaneser; room S10
Date: *terminus ante quem* 612
Location: British Museum, London
Museum Number/Object Number/Field Number:

BM 140359; 1987, 0131.32/ND 8081
Reference: Curtis 1999: 68
Figure Reference: Curtis 1999: 61, fig. 7

Nim94 (Plate 94)
Type: cast-and-cut/rosette inlays (group 3)
Dimensions: w.: 0.8/th.: 0.2–0.3
Description: translucent; dark blue; white filling; round
Context: Fort Shalmaneser; room S10
Date: *terminus ante quem* 612
Location: British Museum, London
Museum Number/Object Number/Field Number: BM 140359; 1987, 0131.32/ND 8081
Reference: Curtis 1999: 68
Figure Reference: Curtis 1999: 61, fig. 7

Nim95 (Plate 95)
Type: cast-and-cut/rosette inlays (group 3)
Dimensions: w.: 0.8/th.: 0.2–0.3
Description: translucent; dark blue; white filling; round
Context: Fort Shalmaneser; room S10
Date: *terminus ante quem* 612
Location: British Museum, London
Museum Number/Object Number/Field Number: BM 140359; 1987, 0131.32/ND 8081
Reference: Curtis 1999: 68
Figure Reference: Curtis 1999: 61, fig. 7

Nim96 (Plate 30)
Type: cast-and-cut/rosette inlays (group 3)
Dimensions: w.: 0.8/th.: 0.2–0.3
Description: translucent; dark blue; white filling; round
Context: Fort Shalmaneser; room S10
Date: *terminus ante quem* 612
Location: British Museum, London
Museum Number/Object Number/Field Number: BM 140359; 1987, 0131.32/ND 8081
Reference: Curtis 1999: 68
Figure Reference: Curtis 1999: 61, fig. 7

Nim97 (Plate 30)
Type: cast-and-cut/rosette inlays (group 3)
Dimensions: w.: 0.8/th.: 0.2–0.3
Description: translucent; dark blue; white filling; round
Context: Fort Shalmaneser; room S10
Date: *terminus ante quem* 612
Location: British Museum, London
Museum Number/Object Number/Field Number: BM 140359; 1987, 0131.32/ND 8081
Reference: Curtis 1999: 68
Figure Reference: Curtis 1999: 61, fig. 7

Nim98 (Plate 30)
Type: cast-and-cut/rosette inlays (group 3)
Dimensions: w.: 0.8/th.: 0.2–0.3
Description: translucent; dark blue; white filling; round
Context: Fort Shalmaneser; room S10
Date: *terminus ante quem* 612
Location: British Museum, London
Museum Number/Object Number/Field Number: BM 140359; 1987, 0131.32/ND 8081
Reference: Curtis 1999: 68
Figure Reference: Curtis 1999: 61, fig. 7

Nim99 (Plate 30)
Type: cast-and-cut/rosette inlays (group 3)
Dimensions: w.: 0.8/th.: 0.2–0.3
Description: translucent; dark blue; white filling; round
Context: Fort Shalmaneser; room S10
Date: *terminus ante quem* 612
Location: British Museum, London
Museum Number/Object Number/Field Number: BM 140359; 1987, 0131.32/ND 8081
Reference: Curtis 1999: 68
Figure Reference: Curtis 1999: 61, fig. 7

Nim100 (Plate 30)
Type: cast-and-cut/rosette inlays (group 3)
Dimensions: w.: 0.8/th.: 0.2–0.3
Description: translucent; dark blue; white filling; round
Context: Fort Shalmaneser; room S10
Date: *terminus ante quem* 612
Location: British Museum, London
Museum Number/Object Number/Field Number: BM 140359; 1987, 0131.32/ND 8081
Reference: Curtis 1999: 68
Figure Reference: Curtis 1999: 61, fig. 7

Nim101 (Plate 30)
Type: cast-and-cut/rosette inlays (group 3)
Dimensions: w.: 0.8/th.: 0.2–0.3
Description: translucent; dark blue; white filling; round
Context: Fort Shalmaneser; room S10
Date: *terminus ante quem* 612
Location: British Museum, London
Museum Number/Object Number/Field Number: BM 140359; 1987, 0131.32/ND 8081
Reference: Curtis 1999: 68
Figure Reference: Curtis 1999: 61, fig. 7

Nim102 (Plate 30)
Type: cast-and-cut/rosette inlays (group 3)
Dimensions: w.: 0.8/th.: 0.2–0.3
Description: translucent; dark blue; white filling; round
Context: Fort Shalmaneser; room S10
Date: *terminus ante quem* 612
Location: British Museum, London
Museum Number/Object Number/Field Number: BM 140359; 1987, 0131.32/ND 8081

Reference: Curtis 1999: 68
Figure Reference: Curtis 1999: 61, fig. 7

Nim103 (Plate 30)
Type: cast-and-cut/rosette inlays (group 3)
Dimensions: w.: 0.8/th.: 0.2–0.3
Description: translucent; dark blue; white filling; round
Context: Fort Shalmaneser; room S10
Date: *terminus ante quem* 612
Location: British Museum, London
Museum Number/Object Number/Field Number: BM 140359; 1987, 0131.32/ND 8081
Reference: Curtis 1999: 68
Figure Reference: Curtis 1999: 61, fig. 7

Nim104 (Plate 30)
Type: cast-and-cut/rosette inlays (group 3)
Dimensions: w.: 0.8/th.: 0.2–0.3
Description: translucent; dark blue; white filling; round
Context: Fort Shalmaneser; room S10
Date: *terminus ante quem* 612
Location: British Museum, London
Museum Number/Object Number/Field Number: BM 140359; 1987, 0131.32/ND 8081
Reference: Curtis 1999: 68
Figure Reference: Curtis 1999: 61, fig. 7

Nim105 (Plate 30)
Type: cast-and-cut/rosette inlays (group 3)
Dimensions: w.: 0.8/th.: 0.2–0.3
Description: translucent; dark blue; white filling; round
Context: Fort Shalmaneser; room S10
Date: *terminus ante quem* 612
Location: British Museum, London
Museum Number/Object Number/Field Number: BM 140359; 1987, 0131.32/ND 8081
Reference: Curtis 1999: 68
Figure Reference: Curtis 1999: 61, fig. 7

Nim106 (Plate 30)
Type: cast-and-cut/rosette inlays (group 3)
Dimensions: w.: 0.8/th.: 0.2–0.3
Description: translucent; dark blue; white filling; round
Context: Fort Shalmaneser; room S10
Date: *terminus ante quem* 612
Location: British Museum, London
Museum Number/Object Number/Field Number: BM 140359; 1987, 0131.32/ND 8081
Reference: Curtis 1999: 68
Figure Reference: Curtis 1999: 61, fig. 7

Nim107 (Plate 30)
Type: cast-and-cut/rosette inlays (group 3)
Dimensions: w.: 0.8/th.: 0.2–0.3
Description: translucent; dark blue; white filling; round
Context: Fort Shalmaneser; room S10
Date: *terminus ante quem* 612
Location: British Museum, London
Museum Number/Object Number/Field Number: BM 140359; 1987, 0131.32/ND 8081
Reference: Curtis 1999: 68
Figure Reference: Curtis 1999: 61, fig. 7

Nim108 (Plate 30)
Type: cast-and-cut/rosette inlays (group 3)
Dimensions: w.: 0.8/th.: 0.2–0.3
Description: translucent; dark blue; white filling; round
Context: Fort Shalmaneser; room S10
Date: *terminus ante quem* 612
Location: British Museum, London
Museum Number/Object Number/Field Number: BM 140359; 1987, 0131.32/ND 8081
Reference: Curtis 1999: 68
Figure Reference: Curtis 1999: fig. 7

Nim109 (Plate 29)
Type: cast-and-cut/rosette inlays (group 2)
Dimensions: ht.: 1.4/w.: 1.6/th.: 0.2
Description: opaque; blue; six-petalled rosette without inner circle; square
Context: Fort Shalmaneser; room T10
Date: mid 9th – mid 8th century
Location: British Museum, London
Reference: Curtis *et al.* 1993: 15–16; Curtis 1999: 59
Figure Reference: Curtis et al. 1993: 14, fig. 14, 5; Curtis 1999: fig. 1, 2

Nim110 (Plate 30)
Type: cast-and-cut/rosette inlays (group 3)
Dimensions: ht.: 0.5/w.: 0.5
Description: translucent; dark blue; white filling; square
Date: Neo-Assyrian
Location: British Museum, London
Museum Number/Object Number/Field Number: 140346; 1987, 0131.19/ND 10229
Reference: Curtis 1999: 58
Figure Reference: Curtis 1999: 61, fig. 6
Analysis: Brill 1999a: 46, no. 3249, 49

Nim111 (Plate 27)
Type: cast-and-cut/rosette inlays (group 1)
Dimensions: ht.: 1.7/w.: 1.7/th.: 0.2
Description: opaque; blue; six-petalled rosette without inner circle; square; side edges sloping outwards towards the underside
Context: Fort Shalmaneser
Date: *terminus ante quem* 612
Location: Metropolitan Museum of Art, New York
Museum Number/Object Number/Field Number: 58.31.43/ND 6406; ND 6415
Reference: Saldern 1966a: 632, 633

Figure Reference: Curtis 1999: fig. 4; Saldern 1966a: fig. 593; Saldern 1970: fig. 36c; http://www.metmuseum.org/art/collection/search/324780?sortBy=Relevance&ft=58.31.43&pg=1&rpp=20&pos=1 (accessed: 08.08.2018)

Nim112 (Plate 25)
Type: cast-and-cut/painted inlays
Dimensions: ht.: 3.2–3.6/w.: 4.5/th.: 0.3
Description: transparent; colourless; opaque white corroded; iridescence; rectangular, convex insert; long sides straight, narrow side bevelled; black paint on surface
Context: Fort Shalmaneser, room SW7
Date: 9th – 8th century
Location: British Museum, London
Museum Number/Object Number/Field Number: BM 134900; 1966, 1217.8/ND 7638
Reference: Barag 1985: 71, no. 52; Mallowan 1966, 415–416; Orchard 1978, 2, 3; Saldern 1966a, 632, no. 3; Saldern 1970: 223, no. 35c
Figure Reference: Barag 1985: pl. 7; Brill 1978: 24; Mallowan 1966: 415, no. 344; Orchard 1978: pl. Ic, Iie, IIIe; Saldern 1970: fig. 35

Nim113 (Plate 25)
Type: cast-and-cut/painted inlays
Dimensions: ht.: 1.75/w.: 4.4/th.: 0.2
Description: transparent; colourless; opaque white corroded; iridescence; rectangular, convex insert; long sides straight, narrow side bevelled; black paint on surface
Context: Fort Shalmaneser, room SW7
Date: 9th – 8th century
Location: Metropolitan Museum of Art, New York
Museum Number/Object Number/Field Number: 59.107.25 + 59.107.26/ND 7631
Reference: Brill 1987: 23; Orchard 1978: 17, 18; Saldern 1966a: 632; Saldern 1970: 223, no. 35a, b
Figure Reference: Orchard 1978: pl. IIc. IIIc

Nim114 (Plate 26)
Type: cast-and-cut/painted inlays
Dimensions: ht.: 2.8/w.: 2.1/th.: 0.2
Description: transparent; colourless; opaque white corroded; iridescence; rectangular, convex insert; long sides straight, narrow side bevelled; black paint on surface
Context: Fort Shalmaneser, room SW7
Date: 9th – 8th century
Location: Iraq Museum, Baghdad
Museum Number/Object Number/Field Number: ND 7632
Reference: Orchard 1978: 18
Figure Reference: Orchard 1978: pl. Iia, IIIa

Nim115 (Plate 25)
Type: cast-and-cut/painted inlays
Dimensions: ht.: 3.2/w.: 4.4/th.: 0.2
Description: transparent; colourless; opaque white corroded; iridescence; rectangular, convex insert; long sides straight, narrow side bevelled; black paint on surface
Context: Fort Shalmaneser, room SW7
Date: 9th – 8th century
Location: Iraq Museum, Baghdad
Museum Number/Object Number/Field Number: ND 7639
Reference: Oates and Oates 2001: 239, 240; Orchard 1978: 19, 20
Figure Reference: Oates and Oates 2001: 239, fig. 152

Nim116 (Plate 26)
Type: cast-and-cut/painted inlays
Dimensions: ht.: 3.3/w.: 3.7/th.: 0.1
Description: transparent; colourless; opaque white corroded; iridescence; flat not convex in cut; black painting on surface
Context: Fort Shalmaneser, room SW37
Date: 9th – 8th century
Location: Iraq Museum, Baghdad
Museum Number/Object Number/Field Number: ND 10280b
Reference: Orchard 1978: 21
Figure Reference: Orchard 1978: pl. Iif, IIIf

Nim117 (Plate 26)
Type: cast-and-cut/painted inlays
Dimensions: ht.: 3/w.: 4/th.: 2
Description: transparent; colourless; opaque white corroded; iridescence; almost completely preserved, rectangular inlay; upper left corner missing; flat not convex, black painting on surface
Context: Fort Shalmaneser, room SW37
Date: 9th – 8th century
Location: Iraq Museum, Baghdad
Museum Number/Object Number/Field Number: ND 10280a
Reference: Orchard 1978: 20, 21
Figure Reference: Orchard 1978: pl. Iih, IIIh

Nim118 (Plate 25)
Type: cast-and-cut/painted inlays
Dimensions: ht.: 7.2/w.: 4/th.: 0.2
Description: transparent; colourless; opaque white corroded; iridescence; upper half a rectangular inlay, this is straight; black painting on surface
Date: 9th – 8th century
Museum Number/Object Number/Field Number: ND 7633
Reference: Orchard 1978: 19; Saldern 1970: 223, no. 35d
Figure Reference: Orchard 1978: pl. Ia, Iib, IIIb

Nim119
Type: cast-and-cut/painted inlays
Dimensions: ht.: 2.7/w.: 4/th.: 0.3
Description: transparent; colourless; opaque white corroded; iridescence; flat; black painting on surface

Context: Fort Shalmaneser, room SW37
Date: 9th – 8th century
Location: The Corning Museum of Glass, Corning
Museum Number/Object Number/Field Number: ND 10279b
Reference: Brill 1978: 24; Orchard 1978: 20; Saldern 1966a: 632, no. 4; Saldern 1970: 223, no. 35d
Figure Reference: Orchard 1978: pl. IIg

Nim120 (Plate 37)
Type: cast-and-cut/small monochrome inlays, simple geometric motif
Dimensions: ht.: max. 1.6/w.: 0.7/th.: 0.3
Description: translucent; blue-green; perforated barrel; iridescent; diamond-shaped
Context: NW Palace, room V ('Layard's dump')
Date: *terminus ante quem* 612
Location: British Museum, London
Museum Number/Object Number/Field Number: BM 136778
Reference: Barag 1985: 74, no. 58
Figure Reference: Barag 1985: colour pl. C; pl. 7

Nim121 (Plate 33)
Type: cast-and-cut/small monochrome inlays, wing-shaped
Dimensions: w.: 5.5/th.: 0.2
Description: translucent; dark blue; winged; straight side edges
Date: Neo-Assyrian
Location: Metropolitan Museum of Art, New York
Museum Number/Object Number/Field Number: 58.31.45
Reference: Saldern 1966a: 632, 633; Saldern 1970: 224, no. 39
Figure Reference: Saldern 1966a: 633, no. 594; Saldern 1970: fig. 36B

Nim122 (Plate 33)
Type: cast-and-cut/small monochrome inlays, wing-shaped
Dimensions: ht.: 2.7/th.: 0.4
Description: translucent; dark blue; broken wing; surface slightly convex; underside flat
Date: Neo-Assyrian
Location: Metropolitan Museum of Art, New York
Museum Number/Object Number/Field Number: 58.31.49
Reference: Saldern 1966a: 632, 633; Saldern 1970: 224, no. 39
Figure Reference: Saldern 1966a: 633, no. 594; Saldern 1970: fig. 36aB

Nim123 (Plate 34)
Type: cast-and-cut/small monochrome inlays, scale-shaped
Dimensions: w.: 2.5/th.: 0.2
Description: translucent; dark blue; irregular; scaly; side edges sloping outwards
Date: Neo-Assyrian
Location: Metropolitan Museum of Art, New York
Museum Number/Object Number/Field Number: 58.31.53
Reference: Saldern 1966a: 632, 633; Saldern 1970: 224, no. 39
Figure Reference: Saldern 1966a: 633, no. 594; Saldern 1970: fig. 36D

Nim124 (Plate 34)
Type: cast-and-cut/small monochrome inlays, triangle-shaped
Dimensions: w.: 1.4/th.: 0.4
Description: opaque, green-yellow corroded; triangular inlay with rounded sides
Date: Neo-Assyrian
Location: Metropolitan Museum of Art, New York
Museum Number/Object Number/Field Number: 58.31.55
Reference: http://www.metmuseum.org/art/collection#!?q=58.31.55&sortBy=Relevance&sortOrder=asc&page=1 (accessed: 08.08.2018)

Nim125 (Plate 34)
Type: cast-and-cut/small monochrome inlays, triangle-shaped
Dimensions: w.: 1.7/th.: 0.2
Description: opaque; green-yellow corroded; triangular insert; straight edges
Date: Neo-Assyrian
Location: Metropolitan Museum of Art, New York
Museum Number/Object Number/Field Number: 58.31.56
Reference: http://www.metmuseum.org/art/collection#!?q=58.31.56&sortBy=Relevance&sortOrder=asc&page=1 (accessed: 08.08.2018)

Nim126 (Plate 34)
Type: cast-and-cut/small monochrome inlays, scale-shaped
Dimensions: w.: 1.3/h.: 2.3/th.: 0.2
Description: translucent; dark blue; irregular; scaly; side edges sloping outwards
Date: Neo-Assyrian
Location: Metropolitan Museum of Art, New York
Museum Number/Object Number/Field Number: 62.269.19a
Reference: http://www.metmuseum.org/art/collection#!?q=62.269.19a&sortBy=Relevance&sortOrder=asc&page=1 (accessed: 08.08.2018)

Nim127 (Plate 34)
Type: cast-and-cut/small monochrome inlays, triangle-shaped
Dimensions: w.: 1.9/th.: 0.3–0.4
Description: opaque, green-yellow corroded; triangular inlay with rounded sides
Date: Neo-Assyrian
Location: Metropolitan Museum of Art, New York
Museum Number/Object Number/Field Number:

62.269.16b
Reference: http://www.metmuseum.org/art/collection#!?q=62.269.16a,%20b&sortBy=Relevance&sortOrder=asc&page=1 (accessed: 08.08.2018)

Nim128 (Plate 35)
Type: cast-and-cut/small monochrome inlays, double-triangle shaped
Dimensions: ht.: max. 1.4/w.: 1.8/th.: 0.1
Description: translucent; dark blue; irregular; double-axe shaped; side edges sloping outwards
Date: Neo-Assyrian
Location: Metropolitan Museum of Art, New York
Museum Number/Object Number/Field Number: 62.269.21c
Reference: http://www.metmuseum.org/art/collection#!?q=62.269.21&sortBy=Relevance&sortOrder=asc&page=1 (accessed: 08.08.2018)

Nim129 (Plate 35)
Type: cast-and-cut/small monochrome inlays, double-triangle shaped
Dimensions: ht.: 1.3/w.: 2.2/th.: 0.1
Description: translucent; dark blue; irregular; double-axe shaped; side edges sloping outwards
Date: Neo-Assyrian
Location: Metropolitan Museum of Art, New York
Museum Number/Object Number/Field Number: 62.269.20b
Reference: http://www.metmuseum.org/art/collection#!?q=62.269.20&sortBy=Relevance&sortOrder=asc&page=1 (accessed: 08.08.2018)

Nim130 (Plate 33)
Type: cast-and-cut/small monochrome inlays, wing-shaped
Dimensions: w.: 2.3/th.: 0.5–0.6
Description: translucent; dark blue; broken wing; surface slightly convex; underside flat; slightly bevelled side edges
Date: Neo-Assyrian
Location: Metropolitan Museum of Art, New York
Museum Number/Object Number/Field Number: 58.31.48

Nim131 (Plate 37)
Type: cast-and-cut/small monochrome inlays, simple geometric motif
Dimensions: ht.: 0.4/w.: /th.: 0.4
Description: translucent; dark blue; white centre
Context: Fort Shalmaneser, room S10
Date: *terminus ante quem* 612
Location: British Museum, London
Museum Number/Object Number/Field Number: BM 140359; 1987, 0131.32/ND 8081

Nim132
Type: ingot/round
Description: red; round bars, broken; OF coated with thick corrosion layer
Location: British Museum, London
Museum Number/Object Number/Field Number: 1992, 0701.1
Reference: http://www.britishmuseum.org/research/collection_online/collection_object_details.aspx?objectId=369269&partId=1&searchText=1992,0701.1&page=1 (accessed: 08.08.2018)
Figure Reference: http://www.britishmuseum.org/research/collection_online/collection_object_details/collection_image_gallery.aspx?assetId=151090001&objectId=369269&partId=1 (accessed: 08.08.2018)

Nim133 (Plate 36)
Type: cast-and-cut/small monochrome inlays, floral motif
Dimensions: ht.: 0.7/w.: 0.9/th.: 0.2
Description: translucent; dark blue; top slightly convex, smooth, underside straight; rolled plant part
Context: Fort Shalmaneser, room S10
Date: *terminus ante quem* 612
Location: British Museum, London
Museum Number/Object Number/Field Number: BM 140359; 1987, 0131.32/ND 8081
Reference: http://www.britishmuseum.org/research/collection_online/collection_object_details.aspx?objectId=1638132&partId=1&searchText=140359&page=1 (accessed: 08.08.2018)

Nim134
Type: inlay (group 3)
Dimensions: ht.: 1.2/w.: 0.7/th.: 0.2
Description: translucent; dark blue; white filling; rectangular
Context: Fort Shalmaneser; room S10
Date: *terminus ante quem* 612
Location: British Museum, London
Museum Number/Object Number/Field Number: BM 140359/ND 8081
Reference: Curtis 1999: 68
Figure Reference: Curtis 1999: 61, fig. 7

Nim135 (Plate 68)
Type: waste material
Dimensions: ht.: 1.2/w.: 1.1
Description: translucent; dark blue; irregular surface
Date: Neo-Assyrian
Location: Metropolitan Museum of Art, New York
Museum Number/Object Number/Field Number: 62.269.19f

Nim136 (Plate 37)
Type: cast-and-cut/small monochrome inlays, simple geometric motif
Dimensions: ht.: 0.9/w.: 1.6/th.: 0.2
Description: translucent; dark blue; top and bottom straight, smooth; very irregular edges; leaf-shaped
Context: Fort Shalmaneser; room S10
Date: *terminus ante quem* 612

Location: British Museum, London
Museum Number/Object Number/Field Number: BM 140359; 1987, 0131.32/ND 8081
Reference: http://www.britishmuseum.org/research/collection_online/collection_object_details.aspx?objectId=1638132&partId=1&searchText=140359&page=1 (accessed: 08.08.2018)

Nim137 (Plate 54)
Type: core-formed/large cylindrical bottles
Dimensions: ht.: 1.1/w.: 1.9
Description: opaque; strong weathering resistance
Context: Burnt Palace; east side of the throne room (room 8)
Date: 8th – late 7th century
Reference: Barag 1970: 156, no. 3; Saldern 1966a, 632, no. 5; Turner 1955, 57, specimen E
Figure Reference: Barag 1970: fig. 46

Nim138 (Plate 54)
Type: core-formed/large cylindrical bottles
Dimensions: ht.: 2/w.: 3/th.: 0.3
Description: opaque; white, slightly wavy, horizontal lines; depression in which handles were attached
Context: Burnt Palace, room 23
Date: 8th – late 7th century
Reference: Barag 1970: 155, no. 1; Saldern 1966a: 632, no. 5; Turner 1955: 57, 59, specimen D
Figure Reference: Barag 1970: fig. 45

Nim139 (Plate 37)
Type: cast-and-cut/small monochrome inlays, simple geometric motif
Dimensions: ht.: 1.2/w.: 1.8
Description: translucent; dark blue; top and underside irregular; diamond-shaped, side edges sloping outwards towards underside
Context: Fort Shalmaneser, room S10
Date: *terminus ante quem* 612
Location: British Museum, London
Museum Number/Object Number/Field Number: BM 140359; 1987, 0131.32/ND 8081

Nim140 (Plate 36)
Type: cast-and-cut/small monochrome inlays, floral motif
Dimensions: ht.: 4.5/w.: 0.9
Description: opaque; yellow, green in places; heavy pitting; heavily corroded; iridescence; palm leaf-shaped
Context: Fort Shalmaneser, room S10
Date: *terminus ante quem* 612
Location: British Museum, London
Museum Number/Object Number/Field Number: BM 140359; 1987, 0131.32/ND 8081

Nim141 (Plate 37)
Type: cast-and-cut/small monochrome inlays, simple geometric motif
Dimensions: ht.: 0.8/w.: 1.8/th.: 0.2
Description: translucent; dark blue; top and underside irregular; diamond-shaped, side edges sloping outwards towards underside
Date: Neo-Assyrian
Location: Metropolitan Museum of Art, New York
Museum Number/Object Number/Field Number: 62.269.19b

Nim142 (Plate 35)
Type: cast-and-cut/small monochrome inlays, floral motif
Dimensions: w.: 3.8/th.: 0.4
Description: opaque; blue-green, yellow corroded; patelled-shaped
Date: Neo-Assyrian
Location: Metropolitan Museum of Art, New York
Museum Number/Object Number/Field Number: 62.269.19c

Nim143 (Plate 35)
Type: cast-and-cut/small monochrome inlays, double-triangle shaped
Dimensions: ht.: 1.8/w.: 2.2/th.: 0.2
Description: translucent; dark blue; irregular; double-triangle shaped; side edges tapering outwards
Date: Neo-Assyrian
Location: Metropolitan Museum of Art, New York
Museum Number/Object Number/Field Number: 62.269.20a

Nim144 (Plate 36)
Type: cast-and-cut/small monochrome inlays, floral motif
Dimensions: w.: 1.7/th.: 0.2
Description: opaque; turquoise, slightly yellowish corroded; very strong pitting; tulip shape
Date: Neo-Assyrian
Location: Metropolitan Museum of Art, New York
Museum Number/Object Number/Field Number: 62.269.21b

Nim145 (Plate 36)
Type: cast-and-cut/small monochrome inlays, floral motif
Dimensions: w.: 2.3/th.: 0.1
Description: opaque; turquoise; slightly yellowish corroded; very strong pitting; tulip shape
Date: Neo-Assyrian
Location: Metropolitan Museum of Art, New York
Museum Number/Object Number/Field Number: 62.269.21a

Nim146 (Plate 34)
Type: cast-and-cut/small monochrome inlays, scale-shaped
Dimensions: ht.: 1./w.: 1./th.: 0.1–
Description: translucent; dark blue; irregular; scaly; side edges tapering outwards
Context: Fort Shalmaneser, room S10
Date: *terminus ante quem* 612

Location: British Museum, London
Museum Number/Object Number/Field Number: BM 140359; 1987, 0131.32/ND 8081

Nim147 (Plate 33)
Type: cast-and-cut/small monochrome inlays, wing-shaped
Dimensions: ht.: 1.1–4.3/w.:2.3/th.: 0.3
Description: opaque; dark blue; top slightly convex, smooth; bottom straight, irregular; bevelled side edges
Context: Fort Shalmaneser, room S10
Date: *terminus ante quem* 612
Location: British Museum, London
Museum Number/Object Number/Field Number: BM 140359; 1987, 0131.32/ND 8081
Reference: Brill 1999a: 46, no. 3231, 3246, 3247, 3251, 3252, 3253; Brill 2012: 625

Nim148 (Plate 33)
Type: cast-and-cut/small monochrome inlays, wing-shaped
Dimensions: ht.: 1.1–4.3/w.: 3.1/th.: 0.3
Description: opaque; dark blue; top slightly convex, smooth; bottom straight, irregular; bevelled side edges
Context: Fort Shalmaneser, room S10
Date: *terminus ante quem* 612
Location: British Museum, London
Museum Number/Object Number/Field Number: BM 140359; 1987,0131.32/ND 8081
Reference: Brill 1999a: 46, no. 3231, 3246, 3247, 3251, 3252, 3253; Brill 2012: 625

Nim149 (Plate 33)
Type: cast-and-cut/small monochrome inlays, wing-shaped
Dimensions: ht.: 1.1–4.3/w.: 3.1/th.: 0.3
Description: opaque; dark blue; top slightly convex, smooth; bottom straight, irregular; bevelled side edges
Context: Fort Shalmaneser, room S10
Date: *terminus ante quem* 612
Location: British Museum, London
Museum Number/Object Number/Field Number: BM 140359; 1987, 0131.32/ND 8081
Reference: Brill 1999: 46, no. 3231, 3246, 3247, 3251, 3252, 3253; Brill 2012: 625

Nim150 (Plate 33)
Type: cast-and-cut/small monochrome inlays, wing-shaped
Dimensions: ht.: 1.1–4.3/w.: 2.9/th.: 0.3
Description: opaque; dark blue; top slightly convex, smooth; bottom straight, irregular; bevelled side edges
Context: Fort Shalmaneser, room S10
Date: *terminus ante quem* 612
Location: British Museum, London
Museum Number/Object Number/Field Number: BM 140359; 1987, 0131.32/ND 8081
Reference: Brill 1999a: 46, no. 3231, 3246, 3247, 3251, 3252, 3253; Brill 2012: 625

Nim151 (Plate 33)
Type: cast-and-cut/small monochrome inlays, wing-shaped
Dimensions: ht.: 1.1–4.3/w.:0.9/th.: 0.3
Description: opaque; dark blue; top slightly convex, smooth; bottom straight, irregular; bevelled side edges
Context: Fort Shalmaneser, room S10
Date: *terminus ante quem* 612
Location: British Museum, London
Museum Number/Object Number/Field Number: BM 140359; 1987, 0131.32/ND 8081
Reference: Brill 1999a: 46, no. 3231, 3246, 3247, 3251, 3252, 3253; Brill 2012: 625

Nim152 (Plate 33)
Type: cast-and-cut/small monochrome inlays, wing-shaped
Dimensions: ht.: 1.1–4.3/w.: 1.4/th.: 0.3
Description: opaque; dark blue; top slightly convex, smooth; bottom straight, irregular; bevelled side edges
Context: Fort Shalmaneser, room S10
Date: *terminus ante quem* 612
Location: British Museum, London
Museum Number/Object Number/Field Number: BM 140359; 1987, 0131.32/ND 8081
Reference: Brill 1999a: 46, no. 3231, 3246, 3247, 3251, 3252, 3253; Brill 2012: 625

Nim153 (Plate 33)
Type: cast-and-cut/small monochrome inlays, wing-shaped
Dimensions: ht.: 1.1–4.3/w.: 0.4–0.9/th.: 0.3
Description: translucent; dark blue; top slightly convex, smooth; underside straight, irregular; bevelled side edges; winged
Context: Fort Shalmaneser, room S10
Date: *terminus ante quem* 612
Location: British Museum, London
Museum Number/Object Number/Field Number: BM 140359; 1987, 0131.32/ND 8081
Reference: Brill 1999a: 46, no. 3231, 3246, 3247, 3251, 3252, 3253; Brill 2012: 625

Nim154 (Plate 33)
Type: cast-and-cut/small monochrome inlays, wing-shaped
Dimensions: ht.: 0.8/w.: 4/th.: 0.2
Description: opaque; dark blue; strongly whitish-green corrosion layer; upper side slightly convex, smooth; underside straight, irregular; bevelled side edges
Context: Fort Shalmaneser, room S10
Date: *terminus ante quem* 612
Location: British Museum, London
Museum Number/Object Number/Field Number: BM 140359; 1987, 0131.32/ND 8081

Nim155 (Plate 34)
Type: cast-and-cut/small monochrome inlays, scale-shaped

Dimensions: ht.: 1/w.: 1/th.: 0.1–0.3
Description: translucent; dark blue; irregular; scale-shaped; side edges tapering outwards
Context: Fort Shalmaneser, room S10
Date: *terminus ante quem* 612
Location: British Museum, London
Museum Number/Object Number/Field Number: BM 140359; 1987, 0131.32/ND 8081

Nim156 (Plate 34)
Type: cast-and-cut/small monochrome inlays, scale-shaped
Dimensions: ht.: 1.3/w.: 1.3/th.: 0.1–0.3
Description: translucent; dark blue; irregular; scale-shaped; side edges tapering outwards
Context: Fort Shalmaneser, room S10
Date: *terminus ante quem* 612
Location: British Museum, London
Museum Number/Object Number/Field Number: BM 140359; 1987, 0131.32/ND 8081

Nim157 (Plate 34)
Type: cast-and-cut/small monochrome inlays, scale-shaped
Dimensions: ht.: 1/w.: 1.3/th.: 0.1–0.3
Description: translucent; dark blue; irregular; scale-shaped; side edges tapering outwards
Context: Fort Shalmaneser, room S10
Date: *terminus ante quem* 612
Location: British Museum, London
Museum Number/Object Number/Field Number: BM 140359; 1987, 0131.32/ND 8081

Nim158 (Plate 34)
Type: cast-and-cut/small monochrome inlays, scale-shaped
Dimensions: ht.: 1/w.: 1.1/th.: 0.1–0.3
Description: translucent; dark blue; irregular; scale-shaped; side edges tapering outwards
Context: Fort Shalmaneser, room S10
Date: *terminus ante quem* 612
Location: British Museum, London
Museum Number/Object Number/Field Number: BM 140359; 1987, 0131.32/ND 8081

Nim159 (Plate 34)
Type: cast-and-cut/small monochrome inlays, scale-shaped
Dimensions: ht.: 1.4/w.: 1.2/th.: 0.1–0.25
Description: translucent; dark blue; irregular; scale-shaped; side edges tapering outwards
Context: Fort Shalmaneser, room S10
Date: *terminus ante quem* 612
Location: British Museum, London
Museum Number/Object Number/Field Number: BM 140359; 1987, 0131.32/ND 8081

Nim160 (Plate 34)
Type: cast-and-cut/small monochrome inlays, scale-shaped
Dimensions: ht.: 1.1/w.: 1.4/th.: 0.1–0.3
Description: translucent; dark blue; irregular; scale-shaped; side edges tapering outwards
Context: Fort Shalmaneser, room S10
Date: *terminus ante quem* 612
Location: British Museum, London
Museum Number/Object Number/Field Number: BM 140359; 1987, 0131.32/ND 8081

Nim161 (Plate 35)
Type: cast-and-cut/small monochrome inlays, double-triangle shaped
Dimensions: ht.: max. 1.1/w.: max.2.1/th.: 0.2
Description: translucent; dark blue; irregular; scale-shaped; side edges tapering outwards
Context: Fort Shalmaneser; room S10
Date: *terminus ante quem* 612
Location: British Museum, London
Museum Number/Object Number/Field Number: 1987-1-31; 140359/ND 8081

Nim162 (Plate 35)
Type: cast-and-cut/small monochrome inlays, double-triangle shaped
Dimensions: ht.: max. 1.1/w.: 2.1/th.: 0.2
Description: translucent; dark blue; irregular; scale-shaped; side edges tapering outwards
Context: Fort Shalmaneser; room S10
Date: *terminus ante quem* 612
Location: British Museum, London
Museum Number/Object Number/Field Number: 1987-1-31; 140359/ND 8081

Nim163
Type: cast-and-cut/small monochrome inlays, double-triangle shaped
Dimensions: ht.: 2.1/w.: 1.1/th.: 0.2
Description: translucent; dark blue; irregular; scale-shaped; side edges tapering outwards
Context: Fort Shalmaneser; room S10
Date: *terminus ante quem* 612
Location: British Museum, London
Museum Number/Object Number/Field Number: 1987-1-31; 140359/ND 8081

Nim164
Type: cast-and-cut/small monochrome inlays, double-triangle shaped
Dimensions: ht.: 2.1/w.: 1.1/th.: 0.2
Description: translucent; dark blue; irregular; scale-shaped; side edges tapering outwards
Context: Fort Shalmaneser; room S10
Date: *terminus ante quem* 612
Location: British Museum, London
Museum Number/Object Number/Field Number: 1987-1-31; 140359/ND 8081

Nim165
Type: cast-and-cut/small monochrome inlays, double-

triangle shaped
Dimensions: ht.: 2.1/w.: 1.1/th.: 0.2
Description: translucent; dark blue; irregular; scale-shaped; side edges tapering outwards
Context: Fort Shalmaneser; room S10
Date: *terminus ante quem* 612
Location: British Museum, London
Museum Number/Object Number/Field Number: 1987-1-31; 140359/ND 8081

Nim166
Type: cast-and-cut/small monochrome inlays, double-triangle shaped
Dimensions: ht.: 2.1/w.: 1.1/th.: 0.2
Description: translucent; dark blue; irregular; scale-shaped; side edges tapering outwards
Context: Fort Shalmaneser; room S10
Date: *terminus ante quem* 612
Location: British Museum, London
Museum Number/Object Number/Field Number: 1987-1-31; 140359/ND 8081

Nim167
Type: cast-and-cut/small monochrome inlays, double-triangle shaped
Dimensions: ht.: 2.1/w.: 1.1/th.: 0.2
Description: translucent; dark blue; irregular; scale-shaped; side edges tapering outwards
Context: Fort Shalmaneser; room S10
Date: *terminus ante quem* 612
Location: British Museum, London
Museum Number/Object Number/Field Number: 1987-1-31; 140359/ND 8081

Nim168 (Plate 35)
Type: cast-and-cut/small monochrome inlays, double-triangle shaped
Dimensions: ht. max. 1.1/w.: 2.4/th.: 0.2
Description: translucent; dark blue; irregular; scale-shaped; side edges tapering outwards
Date: Neo-Assyrian
Location: Metropolitan Museum of Art, New York
Museum Number/Object Number/Field Number: 58.31.52
Reference: Saldern 1966a: 632. 633; Saldern 1970: 224 no. 39
Figure Reference: Saldern 1966a: 633, no. 594; Saldern 1970: fig. 36D

Nim169
Type: cast-and-cut/small monochrome inlays, double-triangle shaped
Dimensions: ht.: 2.3/w.: 1.9/th.: 0.2
Description: translucent; dark blue; irregular; scale-shaped; side edges tapering outwards
Context: Fort Shalmaneser; room S10
Date: *terminus ante quem* 612
Location: British Museum, London
Museum Number/Object Number/Field Number: BM 140359/ND 8081

Nim170
Type: cast-and-cut/small monochrome inlays, double-triangle shaped
Dimensions: ht.: 2.3/w.: 1.9/th.: 0.2
Description: translucent; dark blue; irregular; double-axe shaped; side edges sloping outwards
Context: Fort Shalmaneser; room S10
Date: *terminus ante quem* 612
Location: British Museum, London
Museum Number/Object Number/Field Number: BM 140359/ND 8081

Nim171
Type: cast-and-cut/small monochrome inlays, double-triangle shaped
Dimensions: ht.: 2.3/w.: 1.9/th.: 0.2
Description: translucent; dark blue; irregular; double-axe shaped; side edges sloping outwards
Context: Fort Shalmaneser; room S10
Date: *terminus ante quem* 612
Location: British Museum, London
Museum Number/Object Number/Field Number: BM 140359/ND 8081

Nim172
Type: cast-and-cut/small monochrome inlays, double-triangle shaped
Dimensions: ht.: 2.3/w.: 1.9/th.: 0.2
Description: translucent; dark blue; irregular; double-axe shaped; side edges sloping outwards
Context: Fort Shalmaneser; room S10
Date: *terminus ante quem* 612
Location: British Museum, London
Museum Number/Object Number/Field Number: BM 140359/ND 8081

Nim173 (Plate 34)
Type: cast-and-cut/small monochrome inlays, triangle-shaped
Dimensions: w.: 1.6/th.: 0.4
Description: opaque, green-yellow corroded; triangular inlay with rounded sides
Date: Neo-Assyrian
Location: Metropolitan Museum of Art
Museum Number/Object Number/Field Number: 62.269.16a

Nim174 (Plate 33)
Type: cast-and-cut/small monochrome inlays, wing-shaped
Dimensions: ht.: 0.8/w.: 4/th.: 0.2
Description: translucent; dark blue; white-greenish corrosion layer; pitting; iridescence; straight side edges, straight cut; top slightly convex, smooth; underside rough
Context: Fort Shalmaneser, room S10
Date: *terminus ante quem* 612

Location: British Museum, London
Museum Number/Object Number/Field Number: BM 140359; 1987, 0131.32/ND 8081

Nim175 (Plate 36)
Type: cast-and-cut/small monochrome inlays, floral motif
Dimensions: ht.: 4.5/w.: 0.9
Description: opaque; yellow, partly green, light blue corroded; heavy pitting; heavily corroded; palm petalled-shaped
Context: Fort Shalmaneser, room S10
Date: *terminus ante quem* 612
Location: British Museum, London
Museum Number/Object Number/Field Number: BM 140359; 1987, 0131.32/ND 8081
Figure Reference: Curtis 1999: fig. 7

Nim176 (Plate 37)
Type: cast-and-cut/small monochrome inlays, simple geometric motif
Description: translucent; dark blue; rectangular
Context: NW Palace, room V ('Layard's dump')
Date: *terminus ante quem* 612
Location: Iraq Museum, Baghdad
Museum Number/Object Number/Field Number: N. 368
Reference: Mallowan 1952: 51
Figure Reference: Mallowan 1952: pl. XIV

Nim177 (Plate 37)
Type: cast-and-cut/small monochrome inlays, simple geometric motif
Description: translucent; dark blue; rectangular
Context: NW Palace, room V ('Layard's dump')
Date: *terminus ante quem* 612
Location: Iraq Museum, Baghdad
Museum Number/Object Number/Field Number: N. 368
Reference: Mallowan 1952: 51
Figure Reference: Mallowan 1952: pl. XIV

Nim178 (Plate 37)
Type: cast-and-cut/small monochrome inlays, simple geometric motif
Description: translucent; dark blue; rectangular
Context: NW Palace, 'Layard's dump' in room V
Date: *terminus ante quem* 612
Location: Iraq Museum, Baghdad
Museum Number/Object Number/Field Number: N. 368
Reference: Mallowan 1952: 51
Figure Reference: Mallowan 1952: pl. XIV

Nim179 (Plate 37)
Type: cast-and-cut/small monochrome inlays, simple geometric motif
Description: translucent; dark blue; rectangular
Context: NW Palace, room V ('Layard's dump')
Date: *terminus ante quem* 612
Location: Iraq Museum, Baghdad
Museum Number/Object Number/Field Number: N. 368
Reference: Mallowan 1952: 51
Figure Reference: Mallowan 1952: pl. XIV

Nim180 (Plate 37)
Type: cast-and-cut/small monochrome inlays, simple geometric motif
Description: translucent; dark blue; rectangular
Context: NW Palace, 'Layard's dump' in room V
Date: terminus ante quem 612
Location: Iraq Museum, Baghdad
Museum Number/Object Number/Field Number: N. 368
Reference: Mallowan 1952: 51
Figure Reference: Mallowan 1952: pl. XIV

Nim181 (Plate 37)
Type: cast-and-cut/small monochrome inlays, simple geometric motif
Description: translucent; dark blue; rectangular
Context: NW Palace, 'Layard's dump' in room V
Date: terminus ante quem 612
Location: Iraq Museum, Baghdad
Museum Number/Object Number/Field Number: N. 368
Reference: Mallowan 1952: 51
Figure Reference: Mallowan 1952: pl. XIV

Nim182 (Plate 37)
Type: cast-and-cut/small monochrome inlays, simple geometric motif
Description: translucent; dark blue; rectangular
Context: NW Palace, room V 'Layard's dump' in room V
Date: *terminus ante quem* 612
Location: Iraq Museum, Baghdad
Museum Number/Object Number/Field Number: N. 368
Reference: Mallowan 1952: 51
Figure Reference: Mallowan 1952: pl. XIV

Nim183 (Plate 37)
Type: cast-and-cut/small monochrome inlays, simple geometric motif
Description: translucent; dark blue; rectangular
Context: NW Palace, room V 'Layard's dump' in room V
Date: *terminus ante quem* 612
Location: Iraq Museum, Baghdad
Museum Number/Object Number/Field Number: N. 368
Reference: Mallowan 1952: pl. XIV
Figure Reference: Mallowan 1952: pl. XIV

Nim184 (Plate 36)
Type: cast-and-cut/small monochrome inlays, floral motif
Dimensions: ht.: 1.1–1.2/w.: 0.7–0.8/th.: 0.1

Description: translucent; dark blue; white weathering layer; tulip shape
Context: NW Palace, room V 'Layard's dump' in room V
Date: *terminus ante quem* 612
Location: British Museum, London
Museum Number/Object Number/Field Number: 1848, 0720.84; 122100,b/N. 368
Reference: Mallowan 1952: 51; Barag 1985: 73, no. 57
Figure Reference: Barag 1985: colour pl. C, pl. 7; Mallowan 1952: pl. XIV

Nim185 (Plate 36)
Type: cast-and-cut/small monochrome inlays, floral motif
Dimensions: ht.: 1.1–1.2/w.: 0.7–0.8/th.: 0.1
Description: translucent; dark blue; white weathering layer; tulip shape
Context: NW Palace, 'Layard's dump' in room V
Date: *terminus ante quem* 612
Location: British Museum, London
Museum Number/Object Number/Field Number: 1848, 0720.84; 122100, b/N. 368
Reference: Barag 1985: 73, no. 57; Mallowan 1952: 51
Figure Reference: Barag 1985: colour pl. C, pl. 7; Mallowan 1952: pl. XIV

Nim186 (Plate 36)
Type: cast-and-cut/small monochrome inlays, floral motif
Dimensions: ht.: 1.1–1.2/w.: 0.7–0.8/th.: 0.1
Description: translucent; dark blue; white weathering layer; tulip shape
Context: NW Palace, room V 'Layard's dump' in room V
Date: *terminus ante quem* 612
Location: British Museum, London
Museum Number/Object Number/Field Number: 1848, 0720.84; 122100,b/N. 368
Reference: Barag 1985: 73, no. 57; Mallowan 1952: 51
Figure Reference: Barag 1985: colour pl. C, pl. 7; Mallowan 1952: pl. XIV

Nim187 (Plate 36)
Type: cast-and-cut/small monochrome inlays, floral motif
Dimensions: ht.: 1.1–1.2/w.: 0.7–0.8/th.: 0.1
Description: translucent; dark blue; white weathering layer; tulip shape
Context: NW Palace, 'Layard's dump' in room V
Date: *terminus ante quem* 612
Location: British Museum, London
Museum Number/Object Number/Field Number: 1848, 0720.84; 122100, b/N. 368
Reference: Barag 1985: 73, no. 57; Mallowan 1952: 51
Figure Reference: Barag 1985: colour pl. C, pl. 7; Mallowan 1952: pl. XIV

Nim188 (Plate 36)
Type: cast-and-cut/small monochrome inlays, floral motif
Dimensions: ht.: 1.1–1.2/w.: 0.7–0.8/th.: 0.1
Description: translucent; dark blue; white weathering layer; tulip shape
Context: NW Palace, 'Layard's dump' in room V
Date: *terminus ante quem* 612
Location: British Museum, London
Museum Number/Object Number/Field Number: 1848, 0720.84; 122100, b/N. 368
Reference: Barag 1985: 73, no. 57; Mallowan 1952: 51
Figure Reference: Barag 1985: colour pl. C, pl. 7; Mallowan 1952: pl. XIV

Nim189 (Plate 68)
Type: waste material
Dimensions: ht.: 2/w.: 1.3/th.: 0.5
Description: translucent; dark blue; irregular surface
Date: Neo-Assyrian
Location: Metropolitan Museum of Art, New York
Museum Number/Object Number/Field Number: 62.269.19e

Nim190
Type: ingot/round
Dimensions: ht.: 4.1/w.: 3.2/th.: 0.7
Description: opaque; red; greenish corrosion layer; flat fragment, damaged on all sides
Context: South-Eastern Palace
Location: British Museum, London
Museum Number/Object Number/Field Number: N. 1335
Reference: Barag 1985: 111, no. 178
Figure Reference: Barag 1985: pl. 20

Nim191 (Plate 37)
Type: cast-and-cut/small monochrome inlays, figurative motif
Dimensions: ht.: 1.8/w.: c. 1.3
Description: translucent; blue; wig with horizontal and vertical recesses
Context: Fort Shalmaneser, SW 37
Date: *terminus ante quem* 612
Figure Reference: Fiorina 2009: 45, fig. 17

Nim192 (Plate 37)
Type: cast-and-cut/small monochrome inlays, figurative motif
Dimensions: ht.: c. 1/w.: c. 1
Description: translucent; blue; wig with horizontal and vertical recesses
Context: Fort Shalmaneser, SW 37
Date: *terminus ante quem* 612
Figure Reference: Fiorina 2009: 45, fig. 17

Nim193 (Plate 37)
Type: cast-and-cut/small monochrome inlays, figurative motif
Dimensions: ht.: 1.5/w.: c. 1
Description: translucent; blue; part of wig
Context: Fort Shalmaneser, SW 37

Date: *terminus ante quem* 612
Figure Reference: Fiorina 2009: 45, fig. 17

Nim194 (Plate 37)
Type: cast-and-cut/small monochrome inlays, figurative motif
Dimensions: ht.: 1.5/w.: 1
Description: translucent; blue; part of wig
Context: Fort Shalmaneser, SW 37
Date: *terminus ante quem* 612
Figure Reference: Fiorina 2009: 45, fig. 17

Nim195
Type: ingot/round
Description: opaque; red; grey corrosion layer, green in places; broken; round; convex form
Location: British Museum, London
Museum Number/Object Number/Field Number: 1992, 0701.1

Nim196
Type: ingot/round
Dimensions: th.: 2.7
Description: opaque; turquoise; corrosion layer on top and bottom; pitting; iridescence
Location: British Museum, London
Museum Number/Object Number/Field Number: BM 140427; 1987, 0131.100
Reference: http://www.britishmuseum.org/research/collection_online/collection_object_details.aspx?objectId=369268&partId=1&searchText=140427+&page=1 (Accessed: 8.8.2018)
Figure Reference: http://www.britishmuseum.org/research/collection_online/collection_object_details/collection_image_gallery.aspx?partid=1&assetid=151087001&objectid=369268 (accessed: 8.8.2018)

Nim197
Type: ingot/round
Dimensions: ht.: 7.6/w.: 1.9/rim: 22
Description: opaque; dark red; corrosion layer; fragment of a round bar; top flat, bottom convex
Location: British Museum, London
Museum Number/Object Number/Field Number: BM 91563/N. 787; N. 835
Reference: Barag 1985: 109, no. 167
Figure Reference: Barag 1985: pl. 19

Nim198 (Plate 61)
Type: ingot/round
Dimensions: w.: 16.4/th.: 3.6
Description: opaque; red; stress cracks on surface; green weathering layer; convex bottom; upper coating of charcoal or similar charred material
Context: Burnt Palace; room 47
Date: 860–740
Location: British Museum, London
Museum Number/Object Number/Field Number: BM 132163; 1957, 0209.10
Reference: Barber *et al.* 2009: 117; Barag 1985: 108–109, no. 166; Bimson and Freestone 1985: 121, 122, no. 166; Brill 1999a: 45, no. 200; Mallowan 1954: 77, 82, 83
Figure Reference: Barag 1985: fig. 12, pl. 19

Nim199 (Plate 32)
Type: cast-and-cut/rosette inlays
Dimensions: ht.: 1/w.: 1.3/th.: 0.2
Description: opaque; green; square leaves with middle dot and white filling
Date: Neo-Assyrian
Location: Metropolitan Museum of Art, New York

Nineveh

Nin1 (Plate 43)
Type: cast-and-cut/composite attachments and inlays
Dimensions: ht.: 6.5/w.: 6.2/th.: 2.6
Description: opaque; dark blue; thick white corrosion layer; convex front; beard curls; straight back, irregular surface
Date: 10th–8th century
Location: British Museum, London
Museum Number/Object Number/Field Number: 1825, 0503.62
Reference: http://www.britishmuseum.org/research/collection_online/collection_object_details.aspx?objectId=674199&partId=1&searchText=1825,0503.62&page=1 (accessed: 08.08.2018)
Figure Reference: http://www.britishmuseum.org/research/collection_online/collection_object_details/collection_image_gallery.aspx?assetId=528191001&objectId=674199&partId=1 (accessed: 08.08.2018)

Nin2 (Plate 10)
Type: cast-and-cut/mace-heads
Dimensions: ht.: 6.6/w.: max. 5.9/base: 4.1
Description: opaque; dark blue; elongated bubbles; flat bottom; cylindrical shaft hole in the middle of the bottom
Context: Nabû temple; on the bottom of 'Sargon's well', renewed during the reign of Sargon II and remaining in use until 612
Date: 722–612
Location: British Museum, London
Museum Number/Object Number/Field Number: BM 98922; 1905, 0409.428
Reference: Barag 1985: 74, no. 60; Harden 1968: no. 28; Saldern 1970: 224, no. 41a
Figure Reference: Barag 1985: colour pl. C, pl. 7

Nin3 (Plate 68)
Type: waste material
Dimensions: ht.: 4.4/w.: 3.3/th.: 2
Description: opaque; greenish; corrosion layer brown-grey
Context: probably from the burials excavated by E. A. Wallis Budget 1888–1889
Location: British Museum, London
Museum Number/Object Number/Field Number: 41-

7-26, 80
Reference: Barag 1985: 113, no. 187
Figure Reference: Barag 1985: pl. 20

Nin4
Type: raw glass fragment
Dimensions: ht.: 1.8/w.: 1.4
Description: opaque; red-brown; surface greenish eroded
Location: British Museum, London
Museum Number/Object Number/Field Number: S. 2390
Reference: Barag 1985: 112, no. 182
Figure Reference: Barag 1985: pl. 20

Nippur
Nip1 (Plate 47)
Type: core-formed/piriform bottles
Description: opaque; grey; white vertical lines, this partly slightly downward drawn; steep, slightly inward sloping neck; round, outwardly thickened rim; round body tapering downwards; pointed base
Context: Burial 4B 77; locus NT 25 I 2; Level II; possibly juvenile individual in sitting position; glass vessel on left foot
Date: 7th – 5th century
Location: Iraq Museum, Baghdad
Museum Number/Object Number/Field Number: 58275/4N 133
Reference: MacCown 1978: 49, 62, 99
Figure Reference: MacCown 1978: 56, no. 8

Nip2 (Plate 51)
Type: core-formed/bottles with disc-base
Dimensions: ht.: 8.8/rim: 4.8
Description: opaque; dark; white garland decoration on body; horizontal stripes on neck; straight neck; oval body; ring bottom
Date: mid 6th – early 5th century
Location: Istanbul Arkeoloji Müzeleri
Museum Number/Object Number/Field Number: 2200
Reference: Barag 1970: 158, no. 1
Figure Reference: Barag 1970: fig. 55

Nip3 (Plate 54)
Type: core-formed/large cylindrical bottles
Dimensions: ht.: 8.2/rim: 4.5
Description: opaque; grey; white feather decor
Context: Burial; Level IV; Loc. TA 51 IV: TA (Tablet Hill)
Date: 8th – late 7th century
Location: Iraq Museum, Baghdad
Museum Number/Object Number/Field Number: IM 55970/2N 740
Reference: Barag 1970: 158, no. 2; MacCown *et al.* 1967: 79
Figure Reference: Barag 1970: fig. 56; MacCown et al. 1967: pl. 148, 4

Praenestre
Pr1 (Plate 16)
Type: cast-and-cut/hemispherical bowls
Dimensions: ht.: 7.5/w.: 10
Description: translucent; turquoise, greenish tinge; many small hemispherical bubbles; iridescent in places; matt, grey surface; ground edge; varying wall thickness; uneven base
Context: burial 'Tomba Bernardini'
Date: *terminus ante quem* 650
Location: Museo Preistorico e Etnografico L. Pigorini, Rome
Reference: Brill 2012: 274–277; Fossing 1940: 36, 37; Saldern 1970: 225, no. 42; Saldern 2004: 60
Figure Reference: Saldern 1970: fig. 37

Samaria
Sam1 (Plate 27)
Type: cast-and-cut/rosette inlays (group 1)
Description: opaque; blue; six-petalled rosette without inner circle; square
Context: 'Ivory House'
Date: Neo-Assyrian
Location: Rockefeller-Museum, Jerusalem
Reference: Crowfoot and Crowfoot 1938: 44, no. 5
Figure Reference: Crowfoot and Crowfoot 1938: colour pl. 5; pl. 24, 2; Curtis 1999: 63, fig. 8

Sam2 (Plate 27)
Type: cast-and-cut/rosette inlays (group 1)
Description: opaque; blue; six-petalled rosette without inner circle; square
Context: 'Ivory House'
Date: Neo-Assyrian
Location: Rockefeller-Museum, Jerusalem
Reference: Crowfoot and Crowfoot 1938: 44, no. 5
Figure Reference: Crowfoot and Crowfoot 1938: pl. 24, 2

Sam3 (Plate 30)
Type: cast-and-cut/rosette inlays (group 3)
Description: translucent; dark blue; white filling; round
Date: Neo-Assyrian
Context: 'Ivory House'
Location: Rockefeller-Museum, Jerusalem
Reference: Crowfoot and Crowfoot 1938: 45, no. 10
Figure Reference: Crowfoot and Crowfoot 1938: colour pl. 10; pl. 24, 5

Sam4 (Plate 27)
Type: cast-and-cut/rosette inlays (group 1)
Description: opaque; blue; six-petalled rosette without inner circle; square
Context: 'Ivory House'
Date: Neo-Assyrian
Location: Rockefeller-Museum, Jerusalem
Reference: Crowfoot and Crowfoot 1938: 44, no. 5

Figure Reference: Crowfoot and Crowfoot 1938: colour pl. 5; pl. 24, 2; Curtis 1999: 63, fig. 8

Sam5 (Plate 27)
Type: cast-and-cut/rosette inlays (group 1)
Description: opaque; blue; six-petalled rosette without inner circle; square
Context: 'Ivory House'
Date: Neo-Assyrian
Location: Rockefeller-Museum, Jerusalem
Reference: Crowfoot and Crowfoot 1938: 44, no. 5
Figure Reference: Crowfoot and Crowfoot 1938: colour pl. 5; pl. 24, 2; Curtis 1999: 63, fig. 8

Sam6 (Plate 34)
Type: cast-and-cut/small monochrome inlays, triangle-shaped
Description: opaque, green-yellow corroded; triangular inlay with rounded sides
Context: 'Ivory House'
Date: Neo-Assyrian
Location: Rockefeller-Museum, Jerusalem
Reference: Crowfoot and Crowfoot 1938: 44, 2
Figure Reference: Crowfoot and Crowfoot 1938: pl. 24, 2

Sam7 (Plate 37)
Type: cast-and-cut/small monochrome inlays, figurative motif
Description: translucent; blue; lotus-shaped insert; surface slightly convex
Context: 'Ivory House'
Date: Neo-Assyrian
Location: Rockefeller-Museum, Jerusalem
Reference: Crowfoot and Crowfoot 1938: 44, no. 1
Figure Reference: Crowfoot and Crowfoot 1938: colour pl. 1

Sam8 (Plate 37)
Type: cast-and-cut/small monochrome inlays, simple geometric motif
Description: translucent; dark blue; white centre
Context: 'Ivory House'
Date: Neo-Assyrian
Location: Rockefeller-Museum, Jerusalem
Reference: Crowfoot and Crowfoot 1938: 45, no. 11
Figure Reference: Crowfoot and Crowfoot 1938: colour pl. 11; pl. 24, 2

Sam9 (Plate 37)
Type: cast-and-cut/small monochrome inlays, figurative motif
Description: opaque; red; slightly greenish corroded; human leg; slightly convex upper side
Context: 'Ivory House'
Date: Neo-Assyrian
Location: Rockefeller-Museum, Jerusalem
Reference: Crowfoot and Crowfoot 1938: 44, no. 5
Figure Reference: Crowfoot and Crowfoot 1938: colour pl. 4

Sam10 (Plate 37)
Type: cast-and-cut/small monochrome inlays, figurative motif
Description: translucent; blue; irregular
Context: 'Ivory House'
Date: Neo-Assyrian
Location: Rockefeller-Museum, Jerusalem
Figure Reference: Crowfoot and Crowfoot 1938: pl. 24, 11

Sam11 (Plate 34)
Type: cast-and-cut/small monochrome inlays, scale-shaped
Description: translucent; dark blue; irregular; scale-shaped; side edges sloping outwards
Context: 'Ivory House'
Date: Neo-Assyrian
Location: Rockefeller-Museum, Jerusalem
Reference: Crowfoot and Crowfoot 1938: 45, no. 9
Figure Reference: Crowfoot and Crowfoot 1938: colour pl. 9

Sam12 (Plate 33)
Type: cast-and-cut/small monochrome inlays, wing-shaped
Description: translucent; dark blue; wing-shaped
Context: 'Ivory House'
Date: Neo-Assyrian
Location: Rockefeller-Museum, Jerusalem
Museum Number/Object Number/Field Number: IAA 1933-2648
Figure Reference: Crowfoot and Crowfoot 1938: colour pl. 8; pl. 24, 11

Sam13 (Plate 34)
Type: cast-and-cut/small monochrome inlays, triangle-shaped
Description: opaque, green-yellow corroded; triangular inlay with rounded sides
Context: 'Ivory House'
Date: Neo-Assyrian
Location: Rockefeller-Museum, Jerusalem
Reference: Crowfoot and Crowfoot 1938: 44, no. 2
Figure Reference: Crowfoot and Crowfoot 1938: colour pl. 2; pl. 24, 2

Sam14 (Plate 33)
Type: cast-and-cut/small monochrome inlays, wing-shaped
Description: blue; winged
Context: 'Ivory House'
Date: Neo-Assyrian
Location: Rockefeller-Museum, Jerusalem
Figure Reference: Crowfoot and Crowfoot 1938: pl. 24, 11

Sam15 (Plate 37)
Type: cast-and-cut/small monochrome inlays, simple geometric motif
Description: translucent; dark blue; leaf-shaped
Context: 'Ivory House'
Date: Neo-Assyrian
Location: Rockefeller-Museum, Jerusalem
Reference: Crowfoot and Crowfoot 1938: 45, no. 12
Figure Reference: Crowfoot and Crowfoot 1938: pl. 12

Sam16 (Plate 37)
Type: cast-and-cut/small monochrome inlays, figurative motif
Description: curved; convex surface
Context: 'Ivory House'
Date: Neo-Assyrian
Location: Rockefeller-Museum, Jerusalem
Figure Reference: Crowfoot and Crowfoot 1938: pl. 24, 11

Sam17 (Plate 37)
Type: cast-and-cut/small monochrome inlays, figurative motif
Description: translucent, blue, irregular
Context: 'Ivory House'
Date: Neo-Assyrian
Location: Rockefeller-Museum, Jerusalem
Figure Reference: Crowfoot and Crowfoot 1938: pl. 24, 11

Sam18 (Plate 37)
Type: cast-and-cut/small monochrome inlays, figurative motif
Description: opaque; red; slightly greenish corroded; human leg; slightly convex upper side
Context: 'Ivory House'
Date: Neo-Assyrian
Location: Rockefeller-Museum, Jerusalem
Figure Reference: Crowfoot and Crowfoot 1938: pl. 24, 11

Sam19 (Plate 37)
Type: cast-and-cut/small monochrome inlays, simple geometric motif
Description: translucent; dark blue; top and underside irregular; diamond-shaped, side edges sloping outwards towards underside
Context: 'Ivory House'
Date: Neo-Assyrian
Location: Rockefeller-Museum, Jerusalem
Figure Reference: Crowfoot and Crowfoot 1938: pl. 24, 11

Sam20 (Plate 37)
Type: cast-and-cut/small monochrome inlays, figurative motif
Description: translucent, blue, irregular
Context: 'Ivory House'
Date: Neo-Assyrian
Location: Rockefeller-Museum, Jerusalem
Figure Reference: Crowfoot and Crowfoot 1938: pl. 24, 11

Sam21 (Plate 36)
Type: cast-and-cut/small monochrome inlays, floral motif
Description: slightly curved; convex upper side
Context: 'Ivory House'
Date: Neo-Assyrian
Location: Rockefeller-Museum, Jerusalem
Figure Reference: Crowfoot and Crowfoot 1938: pl. 24, 11

Sam22 (Plate 34)
Type: cast-and-cut/small monochrome inlays, scale-shaped
Description: opaque; yellow; irregular; scaly; side edges sloping outwards
Context: 'Ivory House'
Date: Neo-Assyrian
Location: Rockefeller-Museum, Jerusalem
Figure Reference: Crowfoot and Crowfoot 1938: pl. 24, 1

Sultantepe

Su1 (Plate 47)
Type: core-formed/piriform bottles
Dimensions: ht.: 15/w.: 6
Description: opaque; light blue; feather decoration; neck drawn outwards, round, vessel body tapering strongly downwards; pointed base; handle
Context: Room M. 2; 'hoard'
Date: 7th – 5th century
Reference: Barag 1970: 156; Barnett 1953: 50
Figure Reference: Barag 1970: fig. 48; Barnett 1953: pl. VII, f

Susa

Sus1 (Plate 47)
Type: core-formed/piriform bottles
Dimensions: ht.: 16/w.: 5.6
Description: opaque; white-beige; white, yellow feather finish; heavily corroded; straight neck; round edge; oval body; pointed bottom; surface with vertical grooves; 'duck-head' handle on shoulder
Context: burial
Date: late 7th – 6th century
Location: Musée du Louvre, Paris
Museum Number/Object Number/Field Number: SB 3385
Reference: Amiet 1966: 502, no. 377; Barag 1985: 162, no. 1
Figure Reference: Amiet 1966: 502, no. 377; Barag 1985: fig. 61

Sus2 (Plate 47)
Type: core-formed/piriform bottles
Dimensions: ht.: 10.7/w.: 6.5
Description: opaque; green-grey; white feather decor; spiral on neck; wavy on body; strong vertical depression; wide neck; round edge; oval body; pointed bottom; 'duck-head' handle on shoulder
Context: burial
Date: late 7th – 6th century
Location: Musée du Louvre, Paris

Museum Number/Object Number/Field Number: SB 3386
Reference: Barag 1970: 162, no. 5; Mequenem 1931: 334
Figure Reference: Barag 1970: fig. 63; Mequenem 1931: pl. II, 2

Sus3 (Plate 47)
Type: core-formed/piriform bottles
Dimensions: ht.: 12.6/w.: 5.6
Description: opaque; green-grey corroded; yellow feather decor, white corroded in places; long broad neck; oval body; pointed base; 'duck-head' handle on shoulder
Context: burial
Date: late 7th – 6th century
Location: Musée du Louvre, Paris
Museum Number/Object Number/Field Number: SB 455
Reference: Barag 1970: 162, no. 3
Figure Reference: Barag 1970: fig. 62

Sus4 (Plate 47)
Type: core-formed/piriform bottles
Dimensions: ht.: 11.5/w.: 6
Description: opaque; grey corroded; white feather finish; long, wide neck; round, thick rim; oval body; pointed bottom; 'duck-head' handle on shoulder
Context: burial
Date: late 7th – 6th century
Location: Musée du Louvre, Paris
Museum Number/Object Number/Field Number: SB 2809
Reference: Barag 1970: 162–163, no. 7
Figure Reference: Barag 1970: fig. 64

Ur

Ur1
Type: cast-and-cut/composite attachments and inlays
Dimensions: ht.: 3.9/w.: 4.9/th.: 0.9–1.2
Description: opaque; turquoise; two rows of curls; back straight
Location: British Museum, London
Museum Number/Object Number/Field Number: 1930, 1213.675
Reference: Barag 1985: 76, no. 66
Figure Reference: Barag 1985: pl. 8

Ur2 (Plate 11)
Type: cast-and-cut/jars and alabastra
Dimensions: ht.: 7.5/th.: 0.7–1.4
Description: transparent; colourless with greenish tinge; thick yellowish corrosion layer; surface smooth inside; strongly outwardly pulling, pointed round edge; very round wall; round, very thick base; increasing wall thickness towards the base
Context: Enunmah; room 5: 'found let into the brick pavement of Nebuchadnezzar' (Woolley 1962: 110, U. 791)
Date: *terminus ante quem* Nebuchadnezzar II (reign 605–562)
Location: British Museum, London
Museum Number/Object Number/Field Number: BM 116543; 1923, 1110.144/U. 791
Reference: Barag 1985: 67, no. 43; Woolley 1962: 110, U. 791
Figure Reference: Barag 1985: fig. 4; pl. 5

Ur3 (Plate 46)
Type: core-formed/ ovoid bottles with pointed base
Dimensions: ht.: 9.8
Description: opaque; dark; white thread decoration; yellow handle; long, straight neck; oval wall; tapering downwards; handle below the shoulder; round bottom
Date: 8th – early 6th century
Location: Iraq Museum, Baghdad
Museum Number/Object Number/Field Number: IM 65862
Reference: Barag 1970: 157 no. 3
Figure Reference: Barag 1970: fig. 50

Ur4 (Plate 47)
Type: core-formed/piriform bottles
Dimensions: ht.: 10/rim: 4.5
Description: opaque; blue-grey; white feather decor; long neck waved by threads; round wall, very tapered towards the bottom; 'duck-head' handle; pointed, round bottom
Context: foundation of a dwelling
Date: 7th – 5th century
Location: University of Pennsylvania, Museum of Archaeology and Anthropology
Museum Number/Object Number/Field Number: 31-43-231/U. 17062
Reference: Barag 1970: 157, no. 2; Woolley 1965: 78, 106
Figure Reference: Barag 1970: fig. 49; Woolley 1965: pl. 28

Ur5 (Plate 55)
Type: core-formed/ovoid bottles with rounded base
Dimensions: ht.: 8.7/rim: 4.8
Description: opaque; green; white garland decoration on vessel body; horizontal thread decoration on shoulder and neck; broken neck; oval body; tapering downwards; round bottom; 'duck-head' handle.
Context: Burial 26
Date (Context): *terminus post quem* 8th century
Date (Typology): 8th – 5th century
Location: Iraq Museum, Baghdad
Museum Number/Object Number/Field Number: IM 18482/U. 17395
Reference: Barag 1970: 157 no. 50; Kühne 1971: 422; Woolley 1962: 60, 126
Figure Reference: Barag 1970: fig. 50

Ur6 (Plate 50)
Type: core-formed/small jars
Dimensions: ht.: 8/w.: 6
Description: opaque; dark green corroded; white thread decoration irregular zigzag; not sufficiently marbled; thread-decoration on rim; short, wide neck;

spherical body; straight bottom; very irregular vessel surface
Context: dwelling on Site XNCF; Level II
Date: early 7th century
Location: Iraq Museum, Baghdad
Museum Number/Object Number/Field Number: IM 18485/U. 18125
Reference: Barag 1970: 157, no. 4
Figure Reference: Ådahl and Strommenger 1978: 180, no. 147; Barag 1970: fig. 51; Saldern 2004: pl. 9, 44

Ur7 (Plate 55)
Type: core-formed/vessel fragments
Dimensions: ht.: 6/w.: 5.2
Description: opaque; white; blue-grey feather decor; sherd
Location: Iraq Museum, Baghdad
Museum Number/Object Number/Field Number: IM 689/U. 1404
Reference: Barag 1970: 157, no. 6; Woolley 1965: 101
Figure Reference: Barag 1970: fig. 53; Woolley 1965: pl. 37

Uruk
Urk1 (Plate 55)
Type: core-formed/vessel fragments
Dimensions: ht.: 1.8/w.: 0.9
Description: opaque; dark; yellowish-green thread decoration in zigzag pattern
Context: Qa 14-5; pit in Eanna
Date: Neo-Assyrian/Neo-Babylonian
Location: Warka-Sammlung, Heidelberg
Museum Number/Object Number/Field Number: W 18186
Reference: van Ess and Pedde 1992: 160, no. 1188
Figure Reference: van Ess and Pedde 1992: pl. 95, 1188

Urk2 (Plate 55)
Type: core-formed/vessel fragments
Description: fragment of bottom
Context: Burial 'Doppeltopfgrab' 129; +2,32m NNO/SSW
Date: 7th – 6th century
Location: Iraq Museum, Baghdad
Museum Number/Object Number/Field Number: W 20969, 4a
Reference: van Ess and Pedde 1992: 160, no. 1189
Figure Reference: van Ess and Pedde 1992: pl. 95, 1189

Urk3 (Plate 55)
Type: core-formed/vessel fragments
Description: bottom fragment of a small vessel with pierced rim
Context: Burial 'Doppeltopfgrab' 129; +2,32m NNO/SSW
Date: 7th – 6th century
Location: Iraq Museum, Baghdad
Museum Number/Object Number/Field Number: W 20969, 4b
Reference: van Ess and Pedde 1992: 160, no. 1190
Figure Reference: van Ess and Pedde 1992: pl. 95, 1190

Ziyaret Tepe
Ziy1 (Plate 56)
Type: core-formed/vessel fragments
Description: opaque; blue; yellow thread decoration; spiral on neck; feather decor on body; handle broken off; yellow base preserved; slightly outward drawing neck; round edge
Context: burial; 'Brandgrubengrab' N-070
Date: Neo-Assyrian
Location: Tušḫan Archaeological Project

Objects with no provenience (Art Market)

AM1 (Plate10)
Type: cast-and-cut/mace-heads
Dimensions: ht.: 4/w.: max.4.5
Description: opaque; turquoise; grey weathering layer; pitting; round; cylindrical shaft hole
Date: Neo-Assyrian
Location: British Museum, London
Museum Number/Object Number/Field Number: BM 136777
Reference: Barag 1985: 75, no. 61
Figure Reference: Barag 1985: fig. 6, pl. 8

AM2 (Plate9)
Type: cast-and-cut/pallets
Dimensions: ht.: 1.54/rim: 7.8
Description: translucent; blue; silver iridescent; heavy pitting; heavily corroded; straight top; central semicircular recess; straight bottom; groove below edge
Date: Iron Age II
Location: Israel Department of Antiquities and Museums, Jerusalem
Museum Number/Object Number/Field Number: 68-365
Reference: Barag 1982: 11
Figure Reference: Barag 1982: 12 fig. 3a, b, 4

AM3 (Plate9)
Type: cast-and-cut/pallets
Dimensions: ht.: 1.5/rim: 5.5
Description: transparent; colourless, slightly greenish cast; white corrosion layer in places; iridescence; pitting; straight upper side, underside; grooving below the edge
Date: Iron Age II
Location: Private collection
Reference: Barag 1982: 11
Figure Reference: Barag 1982: 12 fig. 5a, b, 6

AM4 (Plate11)
Type: cast-and-cut/jars and alabastra
Dimensions: ht.: max. 7.2 /w.: 8/rim: 5.2/base: 5.4–5.6/ th.: 0.7–1.4
Description: transparent; colourless, light greenish tinge; small round bubbles evenly distributed in the vessel; brownish weathering layer; pitting; very

strongly iridescent surface; rim sloping inwards, tapering to a point; round body, slightly tapering downwards; shoulders slightly rounded; disc base; 'duck-head' handle below the shoulder
Date: Neo-Assyrian
Location: British Museum, London
Museum Number/Object Number/Field Number: BM 91461/N. 2071
Reference: Barag 1985: 61, no. 27; Saldern 1959: 29; Saldern 1966a: 626, 627; Saldern 1970: 218, no. 18
Figure Reference: Barag 1985: fig. 2, 27; pl. B, 27; Saldern 1959: 29, fig. 5; Saldern 1966a, 626, no. 585; Saldern 1970: fig. 18

AM5 (Plate19)
Type: cast-and-cut/shallow, undecorated and ribbed and petalled bowls
Dimensions: w.: 6.9/th.: 0.1–0.4
Description: transparent; colourless, yellowish-greenish tinge; elongated bubbles; white corrosion layer; pitting; iridescence; central base fragment, strongly inward pressed oval centre of a rosette; five petals arranged radially around the centre
Date: Neo-Assyrian
Location: British Museum, London
Museum Number/Object Number/Field Number: BM 127342
Reference: Barag 1985: 66, 67, no. 42
Figure Reference: Barag 1985: fig. 3; pl. 5

AM6 (Plate13)
Type: cast-and-cut/jars and alabastra
Description: translucent-transparent; long body; sharp edges at edge and shoulder; 'duck-head' handle on shoulders
Date: Post-Assyrian
Location: The J. Paul Getty Museum, Los Angeles
Museum Number/Object Number/Field Number: JPGM 2004.16.
Reference: Wight 2011: fig. 12
Figure Reference: Wight 2011: fig. 12

AM7 (Plate12)
Type: cast-and-cut/jars and alabastra
Dimensions: ht.: 21.3/w.: 3.9 /rim: 3.4/th.: 0.3–0.7
Description: transparent; colourless, greenish tinge; very slight pitting; only few very small, round bubbles; surface smooth on the outside; surface smooth on the inside; edge pulling outwards; sloping inwards; neck pulling outwards; shoulder set off by a prominent heel; round bottom; 'duck-head' handle; transition from edge to neck rounded; cannulation is straight, slightly widening outwards
Date: *terminus post quem* 6th century
Location: British Museum, London
Museum Number/Object Number/Field Number: 1869, 0624.16
Reference: Barag 1985: 67, no. 44; Saldern 1970: 227, no. 51
Figure Reference: Barag 1985: fig. 4; colour pl. B; pl. 5; Saldern 1970: fig. 47

AM8 (Plate12)
Type: cast-and-cut/jars and alabastra
Dimensions: ht.: 22.5/w.: 4.1/rim: 3.2/base: 3.9
Description: transparent; turquoise; very large, elongated, vertical bubble in the middle of the vessel; elongated, fine cracks on surface; straight neck; outward drawing edge, sharp edge, sloping obliquely towards the opening; prominent shoulder wrap; straight, widening body; round bottom
Date: *terminus post quem* 6th century
Location: The Corning Museum of Glass, Corning
Museum Number/Object Number/Field Number: 62.1.22
Reference: Saldern 1970: 226, no. 48; Goldstein 1979: 102, no. 200
Figure Reference: Goldstein 1979: 102, no. 200, pl. 12, 37; Saldern 1970: fig. 44

AM9 (Plate12)
Type: cast-and-cut/jars and alabastra
Dimensions: ht.: 18.1/w.: 6.9/rim: 3.3/th.: 0.5–0.8
Description: transparent; colourless; light iridescence; bubbles evenly distributed on handle and neck; slightly outward pulling neck; edge pointed outward pulling; surface smooth; prominent shoulders; elongated round body; round bottom; wall thickness thickening towards the bottom
Date: *terminus post quem* 6th century
Location: Metropolitan Museum of Art, New York
Museum Number/Object Number/Field Number: 74.51.312
Reference: Saldern 1970: 227, no. 52
Figure Reference: Saldern 1970: fig. 47; Caubet 2014: 168, no. 60

AM10 (Plate13)
Type: cast-and-cut/jars and alabastra
Dimensions: ht.: 9/w.: 3.1/rim: 2.7
Description: transparent; colourless; slightly outward sloping neck; outwardly pulling edge, ground; slightly downward thickening body; round bottom; easy handling
Date: *terminus post quem* 6th century
Location: Musée du Louvre, Paris
Museum Number/Object Number/Field Number: CP 9193
Reference: Arveiller and Nenna 2000: 167, no. 195
Figure Reference: Arveiller and Nenna 2000: 167, no. 195

AM11 (Plate13)
Type: cast-and-cut/jars and alabastra
Dimensions: ht.: 14.8/w.: 4
Description: transparent translucent; greenish; heavily corroded; neck outward pulling; round rim; straight, slightly downward widening body; handling

in upper vessel area
Date: *terminus post quem* 6th century
Location: Boston Museum of Fine Arts, Boston
Museum Number/Object Number/Field Number: 87.50
Reference: Saldern 1970: 227, no.53
Figure Reference: Saldern 1970: fig. 48

AM12 (Plate13)
Type: cast-and-cut/jars and alabastra
Dimensions: w.: 14
Description: transparent; colourless; strongly outward drawing, cut edge; slightly round, downwardly strongly spreading body; small handles on side of body
Date: *terminus post quem* 6th century
Location: Erwin Oppenländer collection, Waiblingen
Museum Number/Object Number/Field Number: No.5038
Reference: Saldern 1970: 226, no. 49
Figure Reference: Saldern 1970: fig. 45

AM13 (Plate31)
Type: cast-and-cut/rosette inlays (group 3) with frame
Dimensions: ht.: 1.8/w.: 1.8/th.: 0.2
Description: opaque; formerly red, frame corroded to green; irregular; rectangular; inserted in the middle with rosette (type 3)
Date: 8th century
Location: Badisches Landesmuseum Karlsruhe
Museum Number/Object Number/Field Number: 72/48
Reference: Thimme 1973: no. 36
Figure Reference: Thimme 1973: no. 36

AM14 (Plate 31)
Type: cast-and-cut/rosette inlays (group 3) with frame
Dimensions: ht.: 1.8/w.: 1.8/th.: 0.2
Description: opaque; formerly red, frame corroded to green; irregular; rectangular; inserted in the middle with rosette (type 3)
Date: 8th century
Location: Badisches Landesmuseum Karlsruhe
Museum Number/Object Number/Field Number: 72/48
Reference: Thimme 1973: no. 36
Figure Reference: Thimme 1973: no. 36

AM15 (Plate31)
Type: cast-and-cut/rosette inlays (group 3) with frame
Dimensions: ht.: 1.8/w.: 1.8/th.: 0.2
Description: opaque; formerly red, frame corroded to green; irregular; rectangular; inserted in the middle with rosette (type 3)
Date: 8th century
Location: Badisches Landesmuseum Karlsruhe
Museum Number/Object Number/Field Number: 72/48
Reference: Thimme 1973: no. 36
Figure Reference: Thimme 1973: no. 36

AM16 (Plate31)
Type: cast-and-cut/rosette inlays (group 3) with frame
Dimensions: ht.: 1.8/w.: 1.8/th.: 0.2
Description: opaque; formerly red, frame corroded to green; irregular; rectangular; inserted in the middle with rosette (type 3)
Date: 8th century
Location: Badisches Landesmuseum Karlsruhe
Museum Number/Object Number/Field Number: 72/48
Reference: Thimme 1973: no. 36
Figure Reference: Thimme 1973: no. 36

AM17 (Plate31)
Type: cast-and-cut/rosette inlays (group 3) with frame
Dimensions: ht.: 1.8/w.: 1.8/th.: 0.2
Description: opaque; formerly red, frame corroded to green; irregular; rectangular; inserted in the middle with rosette (type 3)
Date: 8th century
Location: Badisches Landesmuseum Karlsruhe
Museum Number/Object Number/Field Number: 72/48
Reference: Thimme 1973: no. 36
Figure Reference: Thimme 1973: no. 36

AM18 (Plate31)
Type: cast-and-cut/rosette inlays (group 3) with frame
Dimensions: ht.: 1.8/w.: 1.8/th.: 0.2
Description: opaque; formerly red, frame corroded to green; irregular; rectangular; inserted in the middle with rosette (type 3)
Date: 8th century
Location: Badisches Landesmuseum Karlsruhe
Museum Number/Object Number/Field Number: 72/48
Reference: Thimme 1973: no. 36
Figure Reference: Thimme 1973: no. 36

AM19 (Plate31)
Type: cast-and-cut/rosette inlays (group 3) with frame
Dimensions: ht.: 1.8/w.: 1.8/th.: 0.2
Description: opaque; formerly red, frame corroded to green; irregular; rectangular; inserted in the middle with rosette (type 3)
Date: 8th century
Location: Badisches Landesmuseum Karlsruhe
Museum Number/Object Number/Field Number: 72/48
Reference: Thimme 1973: no. 36
Figure Reference: Thimme 1973: no. 36

AM20 (Plate31)
Type: cast-and-cut/rosette inlays (group 3) with frame
Dimensions: ht.: 1.8/w.: 1.8/th.: 0.2
Description: opaque; formerly red, frame corroded to green; irregular; rectangular; inserted in the middle with rosette (type 3)
Date: 8th century

Location: Badisches Landesmuseum Karlsruhe
Museum Number/Object Number/Field Number: 72/48
Reference: Thimme 1973: no. 36
Figure Reference: Thimme 1973: no. 36

AM21
Type: cast-and-cut/rosette inlays (group 3) with frame and bronze frame
Dimensions: ht.: 0.5
Description: irregular; rectangular; centred with rosette (type 3) inlaid; heavily corroded bronze frame
Date: 8th century
Location: Metropolitan Museum of Art, New York
Museum Number/Object Number/Field Number: 57.80.18a
Reference: Bianchi *et al.* 2002: 172
Figure Reference: Bianchi et al. 2002: 172, no. NE-24a

AM22
Type: cast-and-cut/rosette inlays (group 3) with frame and bronze frame
Dimensions: ht.: 0.5
Description: irregular; rectangular; centred with rosette (type 3) inlaid; heavily corroded bronze frame
Date: 8th century
Location: Metropolitan Museum of Art, New York
Museum Number/Object Number/Field Number: 57.80.18b
Reference: Bianchi *et al.* 2002: 172
Figure Reference: Bianchi et al. 2002: 172, no. NE-24a

AM23 (Plate32)
Type: cast-and-cut/rosette inlays (group 3) with frame and bronze frame
Description: irregular; rectangular; centred with rosette (type 3) inlaid; heavily corroded bronze frame
Date: 8th century
Location: Metropolitan Museum of Art, New York
Museum Number/Object Number/Field Number: 57.80.18c
Reference: Bianchi *et al.* 2002: 172
Figure Reference: Bianchi et al. 2002: 172, no. NE-24a

AM24 (Plate32)
Type: cast-and-cut/rosette inlays (group 3) with frame and bronze frame
Description: irregular; rectangular; centred with rosette (type 3) inlaid; heavily corroded bronze frame
Date: 8th century
Location: Metropolitan Museum of Art, New York
Museum Number/Object Number/Field Number: 57.80.18d
Reference: Bianchi *et al.* 2002: 172
Figure Reference: Bianchi et al. 2002: 172, no. NE-24a

AM25 (Plate32)
Type: cast-and-cut/rosette inlays (group 3) with frame and bronze frame
Description: irregular; rectangular; centred with rosette (type 3) inlaid; heavily corroded bronze frame
Date: 8th century
Location: Metropolitan Museum of Art, New York
Museum Number/Object Number/Field Number: 57.80.18e/
Reference: Bianchi *et al.* 2002: 172
Figure Reference: Bianchi et al. 2002: 172, no. NE-24a

AM26 (Plate32)
Type: cast-and-cut/rosette inlays (group 3) with frame and bronze frame
Description: irregular; rectangular; centred with rosette (type 3) inlaid; heavily corroded bronze frame
Date: 8th century
Location: Metropolitan Museum of Art, New York
Museum Number/Object Number/Field Number: 57.80.18f
Reference: Bianchi *et al.* 2002: 172
Figure Reference: Bianchi et al. 2002: 172, no. NE-24a

AM27 (Plate32)
Type: cast-and-cut/rosette inlays (group 3) with frame and bronze frame
Dimensions: ht.: 0.5
Description: irregular; rectangular; centred with rosette (type 3) inlaid; heavily corroded bronze frame
Date: 8th century
Location: Metropolitan Museum of Art, New York
Museum Number/Object Number/Field Number: 57.80.18g
Reference: Bianchi *et al.* 2002: 172
Figure Reference: Bianchi et al. 2002: 172, no. NE-24a

AM28 (Plate31)
Type: cast-and-cut/rosette inlays (group 3) with frame
Dimensions: ht.: 0.7/w.: 0.7/th.:0.3
Description: opaque; formerly red, frame corroded to green; irregular; rectangular; inserted in the middle with rosette (type 3)
Date: 8th century
Location: British Museum, London
Museum Number/Object Number/Field Number: BM 132815; 1960-409.1
Reference: Barag 1985: 72, no. 54; Barnett 1963: 84; Saldern 1966a: 633; Saldern 1970: 224, no. 40
Figure Reference: Barag 1985: pl. 7, 54; Barnett 1963: pl. XXXIId
Analyses: Bimson and Freestone 1985: 121, no. 54

AM29 (Plate31)
Type: cast-and-cut/rosette inlays (group 3) with frame
Dimensions: ht.: 0.6/w.: 0.6/rim: /base: /th.: 0.3
Description: opaque; formerly red, frame corroded to green; irregular; rectangular; inserted in the middle with rosette (type 3)
Date: 8th century
Location: British Museum, London
Museum Number/Object Number/Field Number:

BM 132816; 1960-409.2
Reference: Barag 1985: 72, no. 54; Barnett 1963: 84; Saldern 1970: 224, no. 40
Figure Reference: Barag 1985: pl. 7, 54; Barnett 1963: pl. XXXIId

AM30 (Plate31)
Type: cast-and-cut/rosette inlays (group 3) with frame
Dimensions: ht.: 0.6/w.: /0.7 rim: /base: /th.: 0.3
Description: opaque; formerly red, frame corroded to green; irregular; rectangular; inserted in the middle with rosette (type 3)
Date: 8th century
Location: British Museum, London
Museum Number/Object Number/Field Number: BM 132817; 1960-409.3
Reference: Barag 1985: 72, no. 54; Barnett 1963: 84; Saldern 1970: 224, no. 40
Figure Reference: Barag 1985: pl. 7, 54; Barnett 1963: pl. XXXIId

AM31 (Plate 38)
Type: cast-and-cut/small monochrome inlays, 'Lady at the Window'
Dimensions: ht.: 6.4/w.: 4.5/rim: /base: /th.: 1.2
Description: opaque; formerly red, frame corroded to green; irregular; rectangular; inserted in the middle with rosette (type 3)
Date: 8th century
Location: Metropolitan Museum of Art, New York
Museum Number/Object Number/Field Number: 57.80.12
Reference: Fontan 2007, 16. 212. 377
Figure Reference: http://www.metmuseum.org/art/collection#!?q=57.80.12&sortBy=Relevance&sortOrder=asc&page=1 (accessed: 08.08.2018)

AM32 (Plate33)
Type: cast-and-cut/small monochrome inlays, wing-shaped
Dimensions: ht.: 0.4–0.8/w.: 6.3/th.: 0.3
Description: translucent; dark blue; winged; straight side edges
Date: Neo-Assyrian
Location: Metropolitan Museum of Art, New York
Museum Number/Object Number/Field Number: 57.80.19a
Reference: http://www.metmuseum.org/art/collection#!?q=57.80.19a&sortBy=Relevance&sortOrder=asc&page=1 (accessed: 08.08.2018)

AM33 (Plate38)
Type: cast-and-cut/small monochrome inlays, 'Lady at the Window'
Dimensions: ht.: 4.5/w.: 2.7
Description: translucent; strongly iridescent; rectangular
Date: 8th century
Location: Metropolitan Museum of Art, New York
Museum Number/Object Number/Field Number: 57.80.13
Reference: http://www.metmuseum.org/art/collection#!?q=57.80.13&sortBy=Relevance&sortOrder=asc&page=1 (accessed: 08.08.2018)

AM34 (Plate33)
Type: cast-and-cut/small monochrome inlays, wing-shaped
Dimensions: ht.: 1.8/w.: 4.4/th.: 0.3
Description: translucent; dark blue; top smooth; bottom matt, rough; straight side edges
Date: Neo-Assyrian
Location: Metropolitan Museum of Art, New York
Museum Number/Object Number/Field Number: 57.80.19c

AM35
Type: cast-and-cut/small monochrome inlays, wing-shaped
Dimensions: ht.: 0.8–1/w.: 5.2/th.: 0.3
Description: opaque; blue-grey; dark blue stripes; surface smooth; underside yellowish-white corroded; surface slightly convex
Date: Neo-Assyrian
Location: Metropolitan Museum of Art, New York
Museum Number/Object Number/Field Number: 57.80.19d

AM36 (Plate38)
Type: cast-and-cut/small monochrome inlays, 'Lady at the Window'
Dimensions: ht.: 6.2/w.: 4.2
Description: transparent; colourless; strongly iridescent; rectangular
Date: 8th century
Location: Kunst- und Gewerbemuseum Hamburg
Museum Number/Object Number/Field Number: 1966.26
Reference: http://sammlungonline.mkg-hamburg.de/de/search?s=1966.26&h=0&sort=scoreDesc (accessed: 08.08.2018)
Figure Reference: http://sammlungonline.mkg-hamburg.de/de/search?s=1966.26&h=0&sort=scoreDesc (accessed: 08.08.2018)

AM37 (Plate31)
Type: cast-and-cut/rosette inlays
Description: translucent; dark blue; white filling; square
Date: Neo-Assyrian
Reference: Schlick-Nolte 1994: 59
Figure Reference: Schlick-Nolte 1994: 59, no. 88

AM38 (Plate32)
Type: cast-and-cut/rosette inlays (group 3) with frame and bronze frame
Dimensions: ht.: 0.5
Description: irregular; rectangular; centred with

rosette (type 3) inlaid; heavily corroded bronze frame
Date: 8th century
Location: Metropolitan Museum of Art, New York
Museum Number/Object Number/Field Number: 57.80.18h
Reference: Bianchi *et al.* 2002: 172
Figure Reference: Bianchi et al. 2002: 172, no. NE-24a

AM39 (Plate33)
Type: cast-and-cut/small monochrome inlays, wing-shaped
Dimensions: ht.: 0.8–1/w.: 4.2/th.: 0.3
Description: translucent; dark blue; upper side smooth, glossy; underside yellowish-white corroded
Date: Neo-Assyrian
Location: Metropolitan Museum of Art, New York
Museum Number/Object Number/Field Number: 57.80.19b

AM40 (Plate46)
Type: core-formed/ovoid bottles with pointed base
Dimensions: ht.: 10.8/w.: 3.6/rim: 2.4
Description: opaque; strongly eroded grey-beige; white-yellow thread decoration; in the neck area white feather decoration, spiral in the shoulder area, on body white feather decoration; yellow bottom and handle; yellow glass attacked by pitting; round bubbles on surface; slightly tapering neck; round edge with yellow thread support; gently rounded shoulder; round vessel body, strongly tapering towards the bottom; pointed bottom
Date: 8th – early 6th century
Location: The Corning Museum of Glass, Corning
Museum Number/Object Number/Field Number: 63.1.29
Reference: Barag 1970: 168, no. 5; Goldstein 1979: 105, no. 5, 205
Figure Reference: Barag 1970: fig. 77; Brill and Stapleton 2012: 597; Goldstein 1979: 105, no. 205

AM41 (Plate46)
Type: core-formed/ovoid bottles with pointed base
Dimensions: ht.: 10.5/w.: 3.7
Description: opaque; grey; white feather decor; tool marks on neck; straight neck; round body, slightly tapering downwards; handle broken; pointed base
Date: 8th – early 6th century
Location: Rockefeller-Museum, Jerusalem
Reference: Barag 1970: 168, no. 5.1
Figure Reference: Barag 1970: fig. 78

AM42 (Plate46)
Type: core-formed/ovoid bottles with pointed base
Dimensions: ht.: 10/w.: 3.6
Description: opaque; greenish; white thread decor; feather decor on body; wavy lines on shoulder; yellow light thread decor on neck; spherical body; short neck
Date: 8th – early 6th century
Location: Iraq Museum, Baghdad
Museum Number/Object Number/Field Number: IM 65863
Reference: Barag 1970: 157, no. 5; Kühne 1971: 422
Figure Reference: Barag 1970: fig. 51

AM43 (Plate46)
Type: core-formed/ovoid bottles with pointed base
Dimensions: ht.: 11.5/w.: 3.6
Description: opaque; dark brown; white feather decor on neck and body; near base spiral; long wide neck; elongated oval body; pointed round base; 'duck-head' handle on shoulder
Date: 8th – early 6th century
Location: Collection Foroughi, Tehran, Iran
Reference: Barag 1985: 163
Figure Reference: Barag 1985: fig. 65

AM44 (Plate47)
Type: core-formed/piriform bottles
Dimensions: ht.: 12.5/w.: 5.2
Description: opaque; brown-yellow; white-green feather decor; surface wavy; long, wide neck; transition from neck to shoulder with hollow; 'duck-head' handles differently shaped
Date: 7th – 5th century
Location: Metropolitan Museum of Art, New York
Museum Number/Object Number/Field Number: 17.194.586
Reference: Barag 1970: 168, no. 4
Figure Reference: Barag 1970: fig. 76

AM45 (Plate53)
Type: core-formed/large cylindrical bottles
Dimensions: ht.: 17/w.: /4.9 rim: 2.6/base: 4.2
Description: translucent; green-yellowish, dark blue in places; yellow-white thread decoration spiral around neck and shoulder; surface with vertical furrows; straight neck; round edge; round shoulders; 'duck-head' handle; pointed bottom
Date: 8th – late 7th century
Location: The Corning Museum of Glass, Corning
Museum Number/Object Number/Field Number: 59.1.65
Reference: Barag 1970: 169, no. 8; Goldstein 1979: 104, no. 204
Figure Reference: Barag 1970: fig. 80; Goldstein 1979: 105, no. 204, pl. 12

AM46 (Plate55)
Type: core-formed/ovoid bottles with rounded base
Dimensions: ht.: 8.6/w.: 2.6/rim: 1.3
Description: opaque; beige-brown, dark blue in places; strongly corroded thread decoration; begins below the shoulder; surface wavy; golden iridescence; long neck tapering inwards; round, irregular rim; oval, irregular vessel body; round bottom
Date Typology: 8th – 5th century
Location: Toledo Museum of Art, Toledo
Museum Number/Object Number/Field Number:

1961.39
Reference: Grose 1989: 85, no. 28, 69
Figure Reference: Grose 1989: 69, no. 28; 398, no. 28; Wittman 1969: 15

AM47 (Plate53)
Type: core-formed/large cylindrical bottles
Dimensions: ht.: 15/w.: 4.8/rim: 2.8/base: /th.: 0.4–0.5
Description: opaque; dark blue, white, yellow, turquoise thread decoration; spiral thread decoration on edge and neck, warped to thread decoration on body; white thread on handles; slightly cylindrical neck; straight, slightly cylindrical body; thickened, round edge; slightly convex bottom; two vertical 'duck-head' handle on shoulder
Date: 8th – late 7th century
Location: British Museum, London
Museum Number/Object Number/Field Number: 1860, 0404.97
Reference: Barag 1985: 70, no. 50; Barag 1970: 166, no. 6; Harden 1981: no. 80
Figure Reference: Barag 1985: pl. 7, B; Barag 1970: fig. 70; Harden 1981: pl. 7, 80

AM48 (Plate54)
Type: core-formed/large cylindrical bottles
Dimensions: ht.: 15.5/w.: 4.1/rim: 2.7
Description: opaque; dark blue; yellow-white feather decor; strongly corroded white in places; light iridescence; pitting; surface irregular, especially on neck; long, straight neck; round, irregular rim with thread support; body spreads slightly downwards; pointed base; a 'duck-head' handle
Date: 8th – late 7th century
Location: British Museum, London
Museum Number/Object Number/Field Number: 1860,0201.53
Reference: Barag 1970: 166, no. 7; Harden 1981: 56, no. 81
Figure Reference: Barag 1970: fig. 71; Harden 1981: pl. 7, no. 81

AM49 (Plate 31)
Type: cast-and-cut/rosette inlays (group 3) with frame (QC 4)
Dimensions: ht.: 1.8/w.: 1.8/th.: 0.2
Description: opaque; formerly red, frame corroded to green; irregular; rectangular; inserted in the middle with rosette (type 3)
Date: 8th century
Location: Badisches Landesmuseum Karlsruhe
Museum Number/Object Number/Field Number: 72/48
Reference: Thimme 1973: no. 36
Figure Reference: Thimme 1973: no. 36

AM50 (Plate46)
Type: core-formed/ovoid bottles with pointed base
Dimensions: ht.: 9.9/w.: 3.6/rim: 2.2
Description: opaque; grey-brown, dark blue in places; white, yellow thread decoration in places; broad feather decoration; golden iridescence; straight neck; round edge with yellow thread support; round shoulder area; round wall, tapering downwards; round tapering base; loop handle on shoulder
Date: 8th – early 6th century
Location: The Corning Museum of Glass, Corning
Museum Number/Object Number/Field Number: 74.1.36
Reference: Goldstein 1979: 104, 105, no. 206
Figure Reference: Goldstein 1979: 105, no. 206

AM51 (Plate50)
Type: core-formed/small jars
Dimensions: ht.: 4.5/w.: max. 4/ rim:3.1
Date: early 7th century
Location: Victoria and Albert Museum, London
Reference: Barag 1970: 169
Figure Reference: Barag 1970: fig. 82; Grose 1989: 77 fig. 43

Museum Number/Object Number/Field Number: c-6-1912

AM52 (Plate51)
Type: core-formed/bottles with disc-base
Dimensions: ht.: 7.5/w.: 3.5
Description: opaque; dark blue; white-yellow garland decoration; yellow handles; yellow horizontal bands around neck and bottom; straight neck; round-oval body tapering downwards; 'duck-head' handle; standing ring
Date: mid 6th – early 5th century
Location: Iraq Museum, Baghdad
Museum Number/Object Number/Field Number: IM 14484
Reference: Barag 1970: 170, no. 13
Figure Reference: Barag 1970: fig. 85

AM53 (Plate51)
Type: core-formed/bottles with disc-base
Dimensions: ht.: 8.3/w.: 3.8
Description: opaque; dark blue; white-yellow garland decoration; straight neck; round-oval body tapering downwards; 'duck-head' handles; base ring
Date: mid 6th – early 5th century
Location: Iraq Museum, Baghdad
Museum Number/Object Number/Field Number: IM 14397
Reference: Barag 1970: 170, no. 14
Figure Reference: Barag 1970: fig. 86

AM54 (Plate51)
Type: core-formed/bottles with disc-base
Dimensions: ht.: 7.2/w.: 3.4
Description: opaque; turquoise; yellow threads around neck and base; straight, slightly inward pulling neck; round-oval body tapering downwards; 'duck-head' handle

Date: mid 6th – early 5th century
Location: Iraq Museum, Baghdad
Museum Number/Object Number/Field Number: IM 14485
Reference: Barag 1970: 170, no. 15
Figure Reference: Barag 1970: fig. 87

AM55 (Plate50)
Type: core-formed/ small jars
Dimensions: ht.: 6/ w.: 5,5
Description: opaque; greenish; white thread decor; feather decor on body; wavy lines on shoulder; yellow light thread decor on neck; spherical body; short neck; two eyelets at neck level
Date: early 7th century
Location: Iraq Museum, Baghdad
Museum Number/ Object Number/ Field Number: IM 65863
Reference: Barag 1970, 157 no. 5; Kühne 1971, 422
Figure Reference: Barag 1970, fig. 51; Goldstein 1979:105 no. 206

As 1

S 1:1

0 5 cm

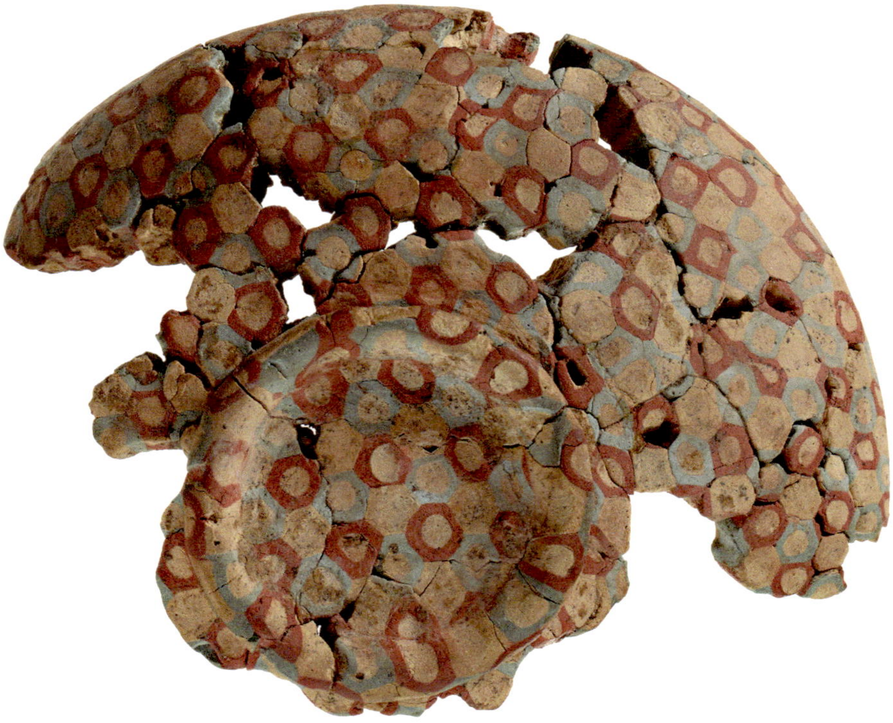

As2

S 1:1

0 5 cm

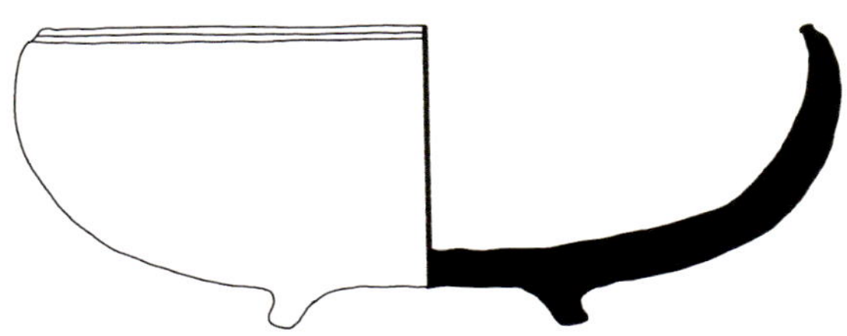

S 1:2

0 5 cm

As2

(not to scale)

BS3

As3

As4

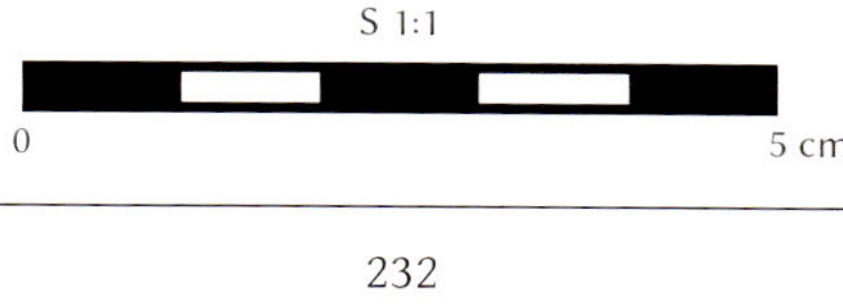

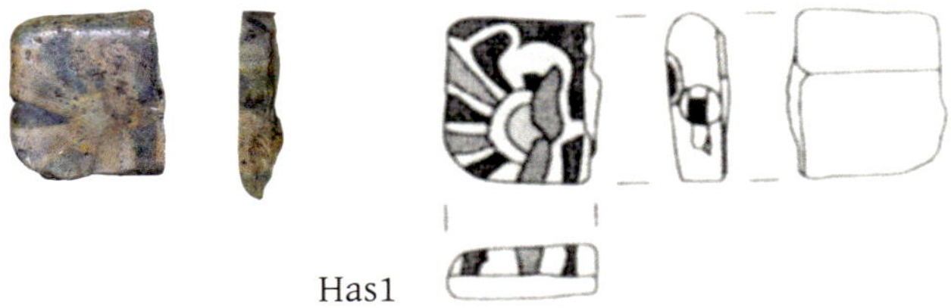

Has1

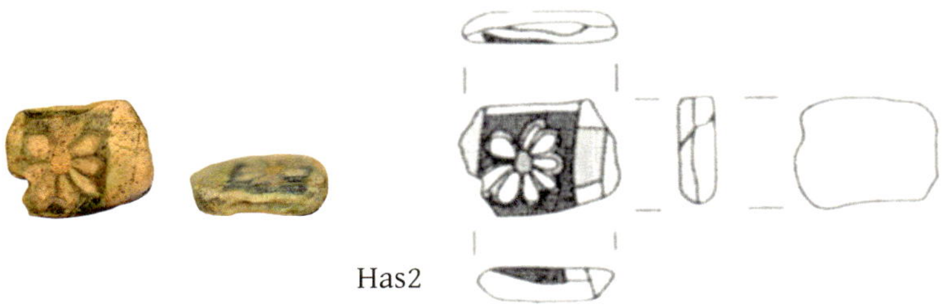

Has2

Has3

As6

As7

S 1:1

0 5 cm

As13

As14

As15

As17

As16

As18

S 1:1

0 5 cm

Meg1

S 1:1

0 5 cm

(not to scale)

(not to scale)

Meg1

S 1:1

0 5 cm

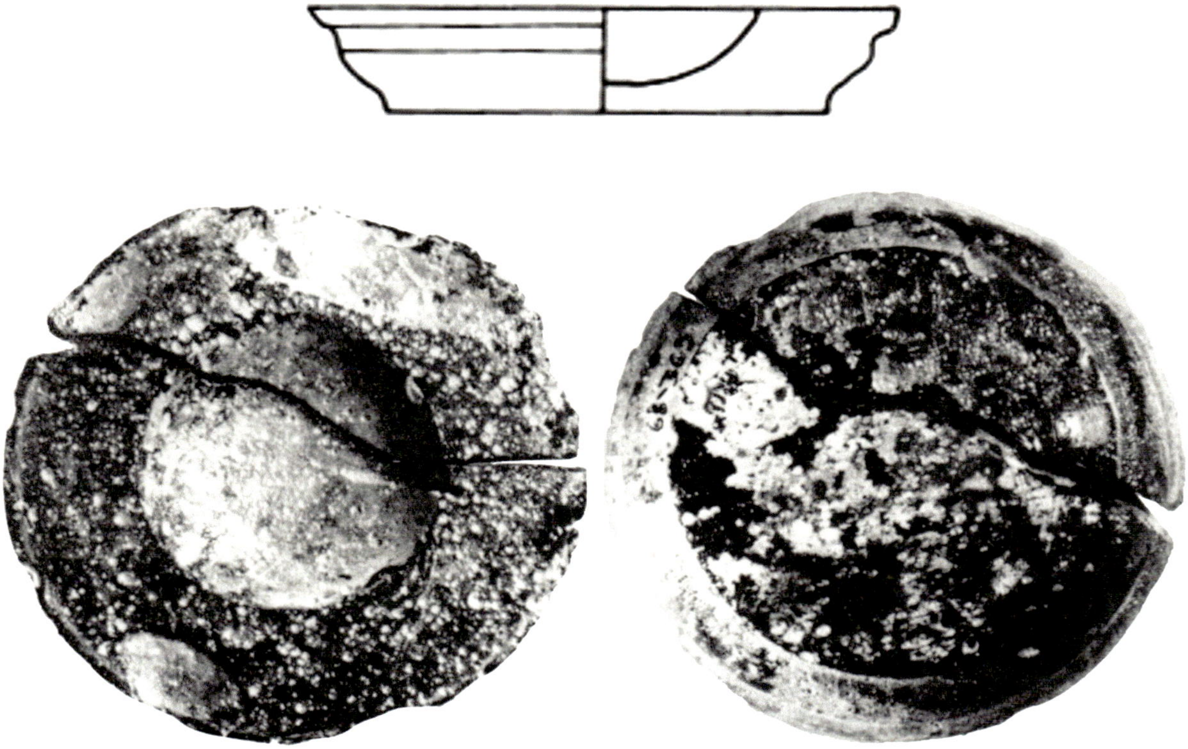

AM2

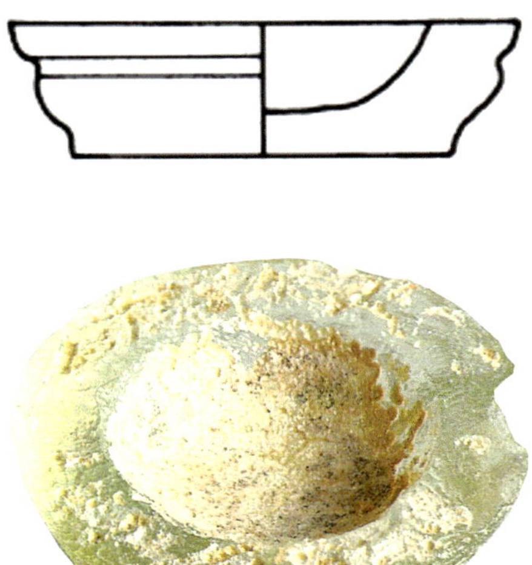

AM3

S 1:1

AM1

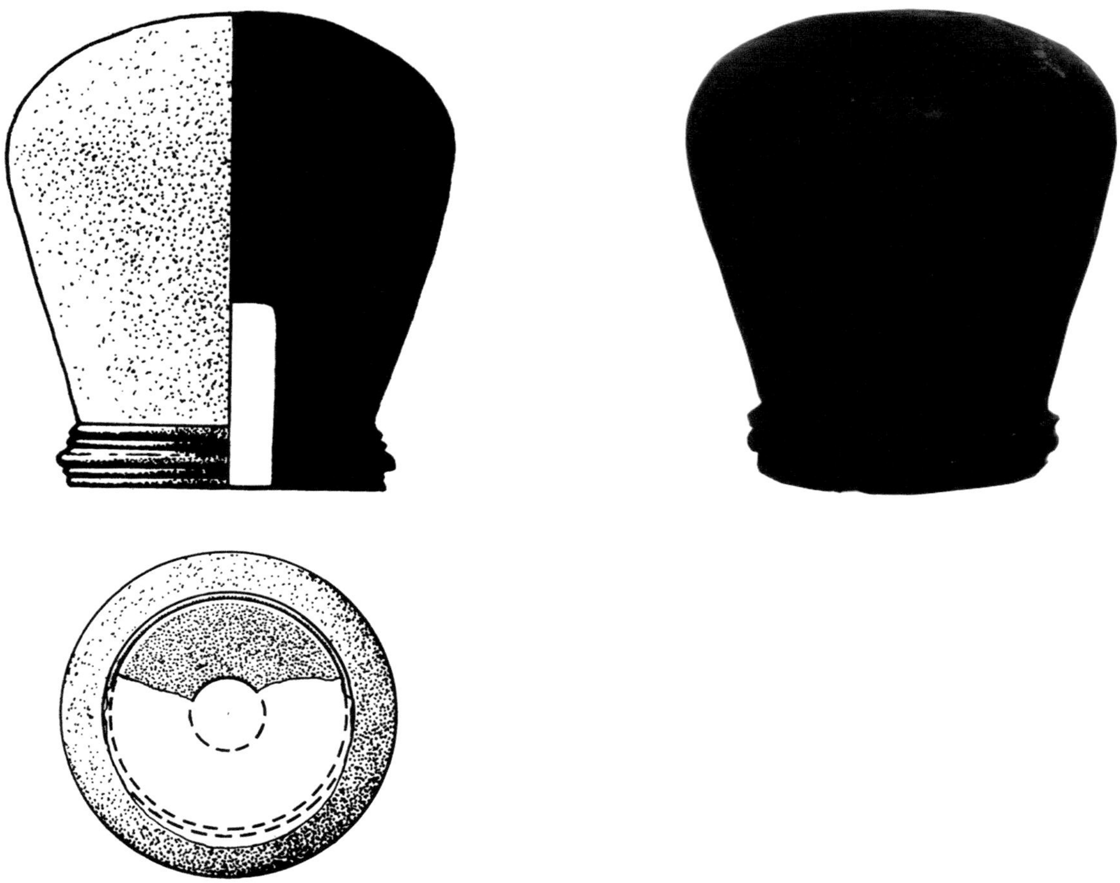

Nin2

S 1:1

0 5 cm

AM4

(not to scale)

Ur2

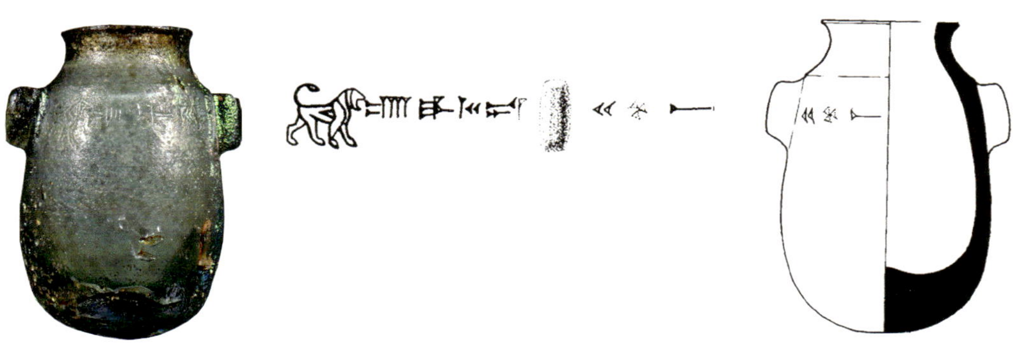

Nim27

S 1:2

0 5 cm

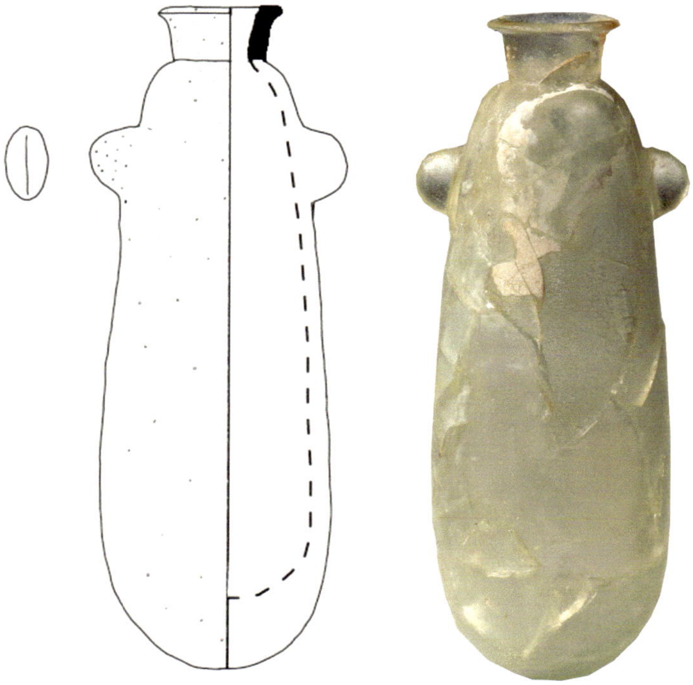

AM9

AM7

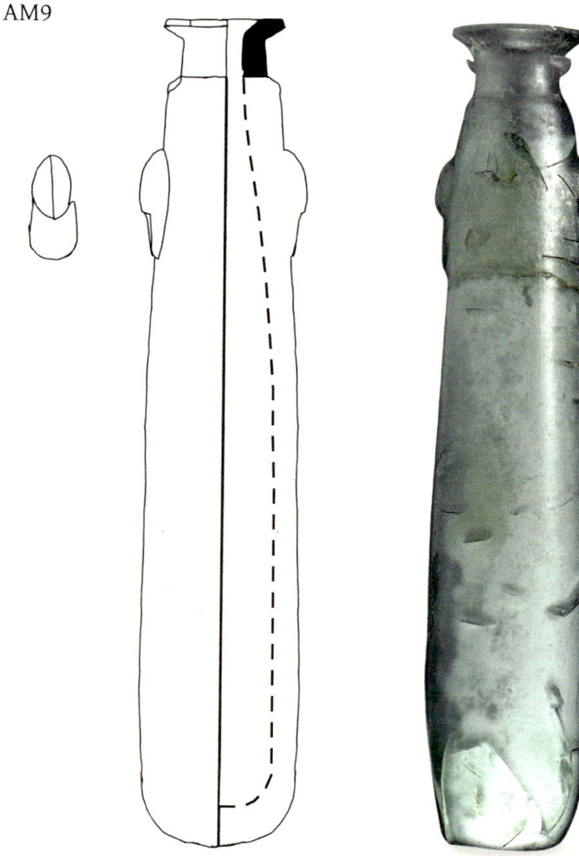

AM8

S 1:2

0 5 cm

AM6

AM11

AM12

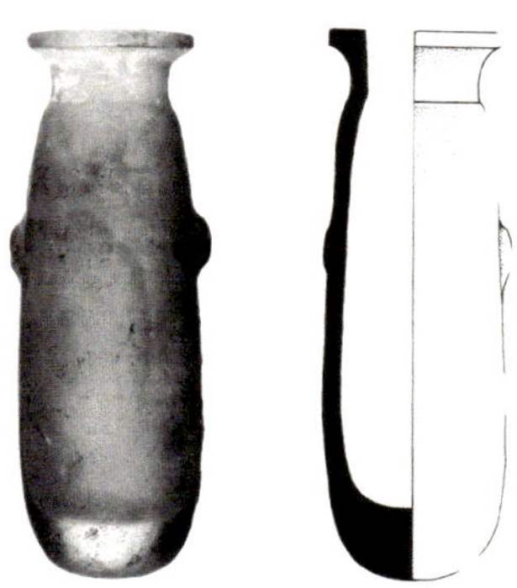

AM10

At1

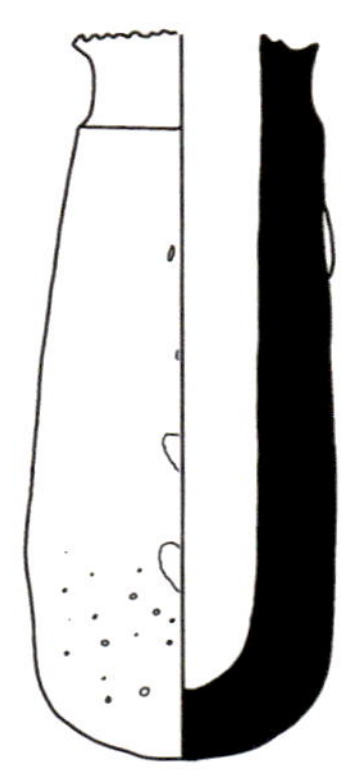

(not to scale)

Id1

S 1:2

0 5 cm

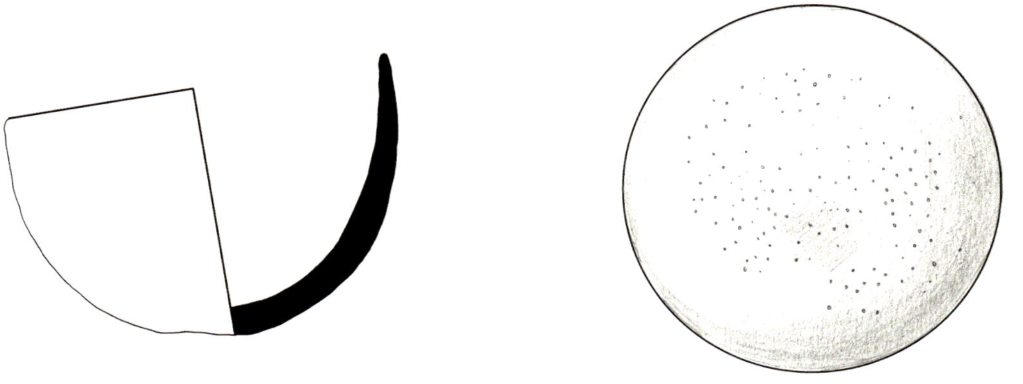

Nim7

Nim11

Nim13

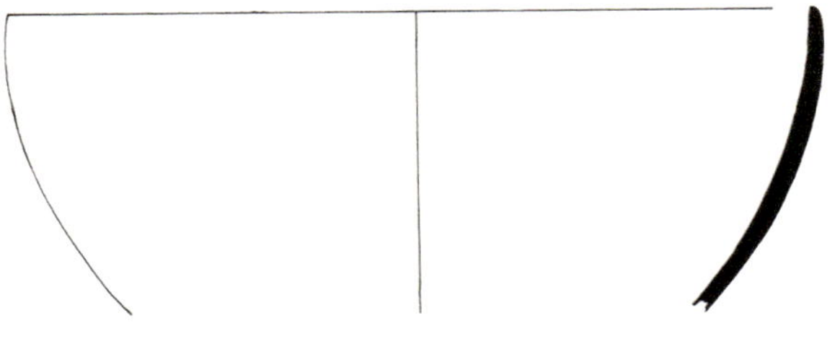

Nim12

Khor1
(not to scale)

S 1:2

0 5 cm

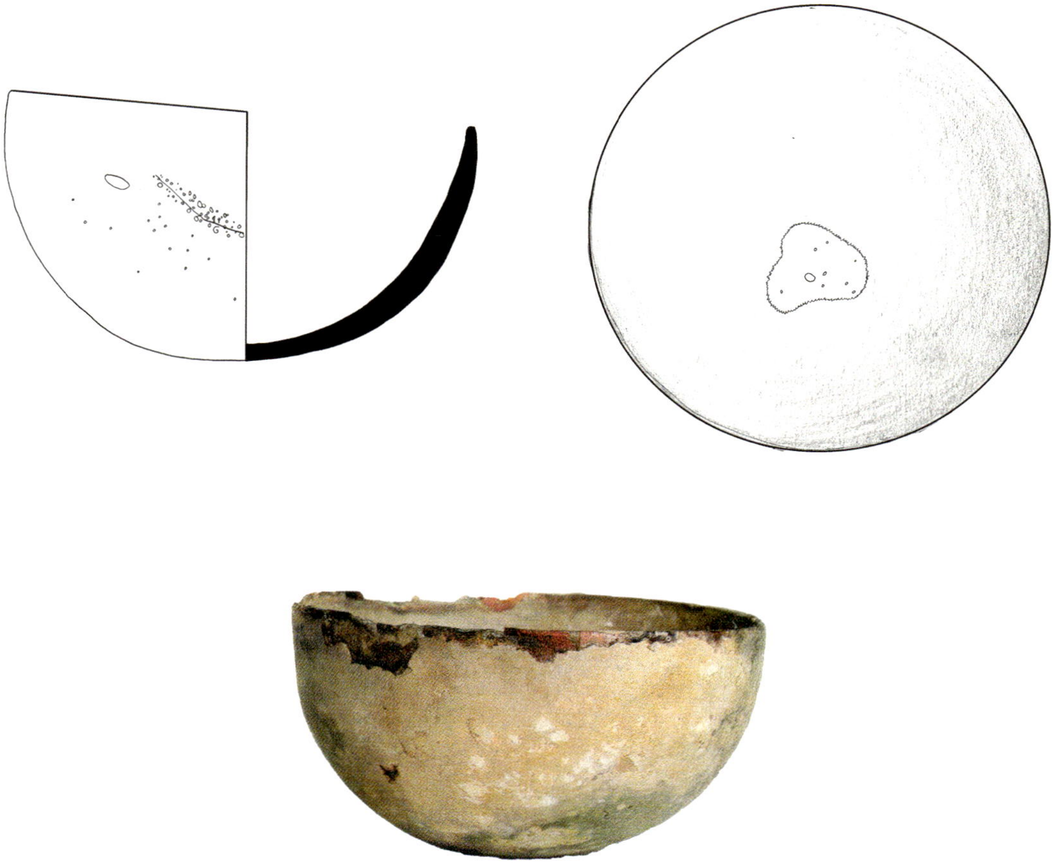

Nim8

Nim9

S 1:2

0 5 cm

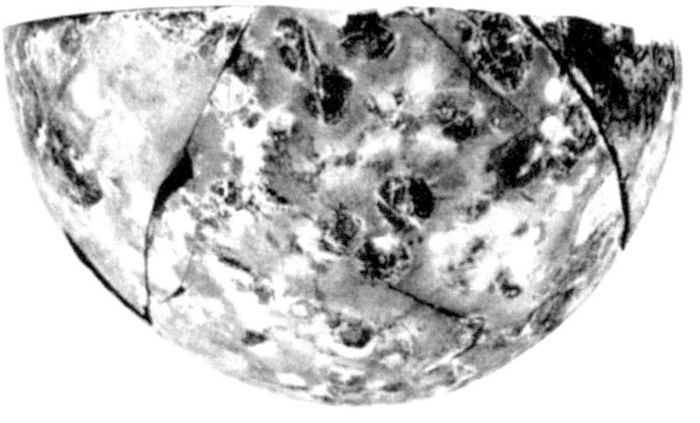

Fo1

Nim10

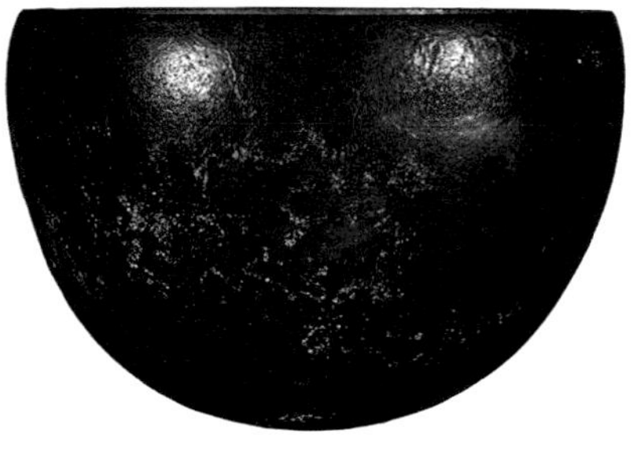

Pr1

S 1:2

0 5 cm

Nim14

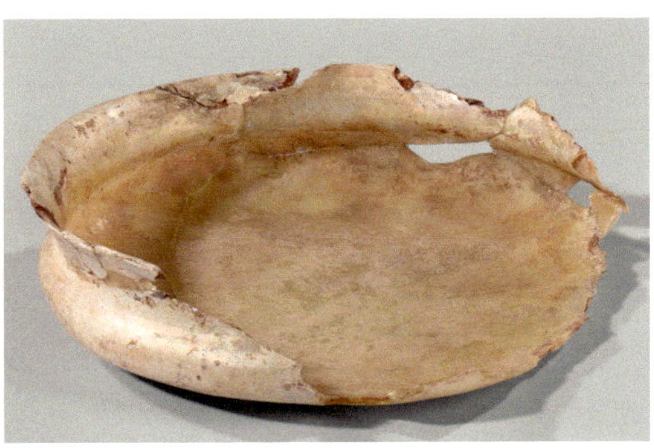

(not to scale)

(not to scale)

Bab3

S 1:2

0 5 cm

(not to scale)

Nim15

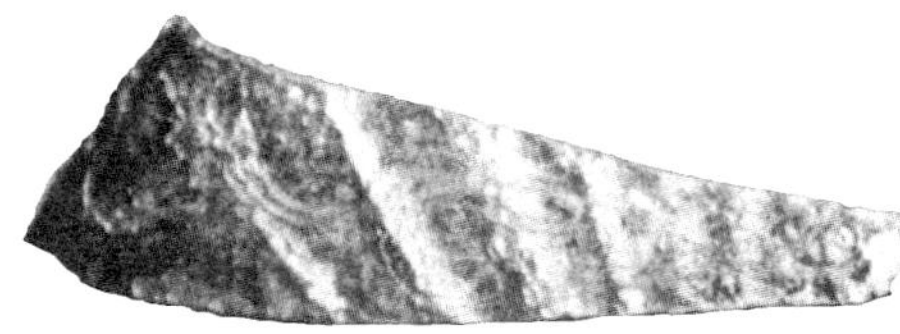

(not to scale)

Nim16

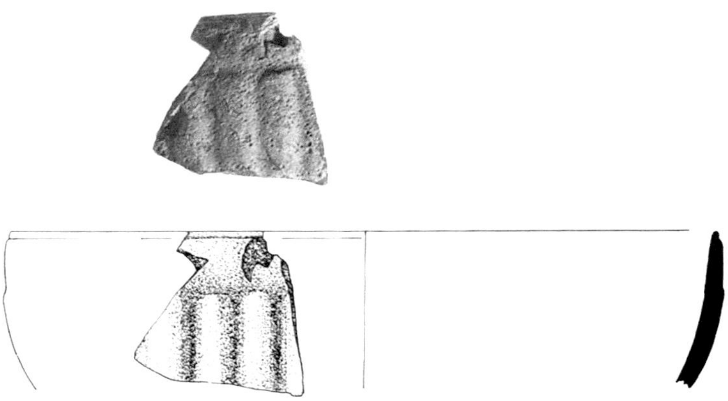

Nim17

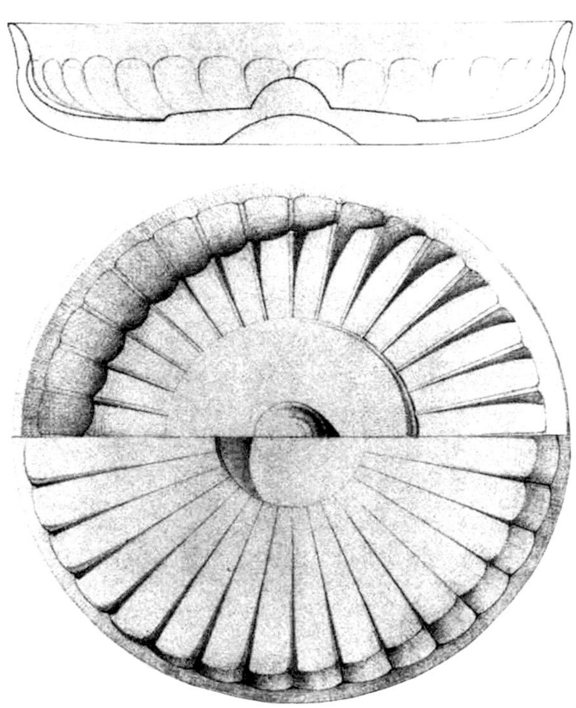

(not to scale)

Gor1

S 1:2

0 5 cm

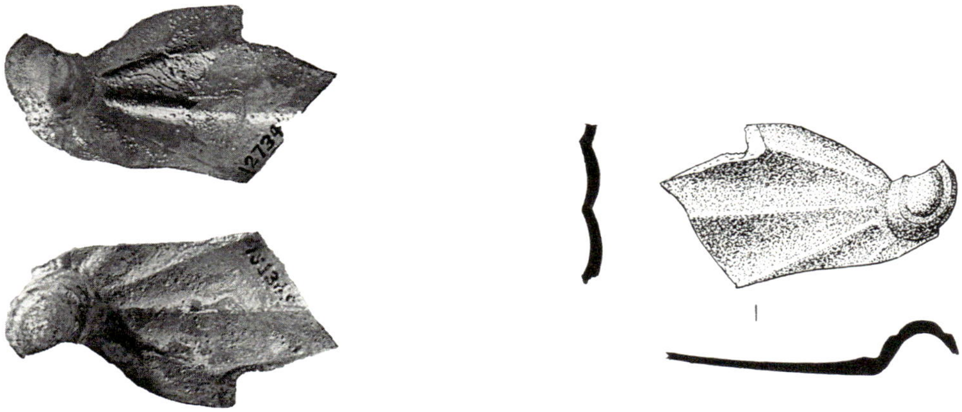

AM5

Has12

S 1:1

0 5 cm

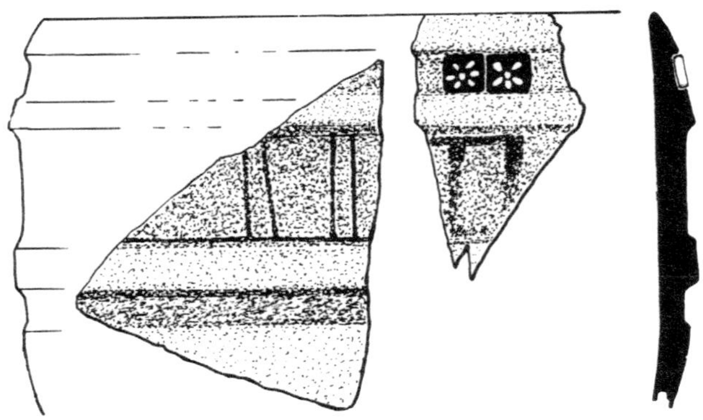

S 1:1

0 5 cm

Nim18

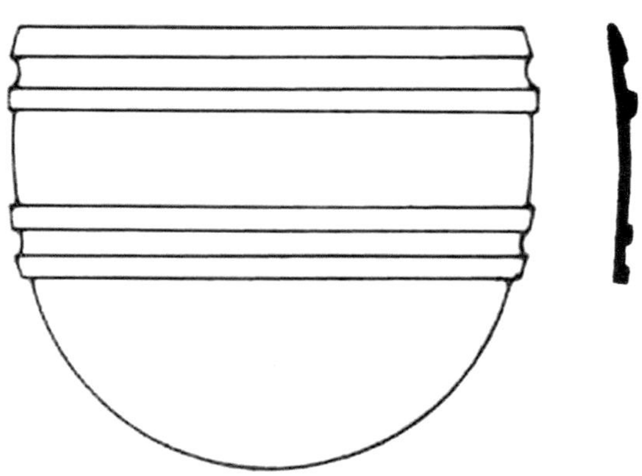

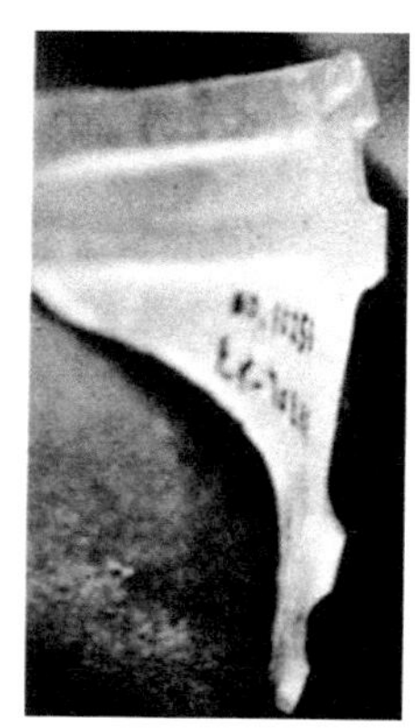

S 1:2

0 5 cm

(not to scale)

Nim20

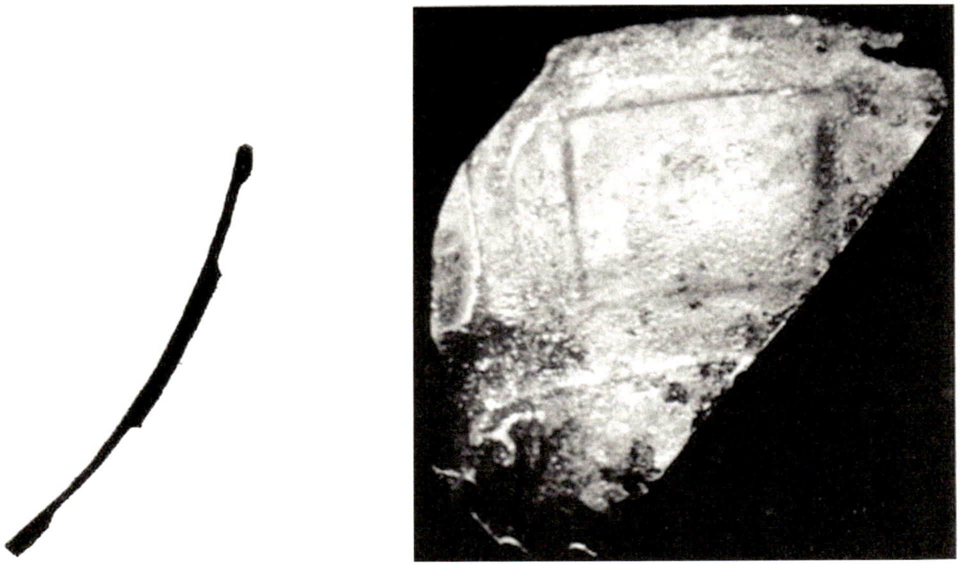

Nim21

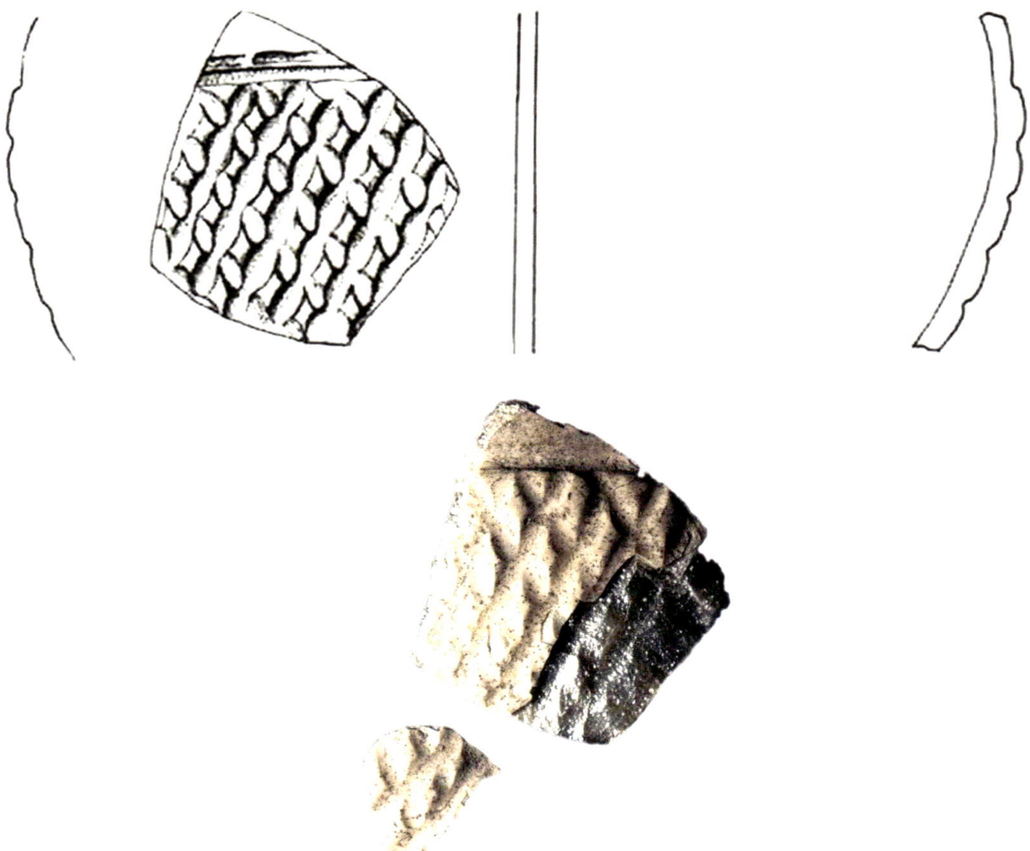

Ar1

S 1:1

0 5 cm

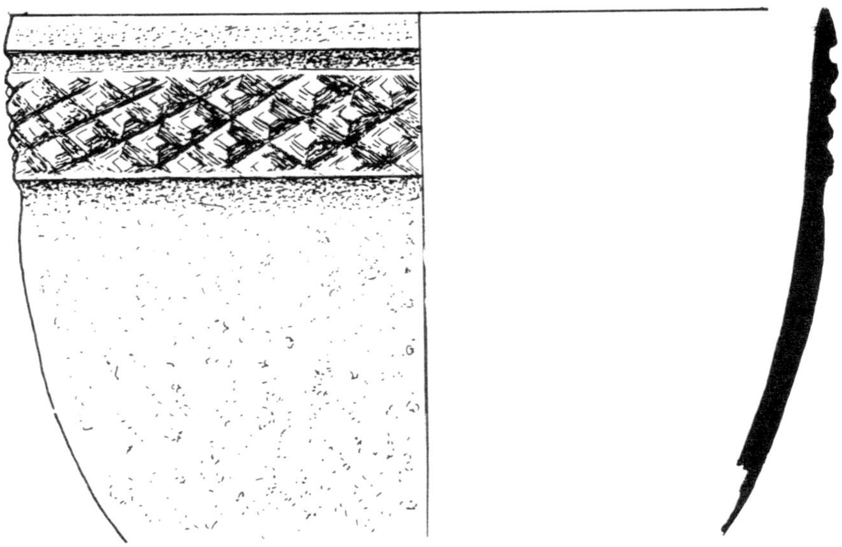

Nim22

Nim22

Nim23

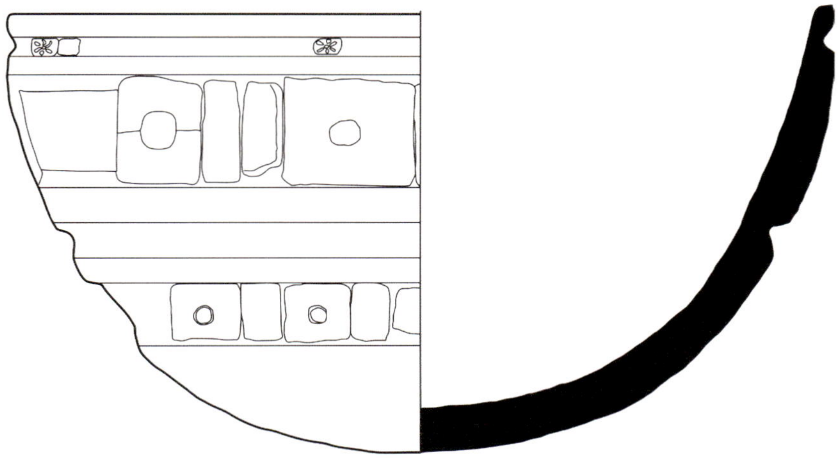

(not to scale)

(not to scale)

(not to scale)

A1

S 1:1

0 5 cm

Nim24

Nim25

Nim26

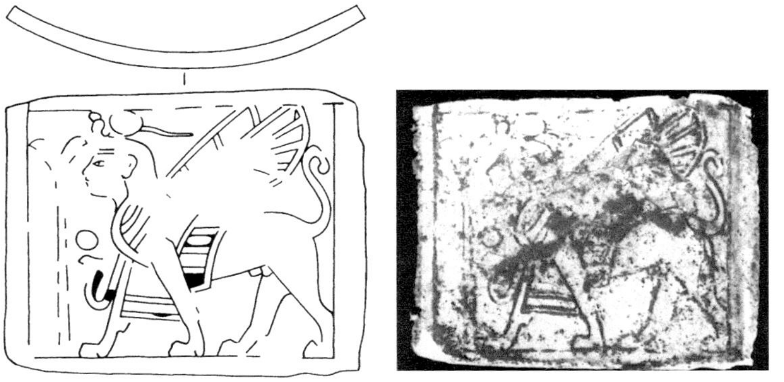

Nim112

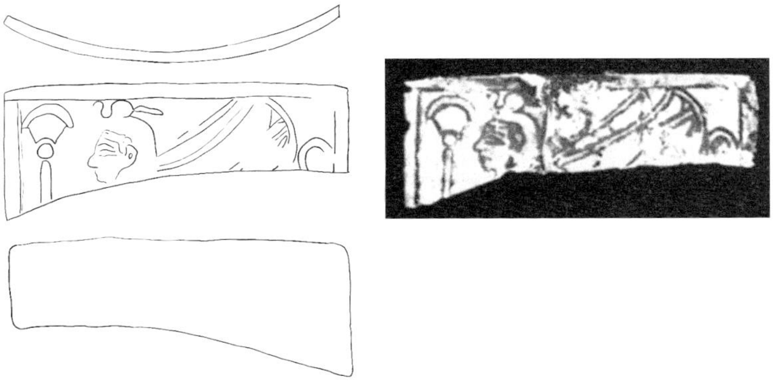

Nim113

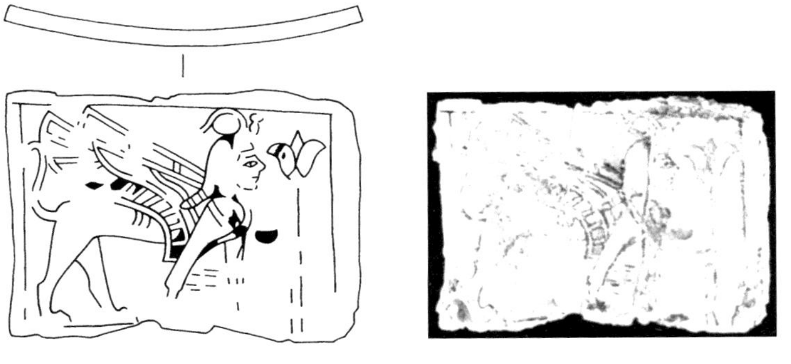

Nim115

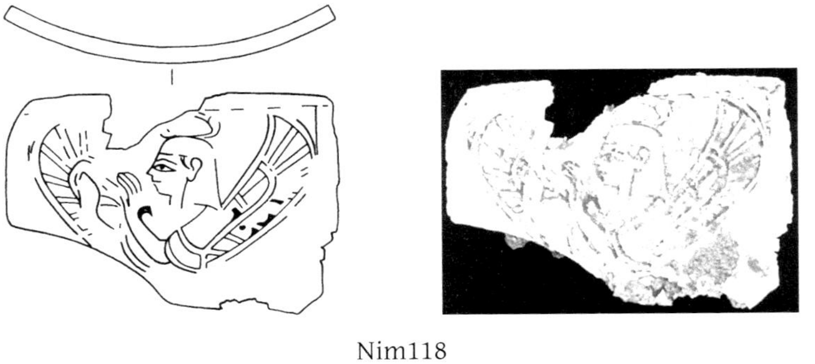

Nim118

S 1:1

0 5 cm

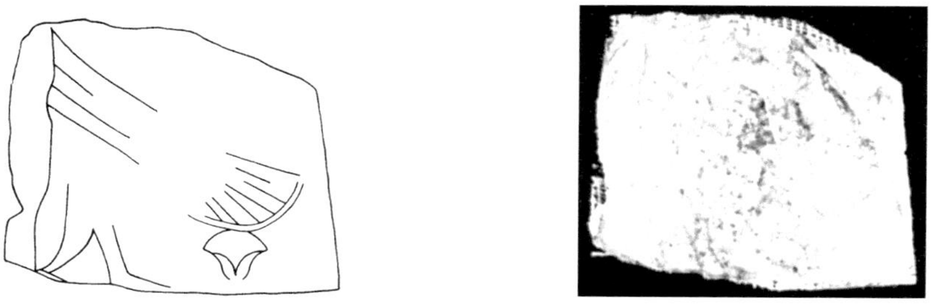

Nim116

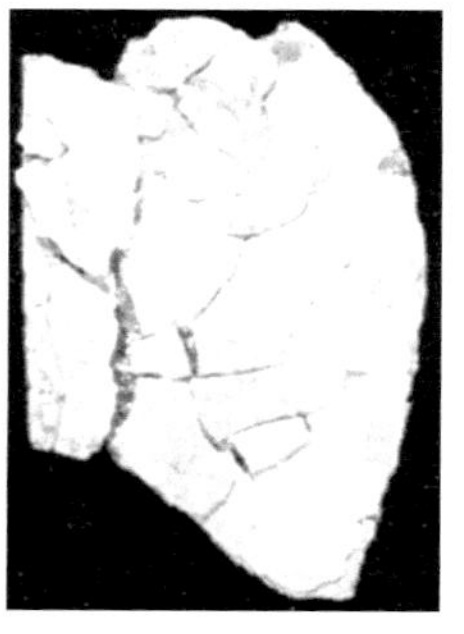

Nim119

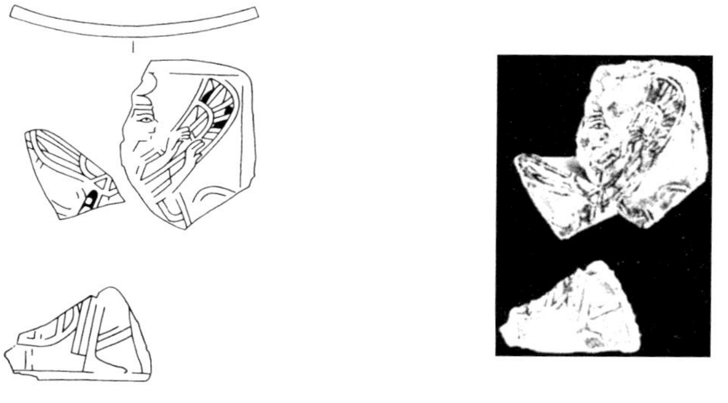

Nim114

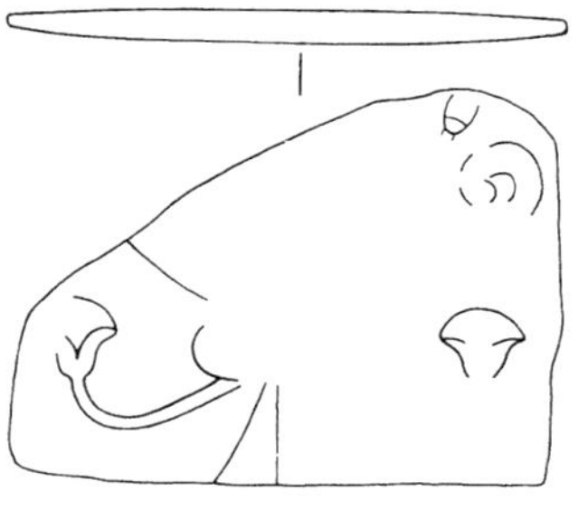

Nim117

Nim34
Nim35
Nim36
Nim59
Nim58
Nim53
Nim54
Nim55
Nim56
Nim29
Nim111
Nim31
Nim32
Nim30
Nim62
Nim63
Nim52
Nim65
Nim70
Nim28
Nim71
Sam1
Sam2
Sam4
Sam5
S 1:1
0
5 cm

Rosette inlays: group 2

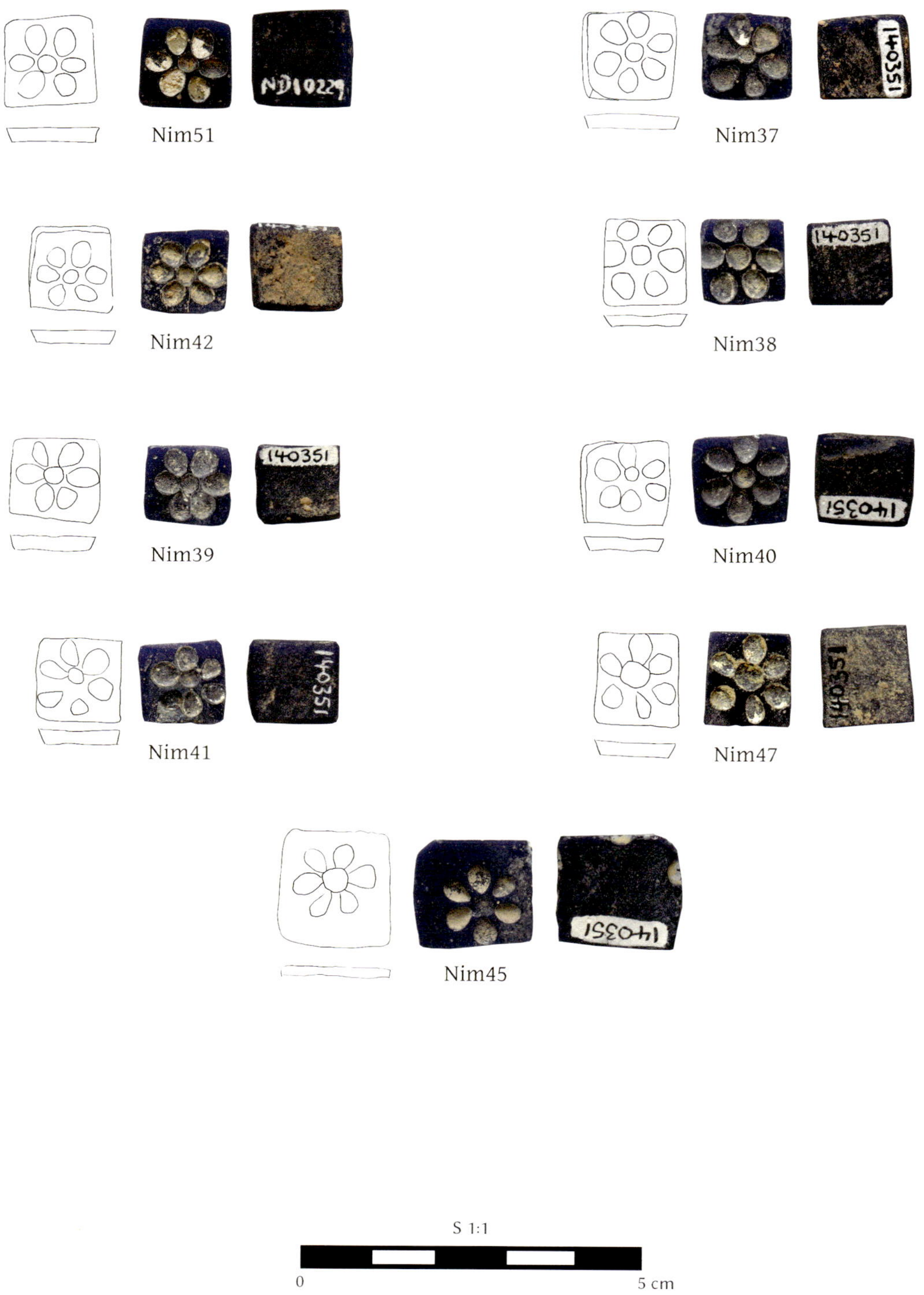

Nim43

Nim44

Nim46

Nim64

Nim61

Nim72

Nim73

Nim74

Nim75

Nim76

Nim77

Nim78

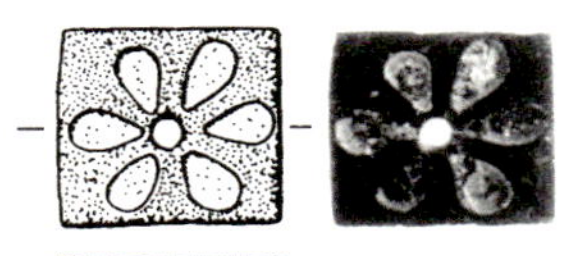

Nim109

TB1

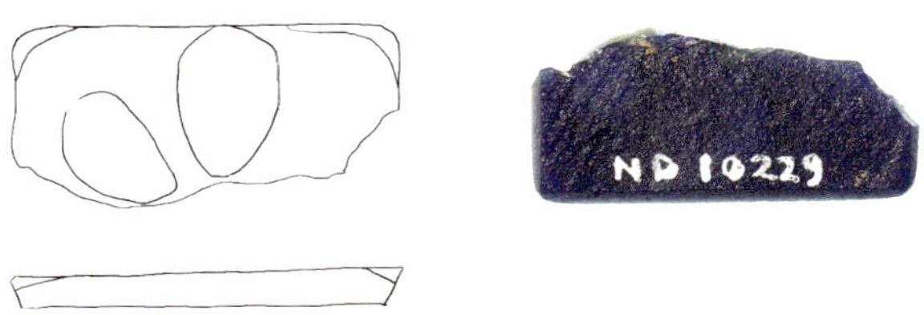

Nim49

Nim50

S 1:1

0 5 cm

Nim60
Nim82
Nim85
Nim84
Nim87
Nim88
Nim79
Nim110
Nim83
Nim86
Nim81
Nim80
Nim92
Nim93
Nim94
Nim95
Nim96
Nim97
Nim98
Nim99
Nim100
Nim101
Nim102
Nim103
Nim104
Nim105
Nim106
Nim107
Sam3
Nim90
Nim108
Nim91
Nim66
Nim67
Nim68
Nim69
S 1:1
0
5 cm

AM37

AM13

AM14

AM15

AM16

AM17

AM18

AM19

AM20

AT1

Nim89

AT5

AM30

AM29

AM28

AM49

S 1:1

0 5 cm

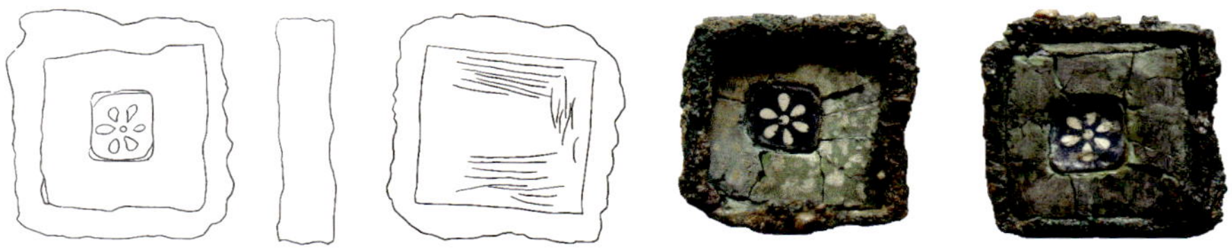

AT2

AT3

AT4

AM21

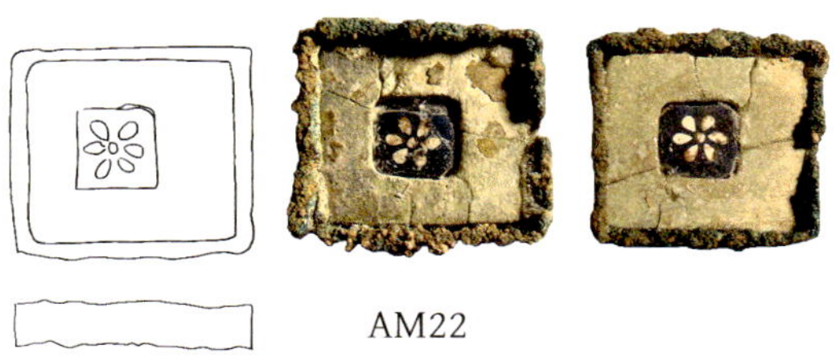

AM22

AM23

AM24

AM25

AM26

AM27

AM38

Nim199

Nim57

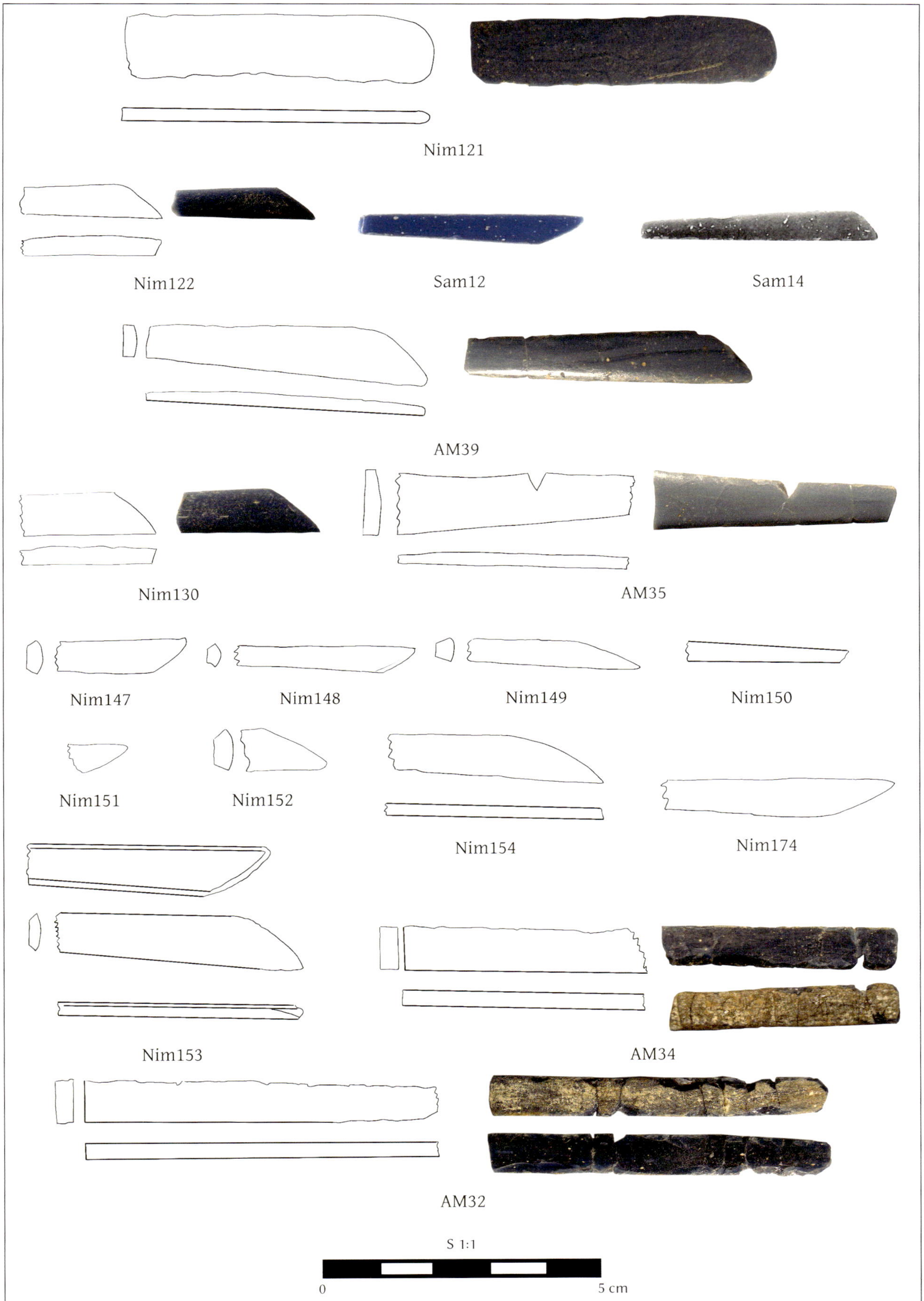
Nim121
Nim122
Sam12
Sam14
AM39
Nim130
AM35
Nim147
Nim148
Nim149
Nim150
Nim151
Nim152
Nim154
Nim174
Nim153
AM34
AM32
S 1:1
0
5 cm

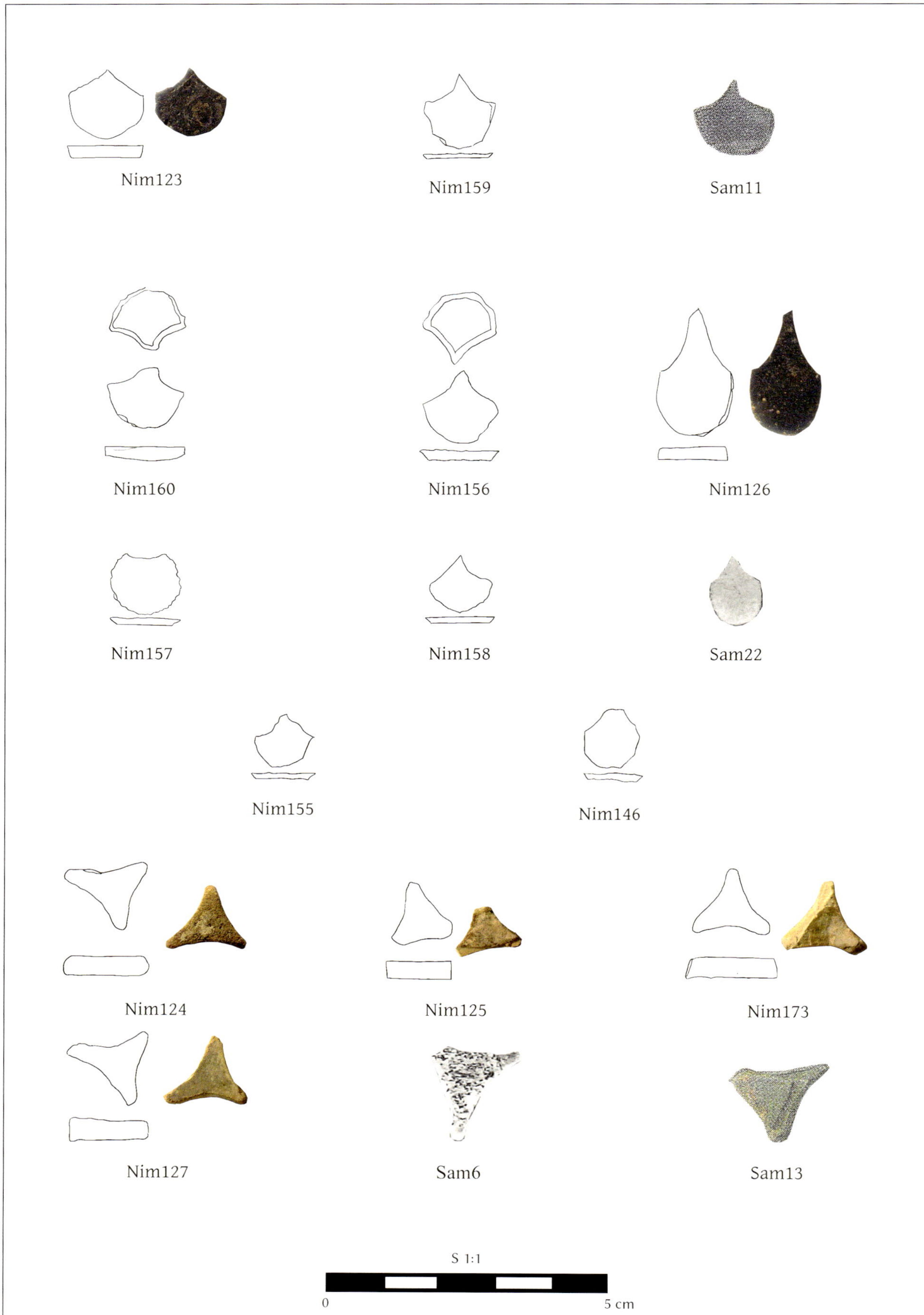
Nim123
Nim159
Sam11
Nim160
Nim156
Nim126
Nim157
Nim158
Sam22
Nim155
Nim146
Nim124
Nim125
Nim173
Nim127
Sam6
Sam13
S 1:1
0
5 cm

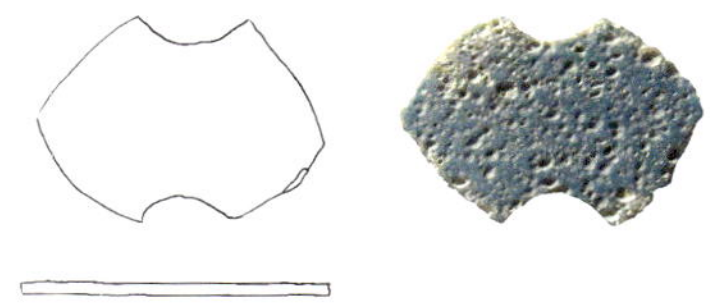

Nim128

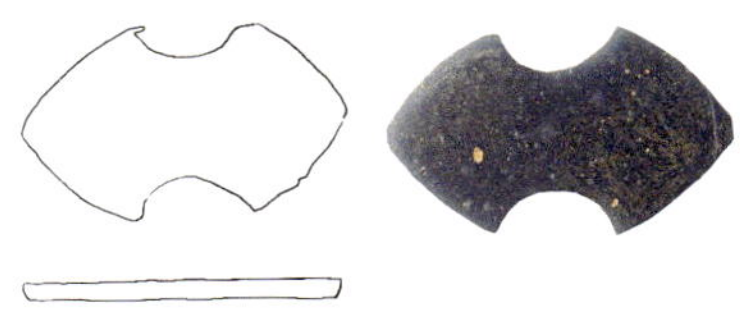

Nim129

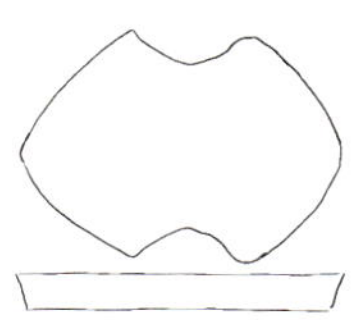

Nim161

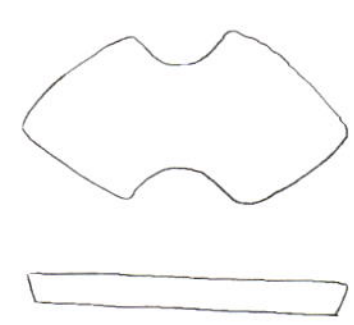

Nim48

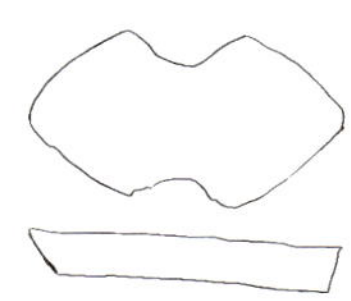

Nim162

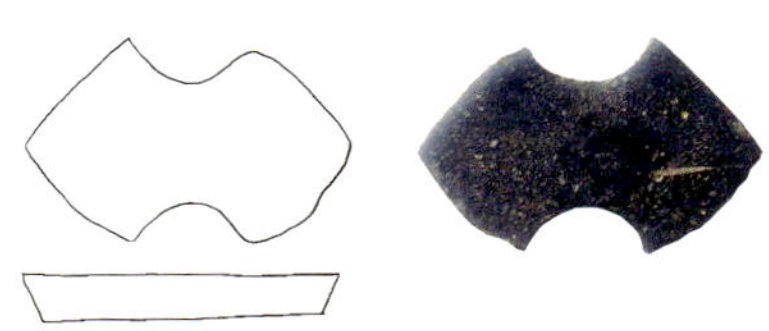

Nim143

Nim168

S 1:1

0 5 cm

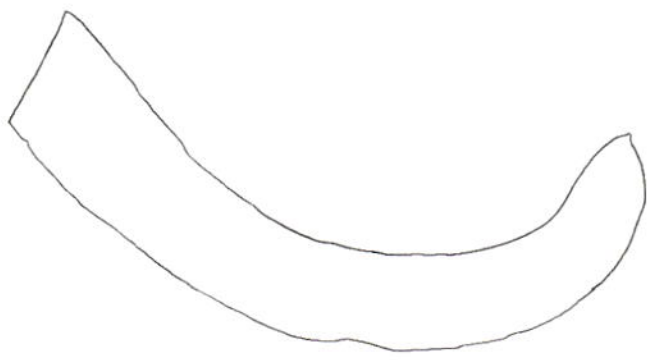

Nim140

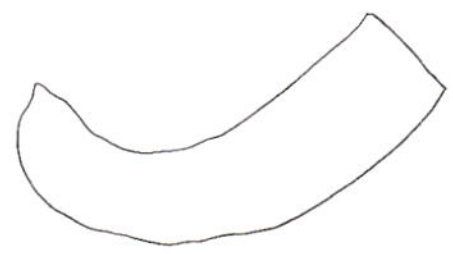

Nim175

Sam21

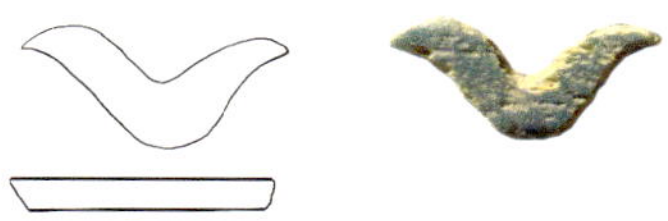

Nim144

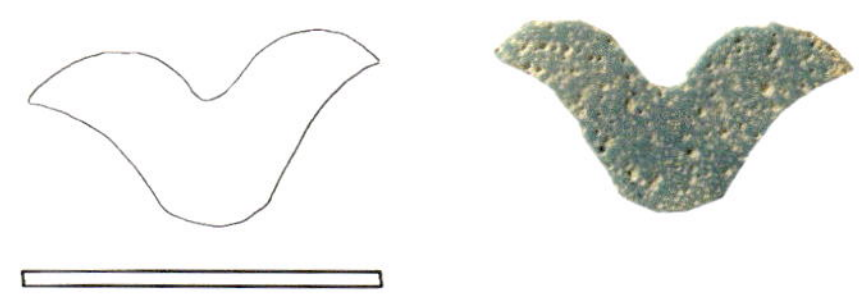

Nim145

Nim184

Nim185

Nim186

Nim187

Nim188

Nim133

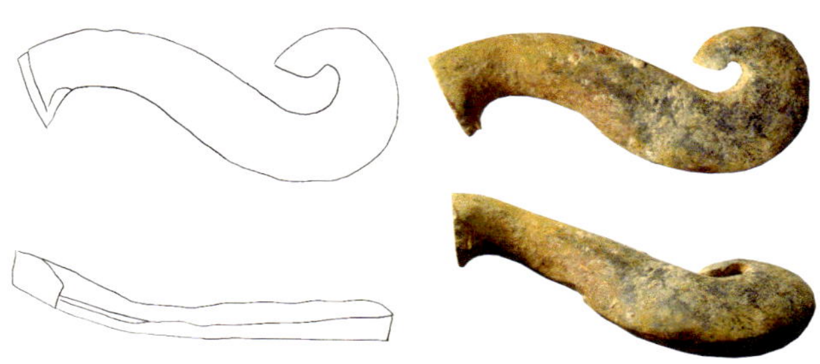

Nim142

S 1:1

0 5 cm

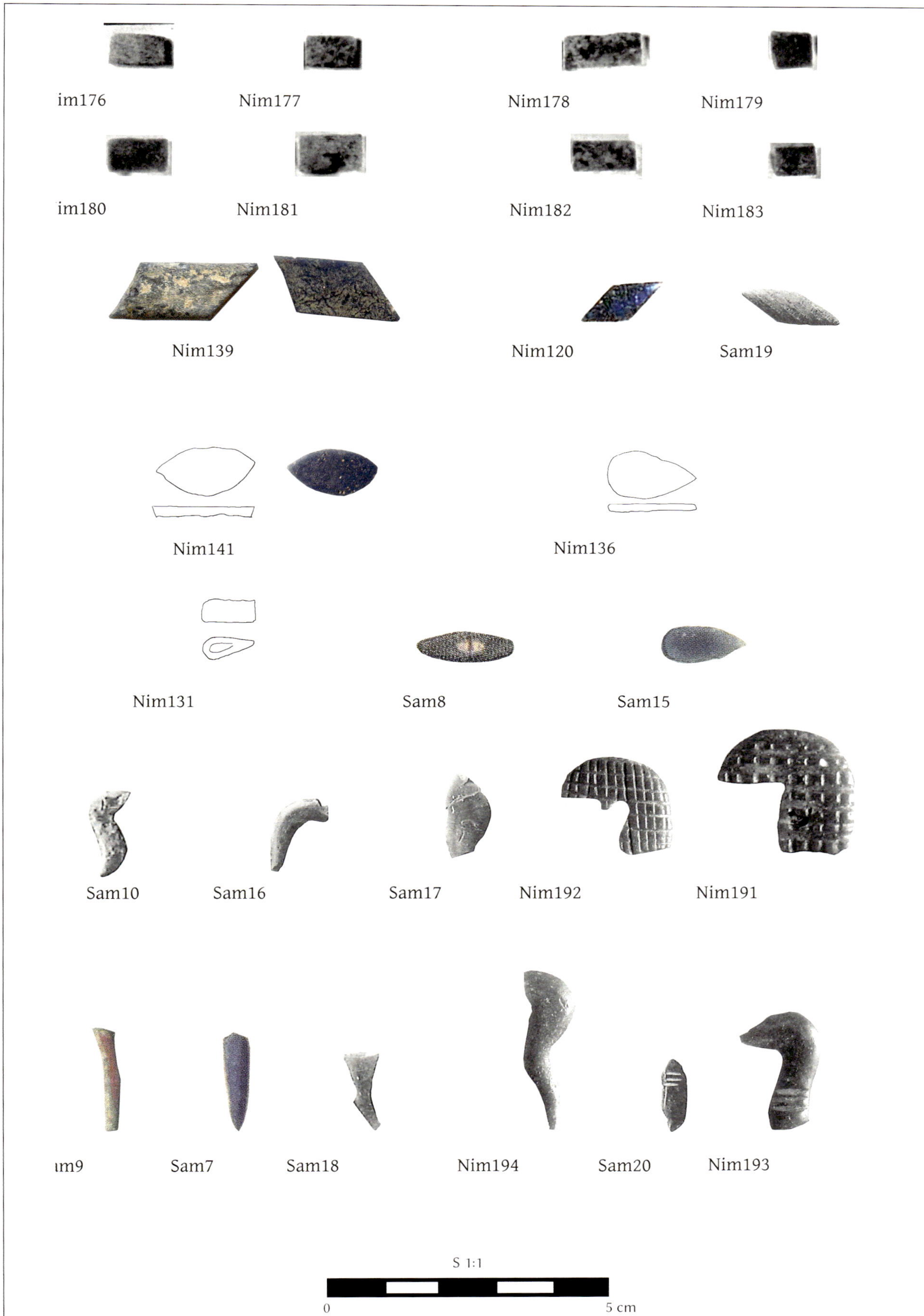
im176
Nim177
Nim178
Nim179
im180
Nim181
Nim182
Nim183
Nim139
Nim120
Sam19
Nim141
Nim136
Nim131
Sam8
Sam15
Sam10
Sam16
Sam17
Nim192
Nim191
ım9
Sam7
Sam18
Nim194
Sam20
Nim193
S 1:1
0
5 cm

AM33

AM36

AM31

S 1:1

0 5 cm

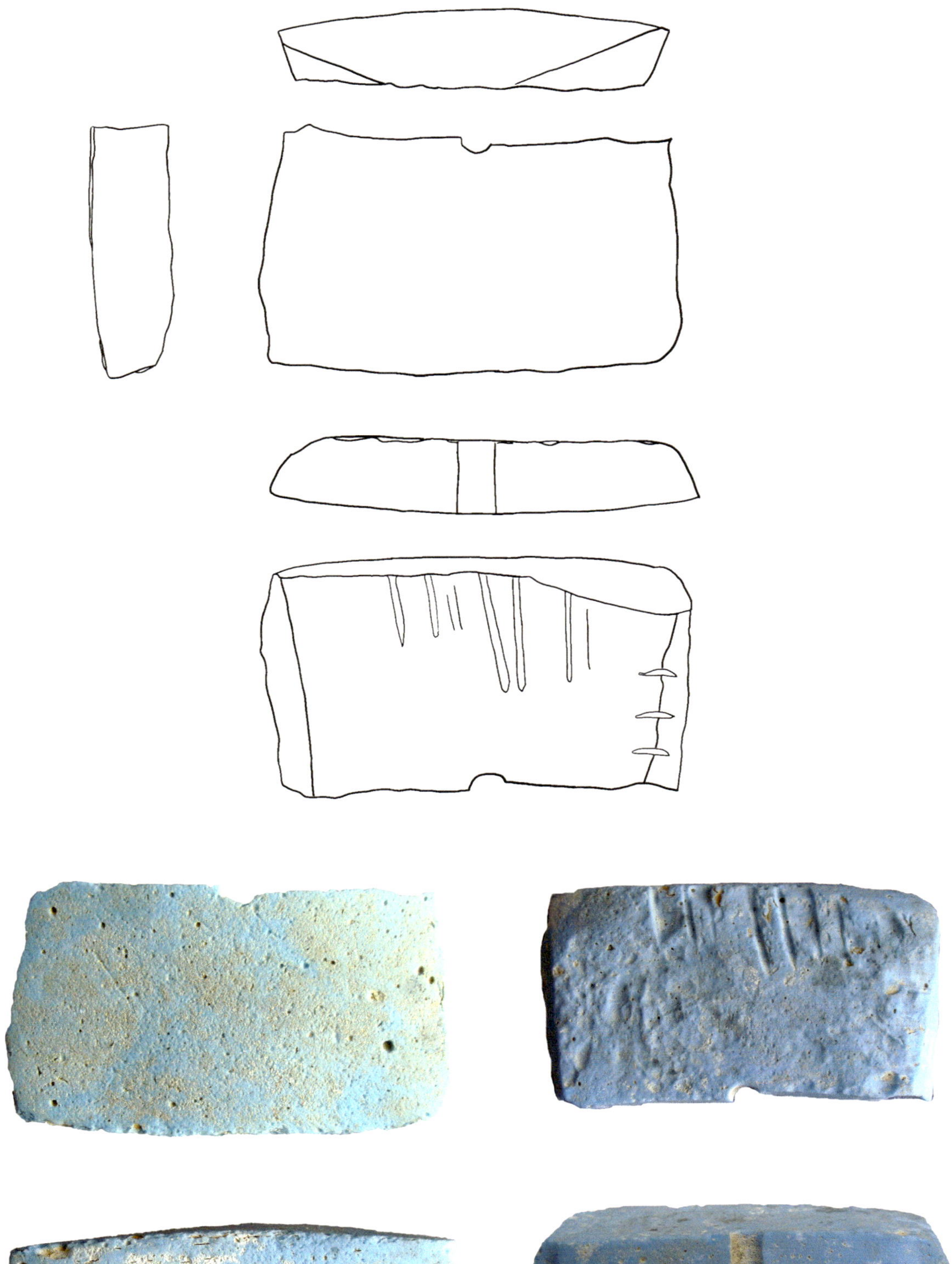

Khor2

S 1:1

0 5 cm

Khor3

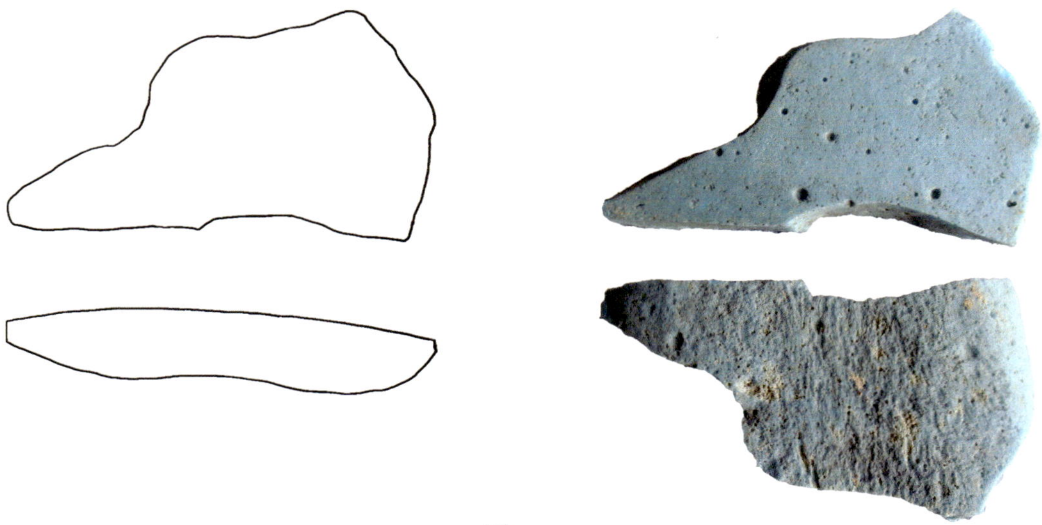

Khor4

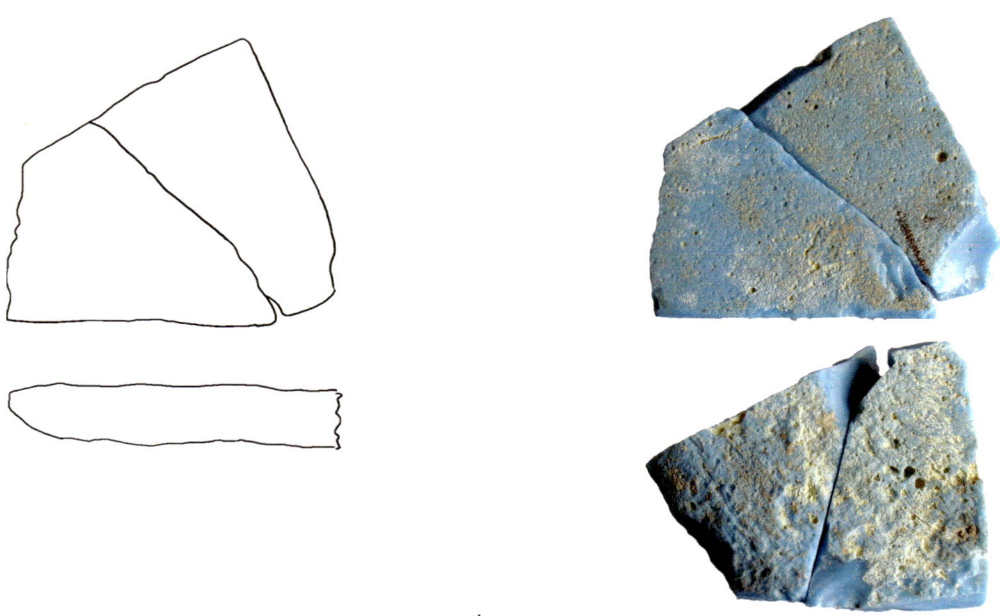

Khor7

S 1:1

0 5 cm

Khor5

Khor8

Khor6

S 1:1

0 5 cm

Bab2

Nim5

Dul1

S 1:1

0 5 cm

Nim1

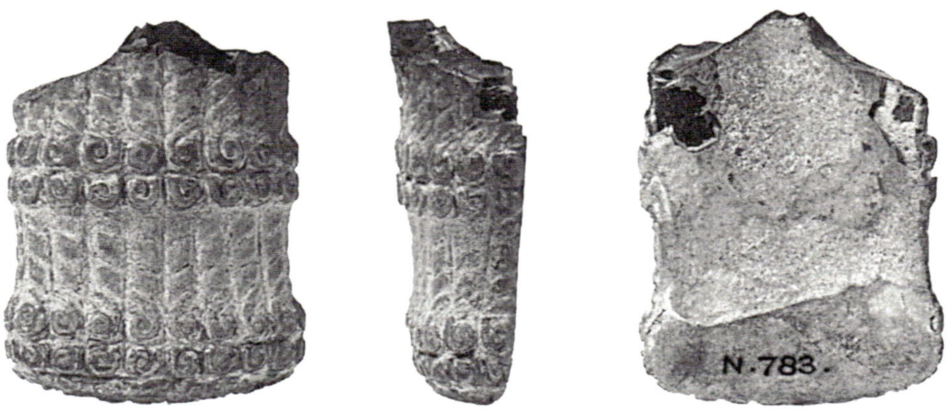

Nim2

(not to scale)

Nin1

S 1:1

0 5 cm

Bab1

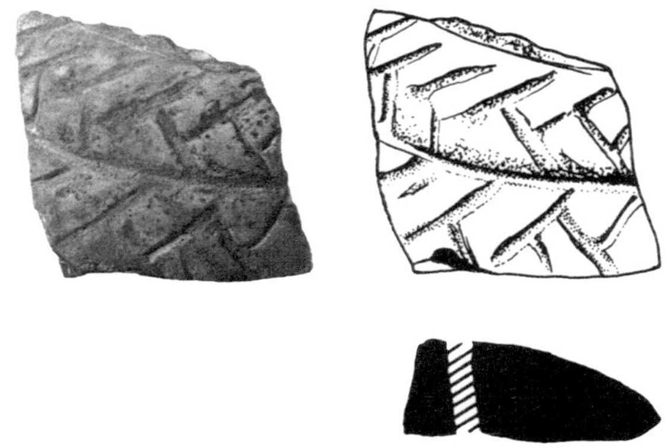

Dul2

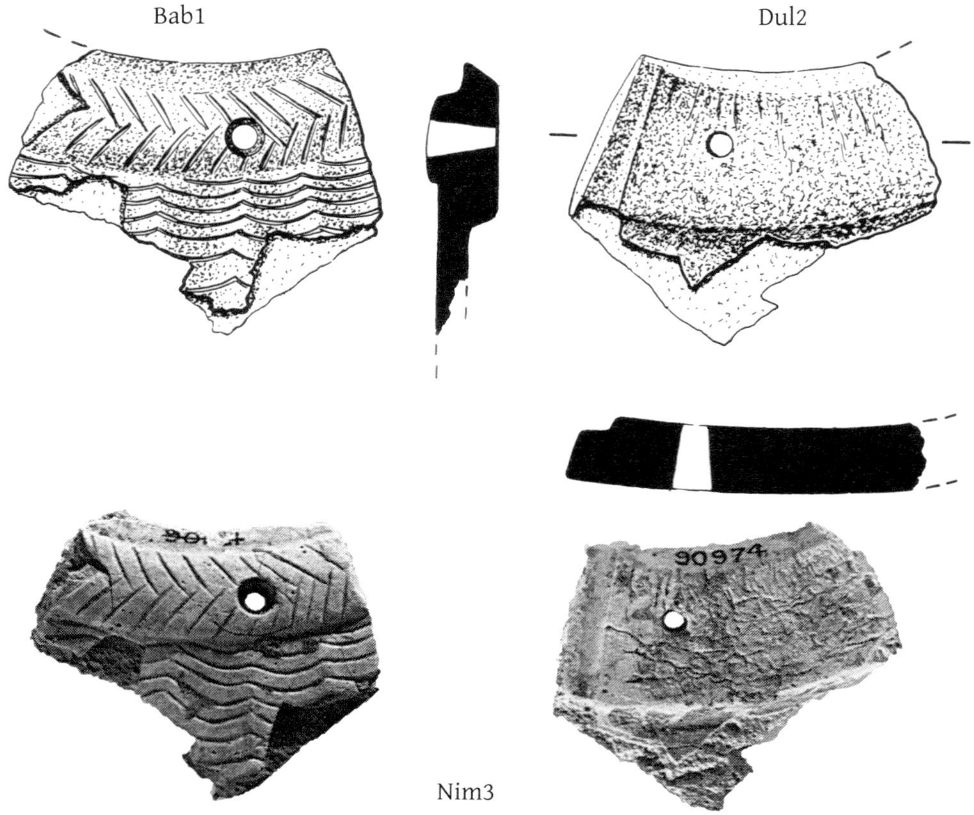

Nim3

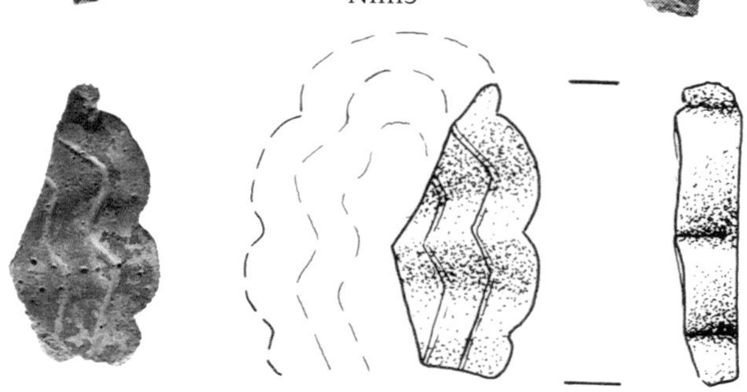

Nim4

S 1:1

0 5 cm

Is1

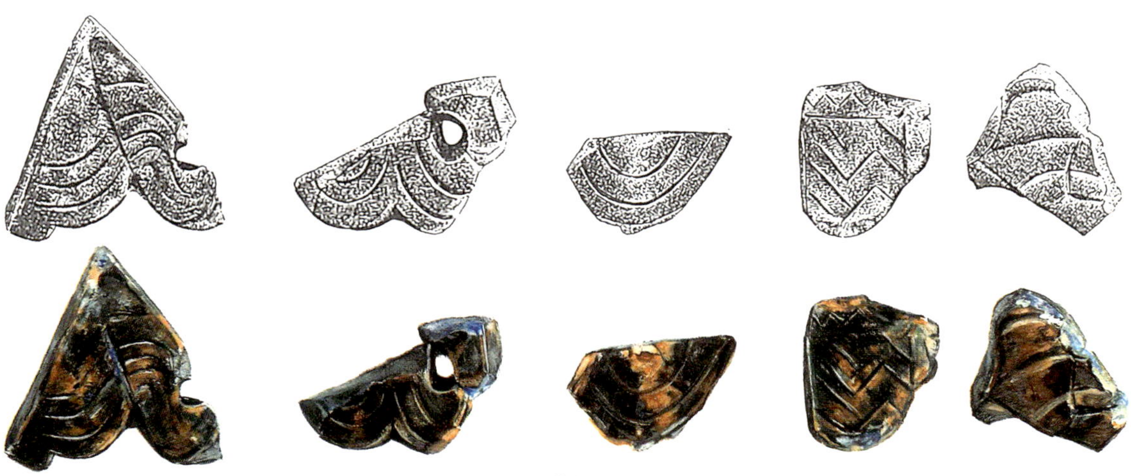

Is2

S 1:1

0 5 cm

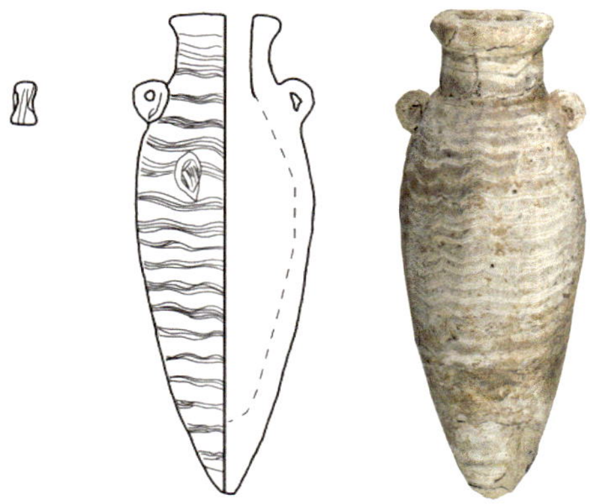

AM50

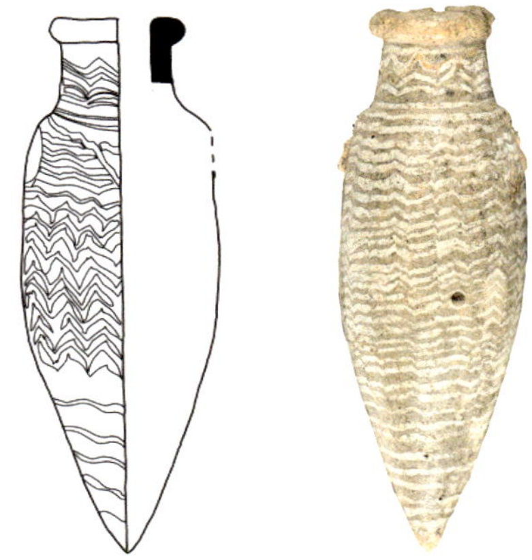

AM40

AM41

AM42

Ur3

Bab4

AM43

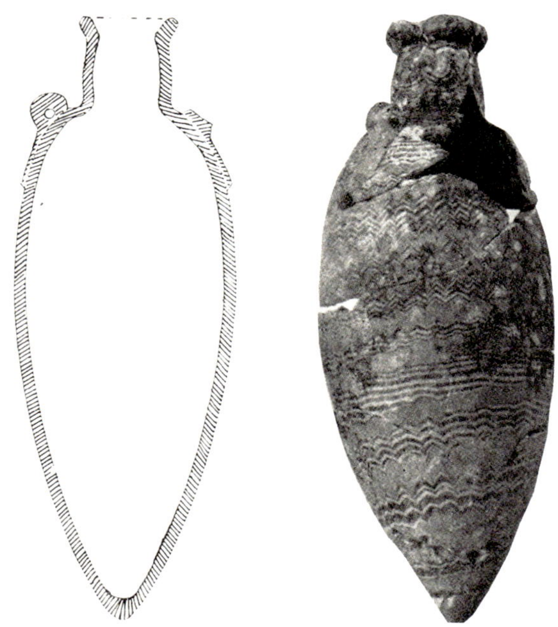

As9

S 1:2

0 5 cm

AM44

Ur4

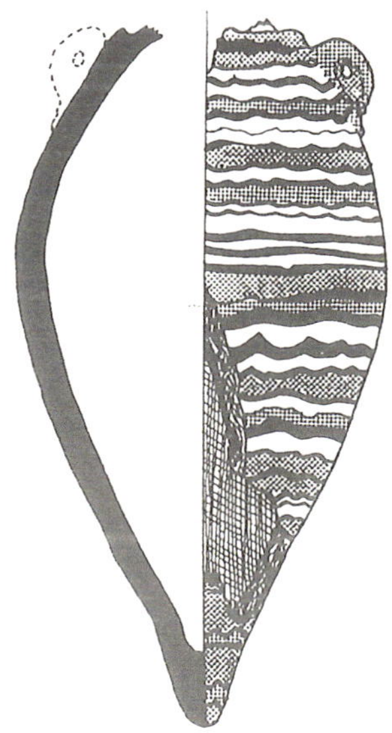
Bus1

Sus1

Sus2

Sus3

Kiš1

Nip1

Sus4

Su1

S 1:2

0 5 cm

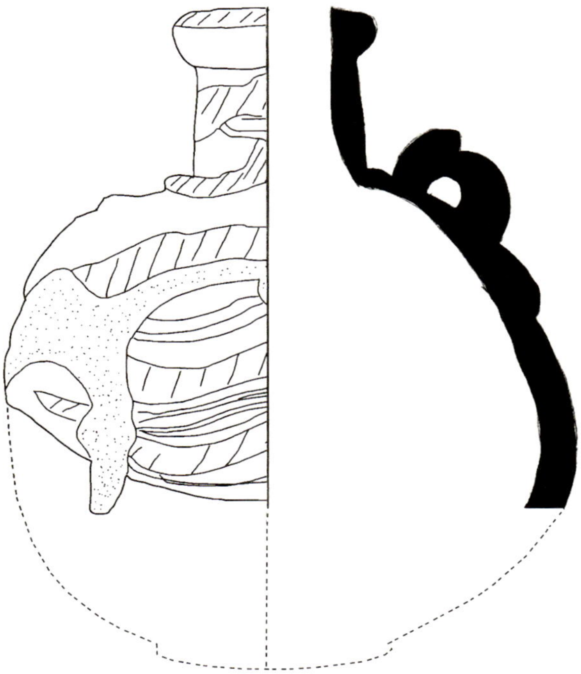

S 1:1

0 5 cm

Bab6

S 1:2

0 5 cm

(not to scale)

(not to scale)

Bab7

Bab5

S 1:1

0 5 cm

TJ1

TJ2

AM51

Ur6

(not to scale)

AM55

S 1:1

0 5 cm

AM52

AM53

AM54

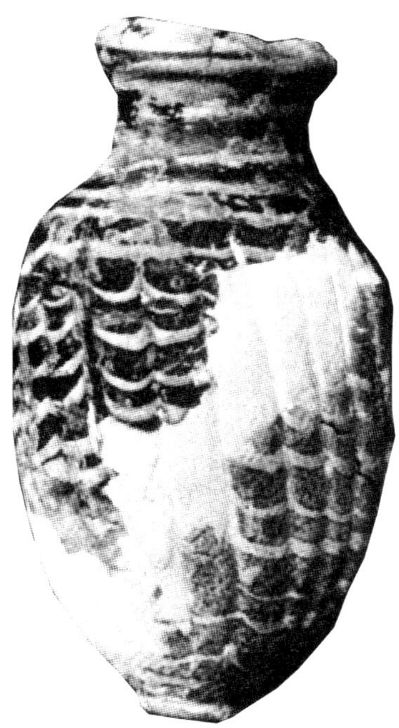

Nip2

S 1:1

0 5 cm

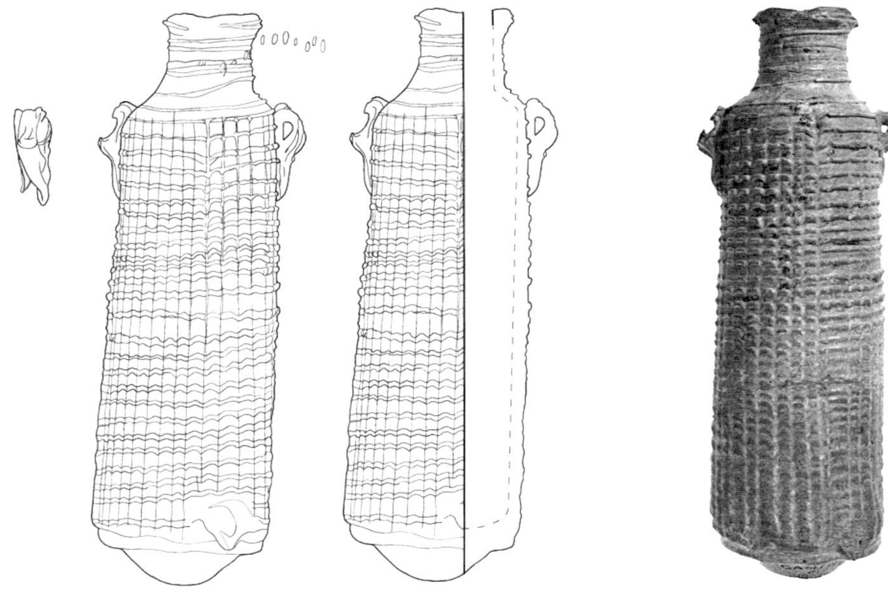

Kam1

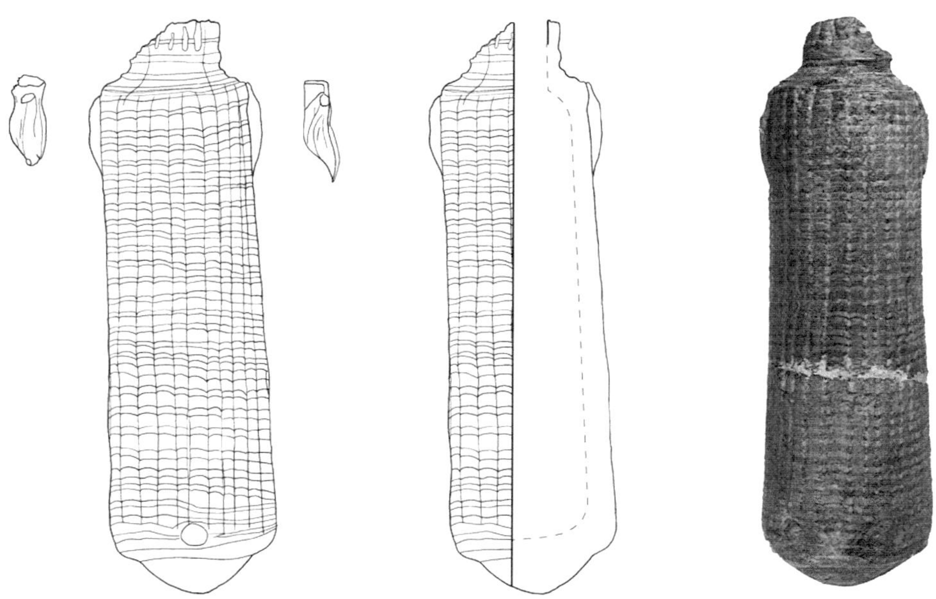

Kam2

S 1:2

0 5 cm

Car1

As10

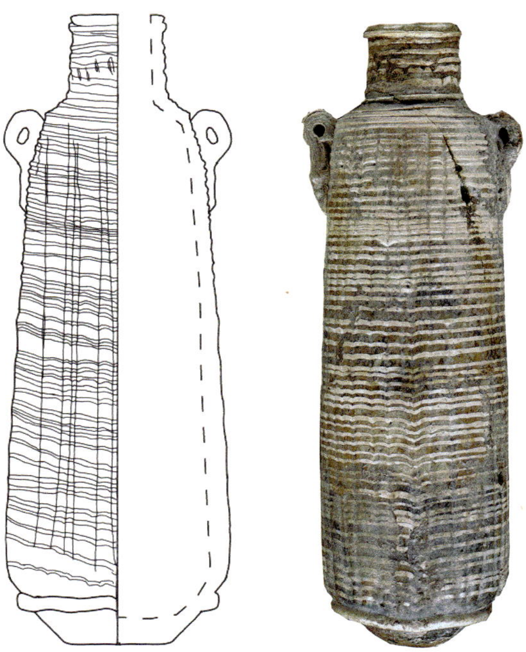

AM45

AM47

S 1:2

0 5 cm

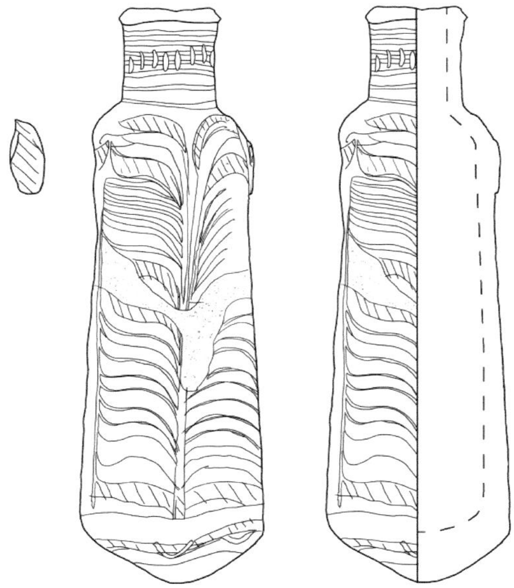

AM48

S 1:2

0 5 cm

Nim138

Nim137

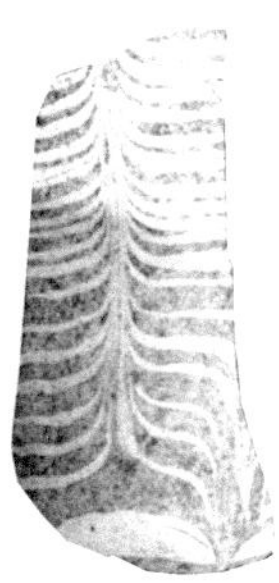

Nip3

S 1:1

0 5 cm

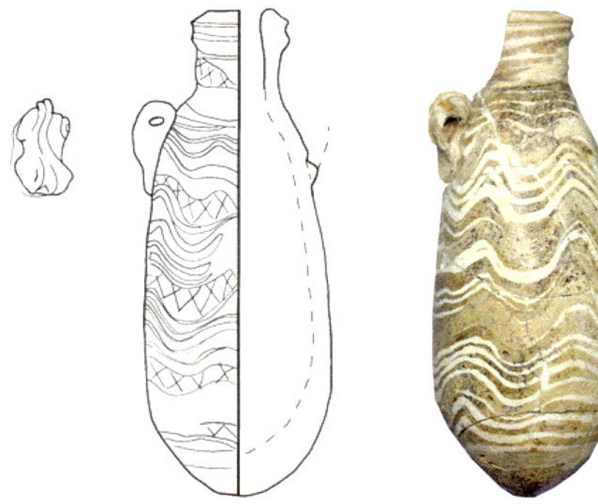

Kiš2

Ur5

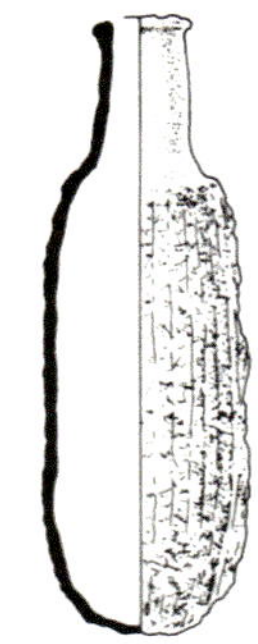

AM46

S 1:2

0 5 cm

BS2

Ur7

Urk1

Urk2

Urk3

S 1:1

0 5 cm

Ziy1

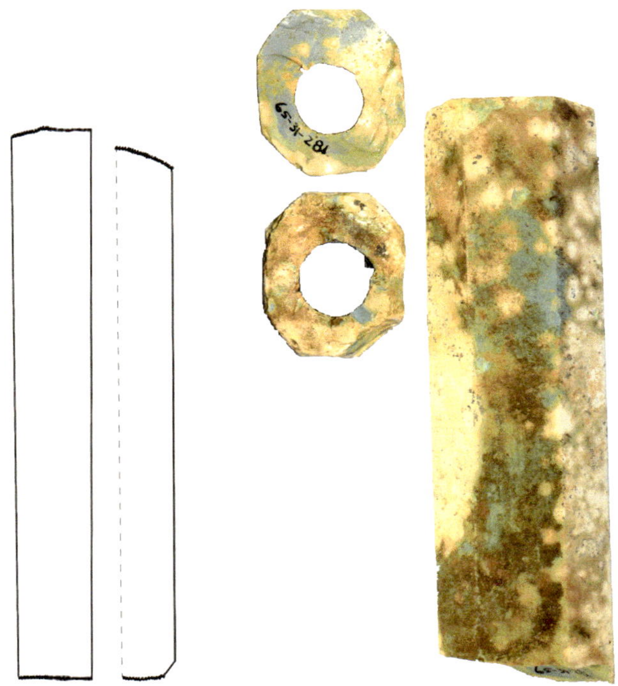

Has4

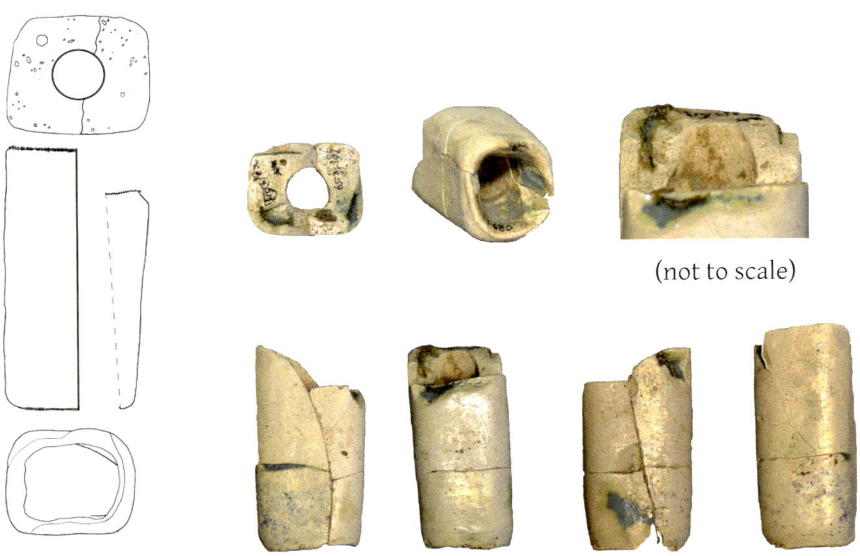

(not to scale)

Has10

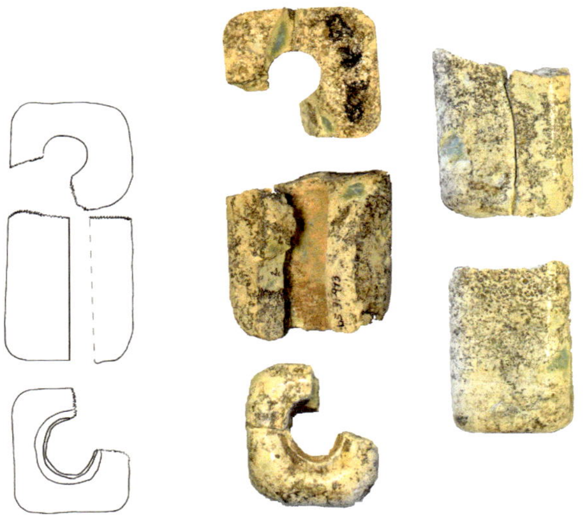

Has11

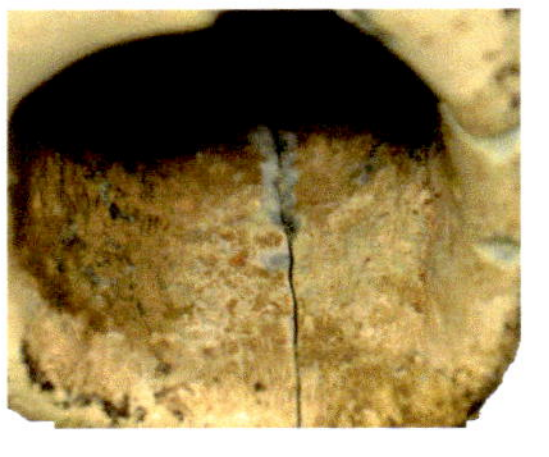

(not to scale)

Has9

S 1:2

0 5 cm

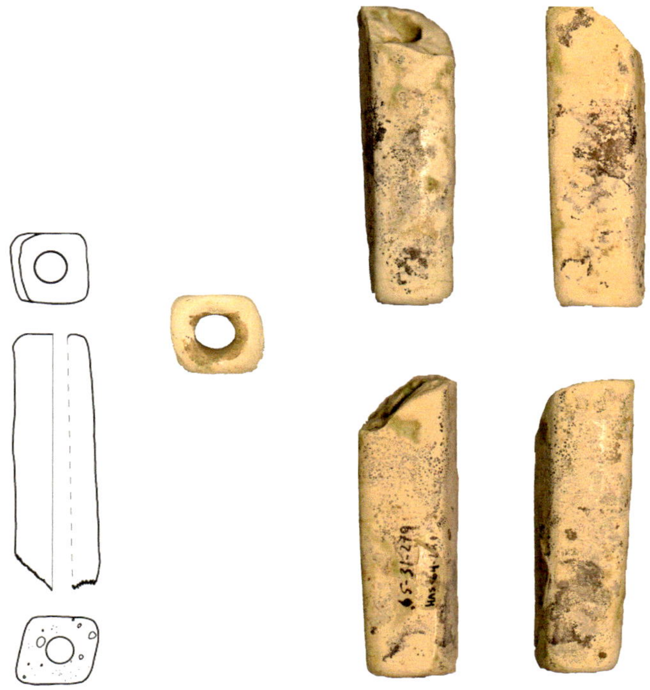

Has8

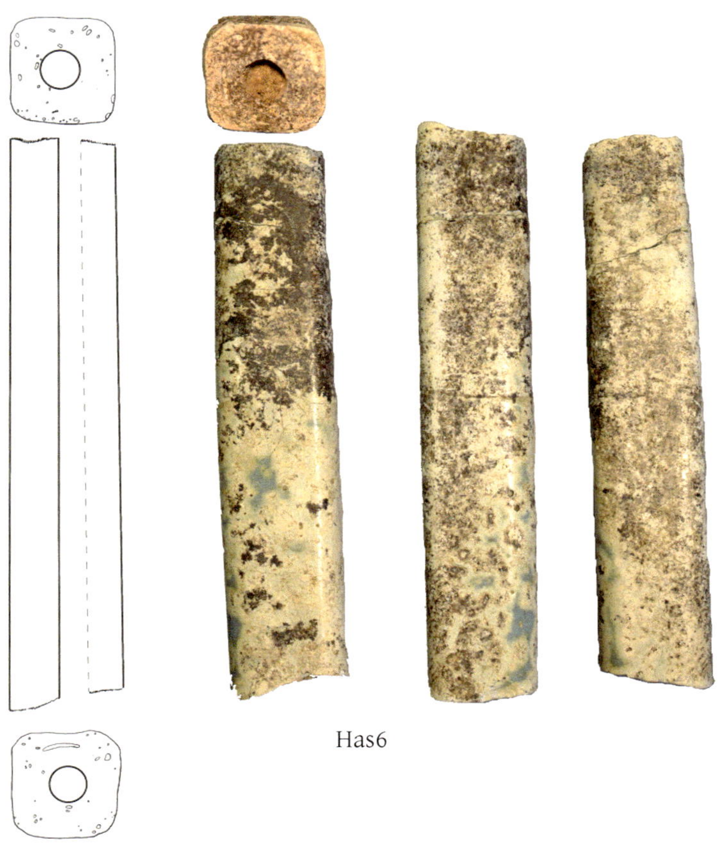

Has6

S 1:2

0 5 cm

Has7

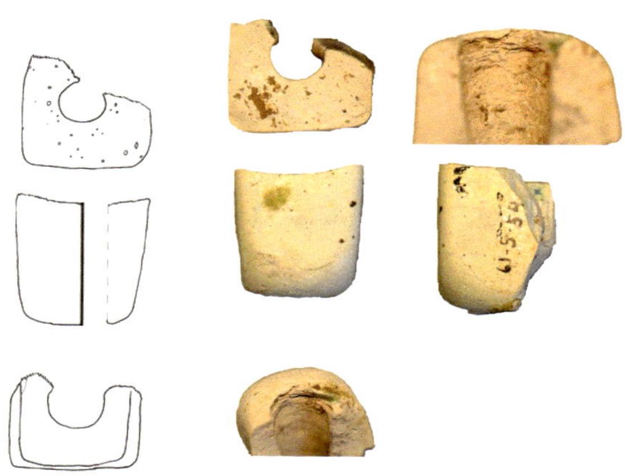

Has5

S 1:2

0 5 cm

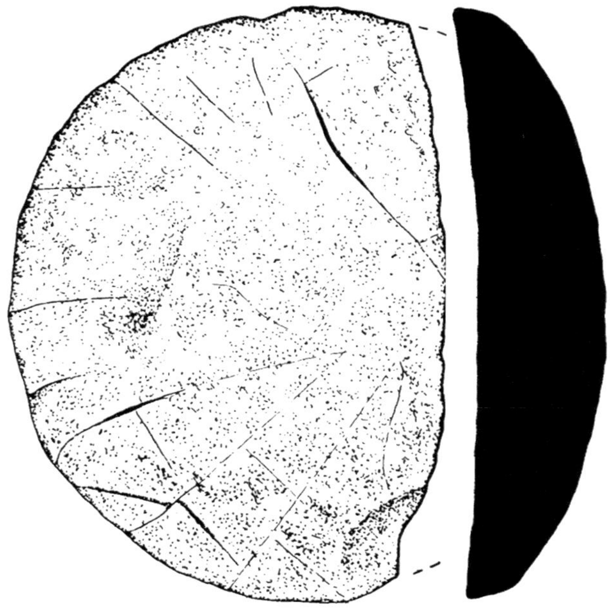

Nim198

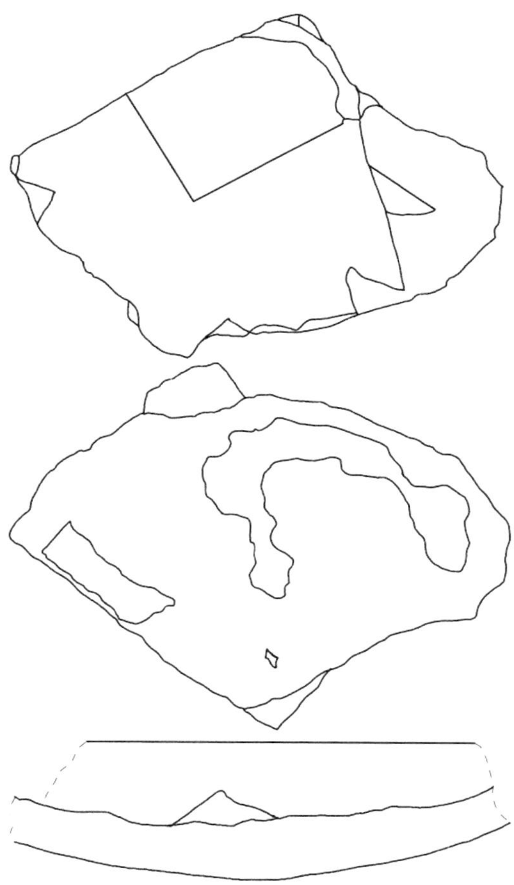

Bab14

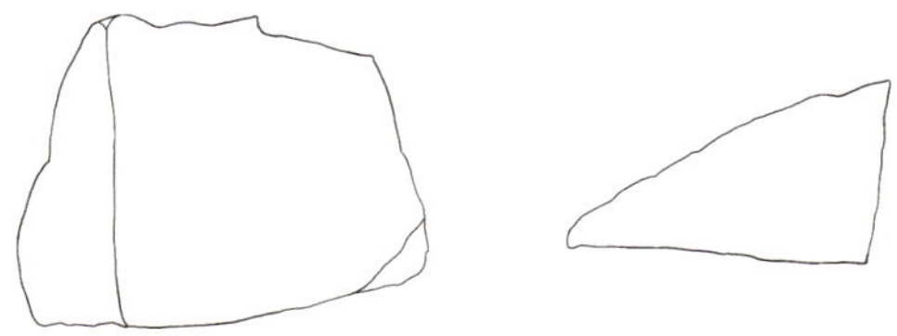

S 1:2

0 5 cm

S 1:1

0 5 cm

BS1

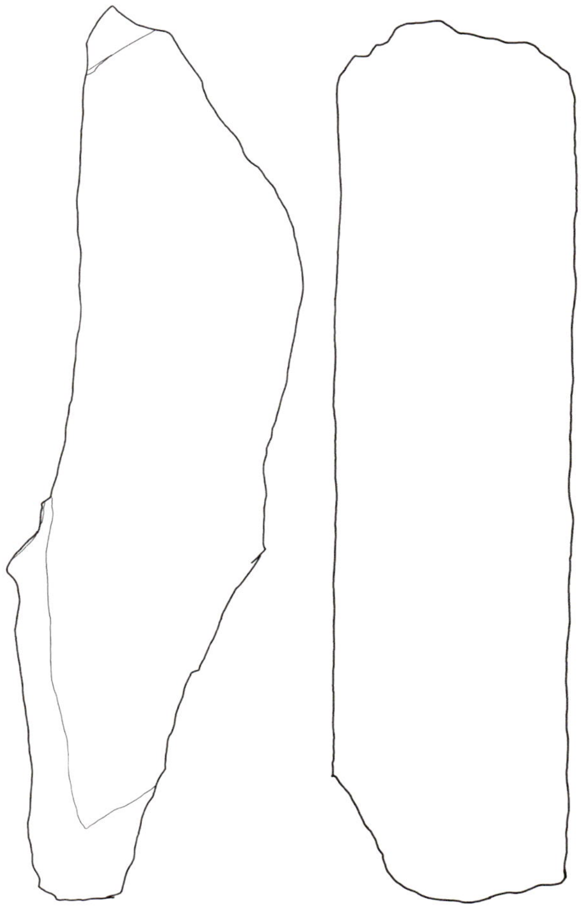

S 1:2

0 5 cm

S 1:1

0 5 cm

Bab9

(not to scale)

Bab9

S 1:1

0 5 cm

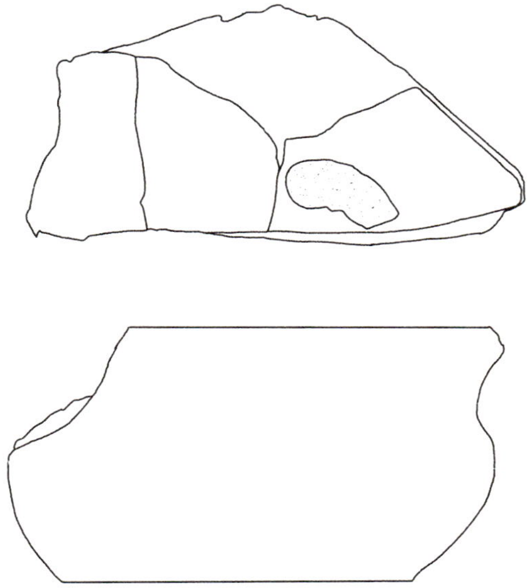

S 1:2

0 5 cm

Bab11

S 1:1

0 5 cm

Has13

Has14

S 1:1

0 5 cm

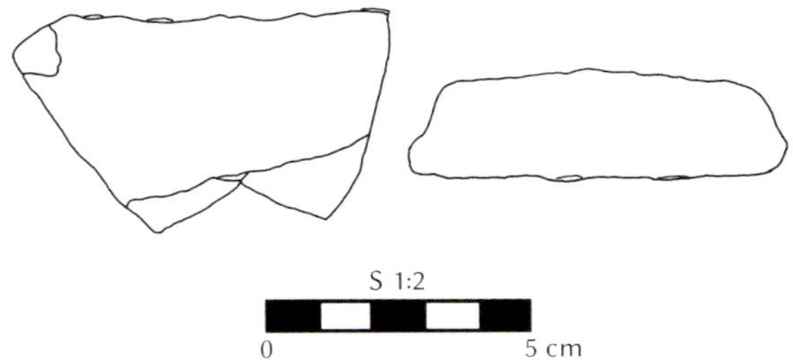

As11

As12

Raw glass fragments / waste material

Bab12

S 1:1

0 5 cm

Nim189

Nim135

Nin3

S 1:1

0 5 cm

Appendix 1

Recipe for blue *zagindurû*-glass

#1.1 When you lay down the foundations of a kiln for glass (lit. 'stone'),

enūma uššī kūri ša abni tanaddû

A:1	*e-nu-ma*	*uš-ši*	*ku-ú-ri*	*šá*	NA$_4$	ŠUB-*du-*[*ú*]
B:1	[*e-n*]*u-ma*	*uš-ši*	*k*[*u-ú-r*]*i*	*šá*	NA$_4$	[*tanaddû*]
C:1	*e-nu-ma*	*uš-šu*	*ku-ú-ri*	*šá*	NA$_4$	[*tanadd*]*u-ú*

#1.2 in a favourable month, you search a propitious day and (then) you lay down the foundations of the kiln.

ina arḫi šalmi ūma magra teštene'e-ma uššī kūri tanamdi

A:2	*i-na*	ITU	*šal-me*	UD	ŠE.G[A	K]IN.KIN-*ma*	*uš-ši*	*ku-ú-ri*	*ta-nam-di*
B:2f	[*i*]-*na*	ITU	*šal-mi*[UD	ŠE.GA	KIN.KIN-*ma*]	(3)*uš-ši*	*ku-ú-ri*	[*ta-nam-di*	x]
C:2				UD	ŠE.G[A	KIN.KIN-*ma*]	*uš-ši*	*ku-ú-ri*	*ta-nam-d*[*i*]

#1.3 Until you have finished building the kiln, (ms. B adds: you ... in(to) the house of the kiln,) you place

adi kūra tuqtettû-ma tētepšu (ms. B adds: *ina bīt kūri* [x])

A:3	*a-di*	*ku-ú-ra*	*tuq-te-et-tu-ma*	*te-te-ep-šú*
B:4	*a-di*	*ku-ú-ra*	*t*[*uq-te-et-tu-m*]*a*	*te-*[*te-ep-šu*]
C:2f	[*a-di*	*ku*]-*ú-ri*	*tu-uq-te-e*[x]	(3)*te-te-ep-*[x]

#1.4 *kūbu*-images. Another person (or) a stranger may not enter, an unclean (person) may not step in front of them.

Kūbē tuš[*eššeb-m*]*a šanû aḫû lā irrub lā ellu ana panīšunu lā igger*

A:4							^{d}KÙ.BU^{MEŠ}	*tu-š*[*e-šeb-m*]*a*	*šá-nu-u*	*a-ḫu-u*	NU	KU$_4$-*ub*	NU	KÙ	*a-na*	*pa-ni-šú-nu*	*la*	GIL
B:5ff	*ina*	É	*ku-ú-r*[*i*	x]-*ri*	*t*	x[...]	(6)^{d}KÙ.BU^{MEŠ}	*tu-še*[*š-šeb-m*]*a*	*šá-nu-u*	*a-ḫ*[*u-u*]	(7)NU	KU$_4$-*ub*	NU	KÙ	[*a-na*	*pa*]-*ni-šú-nu*	*la*	GIL

#1.5 You pour regular strewn offerings (of flour) in front of them.

ginâ sirqa ana panīšunu tasarraq

A:5	*gi-na-a*	*s*[*i-ir-qa*	*a*]-*n*[*a*	*p*]*a-ni-šú-nu*	*ta-sár-raq*					
B:8	*gi-na-a*	*si-ir-qa*	[*a-na*	*pa*]-*ni-šú-nu*	*ta-sár-raq*					
C:4			*a-na*	IGI ^{d}K[Ù(?).BU	*tasarraq*	U$_4$]	NA$_4$	*a-na*	ŠÀ	[*ku-u-ri*]

#1.6 On the day when you put the glass (lit. 'stone') in the middle of the kiln, you make a sheep sacrifice in front of the *kūbu*-images.

ūm abna ana libbi kūri tušerredu nīqa ana maḫar Kūbē teppuš

A:6	U$_4$-*um*	NA$_4$	*ana*	Š[À	*kūri*	*tušerredu*	^{UDU}SISKUR	*a-n*]*a*	IGI ^{d}KÙ.BU^{MEŠ}	DÙ-*uš*
B:9f	U$_4$-*um*	NA$_4$	*ana*	ŠÀ	*k*[*u-ú-r*]*i*	*tu-šèr-re-du*	(10)^{UDU}SISKUR	*ana*	IGI ^{d}[KÙ.BU]	DÙ-*uš*

C:5	*tu-šèr[-re-du*	^{UDU}SISKUR]	*a-na*	IGI ^{d}KÙ.B[U^{MEŠ}	DÙ-UŠ]

#1.7 You put an incense-burner of juniper. (ms. B adds: you sacrifice honey (and) ghee) and you burn a fire in the lower part of the kiln.

nignakki burāši tašakkan (ms. B adds: *dišpa ḫimēta tanaqqī-ma*) *išāta* [*ina*] *šapal kūri tanappaḫ-ma*

A:7	NÍG.NA	ŠIM.LI	GAR-*an*	I[ZI			*ina*	K]I.T[A]	*ku-u-ri*	MÚ-*aḫ-ma*
B:10ff	NÍG.NA	$^{(11)}$ŠIM.LI	GAR-*an*	LÀ[L]	[Ì].N[UN].N[A]	BAL-*qí-ma*	$^{(12)}$*ina*	KI.TA	*ku-ú-ri*	MÚ-*aḫ-ma*

#1.8 You bring down the glass (lit. 'stone') in the middle of the kiln, the persons, whom you bring before the kiln, shall be purified and then you can bring them down to the kiln.

abna [*ana*] *libbi kūri tušerreda amēlū* [*ša ana muḫḫi kū*]*ri tuqarrabu*

A:8	NA$_4$	[*ana*]	ŠÀ	*ku-u-ri*	*tu-šèr-re-da*	LÚ^{MEŠ}	[*šá*	*ana*	UGU	*ku-u*]*ri*	*tu-qar-ra-bu*
B:13f	NA$_4$	*a-na*	ŠÀ	*ku-ú-ri*	[*tu*]-*šèr-r*[*e-d*]*u*	$^{(14)}$LÚ^{MEŠ}	*šá*	*ana*	UGU	*ku-u-r*[*i*]	[*tu-qar-ra-bu*]

[*ūtab*]*babū-ma ana muḫḫi kūri tuše*[*rreda*] (ms. C adds: *ina muḫḫīšun*[*u*])

A:9	[*ú-tab*]-*ba-bu-ma*	*ana*	UGU	*ku-ú-ri*	*tu-šè*[*r-re-da*]	
B:15	*ú-tab-ba-bu-ma*	*ana*	UGU	*ku-ú-r*[*i*	*tušerreda*]	
C:6	[x]	*ina*	UGU-*ḫi-šú-n*[*u*	x]		

#1.9 The wood which you burn in the lower part of the kiln is peeled thick poplar wood, [...], on which no 'knots' are attached, taken from the reed-marsh (ms. C adds: from the forest) (and) cut in the month of *Abu* (August).

[*iṣu*] *ša ina šapal kūri* [*t*]*ašarrapu ṣarbatu kabbartu qaliptu*

A:9f	[GIŠ.ḪI].A	*šá*	*ina*	KI.TA	*ku-u-ri*	$^{(10)}$[*t*]*a-šar-ra-pu*	^{GIŠ}ASAL$_2$	*kab-bar-*[*tu*]	*qa-li-ip-tu*
B:16	GIŠ.ḪI.A	*šá*	*ina*	KI.TA	*ku-u-ri*	*ta-ša*[*r-ra-pu*]	^{GIŠ}ASAL$_2$	*kab-bar-ta*	*qu-li-*[*ip-tu*]
C:7f	[GIŠ.ḪI.A	*ša*	*ina*	KI.TA	*kūri*	*t*]*a-šar-ra-pu*	^{GIŠ}ASAL$_2$	$^{(8)}$[*kabbartu*	*quliptu*]

qurû ša kiṣra lā nadû ina zumur api ṣabtu ina abi naksu (ms. C adds: *ultu qištušu*)

A:11	*qu-ru-u*	*ša*	*ki-iṣ-ra*	*la*	*na-du-u*	*i-na*	SU	*a-pi*	*ṣab-tu*	[*ina*	]TU.NE	KUD-*su*
B:18f	*qu-ru-u*	*šá*	*ki-iṣ-ra*	*la*	*na-d*[*u-u*]	$^{(19)}$*i-na*	SU	*a-pi*	*ṣab-tu*	*ina*	ITU.NE	KU[D-*su*]
C:8f	[*qu-ru-ú*	*š*]*á*	*ki-iṣ-ra*	*la*	*na-*[*du-u*]	$^{(9)}$[*ina*	SU	*api*	*ṣabtu*	?]	*ul-tu*	GIŠ.TIR-*šu*

#1.10 (Only) this wood may go into the lower part of the kiln.

iṣu annû ina šapal kūrīka lillik

A:12	GIŠ.ḪI.A	*an-nu-u*	*ina*	KI.TA	*ku-ri-ka*	[*l*]*il-lik*
B:20	GIŠ.ḪI.A	*an-nu-u*	*ina*	KI.TA	*ku-ú-ri-ka*	*lil-*[*lik*]
C:10	[GIŠ.ḪI.A	*annû*	*ina*	KI.TA	*ku-r*]*i-ka*	*lil*[*lik*]

#2.1 If it is blue *zagindurû*-glass for you to make

šumma zagindurû ana epēšīka

4A:44	[*šu*]*m-ma*	^{NA_4}ZA.GÌN.DURU$_5$	*ana*	DÙ-*ka*
1A:13	*šum-ma*	^{NA_4}ZA.GÌN.DURU$_5$	*a-na*	*e-pe-ši-ka*

1B:21	*šum-ma*	^{NA_4}ZA.GÌN.DURU$_5$	*ana*	DÙ-*ka*
4B:2´	*šum-ma*	^{NA_4}ZA.GÌN.[DURU$_5$	*ana*]	DÙ-*ka*
4C:3	[*šum-ma*	^{NA_4}ZA.GÌN.DURU$_5$]	*a-na*	*e-pe-š*[*i-ka*]

#2.2 you grind separately 10 minas of *immanakku*-stone, 12 minas of salicornia (mss. 1A; 1B add: 1 2/3 minas of 'white plant'), you mix them together.

10 mana immannak[*u*] [*12/15 min*]*a aḫussa* (mss. 1A; 1B add: 1 2/3 *mana šamma peṣâ*) *aḫ*[*ê tamarraq*]

4A:44f	10	MA.NA	*im-ma-na-k*[*a*]	[X	MA.N]A	*a-ḫu-us-su*					*a-ḫ*[*e-e*	*tamarraq*]	(45)*1-niš*	ḪI.ḪI
1A:13f	10	MA.NA	[^{NA_4}]M.MA.NA	(14)[1]5	MA.NA	DÈ	^{Ú}NAGA	1.2/3	MA.NA	Ú BABBAR	*a-ḫe-e*	*ta-*[*mar-raq*]	*1-niš*	ḪI.ḪI
1B:21ff	10	MA.NA	IM.MA.NA	(22)15	MA.NA	DÈ	^{Ú}NAGA	1.2/3	MA.NA	Ú BABBAR	*a-ḫe-e*(23)[*ta-ma*]*r-raq*		*1-niš*	ḪI.ḪI
4B:2´ff	10	MA.N[A	IM.MA.NA]	(3´)12	MA.NA	*a-*[*ḫu-us*]*-su*					*a-ḫe-e*	*t*[*a-mar-raq*]	(4´)*1-niš*	ḪI.ḪI
4C:4f	[10	MA.NA	*am-n*]*a-ku*	12	MA.NA	*a-ḫu-s*[*u*]					(5)[*aḫê*	*ta*]*-mar-raq*	*1-niš*	[ḪI.ḪI]

#2.3 You bring that down into a cold kiln with its four 'eyes' (i.e. openings) and arrange it between the 'eyes'.

ana kūri ša 4 īnātīša kaṣīt[*e tuš*]*erred-ma ina birīt īnāti ta*[*reddi*]

4A:45	*ana*	*ku-ú-ri*		*šá*	4	IGI^{MEŠ}-*šá*	*ka-ṣi-t*[*i*	*tu-š*]*èr-red-ma*	*ina*	*bi-rīt*	IGI^{MEŠ}	*ta-*[*red-di*]
1A:15	[*ana*]	*ku-ú-ri*		*šá*	4	IGI^{MEŠ}-*šá*	*ka-ṣi-ti*	*tu-šèr-re*[*d-ma*	*ina*	*birīt*]	IGI^{MEŠ}	*ta-red-di*
1B:23	*ana*	*ku-ú-ri*		*šá*	4	IGI^{MEŠ}-*šá*	(24)[*ka*]*-ṣi-ti*	*tu-še-red-*[*ma*]	*ina*	*bi-rīt*	IGI^{MEŠ}	(25)*ta-red-di*
4B:4´f	⌜*i*⌝-[*na*	*ku*]*-ú-ri*		*šá*	4	IGI^{MEŠ}-*šá*	*ka-ṣi-*[*te*]	(5´)*tu-šèr-*⌜*red*⌝*-ma*	*ina*	*bi-rīt*	IGI^{MEŠ}	*ta-re*[*d-di*]
4C:6f	[*a-na*]	⌜*ku-ú-ri*⌝	[*ka-ṣi*]*-ti*	*šá*	4	IGI^{MEŠ}-*šú*		*tu-*[*šèr-red-ma*]	[*ina*	*bi*]*-rīt*	*i-*[*n*]*a-a-ti*	*ta-*[*red-di*]

#2.4 You light a good, smokeless fire until your mixture glows (mss. 1A; 1B; 4C instead: 'becomes white').

išāta ṭābta lā qātirta tašarrap adi billuk[*a iraš*]*šušu* (mss. 1A; 1B; 4C instead: *ipeṣṣû*)

4A:46	IZI	*ṭa-ab-ta*	*la*	*qa-tir-ta*	*ta-šár-rap*	*a-di*	*bil-lu-k*[*a*	*i-raš*]*-šu-šu*		
1A:16	[IZ]I	*ṭa-ab-ta*	*la*	*qa-tir-ta*	*ta-šar-rap*	*a-*[*di*	*billuka*			*i*]*-pe-eṣ-ṣu-ú*
1B:25f	IZI	[*ṭa-ab-ta*	*l*]*a*	*qa-tir-ta*	(26)[*ta-š*]*ár-rap*	*a-di*	[*billuka*			*i*]*-pe-eṣ-ṣu-u*
4B:6´f	IZI	*ṭa-ab-ta*	*la*	*qa-tir-ta*	*ta-šár-*[*rap*]	(7´)*a-di*	*bil-lu-ka*			UD^{MEŠ}
4C:8f	[IZI]	*ṭa-ab-ti*			*ta-*[*šár-rap*]	(9)[*a-di*	*b*]*il-lu-ka*	*i-raš-šu-šu*	GIŠ-*ma*	*i-pe-*[x]

#2.5 You lift it (out of the kiln) into daylight, you let it cool and you grind it again. You collect it up for a clean *dabtu*-crucible.

ana ūme tušellâm-ma tu[*kaṣṣ*]*i tatâr-ma tamarraq ana dabti zakūti tessip*

4A:46f	*a-na*	U$_4$-*me*	*tu-šel-lam-ma*	*tu-*[*kaṣ-ṣ*]*i*	(47)GUR-*ma*	*ta-mar-raq*	*a-na*	*da-ab-ti*	*za-ku-ti*	*te-es-sip*
1A:17			*t*[*u-š*]*el-lam-ma*	*tu-kaṣ-ṣi*	GUR-*ma*	*ta-mar-raq*	*a-n*[*a*	*dabti*	*za-ku-t*]*i*	*te-es-si-ip*
1B:27f			[*t*]*u-šel-lam-ma*	*tu-k*[*aṣ-ṣi*	GUR -*ma*	*ta-m*]*ar-*[*ra*]*q*	(28)[*a*]*-na*	*da-ab-ti*	*z*[*a-ku-te*	*te-es-si-ip*]
4B:2´f			*tu-šel-lam-ma*	*tu-k*[*aṣ-ṣi*]	(8´)GUR-*ma*	*ta-mar-raq*	*a-n*[*a*]	*da-ab-ti*	*za-k*[*u-te*	*te-es-sip*]
4C:10f			[*tu-še-lam*]*-ma*	*tu-kaṣ-ṣi*	*ta-ta-ra-am*	*ta-*[*mar-raq*]	(11)[*ana*	*da-a*]*b-ti*	*za-ku-tu$_4$*	*te-es-sip*

#2.6 You put it down into a cold kiln with shelf. You light a good, smokeless fire until it becomes ready. Then you cover it (crucible/kiln-opening). After it has become ready you pour it on baked bricks.

ana kūri ša takkanni kaṣīti tušerred išāta ṭābta lā qātirta tašarrap

4A:47f	*a-na*	*ku-u-ri*	*šá*	*ták-kan-ni*	*ka-ṣi-ti*	*tu-š[e-r]ed*				(48)IZI	*ṭa-ab-tala*	*qa-tir-ta*	*ta-šá[r]-r[ap]*	
1A:18f	[*ana*	*ku-u]-ri*	*ša*	*ták-kan-ni*	*ka-ṣi-ti*	D[U6+DU-*ed*				IZI	*t]a-ab-ta*	(19)[*la*	*qatirta*]	*ta-šár-rap*
1B:29ff	*a-na*	*ku-ú-ri*	*šá*	[*takkanni*	*kaṣiti*]	(30)E11-*ed*				IZI	*ṭa-a[b-t]a*	[*la*	*qa-tir-ta*]	(31)*ta-šár-rap*
4B:9f	*a-na*	*ku-u-ri*	*šá*	*ták-kan-*[*ni*		*tušerred*]				(10)IZI	*ṭa-ab-ta*	*la*	*q*[*a-tir-ta*	*tašarrap*]
4C:11f	*a-na*	*ku-ú-ri*	[*ša*	*takkanni*		*tušerred*]	(12)[*bāb*	*ku*]-*ú-ri*	⌜*tu*⌝-[*kàt-t*]*am*	IZI	*ṭa-a*[*b*]-*ti*	*l*[*a*	*qatirti*]	

adi iḫarraṣu tukattam ultu iḫtarṣu ana muḫḫi agurri tanazzalam-ma

4A:48f	*a-di*	*i-ḫar-ra-ṣu*	*tu-kàt-tam*	(49)*ul-tu*	*iḫ-tar-ṣu*	*ana*	UGU	*a-gur-ri*		[*ta*]-*na-az-za-lam-ma*
1A:19f	*a-di*	*i-*[*ḫar-ra-ṣu*				*ana*	UG]U	*a-gur-ri*		(20)[*ta-na-za-l*]*a-šum-ma*
1B:31f	*a-di*	*i-ḫ*[*ar-ra-ṣu*]				(32)*ana*	UGU	*a-gur-ri*		*ta-*[*n*]*a-az-za-lam-ma*
4B:11f	*a-di*	*i-ḫar-ra-ṣu*	[*tukattam*	*ultu*	*iḫtarṣu*]	(12')*ana*	UGU	*a-gur-ri*		[*tanazzalam-ma*]
4C:13f	[x] NA4		*t*[*u-x*]			(14)[*ana*	UGU	*agurri*	x]	*ta-na-az-za-la-*[*am-ma*]

#2.7 This is called zukû, 'the pure one' (a colourless primary glass).

zukû šumš[*u*]

4A:49	*zu-ku-u*	*šum-š*[*ú*]
1A:20	[NA4*zukû*]	*šum-šu*
1B:32	NA4[*zukû*	*šumšu*]
4B:12	[*zukû*	*šumšu*]
4C:14	[*zukû*	*šumšu*]

#3.1 You collect up (mss. 2A; 2B: instead: put) 10 minas (5 kg) of 'slow' copper in a clean *dabtu*-crucible.

ana erâ nēḫa ana dabti zakūti tessip (mss. 2A; 2B instead: *tašakkan*)

5A:50	[10	M]A.NA	URUDU.ḪI.A	*né-ḫe*	*ana*	*da-ab-ti*	*za-ku-ti*	*te-es-sip*
2A:21	10	[MA.NA	URUDU.ḪI.A	*néḫa*	*ana*	*da-ab-t*]*i*	*za-k*[*u-ti*	*tašakkan*]
2B:33f	10	MA.NA	URUDU.ḪI.A	*né-ḫe*	*ana*	*da-ab-ti*	*za-ku-t*[*i*]	(34)*ta-šak-kan*
5C:15	[10	MA.NA	URUDU.ḪI.A	*néḫa*]	*a-na*	*da-ab-ta*	*-za-ku-tu4* [*te-es*]-⌜*sip*⌝	

#3.2 You put it down into a hot kiln with shelf and you close the door of the kiln.

ana [*k*]*ūri ša takkanni emmeti tu*[*šerr*]*ed bāb kūri tukattam*

5A:50f	*a-na*	[*ku*]-*ú-ri*	*šá*	*ták-kan-ni*	*em-me-ti*	D[U6+D]U	(51)[KÁ	*ku*]-*ú-ri*	DUL-*tam*
2A:21f	[*ana*	*ku-ú*]-*ri*	*šá*	*ták-kan-ni*	*em-*[*me-ti*	*tušerred-ma*	KÁ	*kūritukattam*]	
2B:34f	*a-na*	*ku-ú-ri*	*ša*	*ták-kan-ni*	(35)*em-me-ti*	*tu-šer-red-ma*	KÁ	*ku-u-ri*	*tu-kàt-tam*

5C:16′f	[ana	kūri	ša	ták-ka]n-ni	em-me-ti	t[u-šèr]-red	(17′)[bāb	ku-ú-r]i	[tu-kat-ta]m

#3.3 You light a good, smokeless fire until the copper glows.

išāta danna[ta] lā [qā]tirta tašarrap [adi erû] iraššušu

5A:51f	IZI	dan-na-[ta]	la	[qa]-tir-ta	ta-šár-rap	(52)[adi	URUDU].ḪI	i-ra-áš-šu-šu
2A:22f	[IZI	dan]-na-ta	(23)la	[qatirta	tašarrap	adi	URUDU.ḪI.A	i-r]aš-šu-šu
2B:36f	IZI	dan-na-ta	la	qa-tir-ta	ta-šár-rap	(37)a-di	URUDU.ḪI.A	i-raš-šu-šu
5C:17f	IZI	dan-na-ti	la	qa-tir-ti	(18′)ta-šár-rap	a-[di	URUDU].ḪI.A	i-raš-šu-šu

#3.4 You crush and grind 10 minas (*c.* 5 kg) of the primary colourless glass (*zukû*).

10 mana zakê taḫaššal-ma tamarraq

2A:24	10	MA.[NA	zukê	taḫaššal-ma	tamarraq]
2B:37f	10	MA.NA	za-ku-e	(38)ta-ḫaš-šal-ma	ta-mar-raq
5A:52	10	MA.NA	zu-ke-e	ta-[ḫaš]-šal-ma	ta-mar-raq
5B:17′	10	MA.[NA	zukê	taḫaššal-ma	tamarraq]
5C:18′f	10	MA.NA	zu-ke-e	(19′)[ta-ḫaš]-šal-la(!)-ma	[t]a-mar-raq

#3.5 You open the door of the kiln and (ms. 5C adds: you put *zukû* into it,) you put it on the copper and you close the door of the kiln again until the primary colourless glass (*zukû)* dissolves over the copper (ms. 5C adds: and the glass [lit. 'stone']) and the copper is deposited underneath the glass (lit. 'stone').

[bāb kū]ri tapettī-ma (ms. 5C adds: *[zuk]ê tušerred-ma*) *ana muḫḫi erî tanad[dī]-ma tatâr-ma bā[b] k[ūri tukattam]*

5A:53f	[KÁ	ku-ú]-ri	ta-pe-et-ti-ma			ana		UGU		URUDU.ḪI.A	ŠUB-[di]-ma	GUR-ma	K[Á]	k[u-ú-ri	tukattam]
2A:24f	[KÁ	ku-u-ri	t]a-pe-et-ti			(25)ana	[x	UGU	x	URUDU.ḪI.A	ŠUB-d]i	GUR-ma	KÁ	k[u-u-ri	tukattam
2B:38f	KÁ	ku-ú-ri	(39)ta-pe-et-ti-ma			a-na		UGU		URUDU.ḪI.A	ŠUB-di				
5B:18	KÁ	[ku-ú-ri	tapettima]												
5C:19′ff	KÁ	ku-ú-ri	ta-pét-ti	(20′)[zu-ke]-e	tu-šèr-red-ma	ana		UGU		URUDU	ŠUB-di	(21′)[GUR]-ma	KÁ	ku-ú-ri	tu-kàt-tam

[adi zukû eli] erî (ms. 5C adds: *u abni*) *1-niš immaḫḫaḫū-ma erû in[a šapal abni iššakkunu]* (ms. 2A adds: *imaḫḫaḫu-ma*)

5A:54	[adi	zukû	eli]	URUDU.ḪI.A					im-maḫ-ḫa-ḫu-ma	URUDU.ḪI.A	in[a	šapal	abni	iššakkunu]
2A:25	adi	zukû	UGU	URUDU.ḪI.A]					[i]-maḫ-ḫa-ḫu-ma					
2B:40	a-di	zu-ku-ú	UGU	URUDU.ḪI.A					i-maḫ-ḫa-ḫu-ma	(41)URUDU.ḪI.A	ina	KI.TA	NA4	i-šak-ku-nu
5B:18	KÁ	[ku-ú-ri	tapettima]											
5C:22f	[a-d]i			URUDU	ù	NA4	1-niš		im-maḫ-ḫa-ḫu-ma	(23′)[URUDU.ḪI.A]	i-na	šá-pal	NA4	iš-šak-ku-nu
5d:1f	[x]			[URUDU]ḪI.A		[x]	(2)[NA4]	u		URUDU.ḪI.A	(im)-maḫ (?)	–[ḫa-ḫu]		

#3.6 You stir the copper once, two times, three times with a rake (ms. 5C adds: until the rest of the copper is 'eaten') you arrange it in a new *ḫarāgu*-crucible until you see with the eye a 'crown' on the 'nose' of the rake.

(ms. 2A adds: *erâ*) [*ina muterr*]*i 1-šu 2-šu 3-šu tabeḫḫeš ana ḫa*[*rāgi eššeti*] **(**ms. 5C adds: [*adi r*]*ēḫti erî ikkalu***)**

5A:54f		[*ina*	*mu-ter-r*]*i*	1-*šú*	2-*šú*	3-*šú*	*ta-bé-eḫ-ḫeš*	(55)*a-na*	*ḫa-*[*ra-gi*	*eššeti*]		
2A:27	URUDU.Ḫ[I.A	*ina*	*muterri*	1-*šú*	2-*šú*]	3-*šú*	*ta-bé-eḫ-ḫeš*					
2B:42		*i-na*	*mu-ter-ri*	1-*šú*	2-*šú*	3-*šú*	*ta-be-eḫ-ḫeš*					
5C:24′		[*ina*	*mu-te*]*r-ri*	1-*šú*	2-*šú*	3-*šú*	*ta-be-eḫ-ḫeš*	(25′)[*a-di*		*r*]*e-eḫ-ti*	URUDU	*ik-ka-lu*

ina išāti tareddi ina appi muterri in[*a īni m*]*amm*[*a tammar*] (ms. 5C adds: *tanazalašum-ma*)

5A:56	*ina*	IZI	UŠ-*di*	(56)*ina ap-pi*	*mu-ter-ri*	T[A	IGI *ma*]-*am-m*[*a*		*tammar*]
2A:28	*a-d*[*i*			*ina appi*	*mu-ter-ri*	TA	I]GI^{II} *ma-am-ma*	*ta-mar*	
5C:25′ff	*ina*				*mu-ter-ri*	TA	IGI *mam-ma*	(26′)[*tammar*]	(27′)*ta-na-az-z*[*a*]-*la-áš-šum-ma*
5d:3					[*mu-te*]*r-ri*	TA	IGI *ma-*[*am-ma*		*tammar*]

#3.7 If the glass (lit. 'stone') turns into the colour of ripe grapes you cook the glass in the copper (and) you pour it on baked bricks.

šumma abnu **(**mss. 5C; 5d adds: *pan karāni bašlu/i itta*[*škin*]**)** *t*[*abaššil*]

5A:57	*šum-ma*	NA4	*ina*	⸢URUDU.ḪI.ḪI.A⸣			*t*[*a-ba-ši-il*	*x*]
2A:29	*š*[*um-ma*	*abnu*	*pan*	*karāni*	*bašli*	*ittaškin*	*ta-ba-š*]*i-il*	
5C:26	[*šumma*	NA4]-*ka*	*pa-an*	GEŠTIN	*ba-áš-lu*	*it-taš-kin*		
5d:4f	[*šumma*	NA4	*pa*]-*an*	GIŠ.GEŠTIN	*ba-áš-l*[*i*	*ittaškin*]	(5)[*ta-ba-ši*]-*il*	

ana muḫḫi agurri tan[*a*]*zzala*[*m-m*]*a*

5A:58		*ana*	UGU	*a-gur-ri*	*ta-n*[*a-a*]*z-za-la-a*[*m*]-[*m*]*a*
2A:29f		*a-na*	UGU	*a-gur-ri*	(30)[*tanazzalam-ma*]
5C:27	[NA4	*ina*	UG]U	*a-gur-ru*	
5d:5f		*a-na*	UGU	[*a-gur-ri*]	(6)[*ta-na-a*]*z-za-lam-ma*

#3.8 And this is called *tērsītu* 'preparation' (blue primary glass).

tērsītu šumšu

5A:58	*ter-si-ta*	*šum-š*[*u*]
2A:30	[*ter-s*]*i-tu*	*šum-šu*
5C:27′	[*tērsīta*	*šumšu*]
5d:6	*ter-si-*[*tu*	*šum-šú*]

#4.1 You grind separately 10 minas of the blue primary glass *tērsītu*, 10 minas of the primary glass *būṣu*; prepared salicorina, which does not return (i.e. is lost), and 2/3 minas of the red mother-of-pearl (lit. 'shine from the sea') (ms. 3A adds: and washed *anzaḫḫu*). You mix it together.

10 *mana tērsītu* 10 *mana b*[*ūṣ*]*u aḫussu ḫarṣu lā tayyāru*

6A:59	10	MA.NA	*ter-si-tu*	10	MA.NA	*b*[*u-ṣ*]*u*	ÚNAGA	*ḫar-ṣu*	*la*	*ta-a-a-ru*

3A:31	[*ana*	10	MA.NA	*tērsītu*	10	MA.NA	*būṣu*	(ú)NAGA]	[*ḫar*]-*ṣu*	*la*	*ta-a-a-ru*
6′C:28′f	[*a-n*]*a*	10	MA.NA	*ter-si-tu*	4(?)	MA.NA	*bu-ṣu*		(29′)[*ḫar*]-*ṣu*	*la*	*ta-a-a-r*[*u*
6d:7f	[x		MA.NA	*ter*]-*si-tu*	10	MA.NA	*bu-ṣ*[*u*]		(8)[*ḫar-ṣu*	*la*	*t*]*a-a-a-ru*

2/3 MA.NA *namrūtu ša tâm*[*ti sāndi*] *qalītu aḫê tamarraq* (ms. 3A adds: 1 2/3 *anzaḫḫu mesû*) *ištēniš taballal*

6A:60f	2/3	MA.NA	*nam-ru-tu*$_4$	*šá*	A.AB.[BA	*sa*]-*an-di*	*qa-li-tu*	*a-ḫe-e*	*ta-mar-raq*	(61)1-*niš*	ḪI.ḪI			
3A:32	[x	MA.NA	*nam-ru*]-*tu*$_4$									12/3	NA4AN.ZAḪ	*me-su-u*
6′C:29′f	2/3	M]A.N[A	*n*]*am*-⌜*ru*⌝-*tu*	*šá*	A.AB.BA	(30′)[*sāndi*	*qalītu*	*aḫê*	*tamarraq*]	1-*niš*	ḪI.ḪI			
6d:8f	2/3	MA.NA	[*nam-ru-tu*$_4$]	(9)[*ša*	A.AB.BA	*s*]*a-an-di*	*qa-li-tu*$_4$	⌜*a*⌝-[*ḫe-e*]	x					

#4.2 You collect up (the powder) for a clean (ms. 3A instead: new) *dabtu*-crucible, you put it into a cold kiln with its four 'eyes' (i.e. openings).

ana dabt[*i*] *za*[*kū*]*ti* (ms. 3A instead: *ešēti*) *tessip ana kūri ša 4 īnī kaṣiti* [*tu*]*šerred-ma*

6A:61	*ana*		*da-ab-t*[*i*]	*za*-[*ku*]-*te*	*te-es-sip*	*a-na*	*ku-u-ri*	*šá*	4	IGI^{MEŠ}-*šá*	*ka-ṣi-ti*	(62)[*tu*]-*šèr-red-ma*
3A:33	[*ana*	UGU	*da-a*]*b-ti*	*e-še-ti*	*te-es-sip*	(34)[*ana*	*kūri*	*ša*	4	IGI^{MEŠ} -*ša*		*tu-šèr*]-*red-ma*
6C:29f										[x]-*ku*		(30′)[x]-*ma*

#4.3 You put it on a support between the 'eyes' (i.e. openings). The bottom of the *dabtu*-crucible must not touch the kiln.

i[*na birī*]*ti īnē ina muḫḫi nēmedi tašakkan išid dabti kūra lā ikaššad*

6A:62	*i*-[*na*	*bi-r*]*it*	IGI^{MEŠ}	*ina*	UGU	*ne-me-di*	GAR-*an*	SUḪUŠ	*da-ab-ti*	*ku-u-ra*	NU	KUR
3A:34				*ina*	UGU	*ne-me-di*	GAR-*an*	(35)[SUḪUŠ	*dabti*	*kūra*	NU	KUR]
6B:32						*ne-me-d*[*i*	GAR-*an* ...]					

#4.4 You light a good, smokeless fire. The fire should come out of the 'eyes' (i.e. openings). (ms. B instead: The fire should be good and smokeless).

[*išāta ṭ*]*ābta lā* [*qā*]*tirta tašarrap*

6A:63	[IZI	*ṭ*]*a-ab-ta*	*la*	[*qa*]-*tir-ta*	*ta-šár-ra-ap*	
3A:35	[IZI	*ṭa-a*]*b-ta*	*la*	*qa-tir-ta*	*ta-šár-rap*	
6B:33′	IZI	*ṭa*-[*ab-tu*	*la*	*qa-tir-ta*	*tašarrap*	x]

išātu ultu libbi īnī uṣṣâ

6A:63	IZI	*ul-tu*	ŠÀ	IGI^{MEŠ}		ki-I	x	È.MEŠ-*a*
6B:34	IZI	*ul*-[*tu*	ŠÀ	IGI^{MEŠ}	x	È.MEŠ		*a* x]

#4.5 As soon as the mixture melts, you draw the fire away (and) you let it (mixture) cool in its kiln. You take it up and you grind (it).

[*adi billūk*]*a iṣuddū išāta tašaddad*

6A:63		[*a-di*	*bil-lu-k*]*a*	*i-ṣu-ud-du*	IZI	*ta-šad-da-ad*
3A:36	[x	*a-d*]*i*	*bil-lu-ka*	*i-ṣu-ud-du*		

6B:35´	a-di	bi[l-lu-ka	iṣuddu	...]	

ina kūrīšu tukaṣṣi tušellâm-ma tamarraq

6A:63	i-na	ku-ú-ri-šu	ŠED$_7$	tu-šel-lam-ma	ta-mar-raq
3A:37			[tukaṣṣi	tu]-šel-lam-ma	ta-mar-raq
6B:36´	ina	ku-⸢ú⸣-[ri	ŠED$_7$	tušellâmma	tamarraq]

#4.6 You collect up (the powder) for a clean *dabtu*-crucible and you put it down into a cold kiln with shelf.

[ana dapt]i zakūti tessip-ma ana kūri ša takkanni kaṣīti tušerre[d]

6A:65	[ana	da-ab-t]i	za-ku-ti	te-es-si-ip-ma	a-na	ku-ú-ri	šá	ták-kan-ni	ka-ṣi-i	DUL+DU-e[d]
3A:38	[	x			ana	ku-u]-ri	šá	ták-kan-ni	ka-ṣi-ti	tu-šèr-red
6B:37	a-[na	x]								

#4.7 You light a good, smokeless fire until the glass (lit. 'stone') becomes red. You do not cover the door of the kiln. After the glass has become red, you cover the door of the kiln.

[išāta ṭā]bta lā qātirta tašarrap adi abnu iraššušū

6A:66	[IZI	ṭa-a]b-ti	la	qa-tir-ta	ta-šár-rap	a-di NA$_4$	i-raš-šu-šu	
3A:39	[IZI	ṭa-ab-ta	la	qa-t]ir-ta	[ta-šár-rap	a]-di	NA$_4$	i-raš-šu-šu
6B:40´						a-di NA$_4$	i-raš-šu-[šu]	

[bāb kū]ri la tukattam ultu abnu irtaššu bāb kūri tukattam

6A:67	[KÁ	ku-u]-ri	la	DUL-tam	ul-tu	NA$_4$	ir-taš-šu	KÁ	ku-u-ri	DUL-ma
3A:39f	KÁ	ku-u-ri	la	DUL-tam	(40)[ul-tu		ir-taš]-šu	KÁ	ku-u-ri	D[UL-ma]
6B:40´f	bāb	kūri	la	tukattam]	(41´)ul-tú	NA$_4$	ir-taš-šu	[bāb	kūri	tukattam]

#4.8 Until it becomes ready, you stir it once in front of you. After it has become ready, you will see a 'crown' with the eye. If the glass (lit. 'stone') is grown over (the rake) you pour it into a new *dabtu*-crucible (inside the kiln).

adi iḫarraṣu malani ina panīka tabeḫḫiš ultu iḫtarṣu

6A:68	[a-di	i-ḫar-r]a-ṣu	ma-la-ni	ina	pa-ni-ka	ta-be-eḫ-ḫiš	ul-tu	iḫ-tar-ṣu	
New:1f	a-di	i-ḫar-[x]					(2)ul-tu	iḫ-tar[x]	
6B:42´f	a-di	i-ḫar-ra-ṣu	[malani	ana	panīka	tabeḫḫeš]	(43´)ul-tú	iḫ-tar-ṣu-m[a	x]

[ina īni ma]m[ma] tam[m]ar-ma šumma abnu uppuq ana dabti eššeti [ta]nazza[lam-m]a

6A:69	[TA	IGI	ma]-am-[ma]	⸢tam⸣-[m]ar-ma	šum-ma	NA$_4$	up-pu-uq	ana	da-ab-ti	e-ši-ti	(70)[ta]-na-az-za-[lam-m]a
New:2f					šum-ma	NA$_4$	u[p	x]			(3)ta-na-az-a[x]
3C:1	[	]									ta-n[a-az-za-lam-ma]
6B:44´f					šum-ma	NA$_4$	up-pu-u[q	ana	dabti	eššeti]	(45´)ta-na-az-za-lam-ma

#4.9 You let it cool off in the kiln and (out of the kiln) rises blue *zagindurû*-glass.

	ina kūri tukaṣṣi illâm-ma zagindurû						
6A:70	*ina*	*ku-ri*	ŠED7	*il-lam-ma*	NA4ZA.GÌN.DURU5		
3A:42f	*ina*	*ku-ri*	[ŠED7	*il*]-*lam-ma*	NA4ZA.G[ÌN.DURU5]	(43)x-*še-ep*	x
New:4			[...]	*il-lam-ma*	[NA4ZA.GÌN.DURU5]		
6B:45f	[*ina*	*kūri*	ŠED7]	(46)*il-lam-ma*	[NA4ZA.GÌN.DURU5		
3C:1f	[*ina*	*kūri*]	(2)[ŠED7	*i*]*l-lam-ma*	[x]		

#4.10 copy [x]

gabarû

new:5 GABA.RI

Appendix 2: Chemical raw data of different sites discussed

Site	Obj No.	Sample No	Object	Prop.	n	SiO_2	TiO_2 ppm	Al_2O_3	FeO	Fe_2O_3	MnO	MgO	CaO	Na_2O	K_2O	P_2O_5	Cl	Sb_2O_5	Sb_4O_6	Cl	CuO	CoO	PbO	SrO	NiO	ZnO	total	Literature
Has		61-5-89h2	bead	transl., blue	10.0	59.0	2100.0	5.0	0.5	0.0	0.0	4.7	5.5	20.2	3.9	0.3	0.7	0.3	0.0	0.0	0.9	0.0	0.0	0.0	0.0	0.0	101.6	Stapleton 2003: 107
Has		61-5-95a	bead	transl., blue	3.0	64.3	2700.0	2.5	1.4	0.0	0.3	3.4	7.4	18.1	2.0	0.3	0.6	0.0	0.0	0.0	1.4	0.0	0.0	0.0	0.0	0.0	102.0	Stapleton 2003: 107
Has		61-5-95b	bead	transl., blue	3.0	66.7	2300.0	2.3	0.8	0.0	0.0	3.4	6.4	18.2	2.1	0.0	0.8	0.0	0.0	0.0	0.7	0.0	0.0	0.0	0.0	0.0	101.5	Stapleton 2003: 107
Has		63-5-270a	bead	transl., blue	6.0	61.2	3100.0	3.3	1.7	0.0	0.0	4.6	7.2	20.3	1.7	0.4	0.3	0.0	0.0	0.0	1.3	0.0	0.0	0.0	0.0	0.0	102.8	Stapleton 2003: 107
Has		63-5-276a	bead	transl., blue	9.0	63.9	3100.0	3.3	1.4	0.0	0.0	2.8	8.3	17.4	2.2	0.5	0.5	0.0	0.0	0.0	1.0	0.0	0.0	0.0	0.0	0.0	101.9	Stapleton 2003: 107
Has		63-5-280a	bead	transl., blue	8.0	67.9	2300.0	3.7	0.6	0.0	0.6	1.4	2.6	19.3	1.7	0.4	0.8	0.0	0.0	0.0	2.2	0.0	0.0	0.0	0.0	0.0	101.7	Stapleton 2003: 107
Has		63-5-559a	bead	transl., blue	4.0	65.1	3700.0	4.4	0.9	0.0	0.0	1.1	3.4	20.1	2.3	0.0	0.4	0.6	0.0	0.0	4.8	0.0	0.0	0.0	0.0	0.0	103.9	Stapleton 2003: 107
Has		63-5-559c	bead	transl., blue	5.0	65.5	3700.0	2.8	0.8	0.0	0.0	2.4	5.8	19.6	1.4	0.4	0.6	0.0	0.0	0.0	1.2	0.0	0.0	0.0	0.0	0.0	101.2	Stapleton 2003: 107
Has		65-31-267a	bead	transl., blue	5.0	63.7	2300.0	1.9	1.2	0.0	0.0	4.1	5.7	19.7	4.2	0.4	0.8	0.3	0.0	0.0	0.6	0.0	0.0	0.0	0.0	0.0	103.5	Stapleton 2003: 107
Has		65-31-273a	bead	transl., blue	5.0	61.9	2100.0	2.2	0.9	0.0	0.0	3.3	6.2	20.4	1.8	0.4	0.7	0.7	0.0	0.0	1.5	0.0	0.0	0.0	0.0	0.0	100.6	Stapleton 2003: 107
Has		65-31-396d	bead	transl., blue	5.0	68.6	0.0	5.0	0.0	0.0	0.6	1.4	2.0	18.3	4.0	0.4	0.5	0.0	0.0	0.0	1.2	0.0	0.0	0.0	0.0	0.0	102.3	Stapleton 2003: 107
Has		73-5-293b	bead	transl., blue	5.0	65.9	2700.0	2.6	1.5	0.0	0.0	3.1	7.1	18.7	1.5	0.4	0.3	0.0	0.0	0.0	1.3	0.0	0.0	0.0	0.0	0.0	103.1	Stapleton 2003: 107
Has		73-5-350b	bead	transl., blue	3.0	64.4	3100.0	2.6	1.2	0.0	0.0	3.1	6.9	18.9	1.6	0.5	0.4	0.0	0.0	0.0	1.3	0.0	0.0	0.0	0.0	0.0	101.6	Stapleton 2003: 107
Has		73-5-559d	bead	transl., blue	5.0	64.3	2500.0	2.4	1.1	0.0	0.4	3.0	6.2	19.7	1.9	0.4	0.7	0.0	0.0	0.0	1.3	0.0	0.0	0.0	0.0	0.0	102.0	Stapleton 2003: 107
Has		73-5-559e	bead	transl., blue	4.0	70.5	0.0	3.1	0.5	0.0	1.0	0.9	2.0	19.5	1.4	0.3	0.5	0.0	0.0	0.0	0.8	0.0	0.0	0.0	0.0	0.0	100.8	Stapleton 2003: 107
Has		75-29-199a	bead	transl., blue	4.0	72.0	0.0	0.8	0.6	0.0	0.0	0.9	4.7	19.0	1.0	0.3	0.3	2.6	0.0	0.0	1.1	0.0	0.0	0.0	0.0	0.0	103.6	Stapleton 2003: 107
Has		75-29-307a	bead	transl., blue	8.0	64.7	2700.0	2.3	0.9	0.0	0.0	3.4	5.9	17.9	2.3	0.5	0.4	0.3	0.0	0.0	1.8	0.0	0.0	0.0	0.0	0.0	101.1	Stapleton 2003: 107
Has			bead	transl., blue	10.0	69.3	3200.0	3.0	1.2	1.3	0.1	2.6	4.9	15.5	2.5	0.0	0.0	0.0	0.0	0.0	0.0	0.0	0.0	0.0	0.0	0.0	99.2	Brill 1999: 43
Has			inlay	transl., blue	10.0	65.6	2000.0	1.6	0.7	0.8	0.0	3.8	4.9	19.3	1.4	0.0	0.0	0.0	0.0	0.0	1.8	0.0	0.0	0.0	0.0	0.0	97.9	Brill 1999: 43
Has			bead	transl., blue	10.0	69.0	1500.0	2.2	0.6	0.7	0.6	1.9	3.4	17.3	1.5	0.3	0.0	0.0	0.0	0.0	2.5	0.0	0.0	0.0	0.0	0.0	96.4	Brill 1999: 43
Has			bead	transl., blue	10.0	73.3	800.0	4.2	0.9	1.0	0.0	0.3	0.6	18.3	1.3	0.1	0.0	0.0	0.0	0.0	0.2	0.1	0.0	0.0	0.0	0.0	99.5	Brill 1999: 43
Has		58-4-61b	bead	black	3.0	66.9	2000.0	2.8	7.8	0.0	0.0	1.3	1.8	20.0	1.5	0.4	0.7	0.0	0.0	0.0	0.0	0.0	0.0	0.0	0.0	0.0	103.4	Stapleton 2003: 107
Has		61-5-89-1	bead	black	3.0	67.2	0.0	1.6	10.9	0.0	0.0	1.3	2.1	18.0	1.0	0.3	0.5	0.4	0.0	0.0	0.0	0.0	0.0	0.0	0.0	0.0	103.3	Stapleton 2003: 107
Has		61-5-89h3	bead	black	6.0	64.6	2300.0	2.5	11.0	0.0	0.3	2.4	2.0	18.0	1.2	0.5	0.4	0.0	0.0	0.0	0.0	0.0	0.0	0.0	0.0	0.0	103.5	Stapleton 2003: 107
Has		61-5-89i-3	bead	black	3.0	70.5	2500.0	1.1	8.0	0.0	0.0	1.3	1.7	17.3	1.2	0.4	0.9	0.0	0.0	0.0	0.0	0.0	0.0	0.0	0.0	0.0	103.2	Stapleton 2003: 107
Has		61-5-95a	bead	black	8.0	58.6	4100.0	4.3	1.1	0.0	0.0	5.3	5.4	21.1	4.2	0.6	0.7	0.6	0.0	0.0	0.0	0.0	0.0	0.0	0.0	0.0	103.1	Stapleton 2003: 107
Has		61-5-95c	bead	black	5.0	63.4	2700.0	2.3	0.9	0.0	0.0	3.4	7.5	19.4	3.1	0.6	1.2	0.3	0.0	0.0	0.0	0.0	1.4	0.0	0.0	0.0	103.4	Stapleton 2003: 107
Has		61-5-95d	bead	black	10.0	62.0	0.0	3.0	0.4	0.0	0.0	3.0	9.3	18.1	2.2	0.4	0.6	0.0	0.0	0.0	0.0	0.0	0.0	0.0	0.0	0.0	100.0	Stapleton 2003:108
Has		63-5-269a	bead	black	20.0	68.6	2200.0	2.6	0.9	0.0	0.0	3.6	6.3	20.8	2.0	0.4	1.0	0.0	0.0	0.0	0.0	0.0	0.0	0.0	0.0	0.0	107.2	Stapleton 2003:108
Has		63-5-280a	bead	black	10.0	57.3	3200.0	5.1	2.3	0.0	0.0	6.0	5.5	20.2	2.7	0.6	0.8	2.4	0.0	0.0	0.9	0.0	0.0	0.0	0.0	0.0	105.2	Stapleton 2003:108
Has		65-31-247a	bead	black	3.0	67.4	2400.0	4.1	1.1	0.0	0.0	2.6	6.3	19.4	2.3	0.4	1.1	0.3	0.0	0.0	0.0	0.0	0.0	0.0	0.0	0.0	105.7	Stapleton 2003:108
Has		65-31-273a	bead	black	5.0	60.5	2100.0	1.8	8.1	0.0	0.0	2.6	5.0	18.1	2.2	0.4	0.7	0.0	0.0	0.0	0.0	0.0	0.0	0.0	0.0	0.0	100.3	Stapleton 2003:108
Has		65-31-396e	bead	black	4.0	67.2	1900.0	1.4	10.3	0.0	0.0	1.5	2.7	17.9	0.9	0.0	0.6	0.0	0.0	0.0	0.0	0.0	0.0	0.0	0.0	0.0	102.6	Stapleton 2003:108
Has		65-31-396f	bead	black	4.0	65.7	2000.0	2.0	13.0	0.0	0.0	2.1	1.7	16.4	0.8	0.0	0.3	0.0	0.0	0.0	0.0	0.0	0.0	0.0	0.0	0.0	102.1	Stapleton 2003:108
Has		65-31-396g	bead	black	3.0	62.3	2400.0	1.3	14.8	0.0	0.0	0.7	4.0	15.1	0.8	0.4	0.3	0.0	0.0	0.0	0.0	0.0	0.0	0.0	0.0	0.0	99.8	Stapleton 2003:108
Has		65-31-396h	bead	black	3.0	63.0	0.0	1.4	15.2	0.0	0.0	0.7	4.0	15.1	0.7	0.0	0.3	0.0	0.0	0.0	0.0	0.0	0.0	0.0	0.0	0.0	100.3	Stapleton 2003:108
Has		65-31-397a	bead	black	5.0	65.6	3100.0	3.2	1.2	0.0	0.0	3.0	6.8	17.7	3.3	0.4	0.5	0.4	0.0	0.0	0.0	0.0	0.0	0.0	0.0	0.0	102.8	Stapleton 2003:108
Has		65-31-728c	bead	black	3.0	62.2	2300.0	2.2	4.9	0.0	0.0	4.3	6.0	19.9	1.5	0.8	0.4	0.3	0.0	0.0	0.0	0.0	0.0	0.0	0.0	0.0	103.0	Stapleton 2003:108
Has		73-5-305a	bead	black	6.0	65.7	3400.0	3.4		0.0	0.0	3.3	6.7	19.5	1.9	0.3	0.5	1.3	0.0	0.0	0.0	0.0	1.1	0.0	0.0	0.0	105.9	Stapleton 2003:108
Has		73-5-341 a	bead	black	5.0	68.4	2500.0	2.7	6.3	0.0	0.0	2.6	2.6	17.9	1.3	0.4	0.4	0.4	0.0	0.0	2.2	0.0	0.0	0.0	0.0	0.0	106.5	Stapleton 2003:108
Has		73-5-350a	bead	black	4.0	69.1	2500.0	2.7	0.9	0.0	0.0	2.2	5.7	18.9	2.2	0.6	0.7	0.0	0.0	0.0	0.0	0.0	0.0	0.0	0.0	0.0	103.3	Stapleton 2003:108
Has		73-5-559c	bead	black	4.0	63.8	4700.0	4.2	1.6	0.0	0.0	2.4	5.6	19.8	1.4	0.7	0.7	0.0	0.0	0.0	0.8	0.0	0.0	0.0	0.0	0.0	101.9	Stapleton 2003:108
Has		73-5-559f	bead	black	4.0	68.4	4000.0	2.2	0.9	0.0	0.0	2.9	5.1	18.2	2.5	0.5	0.6	0.0	0.0	0.0	0.0	0.0	0.0	0.0	0.0	0.0	102.2	Stapleton 2003:108

Site	Obj No.	Sample No	Object	Prop.	n	SiO_2	TiO_2 ppm	Al_2O_3	FeO	Fe_2O_3	MnO	MgO	CaO	Na_2O	K_2O	P_2O_5	Cl	Sb_2O_5	Sb_4O_6	Cl	CuO	CoO	PbO	SrO	NiO	ZnO	total	Literature
Has		75-29-307a	bead	black	5.0	65.2	3200.0	2.0	0.7	0.0	0.0	2.7	5.2	19.5	2.9	0.5	0.9	0.5	0.0	0.0	0.0	0.0	0.0	0.0	0.0	0.0	100.9	Stapleton 2003:108
Has		5429	bead	black	10.0	61.2	1500.0	4.5	5.8	6.5	0.0	3.6	5.0	15.9	2.7	0.0	0.0	0.0	0.0	0.0	0.0	0.0	0.0	0.1	0.0	0.0	99.7	Brill 1999: 43
Has		5431	bead	black	10.0	63.8	33200.0	0.8	10.1	11.2	0.0	0.9	1.6	17.1	0.8	0.0	0.0	0.0	0.0	0.0	0.0	0.0	0.0	0.0	0.0	0.0	96.5	Brill 1999: 43
Has		5433	bead	black	10.0	59.7	1500.0	2.3	7.1	7.9	0.0	3.8	5.0	18.4	1.5	0.3	0.0	0.2	0.0	0.0	0.1	0.0	0.1	0.0	0.0	0.0	99.0	Brill 1999: 43
Has		5422	bead	black	10.0	66.6	600.0	1.0	0.5	0.5	0.0	6.4	7.0	14.7	3.4	0.2	0.0	0.0	0.0	0.0	0.0	0.0	0.0	0.1	0.0	0.0	99.6	Brill 1999: 43
Has		5423	bead	black	10.0	73.3	1500.0	1.6	0.6	0.7	0.0	3.0	5.9	13.0	2.3	0.0	0.0	0.0	0.0	0.0	0.1	0.0	0.0	0.0	0.0	0.0	99.7	Brill 1999: 43
Has		5424	bead	black	10.0	79.0	2000.0	3.0	0.5	0.5	0.0	2.0	4.0	5.0	6.0	0.0	0.0	0.5	0.0	0.0	0.0	0.0	0.0	0.0	0.0	0.0		Brill 1999: 43
Has		5430	bead	black	10.0	83.0	1500.0	3.0	0.7	0.8	0.0	0.8	2.0	3.0	6.0	0.0	0.0	1.5	0.0	0.0	0.1	0.0	0.0	0.1	0.0	0.0		Brill 1999: 43
Has		5436	bead	black	10.0	85.0	1500.0	0.8	0.5	0.5	0.0	0.5	2.0	6.0	4.0	0.0	0.0	0.5	0.0	0.0	0.0	0.0	0.2	0.0	0.0	0.0		Brill 1999: 43
Has		5427	bead	black	10.0	72.5	1500.0	2.5	0.9	1.0	0.0	2.4	5.3	13.6	2.4	0.0	0.0	0.0	0.0	0.0	0.0	0.0	0.0	0.0	0.0	0.0	99.8	Brill 1999: 43
Has	Has15	61-5-54	tube	opaque, blue	4.0	64.4	0.0	2.6	0.5	0.0	0.0	4.4	4.9	18.8	1.8	0.4	0.3	2.4	0.0	0.0	3.5	0.0	0.0	0.0	0.0	0.0	104.6	Stapleton 2003: 108
Has	Has16	65-31-280	tube	opaque, blue	4.0	59.4	4100.0	9.4	3.1	0.0	0.0	2.7	4.1	18.1	2.4	0.0	0.6	0.0	0.0	0.0	0.9	0.0	0.0	0.0	0.0	0.0	101.5	Stapleton 2003: 108
Has	Has16	65-31-280	tube	opaque, blue	2.0	62.8	0.0	2.6	0.5	0.0	0.0	4.1	6.5	19.4	1.9	0.4	0.4	1.2	0.0	0.0	1.2	0.0	0.0	0.0	0.0	0.0	101.4	Stapleton 2003: 108
Has	Has13	65-31-282	tube	opaque, blue	1.0	59.8	0.0	2.0	0.5	0.0	0.0	6.4	7.0	20.1	1.4	0.0	0.4	2.3	0.0	0.0	0.0	0.0	0.0	0.0	0.0	0.0	100.6	Stapleton 2003: 108
Has	Has17	65-31-283	tube	opaque, blue	4.0	59.1	0.0	1.4	0.6	0.0	0.0	5.6	6.5	18.5	3.7	0.4	0.7	3.9	0.0	0.0	1.8	0.0	0.0	0.0	0.0	0.0	102.1	Stapleton 2003: 108
Has		65-31-728e	tube	opaque, blue	3.0	65.7	0.0	2.6	0.4	0.0	0.0	4.4	5.1	18.2	1.8	0.5	0.3	2.2	0.0	0.0	3.2	0.0	0.0	0.0	0.0	0.0	105.0	Stapleton 2003: 108
Has		65-31-728f	tube	opaque, blue	4.0	61.4	0.0	2.9	0.7	0.0	0.0	4.5	5.7	19.1	1.7	0.3	0.3	1.8	0.0	0.0	2.2	0.0	0.0	0.0	0.0	0.0	100.9	Stapleton 2003: 108
Has		65-31-992	tube	opaque, blue	3.0	67.3	2100.0	2.4	0.6	0.0	0.0	4.6	5.2	17.5	1.6	0.2	0.3	1.8	0.0	0.0	2.0	0.0	0.0	0.0	0.0	0.0	104.2	Stapleton 2003: 108
Has		65-31-993	tube	opaque, blue	3.0	62.7	0.0	2.1	0.6	0.0	0.0	5.9	6.5	19.6	1.5	0.4	0.5	1.5	0.0	0.0	0.6	0.0	0.0	0.0	0.0	0.0	102.4	Stapleton 2003: 108
Has		73-5-781	tube (?)	opaque, blue	3.0	62.7	0.0	1.5	0.7	0.0	0.0	5.6	6.9	19.4	2.2	0.4	0.4	1.6	0.0	0.0	1.0	0.0	0.0	0.0	0.0	0.0	102.9	Stapleton 2003: 108
Has		73-5-782	tube	opaque, blue	3.0	59.7	0.0	0.9	0.6	0.0	0.0	6.4	6.2	18.1	3.1	0.5	0.4	2.7	0.0	0.0	2.0	0.0	0.0	0.0	0.0	0.0	101.0	Stapleton 2003: 108
Has		367	tube	opaque, blue	10.0	62.8	500.0	0.8	0.4	0.5	0.0	5.8	5.5	17.7	3.2	0.0	0.0	2.5	0.0	0.0	2.2	0.0	0.0	0.0	0.0	0.0	95.2	Brill 1999: 43
Has		5425	bead	opaque, blue	10.0	73.3	3000.0	1.2	0.8	0.9	0.0	1.9	4.7	14.6	1.4	0.0	0.0	0.0	0.0	0.0	1.3	0.0	0.0	0.0	0.0	0.0	98.3	Brill 1999: 43
Has		714	vessel	opaque, blue	10.0	64.8	1000.0	1.4	0.6	0.6	0.1	5.1	6.1	17.2	1.7	0.0	0.0	1.5	0.0	0.0	0.8	0.0	0.0	0.0	0.0	0.0	97.4	Brill 1999: 43
Has		61-5-95a	bead	yellow	7.0	60.6	3100.0	2.2	1.0	0.0	0.0	3.1	6.1	15.1	1.5	0.4	0.4	3.0	0.0	0.0	0.0	0.0	8.6	0.0	0.0	0.0	102.6	Stapleton 2003: 108
Has		61-5-95c	bead	yellow	9.0	60.0	2400.0	2.2	9.9	0.0	0.3	4.6	6.3	16.1	2.7	0.5	0.4	2.0	0.0	0.0	0.0	0.0	7.2	0.0	0.0	0.0	103.8	Stapleton 2003: 108
Has		63-5-280a	bead	yellow	4.0	51.8	0.0	6.0	2.2	0.0	0.0	4.7	4.3	15.6	1.7	0.4	0.3	7.7	0.0	0.0	0.0	0.0	9.9	0.0	0.0	0.0	104.9	Stapleton 2003: 108
Has		73-5-305a	bead	yellow	5.0	53.1	3000.0	2.7	1.5	0.0	0.0	3.4	6.8	12.4	1.3	0.0	0.3	1.8	0.0	0.0	0.0	0.0	16.7	0.0	0.0	0.0	100.9	Stapleton 2003: 108
Has		73-5-559d	bead	yellow	5.0	59.4	0.0	4.5	1.4	0.0	0.0	3.2	5.4	15.2	1.7	0.5	0.6	1.0	0.0	0.0	0.0	0.0	9.1	0.0	0.0	0.0	102.4	Stapleton 2003: 108
Has		73-5-559f	bead	yellow	4.0	52.4	0.0	2.8	1.1	0.0	0.0	2.8	6.7	12.0	1.5	0.5	0.3	4.6	0.0	0.0	0.0	0.0	16.1	0.0	0.0	0.0	101.2	Stapleton 2003: 108
Has		75-29-307a	bead	yellow	7.0	57.3	0.0	3.0	1.5	0.0	0.0	5.0	6.1	16.0	2.2	0.6	0.4	2.3	0.0	0.0	0.0	0.0	8.8	0.0	0.0	0.0	103.7	Stapleton 2003: 108
Has		63-5-276a	bead	yellow	1.0	60.4	2900.0	3.7	1.3	0.0	0.0	4.0	8.1	20.3	1.9	0.0	0.5	2.3	0.0	0.0	0.0	0.0	0.0	0.0	0.0	0.0	103.2	Stapleton 2003: 108
Has		63-5-280a	bead	white	1.0	53.8	2900.0	6.2	1.3	0.0	0.0	4.9	3.2	22.2	1.6	0.5	0.6	6.5	0.0	0.0	0.6	0.0	0.0	0.0	0.0	0.0	101.7	Stapleton 2003: 108
Has		73-5-559e	bead	white	4.0	66.6	2100.0	3.3	0.8	0.0	0.0	1.8	2.9	16.8	1.4	0.0	0.3	7.4	0.0	0.0	0.0	0.0	0.0	0.0	0.0	0.0	102.0	Stapleton 2003: 108
Has		5432	bead	white	10.0	74.0	8000.0	1.5	1.8	2.0	0.0	2.0	3.0	7.0	6.0	0.0	0.0	2.0	0.0	0.0	2.0	0.0	0.0	0.0	0.0	0.0		Brill 1999: 43
Has		5434	bead	white	10.0	49.0	1500.0	3.0	7.2	8.0	0.0	3.0	9.0	10.0	6.0	0.0	0.0	0.8	0.0	0.0	1.0	0.0	0.0	0.1	0.0	0.0		Brill 1999: 43
Has		730	bead	aqua	10.0	70.5	1600.0	1.1	0.4	0.5	0.0	3.1	4.2	18.3	1.2	0.0	0.0	0.3	0.0	0.0	0.0	0.0	0.0	0.0	0.0	0.0	99.3	Brill 1999: 43
Has		706.0	mosaic beaker	red	10.0	53.2	0.1	1.5		2.4	0.0	3.3	4.8	15.4	2.7			2.0			12.5		2.8	0.0	0.1	0.0	83.3	Brill 1999: 43
Nim	Nim119	1712	painted inlay	transparent		63	1500	1.08	0.49	0.55	0.01	3.22	7.52	18.6	1.41	0	0	0.52	0	0	0	0	0	0.03	0	0		Brill 1978, 24
Nim		Specimen A1	bowl	transparent		67.92	0	1.79	0.44	0.49	0	4.34	8.2	14.72	1.89	0.3	0	0	0	0	0	0	0	0	0	0		Turner 1955, 61 Specimen A
Nim		Specimen A 2	bowl	transparent		68	0	2.48	0	0	0	4.2	8.28	14.02	2.7	0	0	0	0	0	0	0	0	0	0	0		Turner 1955, 61 Specimen A
Nim		Specimen B	bowl	transparen		71.54	1900	0.48	0.82	0.91	0	3.07	4.82	12.7	2.7	0.11	0	0	0.25	0	0	0	0	0	0	0		Turner 1955, 61 Specimen B
Nim		545	bowl	transparent	29	70.05	1100	0.87	0.5	0.56	0.03	3.25	9.12	13.7	1.54	0	0	0.41	0	0	0.02	0	0.026	0	0.001	0.008	99.01	Brill 1999: 45 no. 545, 47, 312

Site	Obj No.	Sample No	Object	Prop.	n	SiO_2	TiO_2 ppm	Al_2O_3	FeO	Fe_2O_3	MnO	MgO	CaO	Na_2O	K_2O	P_2O_5	Cl	Sb_2O_5	Sb_4O_6	Cl	CuO	CoO	PbO	SrO	NiO	ZnO	total	Literature
Nim		546	bowl	transparent	29	69.83	1500	1.01	0.35	0.39	0.2	2.92	9.9	13.7	1.5	0	0	0.08	0	0	0.005	0	0	0.05	0	0.006	99.25	Brill 1999: 45 no. 546, 46; Brill 2012: 625
Nim		547	bowl	transparent	29	70.01	1200	0.95	0.55	0.61	0.076	3.21	8.85	14	1.41	0	0	0.22	0	0	0.015	0	0.038	0.2	0.001	0.008	99.04	Brill 1999: 45 no. 547, 47
Nim		548	bowl	transparent	29	68.75	1800	1.12	0.48	0.53	1.55	3.33	10	13	1.25	0	0	0	0	0	0.005	0	0	0.03	0	0.009	97.98	Brill 1999: 45 no. 548, 47; Brill 2012: 625
Nim		3254	bowl	transparent	29	68.89	500	0.5	0.27	0.3	0.02	2.3	7.69	16.5	1.11	0	0	0.78	0	0	1.6	0	0.001	0.05	0	0.01	97.29	Brill 1999: 45 no. 3254, 47
Nim		1712	inlay, painted	transparent	29	64.18	1500	1.08	0.49	0.55	0.01	3.22	7.52	18.6	1.41	0	0	0.52	0	0	0.005	0	2.63	0.03	0	0	96.56	Brill 1999: 46 no. 1712
Nim	Nim28	3230	inlay, rosette (Foto)	opaque blue	29	68.78	1500	4.07	0.86	0.96	0.23	2.04	4.65	15.9	0.25	0	0	2.49	0	0	0.05	0.12	0.05	0.02	0.076	0.062	96.65	Brill 1999: 46 no. 3229, 3230; 48; Brill 2012: 625
Nim	Nim153	3231	inlay, wing (Foto)	opaque blue	29	61.02	1200	6.06	1.11	1.24	0.31	5.91	2.97	15.4	0.87	0.096	0	2.4	0	0	0.05	0.12	3.06	0.01	0.15	0.18	93.47	Brill 1999: 46 no. 3231, 3246, 3247, 3251, 3252,3253; Brill 2012: 625
Nim		3232	inlay, 'figure of eight'	opaque blue	29	63.76	1400	6.91	1.55	1.72	0.5	4.8	4.17	14.9	0.77	0.095	0	1.85	0	0	0.08	0.1	0.005	0.01	0.077	0.069	97.03	Brill 1999: 46 no. 3232,
Nim		3233	inlay, rectangular strip	opaque blue	29	58.65	1200	6.72	1.08	1.2	0.34	4.32	3.82	15.4	0.74	0.072	0	2.05	0	0	0.05	0.21	5.78	0.01	0.21	0.27	90.85	Brill 1999: 46 no. 3233, 47
Nim		3234	inlay, semi-circle	opaque blue	29	57.97	1200	7.29	0.05	1.17	0.39	4.26	3.82	15.8	0.78	0.06	0	2.08	0	0	0.05	0.21	5.47	0.01	0.21	0.27	91.09	Brill 1999: 46 no. 3234, 47
Nim	Nim28	3229	inlay, rosette (Foto)	opaque blue	29	66.02	500	5.02	0.45	0.5	0.5	3.29	2.88	18.5	0.41	0	0	2.13	0	0	0.03	0.18	0.02	0.02	0.16	0.26	96.62	Brill 1999: 46 no. 3229,
Nim		3235	inlay, strip	transl. blue	29	67.21	1200	0.56	0.51	0.57	0.02	2.83	6.19	15.3	1.35	0	0	3.25	0	0	2.09	0	0.02	0.05	0	0.021	94.01	Brill 1999: 46 no. 3235, 47
Nim		3236	inlay, round strip	transl. blue	29	67.6	1200	1.12	0.64	0.71	0.03	2.89	6.23	14.3	1.47	0.12	0	2.89	0	0	1.75	0	0.67	0.05	0	0.009	94.32	Brill 1999: 46 no. 3236, 48; Brill 2012: 625
Nim		1121	fragments	transl. blue	29	63.58	1200	0.72	0.45	0.5	0.01	1.84	7.36	13	0.8	0.11	0	6.06	0	0	5.54	0	0.02	0.02	0.005	0.012	87.8	Brill 1999: 46 no. 1121, 48; Brill 2012: 625
Nim		3251	centre of inlay rosette	white	29	0	0	0	0	0	0	0	21.3	0	0	0	0	31.5	0	0	0	0	0	0	0	0	0	Brill 1999: 46 no. 3251
Nim		3252	centre of inlay rosette	white	29	0	0	0	0	0	0	0	8.21	0	0	0	0	26.8	0	0	0	0	0	0	0	0	0	Brill 1999: 46 no. 3252
Nim		3253	centre of inlay rosette	white	29								23.5					46.9										Brill 1999: 46 no. 3253
Nim	Nim198		ingot	red		42.28	0	0.68	0.43	0	0	2.84	3.82	9.46	1.43	0	0	4.19	0	0.45	9.61	0	24.96	0	0	0		Barber - Freestone 2009, 117
Nim		3237	inlay, plain strip	red	29	72.45		1.09	0.81	0.9		3.55	5.81	15.11	1.09												100	Brill 1999: 46 (reduced values)
Nim		3239	inlay, disc	red	29	72.97		1.64	0.52	0.58		1.81	3.54	18.2	1.27												100	Brill 1999: 46 (reduced values)
Nim		3241	inlay, human male head	red	29	79.91		1.12	0.49	0.54		2.16	4.74	10.7	0.84												100	Brill 1999: 46 (reduced values)
Nim		3245	inlay, small wing	yellow	29	76.2		0.85	2.93	3.26		0.48	1.67	16.54	1												100	Brill 1999: 46 (reduced values)
Nim		3246	inlay, rectangular	yellow	29	77.52		1.01	4.44	4.93		1.63	2.61	11.38	0.91												100	Brill 1999: 46 (reduced values)

Site	Obj No.	Sample No	Object	Prop.	n	SiO_2	TiO_2 ppm	Al_2O_3	FeO	Fe_2O_3	MnO	MgO	CaO	Na_2O	K_2O	P_2O_5	Cl	Sb_2O_5	Sb_4O_6	Cl	CuO	CoO	PbO	SrO	NiO	ZnO	total	Literature
Nim		3247	inlay, irregular shape	yellow	29	83.15		0.62	6.62	7.35		0.86	2.14	5.41	0.46												100	Brill 1999: 46 (reduced values)
Nim		3248	inlay, irregular shape	yellow	29	83.5		0.63	6.53	7.25		0.87	2.16	5.13	0.45												100	Brill 1999: 46 (reduced values)
Nip		B2496/3(5)		Cu blue		63.45	0.02	0.59	0.24		0.03	5.24	5.68	17.02	3.73	0.00		1.69			1.38	0.00	0.00	0.04	0.01	0.02		Walton et al. 2012: 842-843
Nip		B2496/3(6)		Cu blue		62.77	0.02	0.62	0.22		0.03	5.08	5.76	18.25	3.18	0.00		1.60			1.52	0.00	0.00	0.04	0.02	0.01		Walton et al. 2012: 842-844
Nip		B2486/1(3)		Cu blue		63.05	0.04	0.60	0.25		0.03	5.28	5.70	18.20	2.79	0.00		1.71			1.41	0.00	0.00	0.04	0.01	0.01		Walton et al. 2012: 842-845
Nip		B2496/2(2)		Cu blue		70.92	0.02	0.66	0.27		0.04	5.69	5.92	9.15	2.75	0.00		1.94			1.58	0.00	0.00	0.04	0.02	0.01		Walton et al. 2012: 842-846
Nip		B2496/3(10)		Cu blue		72.05	0.02	0.63	0.29		0.04	4.86	5.91	9.11	2.04	0.00		2.06			1.83	0.00	0.00	0.04	0.02	0.01		Walton et al. 2012: 842-847
Nip		B2496/3(3)		Cu-Co blue		65.04	0.03	1.02	0.47		0.03	4.57	4.78	18.17	3.30	0.00		0.98			0.74	0.13	0.00	0.02	0.03	0.02		Walton et al. 2012: 842-848
Nip		B2496/3(4)		Cu-Co blue		64.68	0.02	1.02	0.50		0.03	4.83	5.48	17.37	3.49	0.00		1.01			0.68	0.13	0.00	0.02	0.02	0.02		Walton et al. 2012: 842-849
Nip		B2496/3(7)		Cu-Co blue		63.97	0.03	1.27	0.45		0.03	5.23	6.43	17.34	2.90	0.01		0.90			0.61	0.02	0.00	0.03	0.02	0.01		Walton et al. 2012: 842-850
Nip		B2496/1(5)		Cu-Co blue		64.66	0.03	1.00	0.49		0.03	4.60	4.85	17.83	3.58	0.00		1.23			0.74	0.03	0.00	0.02	0.02	0.02		Walton et al. 2012: 842-851
Nip		B2496/2(1)		Cu-Co blue		65.97	0.02	1.65	0.57		0.02	4.60	5.51	15.53	3.69	0.00		1.09			0.43	0.16	0.00	0.03	0.02	0.01		Walton et al. 2012: 842-852
Nip		B2496/2(5)		Cu-Co blue		63.58	0.02	1.04	0.49		0.03	4.83	5.61	17.46	3.48	0.00		1.78			0.66	0.12	0.00	0.02	0.02	0.02		Walton et al. 2012: 842-853
Nip		B2496.2		Cu-Co blue		64.48	0.02	0.99	0.53		0.03	4.27	5.20	18.88	2.81	0.00		1.08			0.76	0.13	0.00	0.02	0.03	0.02		Walton et al. 2012: 842-854
Nip		B2496.8		Cu-Co blue		63.83	0.02	0.96	0.51		0.03	4.65	5.97	18.21	2.94	0.00		1.22			0.73	0.14	0.00	0.02	0.02	0.01		Walton et al. 2012: 842-855
Nip		B2496.10		Cu-Co blue		63.66	0.03	1.43	0.53		0.02	3.88	5.13	17.86	4.98	0.00		1.11			0.46	0.16	0.01	0.02	0.02	0.01		Walton et al. 2012: 842-856
Nip		B2496.22		Cu-Co blue		68.49	0.02	1.07	0.56		0.03	4.85	6.17	12.84	3.30	0.00		0.91			0.78	0.16	0.00	0.03	0.03	0.01		Walton et al. 2012: 842-857
Nip		B2496.27		Cu-Co blue		64.08	0.02	1.06	0.51		0.03	4.14	5.56	18.92	3.02	0.00		1.01			0.74	0.13	0.00	0.02	0.02	0.01		Walton et al. 2012: 842-858
Nip		B2496.42		Cu-Co blue		63.27	0.02	1.02	0.51		0.03	5.25	6.31	18.78	2.43	0.00		0.79			0.71	0.16	0.00	0.03	0.02	0.01		Walton et al. 2012: 842-859
Nip		B2496.43		Cu-Co blue		65.85	0.02	1.01	0.54		0.03	4.77	5.92	15.35	3.10	0.00		1.60			0.74	0.14	0.00	0.02	0.03	0.01		Walton et al. 2012: 842-860
Nip		B2496/3(12)		Cu-Co blue		64.49	0.02	1.51	0.56		0.02	4.21	5.22	17.91	3.37	0.00		1.25			0.46	0.17	0.01	0.02	0.02	0.01		Walton et al. 2012: 842-861
Nip		B2496.37		Cu-Co blue		60.10	0.02	0.97	0.48		0.02	3.95	5.96	17.10	8.59	0.00		1.17			0.72	0.12	0.00	0.02	0.02	0.01		Walton et al. 2012: 842-862
Nip		B8686		Cu-Co blue		62.16	0.02	1.22	0.43		0.03	5.96	5.73	20.22	1.68	0.00		1.21			0.45	0.12	0.06	0.03	0.02	0.00		Walton et al. 2012: 842-863
Nip		B2496X		Cu-Co blue		66.55	0.05	1.27	0.51		0.03	6.20	5.89	14.66	1.78	0.00		1.53			0.50	0.00	0.07	0.03	0.03	0.03		Walton et al. 2012: 842-864
Nip		B2496D		Cu-Co blue		77.57	0.03	1.59	0.78		0.06	8.22	4.07	2.23	3.40	0.00		0.49			0.53	0.20	0.03	0.04	0.03	0.01		Walton et al. 2012: 842-865

Site	Obj No.	Sample No	Object	Prop.	n	SiO_2	TiO_2 ppm	Al_2O_3	FeO	Fe_2O_3	MnO	MgO	CaO	Na_2O	K_2O	P_2O_5	Cl	Sb_2O_5	Sb_4O_6	Cl	CuO	CoO	PbO	SrO	NiO	ZnO	total	Literature
Nip		B3738		Cu-Co blue		61.69	0.02	1.09	0.65		0.03	5.00	9.57	15.76	3.42	0.00		0.36			0.94	0.23	0.24	0.04	0.07	0.01		Walton et al. 2012: 842-866
Nip		B4558		Cu-Co blue		68.85	0.02	1.02	0.54		0.03	5.86	4.91	11.77	3.49	0.00		0.75			1.19	0.17	0.43	0.03	0.07	0.01		Walton et al. 2012: 842-867
Nip		B8681a		Cu-Co blue		61.48	0.02	1.03	0.62		0.03	4.94	6.74	18.74	3.37	0.00		0.39			1.10	0.25	0.33	0.03	0.08	0.01		Walton et al. 2012: 842-868
Nip		B2496/1(1)		Cu-Co blue		60.10	0.05	0.82	0.49		0.03	5.23	8.54	18.56	2.28	0.00		0.91			1.25	0.20	0.49	0.04	0.06	0.01		Walton et al. 2012: 842-869
Nip		B2496/1(2)		Cu-Co blue		67.67	0.03	1.02	0.53		0.03	5.53	5.91	11.93	3.78	0.00		0.73			1.26	0.15	0.44	0.03	0.06	0.01		Walton et al. 2012: 842-870
Nip		B2496/1(4)		Cu-Co blue		60.50	0.05	0.83	0.50		0.03	5.22	7.44	18.91	2.49	0.00		0.94			1.31	0.20	0.51	0.04	0.07	0.01		Walton et al. 2012: 842-871
Nip		B2496/1(8)		Cu-Co blue		61.41	0.03	0.99	0.48		0.03	5.23	6.53	18.23	3.91	0.00		0.61			1.14	0.14	0.37	0.03	0.05	0.01		Walton et al. 2012: 842-872
Nip		B2496/3(1)		Cu-Co blue		63.03	0.06	1.07	0.53		0.03	5.68	5.62	18.28	2.46	0.00		0.68			1.08	0.15	0.40	0.03	0.05	0.01		Walton et al. 2012: 842-873
Nip		B2496Q		Cu-Co blue		60.22	0.02	0.92	0.49		0.03	5.15	7.82	17.94	4.52	0.00		0.60			0.99	0.14	3.37	0.03	0.05	0.01		Walton et al. 2012: 842-874
Nip		B2496/3(11)		Cu-Co blue		63.26	0.02	1.15	0.51		0.03	5.42	6.70	15.82	3.75	0.00		0.75			1.13	0.15	0.38	0.03	0.05	0.01		Walton et al. 2012: 842-875
Nip		B2496/3(13)		Cu-Co blue		60.08	0.02	0.83	0.49		0.03	5.05	7.04	19.06	3.38	0.00		0.90			1.35	0.21	0.50	0.04	0.06	0.01		Walton et al. 2012: 842-876
Nip		B2496/3(14)		Cu-Co blue		61.12	0.02	0.99	0.50		0.03	5.40	6.82	18.36	3.61	0.00		0.06			1.07	0.15	0.73	0.03	0.05	0.01		Walton et al. 2012: 842-877
Nip		B2496/3(15)		Cu-Co blue		61.66	0.02	0.97	0.48		0.03	5.06	6.43	18.71	3.47	0.00		0.05			1.04	0.15	0.37	0.03	0.05	0.01		Walton et al. 2012: 842-878
Nip		B2496/3(16)		Cu-Co blue		61.15	0.02	1.02	0.48		0.03	5.11	6.98	18.68	3.48	0.00		0.05			1.05	0.14	0.36	0.03	0.05	0.01		Walton et al. 2012: 842-879
Nip		B2496/3(17)		Cu-Co blue		61.86	0.02	0.96	0.49		0.03	5.22	6.31	18.55	3.52	0.00		0.06			1.03	0.14	0.06	0.03	0.04	0.01		Walton et al. 2012: 842-880
Nip		B8685		Cu-Co blue		69.28	0.03	1.68	0.59		0.04	9.38	3.28	9.61	2.87	0.00		0.47			0.94	0.29	0.57	0.03	0.08	0.01		Walton et al. 2012: 842-881
Nip		B2496G		Cu-Co blue		69.50	0.02	1.15	0.61		0.04	5.45	5.84	9.44	4.26	0.00		0.74			1.30	0.17	0.45	0.03	0.06	0.01		Walton et al. 2012: 842-882
Nip		B8762.a1		Cu-Co blue		62.61	0.03	1.27	0.61		0.04	5.88	5.98	16.36	4.16	0.00		0.18			1.26	0.15	0.50	0.04	0.05	0.01		Walton et al. 2012: 842-883
Nip		B4550/1		Cu-Co blue		70.61	0.03	1.38	0.56		0.04	6.50	4.06	10.41	3.30	0.00		0.20			1.31	0.16	0.51	0.03	0.05	0.01		Walton et al. 2012: 842-884
Nip		B4550/2		Cu-Co blue		62.36	0.03	1.31	0.56		0.04	5.97	6.71	15.96	3.98	0.00		0.19			1.30	0.16	0.51	0.03	0.05	0.02		Walton et al. 2012: 842-885
Nip		B4544		Cu-Co blue		65.53	0.03	1.51	0.70		0.04	6.33	3.18	13.77	5.14	0.00		0.22			1.61	0.19	0.62	0.03	0.06	0.02		Walton et al. 2012: 842-886
Nip		B2496.5		Cu-Co blue		61.31	0.03	1.56	0.56		0.04	6.16	4.93	17.47	4.82	0.00		0.19			1.29	0.16	0.53	0.04	0.05	0.01		Walton et al. 2012: 842-887
Nip		B2496/1(3)		Cu-Co blue		61.86	0.03	1.51	0.60		0.04	6.22	5.44	15.64	5.55	0.00		0.20			1.30	0.15	0.53	0.04	0.05	0.02		Walton et al. 2012: 842-888
Nip		B2496/1(7)		Cu-Co blue		61.75	0.03	1.36	0.58		0.04	6.03	5.53	16.31	5.28	0.00		0.20			1.29	0.15	0.52	0.04	0.05	0.02		Walton et al. 2012: 842-889
Nuzi		1930.67.45						0.15	0.09		173							6.88			11,998	0	5.5	380	10.6	15.5		Shortland et al. 2007: 784
Nuzi		1930.82.54						0.15	0.10		243							2.63			10,912	0	11.2	450	21.5	21.2		Shortland et al. 2007: 784
Nuzi		1930.82.10						0.40	0.26		318							4.23			7318	0	7.9	324	13.7	19.5		Shortland et al. 2007: 784

Site	Obj No.	Sample No	Object	Prop.	n	SiO_2	TiO_2 ppm	Al_2O_3	FeO	Fe_2O_3	MnO	MgO	CaO	Na_2O	K_2O	P_2O_5	Cl	Sb_2O_5	Sb_4O_6	Cl	CuO	CoO	PbO	SrO	NiO	ZnO	total	Literature
Nuzi		1930.68.26	bead					0.23	0.17		202							2.15			9510	0	3.1	419	12.5	18.7		Shortland et al. 2007: 784
Nuzi		1930.68.15	bead					0.17	0.13		283							3.1			8846	0	14.6	536	21.8	20.7		Shortland et al. 2007: 784
Nuzi		1930.68.27	bead					0.21	0.14		140							14,230			5810	0	13.3	333	20.4	10.9		Shortland et al. 2007: 784
Nuzi		1930.82.50						0.17	0.11		169							12.5			13,397	0	6.3	432	9.07	16.8		Shortland et al. 2007: 784
Nuzi		1930.82.55	bead					0.33	0.16		240							1.91			6332	0	5.9	374	12	18.6		Shortland et al. 2007: 784
Nuzi		1930.67.82						0.23	0.15		132							13614			7342	0	85.2	420	34.7	17.9		Shortland et al. 2007: 784
Nuzi		1930.82.15a	vessel					0.21	0.14		187							504			20,800	0	20.5	430	12.5	23.5		Shortland et al. 2007: 784
Nuzi		1930.66.90						0.15	0.12		162							10.3			6617	0	51.4	297	26.7	12.1		Shortland et al. 2007: 784
Nuzi		1930.82.15b	vessel					0.25	0.16		186							486			22,117	0.01	7	443	11.2	23.2		Shortland et al. 2007: 784
Nuzi		1930.82.17a	bead					0.44	0.26		326							4.05			7688	0	8	339	14.8	18.1		Shortland et al. 2007: 784
Nuzi		1930.82.17b	bead					0.43	0.26		318							3.95			7488	0	7.9	327	14.8	16.9		Shortland et al. 2007: 784
Brak		HH42.1698	ingot					0.10	0.05		124							15,530			9653	0	3.7	413	16.7	12.2		Shortland et al. 2007: 784
Brak		HH581.4360	ingot					0.48	0.31		223							14,133			6227	0	685	469	25.2	24.2		Shortland et al. 2007: 784
Brak		HH204	vessel					0.26	0.18		202							455			16,785	0.01	43	488	46.8	25.1		Shortland et al. 2007: 784
Brak		HH581.4363	ingot					0.10	0.35		84							18,700			12,511	0	36.5	315	4.9	27.3		Shortland et al. 2007: 784
Brak		HH581.4366	ingot					0.27	0.20		132							12,887			5535	0	56.4	415	29.6	15.1		Shortland et al. 2007: 784
Brak		HH427.3067	ingot					0.51	0.21		250							23,547			6582	0	24.2	418	17	21.3		Shortland et al. 2007: 784
Brak		HH545.30	vessel					0.72	0.19		84							<0.10			2	0	0.5	356	0.83	5		Shortland et al. 2007: 784
Brak		HH572.4434	vessel					0.24	0.32		178							284			6487	0.01	2.6	361	14.9	19.9		Shortland et al. 2007: 784
Brak		HH572.4435	vessel					0.29	0.31		210							11			6512	0.01	55.3	287	17.4	16.3		Shortland et al. 2007: 784
Brak		HH224	ingot					0.12	0.10		34							33,658			8784	0	33.3	524	14.2	13		Shortland et al. 2007: 784
Brak		HH85	ingot					0.22	0.30		296							<0.26			4743	0.7	3.7	618	61.7	19.8		Shortland et al. 2007 784

Index